Engineering Materials

AF181242

For further volumes:
http://www.springer.com/series/4288

Alexandru Aldea · Victor Bârsan
Editors

Trends in Nanophysics

Theory, Experiment and Technology

With 164 Figures and 11 Tables

Springer

Editors
Prof. Dr. Alexandru Aldea
National Institute for Materials
Physics (INFM)
Atomistilor St. 105
077125 Magurele-Bucharest
Romania
aldea@infim.ro

Dr. Victor Bârsan
Horia Hulubei National Institute for Physics
and Nuclear Engineering (IFIN-HH)
Atomistilor St. 407
077125 Magurele-Bucharest
Romania
vbarsan@theory.nipne.ro

ISSN 1612-1317 e-ISSN 1868-1212
ISBN 978-3-642-26445-0 ISBN 978-3-642-12070-1(eBook)
DOI 10.1007/978-3-642-12070-1
Springer Heidelberg Dordrecht London New York

Cover design: deblik, Berlin

Printed on acid-free paper

Springer is part of Springer Science+Business Media (www.springer.com)

Preface

This volume is based on some representative contributions presented in the workshop: "Trends in nanophysics: theory, experiment, technology", which took place in Sibiu, Romania, 23–29 August 2009, being organized by ICTP-Trieste, IAEA, IFIN-HH – Bucharest and ULB – Sibiu. The aim of this workshop was to facilitate experts and active researchers to exchange ideas and information on the most recent results in nanophysics and nanotechnology. It was also an opportunity for young researchers and for researchers from developing countries to enlarge their knowledge and to approach new themes in this area.

In fact, the articles contained in this book represent written and enriched versions of the workshop oral presentations. The topics covered by them are the following:

1. Ordered atomic-scale structures
2. Nanowires: growth and properties
3. Transport phenomena in nanostructures
4. Optical properties of nanostructures
5. Magnetic nanophases; magnetic and non-magnetic nanocomposites
6. Nanofluids and flows at nanoscale

1 Ordered Atomic-Scale Structures

The quest of a reliable method for fabricating ordered atomic-scale structures is a prerequisite for future atomic-scale technology – the ultimate goal of nanosciences. In his lecture devoted to this subject, Schneider reviews selected examples concerning atomic and supra-molecular self-assembly investigated by low temperature scanning tunneling microscopy: two-dimensional arrays of individual Ce atoms on a metal surface; the behaviour of the superconductor energy gap in ultra-thin Pb islands and the conservation of chirality in a hierarchical supra-molecular self-assembly of pentagonal symmetry of rubene on an Au surface. Another key issue for the success of many nanotechnologies is our ability to understand the mechanics of nano-objects, such as nanotubes and nanobelts. Dumitrica's contribution is devoted to an ingenious symmetry-adapted atomistic scheme, based on a quantum-mechanical description of chemical bonding, that performs calculations under helical boundary conditions. As an application, the nanomechanical response of carbon nanotubes and thermodynamical stability of silicon nanowires are obtained.

2 Nanowires: Growth and Properties

Growths and properties of nanowires are a central issue of nanoscience and nanotechnology. Lakhtakis's paper presents a new class of assemblies of nanowires, named sculptured thin films, that can be fabricated typically via physical vapor deposition onto rotating substrates. Their optical properties can be tailored by varying their morphology. The optical, thermal, chemical, and biological applications of sculptured thin films are reviewed. Stoica's contribution is focused on growths and optical properties of an important class of semiconductor nanowires – GaN and InN ones, obtained by self-assambling, with a catalist-free molecular beam epitaxy technique. The optoelectronic properties, as well as the influence of surface effects on the growth and properties of these nanowires are carefully analyzed.

3 Transport Phenomena in Nanostructures

The permanent requirement of shrinking the semiconductor devices in integrated circuits request a good understanding of transport phenomena in nanostructures. A comprehensive review of such topics is given in the presentation of Kuhn and Paraoanu, devoted to electronic and thermal sequential transport in metallic and superconducting two-junction arrays. The authors analyse Coulomb-blockade thermometers, superconductor-insulator-normal-insulator-superconductor structures, and superconducting single-electron transistors. Racec et al. present a general theory of multi-channel scattering for a general two-dimensional potential, based on the R-matrix formalism; it allows a semi-analytical treatment of the problem, and yields a powerful and efficient numerical method, with applications to nanostructures with quantum dots. In the review of Nemnes et al., planar nanoscale transistors and cylindrical nanowire transistors are analyzed in the framework of coherent transport. The Landauer-Buttiker formalism is efficiently implemented using also a R-matrix approach. The advantages of new geometries, like the cylindrical nanowire transistors, are discussed. As charge fractionalization has been observed experimentally in quantum wires, this fundamental phenomenon deserves special attention. In his lecture, Leinaas discusses the issue of fractional charge and statistics in Luttinger liquids – one of the most popular models describing one-dimensional systems of fermions.

4 Optical Properties of Nanostructures

If trapping and moving of dielectric nanoparticles with laser beams constitute a well understood issue, the situation of metallic nanoparticle is quite different. In Prof. Crozier's contribution, it is explained how the propulsion of gold nanoparticles by surface plasmon polaritons was demonstrated experimentally. The optical forces are enhanced, due to the field enhancement provided by plasmon polaritons and near-field coupling between the gold particle and the film. The plasmon spectra of plasmons excited in metallic nanoparticles and nanowires are discussed also in a theoretical paper, by Villo-Perez, Mišković, and Arista. They apply Bloch's hydrodynamic model of an electron gas to describe plasma excitations in thin metallic films, obtaining a good description of the excitation, propagation and decay of bulk and surface modes, in different geometries. A two-fluid model, in which the σ and π electrons of carbon are the constituents of these fluids, is used in order to obtain the plasmon spectra in carbon nanotubes.

5 Magnetic Nanophases; Magnetic and Non-Magnetic Nanocomposites

Nanomagnetism is important for both fundamental and applicative reasons. Functionalized nanocomposites consisting of magnetic nanoparticles (Co, Fe), embedded in dielectric matrices, have a significant potential for the electronics industry. In the contribution of Timonen et al., the theory of such materials is reviewed; also, the authors present a novel measurement method used for the characterization of the electromagnetic properties of composites with nano-magnetic insertions. The article of Tolea et al. is devoted to spring magnets, consisting of interfaced hard (containing rare earths, iron and boron) and soft (containing iron and boron) magnetic nanophases, coupled by exchange interactions. Their magnetic properties depend on the thermal treatment and of amount of added iron, the optimal situation corresponding when hard and soft magnetic phases coexist with a small amorphous phase. Kuncser et al. describe how Mossbauer spectroscopy, applyed in complementarity with magnetic and structural techniques, can be used in order to obtain a comprehensive characterization of the magnetic configuration and magnetic relaxation of nanoparticles. The contribution of Jovanovic et al. is devoted to non-magnetic nanocomposites: silver nanoparticles embedded in a hydrogel, synthetized in situ by gamma iradiation. The plasmon spectra of nanoparticles are described and the biomedical applications are discussed.

6 Nanofluids and Flows at Nanoscale

The characterization of nanofluids, consisting of dispersed magnetic nanoparticles in a liquid carrier, is important mainly due to the the specific applications of such complex magnetic systems. Prof. Chicea's contribution is focused on magnetite nanoparticle aggregation dynamics in an aqueous suspension and on its effects on the modification of the rheological properties of the fluid.The time variation of the average diameter of the aggregates is obtained using light scattering techniques. Last but not the least, Prof. Niemela's contribution reviews some of the nano-physics appearing in the turbulent flow of classical and quantum fluids.

We hope that the diversity of themes and the clarity of contributions, written by leading experimental and theoretical researchers in these fields, recommend this volume as a useful and attractive lecture for researchers or students interested in nanophysics.

Bucharest, Romania

Alexandru Aldea
Victor Bârsan

Acknowledgements

The editors are grateful to EOARD and ANCS for their financial support, which was essential for the organization of the workshop. They are also grateful to Springer Publishing House, for its interest in the publication of these proceedings.

Bucharest, Romania

Alexandru Aldea
Victor Bârsan

Contents

Contributors

R. Alexandrescu National Institute for Lasers, Plasma and Radiation Physics (INFLPR), 409 Atomistilor St., 077125 Bucharest-Magurele, Romania, ralexandrescu2001@yahoo.co.uk

S. Antohe Faculty of Physics, University of Bucharest, "Materials and Devices for Electronics and Optoelectronics" Research Center, PO Box MG-11, 077125 Magurele-Ilfov, Romania, santohe@solid.fizica.unibuc.ro

N.R. Arista División Colisiones Atómicas, Instituto Balseiro and Centro Atómico Bariloche, Comisión Nacional de Energía Atómica, 8400 Bariloche, Argentina, arista001@yahoo.com

A. Birsan National Institute of Materials Physics (INFM), 105 Atomistilor Street, 077125 Magurele-Bucharest, Romania, ancutabirsan@yahoo.com

Raffaella Calarco Institute of Bio- and Nanosystems (IBN-1) and JARA-FIT Jülich-Aachen Research Alliance, Forschungszentrum Jülich, 52425 Jülich, Germany, R.Calarco@fz-juelich.de

Khattiya Chalapat Low Temperature Laboratory, School of Science and Technology, Aalto University, PO Box 15100, FI-00076 Aalto, Finland, khattiya.chalapat@ltl.tkk.fi

Dan Chicea Department of Physics, University Lucian Blaga of Sibiu, Dr. Ion Ratiu Str. 7-9, Sibiu 550012 Romania, dan.chicea@ulbsibiu.ro

Kenneth B. Crozier School of Engineering and Applied Science, Harvard University, Cambridge, MA 02138, USA, kcrozier@seas.harvard.edu

M. Dragašević Institute of Endocrinology, Diabetes and Metabolic Diseases, University Clinical Centre of Serbia, Dr Subotića 13, 1100 Belgrade, Serbia, mirdraga@yahoo.com

Traian Dumitrica Department of Mechanical Engineering, University of Minnesota, Minneapolis, MN, USA, td@me.umn.edu

G. Filoti National Institute for Materials Physics (INFM), 105 Atomistilor Street, 077125 Magurele-Bucharest, Romania, filoti@infim.ro

Joseph B. Geddes III Beckman Institute, University of Illinois at Urbana–Champaign, Urbana, IL 61801, USA, jbgeddes3@jbg3.net

Olli Ikkala Molecular Materials, Department of Applied Physics, School of Science and Technology, Aalto University, PO Box 15100, FI-00076 Aalto, Finland, Olli.Ikkala@tkk.fi

L. Ion Faculty of Physics, University of Bucharest, "Materials and Devices for Electronics and Optoelectronics" Research Center, PO Box MG-11, 077125 Magurele-Ilfov, Romania

Ž. Jovanović Faculty of Technology and Metallurgy, University of Belgrade, PO Box 3503, 11120 Belgrade, Serbia, zjovanovic@tmf.bg.ac.rs

Z. Kačarević-Popović Vinča Institute of Nuclear Sciences, University of Belgrade, PO Box 522, 11001 Belgrade, Serbia, zkacar@vinca.rs

A. Krklješ Vinca Institute of Nuclear Sciences, University of Belgrade, PO Box 522, 11001 Belgrade, Serbia, krkljes@vinca.rs

T. Kühn NanoScience Center and Department of Physics, University of Jyväskylä, Jyväskylä PO Box 35 (YFL), FIN-40014, Finland, kuehn@cc.jyu.fi

V. Kuncser National Institute for Materials Physics (INFM), 105 Atomistilor Street, 077125 Magurele-Bucharest, Romania, kuncser@infim.ro

Akhlesh Lakhtakia Department of Engineering Science and Mechanics, Pennsylvania State University, University Park, PA 16802, USA, axl4@psu.edu

Jon Magne Leinaas Department of Physics, University of Oslo, N-0316 Oslo, Norway, j.m.leinaas@fys.uio.no

Jian Li Low Temperature Laboratory, School of Science and Technology, Aalto University, PO Box 15100, FI-00076 Aalto, Finland, jianli@ltl.tkk.fi

Z.L. Mišković Department of Applied Mathematics, University of Waterloo, Waterloo, ON, Canada, N2L 3G1, zmiskovic03@gmail.com

V. Mišković-Stanković Faculty of Technology and Metallurgy, University of Belgrade, PO Box 3503, 11120 Belgrade, Serbia, vesna@tmf.bg.ac.rs

I. Morjan National Institute for Lasers, Plasma and Radiation Physics (INFLPR), 409 Atomistilor St., 077125 Magurele-Bucuresti, Romania, Ion.Morjan@inflpr.ro

H. Neidhardt Weierstraß-Institut für Angewandte Analysis und Stochastik, Mohrenstr. 39, 10117 Berlin, Germany, neidhard@wias-berlin.de

G.A. Nemnes Faculty of Physics, University of Bucharest, "Materials and Devices for Electronics and Optoelectronics" Research Center, PO Box MG-11, 077125 Magurele-Ilfov, Romania, nemnes@solid.fizica.unibuc.ro

J.J. Niemela The Abdus Salam ICTP, Strada Costiera 11, 34014 Trieste, Italy, niemela@ictp.it

Markku Oksanen Nokia Research Center, Itämerenkatu 11-13, 00180 Helsinki, Finland, markku.a.oksanen@nokia.com

G.S. Paraoanu Low Temperature Laboratory, School of Science and Technology, Aalto University, PO Box 15100, FI-00076 Aalto, Finland, paraoanu@cc.hut.fi

S. Popović Institute of Endocrinology, Diabetes and Metabolic Diseases, University Clinical Centre of Serbia, Dr Subotića 13, 1100 Belgrade, Serbia, prof.srdjan.popovic@gmail.com

E.R. Racec Institut für Physik, Technische Universität Cottbus, Postfach 101344, 03013 Cottbus, Germany; Faculty of Physics, University of Bucharest, PO Box MG-11, 077125 Bucharest Magurele, Romania, roxana@physik.tu-cottbus.de

P.N. Racec Weierstraß-Institut für Angewandte Analysis und Stochastik, Mohrenstr. 39, 10117 Berlin, Germany; National Institute of Materials Physics, PO Box MG-7, 077125 Bucharest Magurele, Romania, racec@wias-berlin.de

Robin H.A. Ras Molecular Materials, Department of Applied Physics, School of Science and Technology, Aalto University, PO Box 15100, FI-00076 Aalto, Finland, Robin.Ras@tkk.fi

G. Schinteie National Institute of Materials Physics (INFM), 105 Atomistilor Street, 077125 Magurele-Bucharest, Romania, schinteie@infim.ro

Wolf-Dieter Schneider Ecole Polytechnique Fédérale de Lausanne (EPFL), Institut de Physique de la Matière Condensée, CH-1015 Lausanne, Switzerland, wolf-dieter.schneider@epfl.ch

Eira Seppälä Nokia Research Center, Itämerenkatu 11-13, 00180 Helsinki, Finland, Eira.Seppala@nokia.com

M. Sofronie National Institute of Materials Physics (INFM), 105 Atomistilor Street, 077125 Magurele-Bucharest, Romania, sofmiha@yahoo.com

Toma Stoica Institute of Bio- and Nanosystems (IBN-1) and JARA-FIT Jülich-Aachen Research Alliance, Forschungszentrum Jülich, 52425 Jülich, Germany, t.stoica@fz-juelich.de

Eli Sutter Brookhaven National Laboratory, Center for Functional Nanomaterials, Upton, NY 11973, USA, esutter@bnl.gov

Jaakko V.I. Timonen Molecular Materials, Department of Applied Physics, School of Science and Technology, Aalto University, PO Box 15100, FI-00076 Aalto, Finland, jaakko.timonen@tkk.fi

F. Tolea National Institute of Materials Physics (INFM), 105 Atomistilor Street, 077125 Magurele-Bucharest, Romania, felicia@infim.ro

S. Tomić Faculty of Technology and Metallurgy, University of Belgrade, PO Box 3503, 11120 Belgrade, Serbia, simonida@tmf.bg.ac.rs

M. Valeanu National Institute of Materials Physics (INFM), Bucharest-Magurele, Romania, valeanu@infim.ro

L. Vekas Centre of Fundamental and Advanced Technical Research, Romanian Academy-Timisoara Division, Timisoara, Romania, vekasladislau@yahoo.co

I. Villo-Perez Departamento de Electrónica, Universidad Politécnica de Cartagena, E-30202 Cartagena, Spain, isidro.villo@upct.es

U. Wulf Technische Universität Cottbus, Fakultät 1, Postfach 101344, 03013 Cottbus, Germany, wulf@physik.tu-cottbus.de

List of Acronyms

Institutions

ANCS	Autoritatea Nationala pentru Cercetare Stiintifica (National Authority for Scientific Research)
EOARD	European Office for Aerospace Research and Development
IAEA	International Agency for Atomic Energy
ICTP	International Centre for Theoretical Physics
IFIN-HH	Institutul de Fizica si Inginerie Nucleara Horia Hulubei
ULB	Universitatea Lucian Blaga

Scientific Concepts

2DEG	two-dimensional electron gas
AFM	atomic force microscope
BCC	body centred cubic
BEP	beam-equivalent pressure
CBT	Coulomb-blocade thermometer
CCD	charge coupled device
CNFET	carbon nanotube field-effect transistor
CNT	carbon nanotubes
DGFET	double gate field-effect transistor
DLS	dynamic light scattering
DW	domain walls
EG	electron gas
EMT	effective medium theories
FCC	face centred cubic
FET	field-effect transistor
FMR	ferromagnetic resonance
FT	Fourier transformation
FWHM	full width at half maximum
GAA	gate-all-around

GNR	graphene nano-ribbons
HCP	hexagonal close packed
HDM	hydrodynamic model
HOMO	highest occupied molecular orbital
HRTEM	high-resolution transmission electron microscopy
IR	infrared
kMC	kinetic Monte Carlo
KTHNY	Kosterlitz-Thouless-Halperin-Nelson-Young
LE	light emission
LPS	localized surface plasmon
LUMO	lowest unoccupied molecular orbital
ML	monolayer
MOSFET	metal-oxide-semiconductor field-effect transistor
MWCNT	multi-walled CNT
NA	numerical aperture
NW	nanowire
PAMBE	plasma-assisted molecular beam epitaxy
PL	photoluminescence
PCS	photon correlation spectroscopy
PVP	poly(N-vinyl-2-pyrrolidine)
RIE	reactive ion etching
RKKY	Ruderman-Kittel-Kasuya-Yosida
rf	radio frequency
RT	room temperature
SC	scattering center
SD	single-domain
SET	single-electron transistor
SHF	super high frequency
SINIS	superconducting-insulator-normal-insulator-superconducting
SLS	static light scattering
SNWT	silicon nanowire transistor
SOI	silicon-on-insulator
SPP	surface plasmon polariton
SQUID	Superconducting Quantum Interference Device
STM	scanning tunneling microscopy
SVP	saturated vapor pressure
SWCNT	single-walled CNT
TD-DFT	time-dependent density functional theory
TE	transversal electric
TM	transversal magnetic
UHV	ultra-high vacuum
vdW	van der Waals
VLS	vapor-liquid-solid
XRD	X-ray diffraction

Part I
Ordered Atomic-Scale Structures

Fabrication and Characterization of Ordered Atomic-scale Structures – A Step towards Future Nanoscale Technology

Wolf-Dieter Schneider

Abstract The quest of a reliable method for fabricating ordered atomic-scale structures is a prequisite for future atomic-scale technology. The interest in such nanostructured materials, consisting of building blocks of a small number of atoms or molecules, arises from their promising new optic, catalytic, magnetic and electronic poperties, which are fundamentally different from their macroscopic bulk counterparts: small is different. Here we review selected examples concerning atomic and supramolecular self-assembly investigated by low-temperature scanning tunneling microscopy (STM). (i) The self-assembly and the melting of a two-dimensional array of individual Ce adatoms (the smallest possible building block) on a metal surface based on long-range interactions between adatoms mediated by surface state electrons. Ce is a magnetic atom, and such hexagonal superlattices of magnetic adatoms might be useful for the development of future atomic-scale magnetic devices. (ii) The reduction of the superconducting energy gap in ultrathin Pb islands grown on Si(111), when the thickness is reduced down to a few atomic with monolayers (MLs). (iii) The conservation of chirality in a hierarchical supramolecular self-assembly of pentagonal symmetry of the organic molecule rubrene on a reconstructed Au(111) surface. We show the spontaneous chiral resolution of the racemate into disjoint homochiral complex architectures and demonstrate the ability to monitor directly the evolution of chiral recognition processes on the molecular and supramolecular level. (iv) Taking advantage of inelastic electron tunneling processes, we excite luminescence from C_{60} and C_{70} molecules in the surface layer of fullerene nanocrystals self-assembled on an ultrathin NaCl film on Au(111). The observed fluorescence and phosphorescence spectra are found to be characteristic for the two molecular species, leading to unambiguous chemical recognition on the molecular scale.

W.-D. Schneider (✉)
Ecole Polytechnique Fédérale de Lausanne (EPFL), Institut de Physique de la Matière Condensée, CH-1015 Lausanne, Switzerland
e-mail: wolf-dieter.schneider@epfl.ch

A. Aldea, V. Bârsan (eds.), *Trends in Nanophysics,* Engineering Materials,
DOI 10.1007/978-3-642-12070-1_1, © Springer-Verlag Berlin Heidelberg 2010

1 Introduction

A promising route leading towards the fabrication of ordered nano-scale structures is based on self-assembly of basic building blocks of matter, i.e., of atoms and/or molecules [1]. The key challenges are on one hand, to observe and to characterize the atomic and molecular processes leading to self-assembly and, on the other hand, to determine the electronic and optical properties as well as the chemical identity of the molecular building blocks and of the resulting nanostructures on the molecular level. Here we address both aspects of this challenge employing low-temperature scanning probe techniques. In the first example we present the self-assembly and the melting of 2-D arrays of individual Ce adatoms on Ag(111) and on Cu(111) surfaces, which is based on long-range interactions between adatoms mediated by surface state electrons [2–5]. In the second example we discuss the reduction of the superconducting gap upon the thickness reduction of Pb nano-islands grown on Si(111) [6, 7]. In the third example we discuss the chiral supermolecular self-assembly and the electronic structure of the organic molecule rubrene [8–11], and in the forth example we report on the chemical recognition of C_{60} and C_{70} molecules within the surface layer of self-assembled fullerene nanocrystals by tip-induced fluorescence and phosphorescence [12, 13].

2 Experimental

The experiments were performed with a home-built low-temperature ultra-high-vacuum (UHV) STM operating at a pressure of 10^{-11} mbar and a temperature of 5 K and of 50 K, using cut PtIr tips. Ce atoms were deposited from a thoroughly degassed tungsten filament onto well prepared Ag(111) and Cu(111) surfaces held at 4.8 K. During specific experiments the temperature of the sample was lowered to 3.0 K by pumping on the He-bath [14].

Pb was thermally evaporated on the Si(111)-7×7 or on the Pb-$\sqrt{3} \times \sqrt{3}$/Si(111) substrate [15] kept at room temperature (RT) favoring the growth of Pb single crystals with their (111) axis perpendicular to the surface [16, 15]. Differential conductance (dI/dV) measurements were performed with open feedback loop, using lock-in technique with a modulation voltage of 0.2–0.5 mV_{pp} at $\sim$ 277 Hz, with a typical tunneling current of 1 nA. Radio frequency (rf) noise has been carefully filtered.

The rubrene molecules were deposited at low coverage (0.3 ML) in situ by sublimation from a homebuilt evaporator onto a clean Au(111) substrate at room temperature.

C_{60} and C_{70} nanocrystals were grown on thin insulating NaCl layers deposited onto an atomically flat Au(111) substrate. NaCl was deposited from a resistively heated evaporator onto a clean Au(111) surface at room temperature. Subsequently, the fullerenes were sublimated onto the NaCl covered substrate forming fullerene nanocrystals of hexagonal and truncated triangular shape. Experiments have been performed at a temperature of 50 K, using etched W tips. Photons emitted from the tunnel junction were collected by a plano-convex lens (NA = 0.34) near the tip-sample gap along the direction 60° with respect to the surface normal. The collected beam was then transmitted through a view port outside the UHV chamber and guided simultaneously to (i) a grating spectrometer (50 l/mm) coupled to a liquid-nitrogen-cooled CCD camera for spectral analysis (90% of the signal) and to (ii) an avalanche photodiode to record the total light intensity and to optimize the alignment of the lens with the tunnel junction (10% of the signal). For the light emission measurements, the tip was positioned over a target location with a fixed tunnel resistance. Spectra were not corrected for the wavelength dependent sensitivity of the detection system. The wavelength resolution of the experiment was 8 nm, corresponding to ≈ 20 meV in the energy range of interest. The spectra were acquired with closed feedback loop while tunneling over a defined position on the sample, e.g. over a single molecule, with a typical acquisition time of 300 s. Bias voltages V refer to the sample voltage with respect to the tip.

3 Results and Discussion

3.1 Self-Assembly of Adatom Superlattices

An interesting possibility to achieve long-range self-assembly of individual atoms with interatomic distances of several lattice constants of the supporting substrate is to take advantage of surface-state-mediated adatom interactions [17] that have been found on noble metal (111) surfaces [18–20]. While the experiments of Refs. [19, 20] displayed preferential adatom-adatom separations illustrating the oscillatory behaviour of the potential of the adatom-adatom interaction mediated by the surface state electrons, no large ordered atomic superlattice was produced. This situation changed when Silly et al. [2, 21] reported on the discovery of the formation of a 2D hexagonal superlattice of Ce adatoms on Ag(111). The control of the temperature permits to tune the adatom mobility on the surface, while the control of the adatom concentration influences the adatom-adatom interaction and, most interestingly, the interatomic distance of the adatoms in the superlattice [2, 3]. The surface state electron scattering at the adatom sites sets the scale of the superlattice constant at a temperature of 4.8 K and a concentration of 1% of a monolayer (ML) of Ce adatoms to 3.2 nm, corresponding closely to half of the Fermi wavelength of the suface state electrons.

Figure 1 shows a STM topograph after deposition of about 1% of a ML of Ce onto Ag(111). The Ce adatoms form a hexagonal superlattice covering the Ag(111) surface. The distance between two neighboring adatoms is determined to be 32 ± 2 Å . The superlattice is visible on all large scale STM images but also at different regions of the sample separated by macroscopic distances (2 mm). Such a long-range ordered superlattice was not observed before, neither with Cu on Cu(111) [18, 19] nor with Co adatoms deposited on Ag(111) [20].

Fig. 1 155 × 108 nm^2 STM topography of a Ag(111) surface covered by 1% ML of Ce atadoms at a temperature of 3.9 K ($U_s = -90$ mV, $I_s = 40$ pA). Nearest neighbor adatatom-adatom separations are 3.2 nm. Dark (*bright*) contrast: lower (*upper*) terrace separated by a monatomic Ag step. After Ref. [4]

The two-body interaction energy decays as $1/\rho^2$ where ρ is the distance between two adatoms, and oscillates with a period of $\lambda_F/2$, half of the Fermi wavelength of the surface state electrons. A quantitative analysis of the phase shift δ_0 of the surface state electrons scattered at the adatom sites, which can be obtained independently from an analysis of the standing wave patterns, and of the observed adatom mobility around their superlattice sites enabled the determination of the depth of the first minimum in the oscillating two-body interaction potential [22] to 0.8 meV [2, 21].

Thus the self-assembly of the adatom superlattice is a consequence of a subtle balance between the sample temperature, the surface diffusion barrier, and the concentration-dependent adatom interaction potential generated by the surface state electrons. We note that recent kinetic Monte Carlo (kMC) simulations of the self-organisation of Ce adatoms on Ag(111) confirm quantitatively our conclusions [23, 24]. Variation of support and of adatom element as well as adatom concentration may allow us to tune the lattice constant of the superlattice over a wide range [21]. One recent example is the observation of such a superlattice for the adsorption of Cs on Cu(111) where the adatom distance was found to be 1.1 nm, in agreement with the value of half of the Fermi wavelength of the Cu(111) surface state electrons [25, 21].

3.2 Melting of Adatom Superlattices

The first order phase transition which occurs when regular three-dimensional (3D) crystals melt and the highly ordered crystal structure changes into the irregular order of a liquid is well known. In 2D systems, however, melting is fundamentally different. This difference stems from the radical change of the intrinsic properties of matter in reduced dimensions. The long-range order which defines the periodic structure of a 3D crystal changes in 2D to only quasi-long-range translational (or positional) order [26, 27].

According to the Kosterlitz-Thouless-Halperin-Nelson-Young (KTHNY) theory [28–31], a true 2D solid melts in two steps via two distinct successive phase transitions occurring at temperatures T_m and T_h, respectively. The intermediate hexatic phase, is thereby characterized by the loss of translational order as for the liquid phase and by the subsistence of a certain degree of orientational order, in contrast to the liquid isotropic phase. For the melting of a 2D solid in the presence of a substrate potential, Nelson and Halperin predicted that the existence of the hexatic phase is hindered and melting directly occurs from the solid into the liquid phase [30].

Many experimental and numerical investigations have been devoted to study the melting transitions in 2D and to verify the KTHNY theory (for a review see [32]) but this new class of atomic 2D superlattices allowed us to directly investigate the melting transition in two dimensions on the atomic scale[5, 21].

For Ce on Ag(111) the adatoms diffuse in the minimum of the superlattice potential already at 4.8 K. Therefore, this system is not well adapted for a detailed melting study because the relatively large adatom-adatom distance and the relatively small interaction energy compared to the diffusion barrier lead to a predicted melting temperature of only $T_m \lesssim 5$ K [5]. Thus, we have examined the temperature evolution of the superlattice of Ce adatoms on Cu(111) [5].

STM images obtained at a temperature of 8 K, 9 K and 14 K, respectively, are shown in Fig. 2a–c. At the elevated temperatures, the Ce adatoms become more mobile on the surface, inducing a modification of the apparent contrast between adatoms and aggregates. The gray-scale insets display the 2D Fourier transformation (FT), corresponding to the structure factor of the positional data. Six sharp and distinct Bragg reflexes, signature of a 2D crystalline order, are observed at 8 K. At 9 K and 14 K the Bragg spots become more diffuse, and a ring appears, indicating melting of the 2D crystal. A weak hexagonal symmetry, however, is still visible in the Fourier transformation even at the elevated temperatures.

In a true 2D system the solid, hexatic, and liquid phases can be identified by a characteristic decay behavior of the pair correlation function $f(r)$, the density-density correlation function $g_r(r)$ and the bond-angular correlation function $g_6(r)$:

$$f(r) = \langle \sum_i \sum_{j \neq i} \delta(r - |\mathbf{r}_i - \mathbf{r}_j|) \rangle \quad (1)$$

$$g_r(|\mathbf{r} - \mathbf{r}'|) =< \exp(i\mathbf{b}[\mathbf{u}(\mathbf{r}) - \mathbf{u}(\mathbf{r}')]) >, \quad (2)$$

$$g_6(|\mathbf{r} - \mathbf{r}'|) =< \exp(i6[\theta(\mathbf{r}) - \theta(\mathbf{r}')]) >, \quad (3)$$

where $\mathbf{b}$ denotes a reciprocal lattice vector of the superlattice, $\mathbf{u}(\mathbf{r})$ is the particle displacement field, $\theta(\mathbf{r})$ is the angle (with respect to a fix axis) of the bond centered at position $\mathbf{r}$. The solid phase is characterized by a quasi-long range positional order and a long range orientational order, corresponding to an algebraic decay of $g_r(r)$ and to the absence of decay of $g_6(r)$ for $r \to \infty$. In the hexatic phase, the positional order is only short range, i. e. decays exponential, while the orientational order is

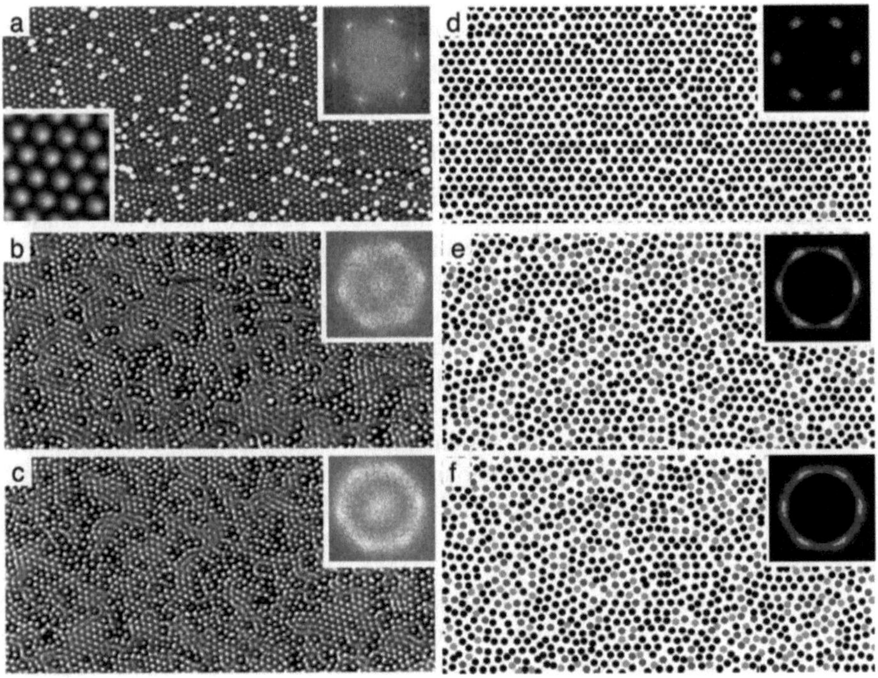

Fig. 2 (**a–c**) STM images (75 × 37 nm^2, $V = -20$ mV, $I = 20$ pA) acquired on Ce/Cu(111) for increasing temperatures T: (**a**) 8 K; (**b**) 9 K; (c) 14 K. About 0.04 ML of Ce adatoms form a macroscopic-ordered superlattice with an interatomic distance of 1.4 nm, as shown in the color inset to (**a**) (9.3 × 9.3 nm^2, $V = +1.8$ V, $I = 20$ pA). The grey-scale insets display the FT of each image. (**d–f**) Results of kMC simulations for about 0.04 ML Ce on Cu(111) at (**a**) 9 K; (**b**) 10 K; (**c**) 13 K. Black atoms are sixfold coordinated, red sevenfold, and green fivefold. The insets show the FT of each snapshot (Adapted from [5])

quasi-long range (algebraic decay of $g_6(r)$). Finally, in the liquid phase both order parameters are short range, i. e. $g_r(r)$ and $g_6(r)$ decay exponentially.

The extraction of $g_r(r)$ from the experimental data is difficult because it would require a large number of STM images for each temperature. However, from our data it was possible to compute the pair correlation function. For $T = 8$ K a power-law decay $f(r) \sim r^{-a}$ with $a \approx 0.9$ was found. For $T = 9$ and 14 K we observed an exponential decay, see Fig. 3a. This finding unambiguously shows that at 8 K the system is in the solid phase, while at $T \geq 9$ K the translational order is destroyed. The results for $g_6(r)$, deduced from the STM images of figure 2(a-c) and summarized in Fig. 3(b), reveal that in the solid state at $T = 8$ K, $g_6(r)$ tends to a finite value in agreement with the prediction of the KTNHY theory. The behavior of $g_6(r)$ for 9 and 14 K, however, is not the one expected for a true 2D system: $g_6(r)$ approaches a constant value for large r, and does not decay to zero despite the fact that the translational order is destroyed.

An extensive theoretical investigation has been performed to clarify the 2D melting process for the Ce superlattice on Cu(111) [5]. The long-range interaction

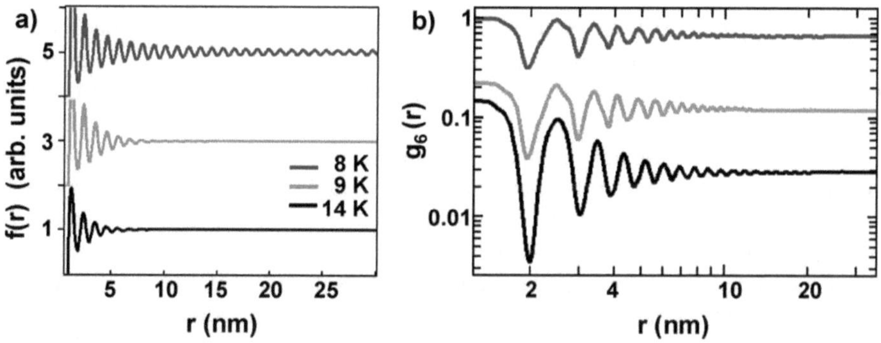

Fig. 3 (**a**) Pair correlation function $f(r)$ and (**b**) bond-angular correlation function $g_6(r)$ extracted from the experimental data shown in figure 2(a-c) (Adapted from [5])

between two single Ce adatoms at different interatomic separations was deduced using ab initio density functional theory. This calculated long-range interaction potential was then used in a large-scale kinetic Monte-Carlo simulation [5] in which the surface was represented by a triangular lattice of equivalent fcc and hcp hollow sites.

The results obtained for different temperatures are presented in Fig. 2d–f. For $T = 8$ K a 2D ordered solid is formed by the Ce adatoms, as confirmed by the Fourier transformation in the inset. Almost all atoms are found to be sixfold coordinated (black color code). In the snapshots at elevated temperatures (Fig. 2e, f), the number of adatoms with six nearest neighbors decreases, while the amount of sevenfold (red) and fivefold (green) coordinated atoms increases. In the Fourier transformation of the images a ring appears, indicating that a melting transition has taken place. However, the hexagonal symmetry persists, with a striking resemblance to the experimental data of Fig. 2a–c.

Figure 4a shows the density-density correlation function $g_r(r)$ obtained for the Ce/Cu(111) kinetic Monte Carlo simulations at different temperatures. At temperatures $T \leq 9$ K, the system exhibits crystalline long-range order, as seen in

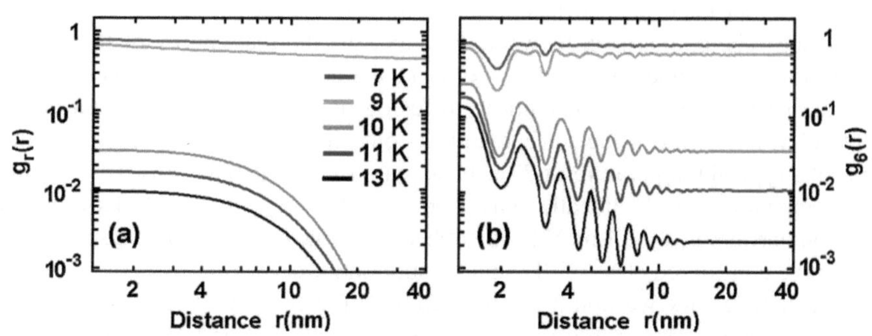

Fig. 4 (**a**) Density-density $g_r(r)$, (**b**) bond-angular $g_6(r)$ correlation functions calculated for about 0.04 ML of Ce on Cu(111) generated by kinetic Monte Carlo simulations (Adapted from [5])

the slow $r^{-\eta}$, $\eta < 0.1$ decay of $g_r(r)$. The melting of the 2D Ce lattice occurs between 9 and 10 K. Between these two temperatures the behavior of $g_r(r)$ changes dramatically – the correlation function decays exponentially above the melting temperature.

In order to detect the hexatic phase, the analysis of the bond-angular correlation function $g_6(r)$ extracted for different temperatures and shown in Fig. 4b is required. In qualitative agreement with the results deduced from the experimental data, surprisingly $g_6(r)$ does not decay to zero at any temperature. In fact, these observations are in complete agreement with the predictions of Nelson and Halperin [30]: for the melting of a 2D crystal on a 'fine mesh' potential, (i) the existence of the hexatic phase is hindered and melting directly occurs from the solid into the liquid phase, (ii) in the liquid phase $g_6(r)$ tends to a finite value at large r and a substrate-induced hexagonal symmetry is present. According to the classification described by Nelson and Halperin [30], the superlattice of Ce/Cu(111) is (i) a commensurate solid at 7 K ($g_r(r) \sim r^{-\eta}$, $\eta = 0$), (ii) a floating solid at 9 K ($g_r(r) \sim r^{-\eta}$, $\eta = 0.09$), (iii) a fluid for $T \geq 10$ K.

For a Ce superlattice on Ag(111), a corresponding theoretical analysis has been performed in which the melting point T_m has been found to be between 4.5 K and 4.9 K [5]. The superlattice of Ce on Ag(111) constitutes a floating solid at 4.0 K ($\eta = 0.14$) and 4.5 K ($\eta = 0.18$), and becomes a fluid for $T \geq 4.9$ K. These findings demonstrate that the behavior of $g_6(r)$ is intrinsic to the Ce/Cu(111) and Ce/Ag(111) superlattices, i. e. the absence of hexatic phase is due to the substrate potential.

The critical parameters defining the surface potential are (i) the adatom diffusion barrier E_{diff} and (ii) the periodicity, i. e. the separation between nearest adatoms in a superlattice d with respect to the mesh density r_0. For Ce on Cu(111) the ratio between $d = 1.3$ nm and $r_0 = 0.256$ nm sets the relative mesh density to ≈ 5. For Ce on Ag(111) the ratio is ≈ 11. In both cases, even in the liquid phase the Ce adatoms occupy discrete positions with respect to each other, and only a limited number of angles between two bonds are possible. Consequently, $g_6(r)$ doesn't decay to zero. In the limit of vanishing relative distance between neighboring adsorption sites (i. e. $d/r_0 \gg 1$), hexatic phase appears.

On the other hand, the diffusion barrier height E_{diff} describes the 'flatness' of the surface. The comparison between E_{diff} and the thermal energy of the adatom k_BT has to be considered. For $E_{\text{diff}} < k_BT$, the adatom is not influenced by the substrate periodic potential, leading to diffusion on a flat surface: the adatom can be found at any point of the surface and not only in the hollow (adsorption) sites. Thus, at a typical temperature of $T = 10$ K, with $k_B = 0.086$ meV/K, the condition for the appearance of the hexatic phase corresponds to $E_{\text{diff}} < 1$ meV. However, as diffusion barriers $E_{\text{diff}} < 1$ meV have not been observed, the hexatic phase does not exist in the considered class of 2D adatom superlattices.

As a final remark we note that the creation of such adatom superlattices with magnetic adatoms, depending on the relative strength of Kondo scattering versus RKKY interaction, may lead to the formation of ferromagnetic or antiferromagnetic adatom pairs at different separations [33]. Such a 2D superlattice may be of interest for electron-spin based quantum computing. The ability to control the long range order of magnetic adatoms opens new possibilities and presents new challenges for the investigation of 2D magnetism and for the development of novel atomic-scale magnetic devices.

3.3 Reduction of the Superconducting Gap of Ultrathin Pb Islands Grown On Si(111)

The fundamental question of how the superconducting properties of a material are modified when its thickness is reduced down to a few atomic monolayers, has stimulated considerable theoretical and experimental interest since the 1960's [34–40]. Today, with the emergence of nanoscience and nanotechnology this question is of special relevance for possible technological applications in superconducting nanodevices. The early model of Blatt and Thompson predicted an increase of the critical temperature (T_C) above the bulk value with decreasing film thickness, together with T_C oscillations due to quantum size effects (QSE) [34]. However, if proper boundary conditions allowing for spill-out of the electronic wave functions in thin films are taken into account, a decreasing T_C with decreasing film thickness was predicted [37, 38, 41–44]. Depending on the material, early experimental results showed either a decrease (Pb) or an increase (Al, Ga, Sn, In) in T_C on film thickness reduction [45, 46], mainly related to disorder in the films composed of small metallic grains [46]. In contrast, pioneering experiments on crystalline Al films reported no T_C enhancement, allowing to address intrinsic thickness-dependent properties of crystalline superconducting films [47].

Recently, Pb films grown on Si(111) attracted much attention in this context [16]. Ex situ resistivity [48] and magnetic susceptibility measurements [49] reported a decrease of T_C with decreasing Pb film thickness. In contrast, recent in situ scanning tunneling spectroscopy (STS) investigations on Pb/Si(111) islands reported no significant change in T_C upon thickness reduction [50, 51], while the very small T_C oscillations observed were attributed to QSE [34, 50]. These contradictory experimental results call for a clarification.

Recently, we reported in situ layer-dependent STS measurements of the energy gap of ultrahigh-vacuum grown single-crystal Pb/Si(111)-7×7 and Pb-$\sqrt{3} \times \sqrt{3}$/Si(111) islands in the thickness range of 5 to 60 monolayers (ML). In contrast to previous STS studies on this system [50, 51], we show that the energy gap decreases with decreasing island thickness d for both, crystalline and disordered interfaces [6]. Corresponding T_C values, estimated using the bulk gap-to-T_C ratio, decrease with a $-1/d$ dependence, in quantitative agreement with ex situ measurements of Ozer et al. [49].

Figure 5 shows an STM image of a flat-top Pb island extending over two Si terraces separated by a single Si(111) step [52]. The island mainly consists of an 8 ML thick Pb area with respect to the Si surface, as determined from the apparent height in the STM topograph [7]. The inset shows a magnified view of the Pb surface lattice with atomic resolution. The observed superstructure, reflects the buried 7×7

Fig. 5 (*Color online*). STM image of a flat-top Pb(111) single-crystal island grown on Si(111)-7×7. The island extends over two Si terraces. Island thickness includes the wetting layer. $V_{bias} = -1.0$ V, I = 100 pA. The inset shows a magnified view, revealing the Pb lattice with atomic resolution ($V_{bias} = 20$ mV, I = 1 nA). After Ref. [6]

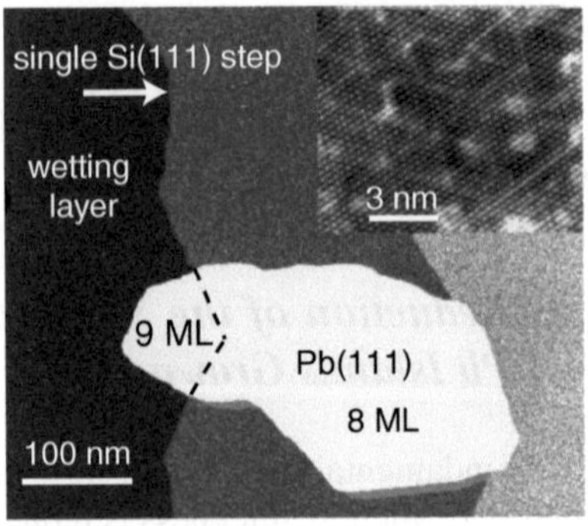

interface [52]. Pb areas of constant thickness with a lateral extension larger than the $T = 0$ K Pb bulk coherence length (~ 80 nm) were selected for this study, excluding transition regions where the island thickness changes.

Figure 6 displays a selected set of measured dI/dV spectra for the indicated island thicknesses at $T = 3.0$ K. Measurements have also been performed at 4.6 K (not shown). Each curve is an average of more than ten individual dI/dV spectra taken at various locations on one island. The background conductance was subtracted and the curves were normalized. The spectra display a clear superconducting energy gap Δ decreasing with decreasing island thickness. The observed small spectral asymmetry between positive and negative bias reflects the limits of the background subtraction on the 7×7 interface. Two phonon modes (indicated by

Fig. 6 (*Color online*). Experimental (*dots*) and calculated (continuous lines) differential conductance spectra for tunneling between a PtIr tip and a large atomically flat Pb island of selected thickness. All spectra are measured at 3.0 K on Pb/Si-7x7, except for the 7ML measured on Pb/Pb-$\sqrt{3} \times \sqrt{3}$/Si. $\Delta_{Bulk}(3.0\text{ K}) = 1.23$ meV. Arrows indicate observed phonon modes. The zero conductance level is indicated for the 6 ML spectrum. The other spectra are vertically displaced for clarity by integers. After Ref. [6]

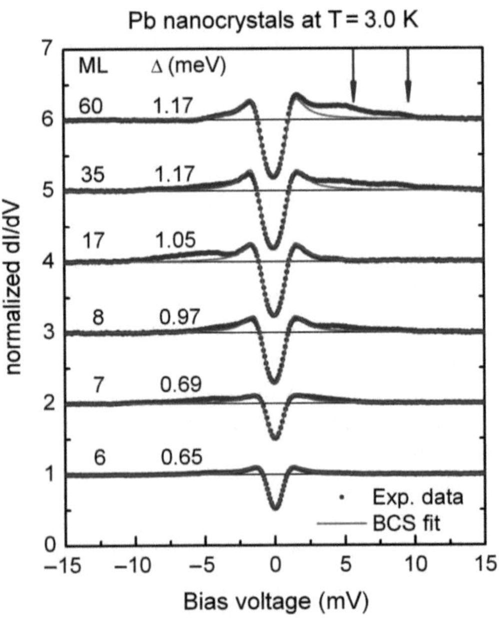

arrows for the 60 ML curve) are clearly detected. Their energies (the difference between the local extrema in d^2I/dV^2 and Δ) are 4.6 $\pm$ 0.2 meV and 8.5 $\pm$0.2 meV, in excellent agreement with the values reported for bulk Pb [53].

The extracted gap values are averaged and plotted in Fig. 7a as a function of inverse film thickness. The energy gaps for both interfaces show a comparable reduction upon decreasing thickness. To allow for a comparison with previous results, the critical temperature T_c was estimated from the measured gap values using the bulk Δ/T_c ratio and assuming the BCS temperature dependence of $\Delta(T)$. The results are displayed in Fig. 7b). The continuous line represents a least squares analysis of the STS data, leading to the relation $T_c(d) = 7.08 \times (1 - d_0/d)$ with $d_0 = 1.88$ ML. Consequently, for $d \leq 2$ ML the superconducting properties of Pb are expected to vanish.

Our estimated T_c's are in quantitative agreement with the ones of Ozer et al. [49], obtained ex situ on Ge capped Pb/Pb$\sqrt{3}\times\sqrt{3}$/Si(111). The trend observed in in situ resistivity measurements [54] is consistent with the present findings, the quantitative differences with respect to our data being most likely caused by the inhomogeneity of the films. Ex situ resistivity measurements [48] on Au capped Pb/Si(111)-7×7 films show a much faster decrease of T_c with decreasing film thickness than in

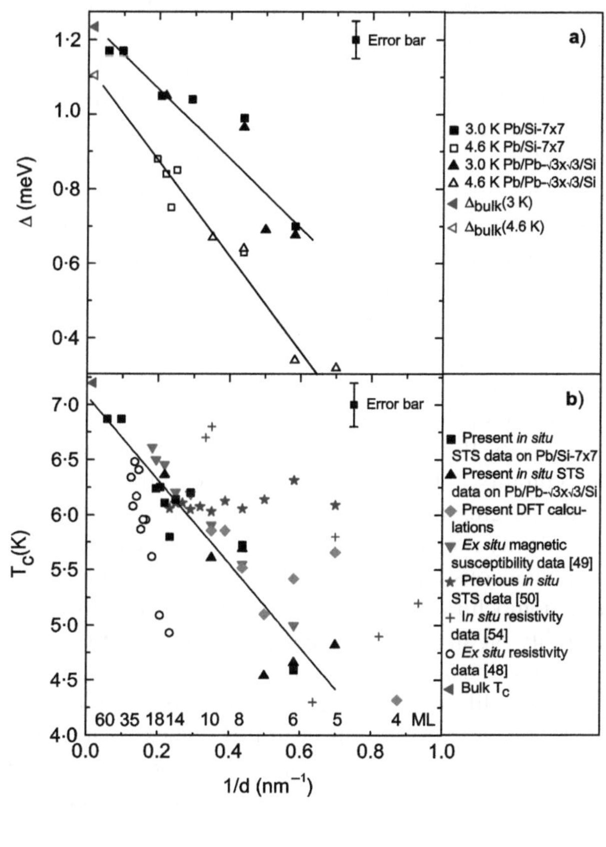

Fig. 7 (*Color online*). (**a**) Superconducting energy gap Δ as a function of inverse Pb island thickness $1/d$, extracted from BCS fits of dI/dV tunneling spectra, for the crystalline (Pb/Pb-$\sqrt{3}\times\sqrt{3}$/Si) and disordered (Pb/Si-7×7) interface. Continuous lines are guides for the eyes. b) Estimated critical temperature T_c as a function of $1/d$, using the bulk Δ/T_c ratio and assuming BCS temperature dependence of $\Delta(T)$, to allow comparison with previously reported results. Continuous line is a fit to the present STS data. For both (**a**) and (**b**) error bars: experimental dispersion and uncertainty in the fit results. After Ref. [6]

our case, probably a consequence of the inverse proximity effect induced in the Pb film by the Au capping layer [55]. However the previous in situ STS results on Pb/Si(111)-7×7, which reported an essentially constant T_c in this thickness range [50, 51], are at variance with our data. Furthermore, our observed energy gap reduction is much larger in amplitude than the very small oscillations of T_c as a function of thickness [50], the amplitude of which is of only 2% of the average T_c.

For thin Pb islands on Si(111) the experimentally observed reduction of the superconducting energy gap with decreasing film thickness is consistent with the first principle results of a thickness-dependent electron-phonon coupling constant λ, where close to the ultrathin Pb film limit the variations of the density of states at E_F play a decisive role. Interestingly, both atomically smooth (Pb/Pb-$\sqrt{3} \times \sqrt{3}$/Si) and disordered (Pb/Si-7×7) interfaces yield similar experimental behavior, in agreement with results showing that both systems are in the diffusive limit [49, 56].

3.4 Chiral Self-Assembly of Rubrene Molecules on Au(111)

Surface nanostructuring by molecular self-organization is a relevant process in the growing field of nanotechnology. Depending on the characteristics of the molecules and on the type of interactions among them and with the substrate, a variety of surface patterns have been observed by means of scanning tunneling microscopy (STM) [57]. The aromatic hydrocarbon rubrene (5,6,11, 12-Tetraphenylnaphthacene, $C_{48}H_{28}$) is chiral and non-planar due to intramolecular steric hindrance [58, 59] which causes a twisting of the central tetracene-backbone around its axis and forces the four phenyl-groups to rotate around their π-bonds. Its highly efficient luminescence established the success of rubrene as a dopant for organic light emitting diodes to improve the efficiency and stability of the devices [60]. Despite this growing interest in the application of rubrene, there have been no nanoscale investigations of the self-assembly process and the molecular and supramolecular chirality of rubrene until recently [8].

Figure 8 shows the surprising hierarchical complexity of the nested 2D supramolecular assembly of rubrene on gold with its three successive generations: single molecules, pentagonal supermolecules and supramolecular decagons. The chirality of the individual molecules is conserved in both assembly-steps. Due to the chirality dependent rotation of the building blocks, both resulting generations of the progression are chiral on their own. Since this self-organization of increasing complexity is enantioselective on the molecular and supramolecular level, we obtain a spontaneous resolution of the original racemic mixture of individual molecules into homochiral architectures. Our findings on the conservation and recognition of chirality on all stages of the supramolecular self-assembly of rubrene may have an impact on the development of chiral molecular electronic and optoelectronic devices and exemplify the working principle of basic processes in nature.

A very interesting aspect in this context is the 3D chirality transfer in rubrene multilayer islands [11]. Multilayer islands up to a thickness of six layers on a Au(111) surface have been investigated. The molecules self-organize in parallel

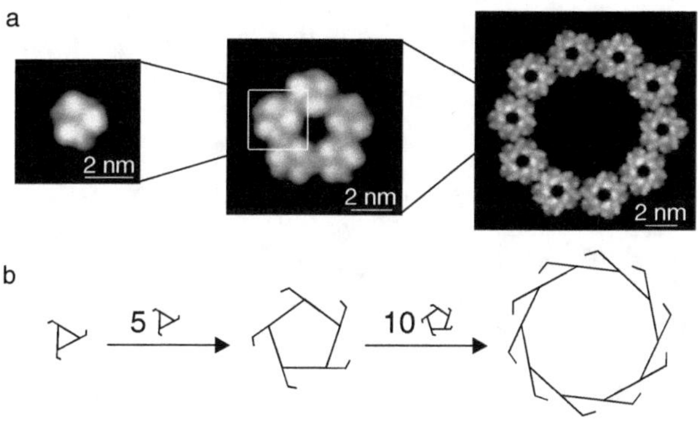

Fig. 8 Hierarchy and conservation of chirality in the spontaneous 2D supramolecular assembly. a, STM images representing the three generations of the nested two-staged self-organization. From left to right: single molecule, pentagonal supermolecule, supramolecular decagon. Typical tunneling parameters are I = 20 pA and V = −0.8 V. b, From left to right: Enantioselective assembly from L-type monomers to L-type pentagons and further on to L-type decagons. After Ref. [8]

twin rows, forming mirror domains of defined local structural chirality. Each layer is composed of twin-row domains of the same structural handedness rotated by 120° with respect to each other. Moreover, this structural chirality is transferred to all successive layers in thc island, resulting in the formation of 3D objects having a defined structural chirality. The centered rectangular surface unit cell differs from the one characteristic for the single-crystal orthorhombic phase.

Two-dimensional (2D) tiling constitutes a fundamental issue in topology [61], with fascinating examples in nature and art, and applications in many domains such as cellular biology, [62], foam physics [63], or crystal growth [64]. Recently, we presented the first observation of surface tiling with both, non-periodic and periodic arrangements of slightly distorted pentagons, hexagons, and heptagons formed by rubrene molecules adsorbed on a Au(111) surface [10]. On adjacent regions of the sample, ordered honeycomb and hexagonal close-packed patterns are found. The existence of manifold arrangements in the supramolecular self-assembly of rubrene on gold originates from the 3D non-planar flexible structure of the molecule, as well as from the nature of the intermolecular bonds.

A detail of the latter non-periodic phase is shown in Fig. 9a. The dashed blue circle surrounds a single rubrene molecule, with the sub-molecular contrast revealing three lobes and a quite regular three-fold symmetry [8, 9]. The self-assembled pattern is composed of supramolecular pentagons, hexagons and filled heptagons, which appear to be randomly distributed over the surface. There exist 11 distinct tilings by regular polygons [61], however, a combination of regular pentagons, hexagons, and heptagons generates empty gaps and overlapping regions, as inferred from the consideration of the angles at the corners of a regular pentagon (108°), hexagon (120°) and heptagon ($\approx$ 129°). Only in the case of joining three hexagons with one common corner, the angular sum is exactly 360°. Nevertheless, there exist three configurations yielding an angular sum close to 360°, schematized in Fig. 9b:

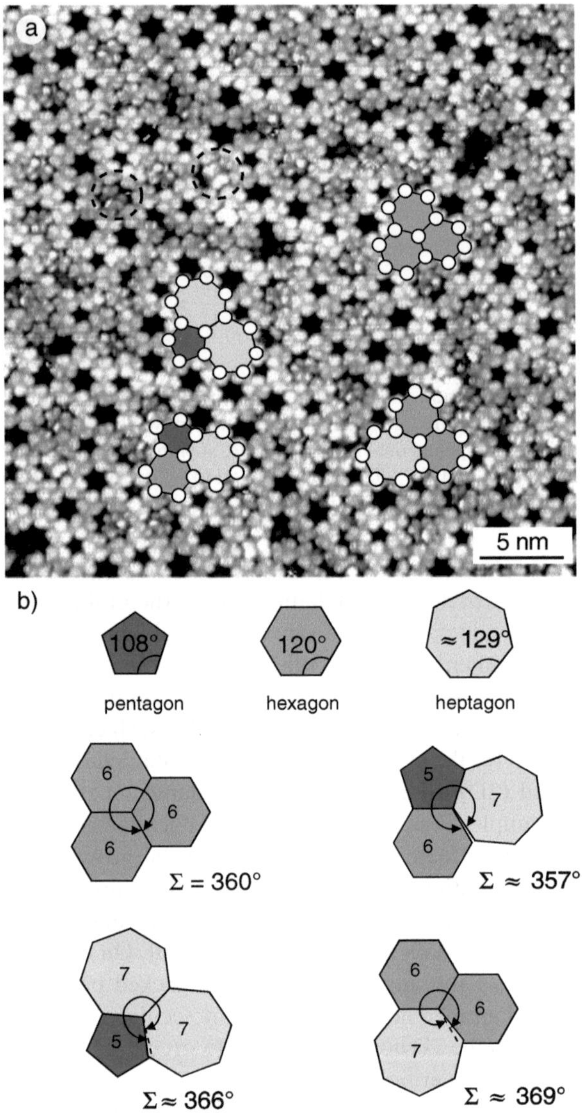

Fig. 9 (**a**) STM image showing a zoom of tiling of type B formed by a non-periodic assembly of pentagons, hexagons, and filled heptagons. (**b**) Schematic representation of the combinations of pentagons, hexagons, and heptagons resulting in three configurations close to the case of an ideal honeycomb arrangement. After Ref. [10]

pentagon-hexagon-heptagon ($\approx 357°$), pentagon-heptagon-heptagon ($\approx 366°$), and hexagon-hexagon-heptagon ($\approx 369°$). The introduction of a slight distortion of the polygons allows a plane-filling tessellation. These three configurations are the most frequently realized in the non-periodic supramolecular tiling shown in Fig. 9a. An example of each combination is superimposed to the STM image. As expected, the fact that the angular sum of the polygons involved in the patterning is either smaller or larger than the required 360° for a true plane-filling leads to small distortions of the side lengths and the angles of the objects, compared to regular polygons. The STM image in Fig. 9a also reveals that the supramolecular heptagons accommodate an additional molecule in their center. These captured specimens present two different appearances, highlighted in Fig. 9a by the two dashed black circles. The two conformations are presumably due to the varying small deformations of the heptagons, offering more circular or more elliptical holes for the additional inserted molecules. An arrangement of pentagons, hexagons and heptagons which is periodic throughout a whole domain is realized in another type of supramolecular phase present in this mixed molecular layer (not shown) [10].

A beautiful singularity existing inside these perfectly periodic domains is shown in the STM image of Fig. 10a. The honeycomb pattern is locally interrupted by a circular replacement of twelve hexagons by a ring of six pentagon-heptagon pairs, surrounding a core of seven hexagons. These rosettes are preferentially formed at elbow sites of the herringbone reconstruction of the substrate. In this image three such rosettes are present, with dashed blue circles highlighting two of them. A schematic representation of the rosette structure superimposed to the third pattern reveals that this symmetric structure with six outer pentagons and six heptagons forms a regular hexagon, which is perfectly insertable into the surrounding honeycomb arrangement. However, the ring of pentagons and heptagons suffer from the constriction from two sides due to the presence of an inner ring as well as an outer ring of hexagons. Moreover, on the outer side of the rosette the combination pentagon- hexagon-hexagon is present which – with a nominal angular sum of 348° – induces a more important deformation of the pentagons. This angular stress leads to a compression of both, the pentagons and the heptagons in opposite directions, as illustrated in Fig. 10b. Again, within the supramolecular rosettes the heptagons are filled with additional molecules, exhibiting mainly the elongated conformation. We note that the replacement of two hexagons with a pentagon-heptagon pair is one of the most common topological defects found in honeycomb structures, such as in graphite and in carbon nanotubes [65]. The replacement of all hexagons would generate the so-called pentaheptite structure, a tiling observed in crystallography [66].

In the sub-monolayer regime, the self-assembly of rubrene on Au(111) results in a variety of distinct phases. As a function of the molecular coverage, honeycomb islets, supramolecular pentagonal chains, and close packed islands have been observed [8, 9]. The periodic and non-periodic tilings presented here demonstrate in an especially fancy manner the conformational flexibility of the rubrene molecules. Supramolecular self-assembly results from the contributions of molecule-substrate and molecule-molecule interactions. Intermolecular bonds are mainly based on

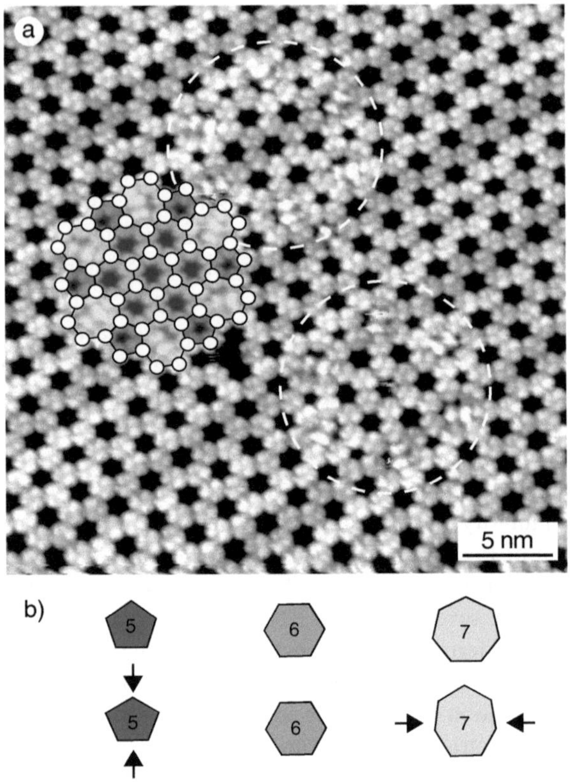

Fig. 10 (**a**) STM image showing three supramolecular rosettes embedded in a well-ordered honeycomb domain. The schematic drawing of the pattern is superimposed to one of the three rosettes. (**b**) Schematic representation of the distortion of the pentagons and heptagons to compose the tiling. After Ref. [10]

weak, non-covalent interactions such as dipole-dipole (or higher order multipole) forces, hydrogen bonds, or van der Waals (vdW) interactions [57, 67]. Given the absence of a dipole moment, the bonding between rubrene molecules might have contributions from electrostatic quadrupole-quadrupole interactions, vdW forces, π-π and CH-π bonding. Among them vdW forces are isotropic, the other interactions are directional. While the π-π interaction favours a parallel arrangement of the π systems [68], the CH-π bond is particularly strong for a CH group pointing perpendicular to a π system of another molecule [68]. These forces relying on the existence of π systems are effectively realized between rubrene molecules, due to the non-planar aromatic tetracene backbone, the four phenyl groups, and the presence of 28 CH bonds pointing into three dimensions. The directional interactions can account for the formation of the ordered hexagonal structures, in which the bond angle between molecules is precisely 120°. Intriguingly, a small modification of the intermolecular bonds or a slight change in the molecular conformation results in the formation of pentagons and heptagons. The adaptive behaviour of rubrene allows these modifications leading to the observed 2D periodic and non-periodic tilings with pentagons, hexagons, and heptagons.

The fact that the supramolecular self-assembly of rubrene on Au(111) results in different adsorption conformations of the physisorbed molecules is reflected in a different electronic structure of the adsorbates. Owing to the 3D geometry of the molecule providing an inherent decoupling of the molecular π states from the substrate the conformers are distinguished by their submolecular appearance in the scanning tunneling microscopy images and by the corresponding position of the highest molecular orbital in the differential conductance (dI/dV) spectra. The application of an electric field induces a switching of the electronic and geometric conformation of the self-assembled molecules [9].

3.5 STM-induced Light Emission from Fullerene Molecules on a Dielectric Substrate

Light emission induced by electrons tunneling through the junction formed by the sample and the tip of a STM has been proposed to characterize the optical properties of nanoscale objects at surfaces [69, 70]. Contrary to conventional non-local techniques, the local character of this method offers the unique possibility to select and probe individual atoms, molecules or clusters on surfaces.

Recently, luminescence from supported molecules has been obtained [71, 72] by successfully decoupling them from the metallic substrate in order to avoid quenching of the radiative transitions [73, 74], using either a thin oxide film [71] or several molecular layers [72]. However, unambiguous chemical identification of single complex molecules requires the observation and identification of *several* vibrational and/or electronic-vibrational transitions, which are the spectroscopic fingerprint of the species. Recently, Ćavar et al. [12] presented the first observation of energy resolved luminescence from selected C_{60} molecules embedded in an ensemble of C_{60} molecules excited by electrons tunneling through a double barrier STM junction.

Figure 11 shows C_{60} islands grown on both, the bare Au(111) (**A**) and the NaCl covered surface (**B–D**). The nanocrystals present a minimum height of two layers of C_{60} molecules (island **B**). The nucleation of the C_{60} nanocrystals starts at defects of the NaCl layer (protrusions or vacancies), monatomic steps of Au(111) (covered with NaCl), or edges of the second layer of NaCl. The C_{60} molecules form hexagonally arranged layers with an intermolecular distance of 1 nm.

The next step in the investigation of these complex samples is the characterization of their plasmonic properties related to the gold substrate. The tip-induced localized surface plasmon (LPS) is excited through an inelastic electron tunneling mechanism and then decays radiatively [75, 76]. Similar spectra are obtained with a reduced intensity on the ultrathin NaCl films and with very low intensity on the fullerene monolayer islands grown on the metal surface, as shown for C_{70}. No intrisic luminescence is observed for the fullerene monolayers on Au(111), as radia-

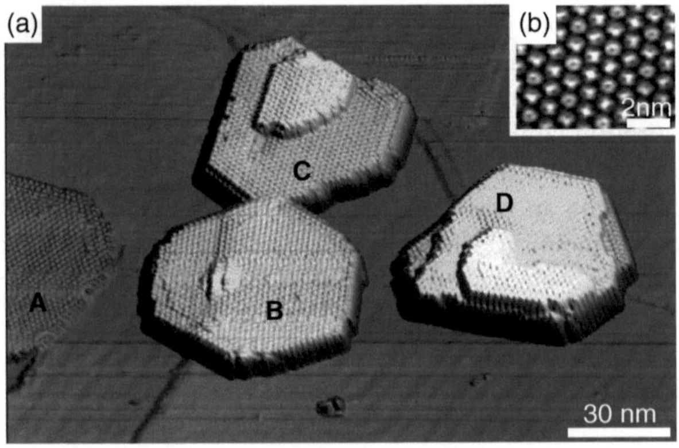

Fig. 11 (**a**) STM image of C_{60} nanocrystals formed on a NaCl ultrathin film grown on Au(111) ($V = -3$ V, $I = 0.02$ nA). Island **A** is a C_{60} monolayer on Au(111), the small blue triangle below **A** is part of the bare Au surface. Hexagonal island **B**, and truncated triangular islands **C** and **D** consist of up to two, three, and four C_{60} molecular layers, respectively, on NaCl. (**b**) Sub-molecular resolution on island **B** ($V = -3$ V, $I = 0.1$ nA). After Ref. [12]

tionless relaxation takes place owing to the direct contact with the metal substrate, a process known in both, photoluminescence [77] and STM-LE experiments [72, 73]. As expected, the LSP is excited for both polarities [76]. However, the excitation parameters for bare Au and for the NaCl layer are not the same in the positive voltage range, in contrast to the negative one, owing to different image potentials [78]. Considering the plasmon enhancement effect, the bias-dependent emission indicates that, if the NaCl layer is used as spacer to decouple the molecules from the metal substrate, a detection of molecular luminescence is only expected for negative voltages larger than -1.8 V and for positive ones between +1.8 V and +3.2 V.

In order to verify the effect of the plasmon amplification on the intrinsic molecular luminescence, different series of STM-LE spectra have been acquired on the fullerene nanocrystals. Figure 12 shows STM-induced light emission spectra acquired over two C_{60} nanocrystals (a) and over their respective NaCl spacer (b), for two different tip conditions (runs 1 and 2). The origin and nature of the molecular emission will be discussed later. The wavelength of the LSP resonance was changed between the two experimental runs by modifying the tip shape with a series of high-voltage pulses across the tunnel junction. Even though the two main peaks are found at similar positions in both molecular spectra, the spectral characteristics of the LSP influence their relative intensity. To substantiate this statement, we divide each molecular spectrum by its corresponding plasmon-emission spectrum. The resulting two normalized curves, shown in Fig. 12c, identical in both spectral features and relative intensities, demonstrate an enhancement of the molecular luminescence in the STM tip-sample gap by coupling with the plasmon modes. This finding indicates that the localized surface plasmons are essential for the detection of molecular light emission.

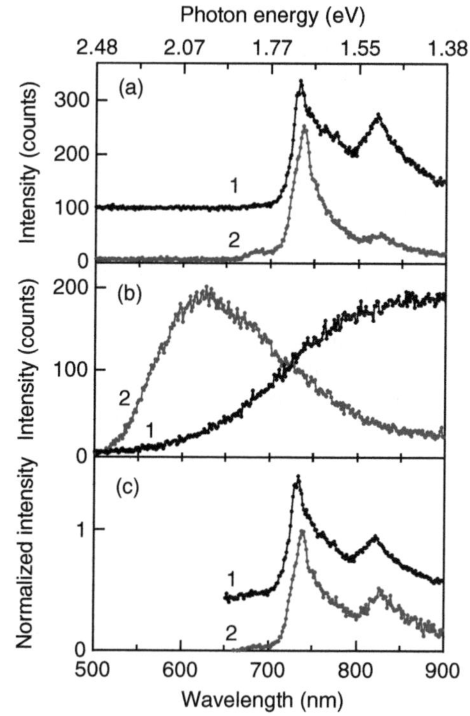

Fig. 12 STM-LE spectra for two different experimental runs (1 and 2). (**a**) Spectra acquired over C_{60} nanocrystals grown on NaCl/Au(111). Spectra 1 ($V = -3$ V, $I = 2$ nA, $t = 300$ s) and 2 ($V = -3$ V, $I = 3$ nA, $t = 300$ s) are vertically shifted for clarity. (**b**) Spectra acquired on NaCl at the same conditions as in (**a**). (**c**) Molecular spectra from (**a**), divided by the corresponding smoothed plasmon-mediated spectra from (**b**) and normalized to fit the same scale. Spectra are vertically shifted for clarity. After Ref. [13]

Figure 13 shows STM-induced light emission spectra acquired over a C_{60} (a) and a C_{70} (b) nanocrystal, as well as over the NaCl spacer layers. Luminescence from the supported fullerenes, clearly distinguishable from the LSP emission, is observed for negative excitation voltages larger than −2.3 V for C_{60} and −2.5 V for C_{70}. For positive voltages up to +5 V, no photon emission was detected for both fullerenes.

The observed bias dependence of the STM-induced light emission from the fullerene molecules is characteristic of a hot electron injection mechanism [71, 72] followed by a radiative decay associated with the highest occupied molecular orbital-lowest unoccupied molecular orbital (HOMO–LUMO) gap of the molecules. This interpretation implies that the energy levels of the emitting molecules are not pinned to the gold substrate, but shift with the applied bias, as illustrated in Fig. 14a. When the substrate is negatively biased with a voltage above the observed threshold value, the LUMO falls at an energy level lower than the Fermi energy (E_F) of the sample and the HOMO at an energy level higher than the Fermi energy of the tip. Electrons tunnel elastically from the gold substrate into the LUMO through the NaCl barrier (hot electron injection) and simultaneously from the HOMO into the tip through the vacuum barrier to create electronically and vibrationally excited states of molecules. The transitions to the ground state give rise to the observed molecular luminescence. The fact that the latter is not observed for positive excitation voltages, in spite of the presence of the plasmon resonance between +1.8 V and +3.2 V, may be due to an asymmetry of the HOMO and LUMO position with respect to E_F

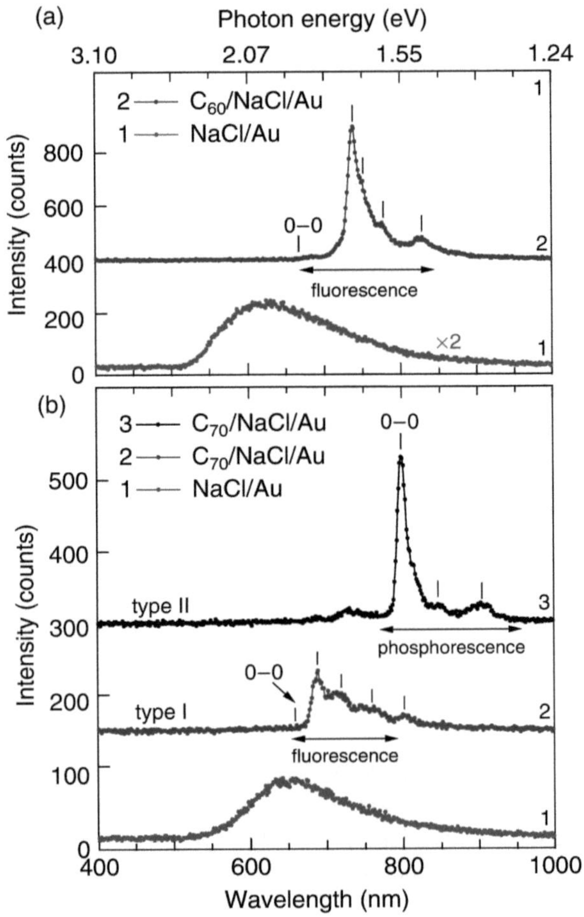

Fig. 13 (**a**) STM-LE spectra acquired over a C_{60} nanocrystal (spectrum 2, $V = -3$ V, $I = 4$ nA, $t = 60$ s) and over the underlying NaCl film (spectrum 1, $V = -3$ V, $I = 2$ nA, $t = 60$ s). (**b**) STM-LE spectra acquired over a C_{70} nanocrystal (spectra 2 and 3) and over the NaCl film (spectrum 1), ($V = -3$ V, $I = 1$ nA, $t = 60$ s). Spectra are vertically shifted for clarity. 0-0 indicates the pure electronic origins. After Ref. [13]

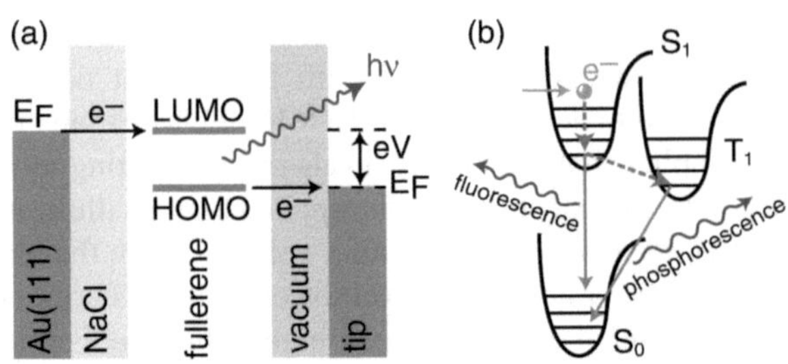

Fig. 14 (**a**) Schematic energy diagram of a double-barrier tunnel junction at negative bias voltage, corresponding to the conditions for luminescence. The simultaneous population and depopulation of the LUMO and HOMO by tunneling electrons creates an electronically and vibrationally excited state of the molecule. (**b**) Fluorescence channel: the molecule relaxes to the ground vibrational state of the lowest electronic excited singlet state S_1 followed by a radiative transition to the singlet ground state S_0. Phosphorescence channel: a radiationless transition occurs between the singlet state S_1 and triplet state T_1 with subsequent radiative molecular transition to the ground state S_0. After Ref. [13]

and to different characteristics for the tip-molecule and molecule-substrate double barrier [79, 12]. As a consequence, an energy level alignment leading to luminescence can not be realized in the range of positive bias corresponding to the plasmon enhancement.

The luminescence spectrum from a C_{60} nanocrystal shown in Fig. 13a (curve 2) is characterized by a peak at 738 nm followed by two shoulders, and a second peak at 828 nm. On the other hand, the excitation of a C_{70} molecular nanocrystal gives rise to two substantial different types of optical spectra shown in (b): (i) one characterized by an intense onset at 690 nm and small broad peaks up to 800 nm (type I) and (ii) a second one covering the spectral range between 700 and 950 nm with an intense line at 800 nm (type II). It is important to mention that the plasmon resonance, centered at 620 nm for C_{60} measurements (curve 1 in a) and at 650 nm (curve 1 in b) in the case of C_{70}, was unchanged during each set of measurements. Thus, the existence of the two types of spectra observed for C_{70} can not be attributed to a different plasmon enhancement. For measurements on both fullerenes, a rigid shift of the entire spectrum has been observed, independently of the excitation voltages, covering to the following wavelength range for the main peaks: between 718 and 745 nm for C_{60}; between 682 and 709 nm for C_{70} spectra of type I; between 800 and 813 nm for C_{70} spectra of type II. These shifts are independent of the corresponding plasmon-mediated emission.

In order to identify the electronic transitions giving rise to the observed molecular light emission spectra shown in Fig. 13, we compare our results with laser-induced photoluminescence data from fullerene molecules in different media.

Photoluminescence spectra acquired for C_{60} in rare gas matrices [80, 81], thin films [82, 83], and single crystals [84, 85] have been considered. Although extensively investigated, the importance of solid state effects in the description of the luminescence process for C_{60} solids has not been completely clarified. Nevertheless, this comparison permits to attribute the observed spectra to fluorescence from the $S_1 \rightarrow S_0$ transition. In the light of the present results on the plasmon-mediated amplification, the previously reported STM-induced luminescence from C_{60} molecules [12] has to be reinterpreted. Two different types of molecular spectra, assigned to fluorescence and phosphorescence, were claimed to be observed. The dissimilarity between both spectra is now understood as arising mainly from differently structured LSP spectra, originating from cut PtIr tips with unstable and ill-defined shape used in those measurements. We interpret the observed luminescence spectra as fluorescence from C_{60}, although a contribution from a phosphorescence channel, corresponding to the $T_1 \rightarrow S_0$ transition, cannot be excluded in the low energy part of the spectra.

Intriguingly, the observation of both, fluorescence and phosphorescence is realized in the case of C_{70}, as deduced from the comparison of our results with laser-induced photoluminescence data from dispersed C_{70} molecules in different media [86–88] and from C_{70} solids [89–91]. The observed spectra of type I are attributed to fluorescence from the $S_1 \rightarrow S_0$ transition and the spectra of type II to phosphorescence from the $T_1 \rightarrow S_0$ transition. Because of a relatively low spectral resolution in the present experiment (8 nm), only the pure electronic origin (0–0) corresponding to the triplet-to-singlet ground state transition is identified at about 800 nm. The other spectral features correspond to unresolved multiplet vibronic structures. Note that the broad peak at 800 nm in the spectrum of type I shown in Fig. 13b is assigned to the phosphorescence channel rather than to the fluorescence one. We indeed observed several times emission spectra consisting of a superposition of components from both radiative transitions.

The spectral shifts discussed above for both C_{60} and C_{70} which have been also observed in laser-induced photoluminescence spectra are attributed to a site dependence [84, 88]. The local environment or the presence of defects may also be at the origin of the two types of relaxation channels, fluorescence and phosphorescence, observed for C_{70}.

In summary, light emission induced by tunneling electrons was observed from C_{60} and C_{70} molecules that were electronically decoupled from the metal substrate by a thin NaCl film. The molecular transitions giving rise to light emission were identified by comparison with laser-induced photoluminescence data from dispersed molecules or molecular crystals reported in the literature. The key role of localized surface plasmons in the enhancement of molecular emission is unambiguously demonstrated. The present observation of local fluorescence and phosphorescence represents a crucial step towards chemical recognition on the nanoscale.

4 Conclusions

In this short review experiments at the scale of atoms, molecules, and supermolecules have been presented which exhibit exciting physical effects. These include light emission from individual molecules, reduction of the supraconducting gap in lead nano-islands, 2D self-assembly and melting of adatoms, and the formation of chiral supramolecular structures. The common link between these different topics is the characterization of the structural and electronic properties of supported nanostructures at atomic-scale spatial and meV-scale spectral resolution. The opportunities of this concept in the emerging field of nanotechnology, where size matters, are evident. For example, a reduction in size often provides an increase in speed for electronic or magnetic devices. Therefore, a further advancement towards a future nanotechnology will depend crucially on the precise control, characterization, and, ultimately, functionalization, of matter at the atomic and molecular level. This prospect remains an inciting scientific and technological challenge.

Acknowledgments I would like to thank M. Ternes, C. Weber, M.-C. Blüm, E. Ćavar, M. Pivetta, F. Patthey, F. Silly, J. P. Pelz, T. Giamarchi, F. Mila, M. Chergui, C. Brun, I. P. Hong, C. Rossel, N. N. Negulyaev, V. S. Stepanyuk, L. Niebergall, P. Bruno, I. Yu Sklyadneva, X. Zubizarreta, R. Heid, V. M. Silkin, P. M. Echenique, K. P. Bohnen, and E. V. Chulkov for a very fruitful and stimulating collaboration. This work has been supported by the Swiss National Science Foundation.

References

1. G.M. Whitesides, B. Grzybowski, Science **295**, 2418 (2002)
2. F. Silly, M. Pivetta, M. Ternes, F. Patthey, J.P. Pelz, W.-D. Schneider, Phys. Rev. Lett. **92**, 016101 (2004)
3. F. Silly, M. Pivetta, M. Ternes, F. Patthey, J.P. Pelz, W.-D. Schneider, New J. Phys. **6**, 16 (2004)
4. M. Ternes, C. Weber, M. Pivetta, F. Patthey, J.P. Pelz, T. Giamarchi, F. Mila, W.-D. Schneider, Phys. Rev. Lett. **93**, 14805 (2004)
5. N.N. Negulyaev, V.S. Stepanyuk, L. Niebergall, P. Bruno, M. Pivetta, M. Ternes, F. Patthey, W.-D. Schneider, Phys. Rev. Lett. **102**, 246102 (2009)
6. C. Brun, I.-P. Hong, F. Patthey, I. Yu Sklyadneva, R. Heid, P.M. Echenique, K.P. Bohnen, E.V. Chulkov, W.-D. Schneider, Phys. Rev. Lett. **102**, 207002 (2009)
7. I.-P. Hong, C. Brun, F. Patthey, I. Yu Sklyadneva, X. Zubizarreta, R. Heid, V.M. Silkin, P.M. Echenique, K.P. Bohnen, E.V. Chulkov, W.-D. Schneider, Phys. Rev. B **80**, 081409(R) (2009)
8. M.-C. Blüm, E. Ćavar, M. Pivetta, F. Patthey, W.-D. Schneider, Angew. Chem. Int. Ed. **44**, 5334 (2005)
9. M.-C. Blüm, M. Pivetta, F. Patthey, W.-D. Schneider, Phys. Rev. B **73**, 195409 (2006)
10. M. Pivetta, M.-C. Blüm, F. Patthey, W.-D. Schneider, Angew. Chem. Int. Ed. **47**, 1076 (2008)
11. M. Pivetta, M.-C. Blüm, F. Patthey, W.-D. Schneider, J. Phys. Chem B **113**, 4578 (2009)
12. E. Ćavar, M.-C. Blüm, M. Pivetta, F. Patthey, M. Chergui, W.-D. Schneider, Phys. Rev. Lett. **95**, 196102 (2005)
13. F. Rossel, M. Pivetta, F. Patthey, W.-D. Schneider, Optics Express **17**, 2714 (2009)
14. R. Gaisch, J.K. Gimzewski, B. Reihl, R.R. Schlittler, M. Tschudy, W.-D. Schneider, Ultramicroscopy **42–44**, 1621 (1992)
15. H.H. Weitering, D.R. Heslinga, T. Hibma, Phys. Rev. B **45**, 5991 (1992)
16. M. Jalochowski, E. Bauer, Phys. Rev. B **38**, 5272 (1988)
17. K.H. Lau, W. Kohn, Surf. Sci. **75**, 69 (1978)
18. E. Wahlström, I. Ekvall, H. Olin, L. Wallden, Appl. Phys. A **66**, S1107 (1998)
19. J. Repp, F. Moresco, G. Meyer, K.H. Rieder, P. Hyldgaard, M. Persson, Phys. Rev. Lett. **85**, 2981 (2000)
20. N. Knorr, H. Brune, M. Epple, A. Hirstein, M.A. Schneider, K. Kern, Phys. Rev. B **65**, 115420 (2002)
21. M. Ternes, M. Pivetta, F. Patthey, W.-D. Schneider, Progr. Surf. Sci. **85**, 1–27 (2010)
22. P. Hyldgaard, M. Persson, J. Phys. Condens. Matter **12**, 13 (2000)
23. N.N. Negulyaev, V.S. Stepanyuk, W. Hergert, H. Fangohr, P. Bruno, Sur. Sci. Lett. **600**, L58 (2006)
24. N.N. Negulyaev, V.S. Stepanyuk, L. Niebergall, W. Hergert, H. Fangohr, P. Bruno, Phys. Rev. B **74**, 035421 (2006)
25. Th. von Hofe, J. Kröger, R. Berndt, Phys. Rev. B **73**, 245434 (2006)
26. N.D. Mermin, H. Wagner, Phys. Rev. Lett. **17**, 1133 (1966)
27. N.D. Mermin, Phys. Rev. **176**, 250 (1968)
28. J.M. Kosterlitz, N.J. Thouless, J. Phys. C **6**, 1181(1973)
29. B.I. Halperin, D.R. Nelson, Phys. Rev. Lett. **41**, 121(1978)
30. D.R. Nelson, B. Halperin, Phys. Rev. B **19**, 2484 (1979)
31. A.P. Young, Phys. Rev. B **19**, 1855 (1979)
32. K.J. Strandburg, Rev. Mod. Phys. **60**, 161 (1988)
33. V.S. Stepanyuk, L. Niebergall, R.C. Longo, W. Hergert, P. Bruno, Phys. Rev. B **70**, 075414 (2004)
34. J.M. Blatt, C.J. Thompson, Phys. Rev. Lett. **10**, 332 (1963)
35. F. Yu, Komnik, E.I. Bukhshtab, K.K. Mankovskii, JETP **30**, 807 (1970)
36. M. Strongin, R.S. Thompson, O.F. Kammerer, J.E. Crow, Phys. Rev. B **1**, 1078 (1970)
37. R.E. Allen, Phys. Rev. B **12**, 3650 (1975)
38. M. Yu, M. Strongin, A. Paskin, Phys. Rev. B **14**, 996 (1976)
39. B.G. Orr, H.M. Jaeger, A.M. Goldman, Phys. Rev. Lett. **53**, 2046 (1984)

40. D.B. Haviland, Y. Liu, A.M. Goldman, Phys. Rev. Lett. **62**, 2180 (1989)
41. F.K. Schulte, Surf. Sci. **55**, 427 (1976)
42. J. Simonin, Phys. Rev. B **33**, 7830 (1986)
43. B. Chen, Z. Zhu, X.C. Xie, Phys. Rev. B **74**, 132504 (2006)
44. A.A. Shanenko, M.D. Croitoru, F.M. Peeters, Phys. Rev. B **75**, 014519 (2007)
45. B. Abeles, R.W. Cohen, G.W. Cullen, Phys. Rev. Lett. **17**, 632 (1966)
46. A.M. Goldman, N. Markovic, Phys. Today **51**, 11, 39 (1998)
47. M. Strongin, O.F. Kammerer, H.H. Farrel, D.L. Miller, Phys. Rev. Lett. **30**, 129 (1973)
48. Y. Guo et al., Science **306**, 1915 (2004)
49. M.M. Ozer, J.R. Thompson, H.H. Weitering, Nature Phys. **2**, 173 (2006)
50. D. Eom, S. Qin, M.-Y. Chou, C.K. Shih, Phys. Rev. Lett. **96**, 027005 (2006)
51. T. Nishio et al., Appl. Phys. Lett. **88**, 113115 (2006)
52. I.B. Altfeder, K.A. Matveev, D.M. Chen, Phys. Rev. Lett. **78**, 2815 (1997)
53. J.M. Rowell, L. Kopf, Phys. Rev. B **137**, A907 (1965)
54. O. Pfennigstorf, A. Petkova, H.L. Guenter, M. Henzler, Phys. Rev. B **65**, 045412 (2002)
55. P.G. De Gennes, Rev. Mod. Phys. **36**, 225 (1964)
56. T. Nishio et al., Phys. Rev. Lett. **101**, 167001 (2008)
57. S. De Feyter, F.C. De Schryver, Chem. Soc. Rev. **32**,139 (2003); J.V. Barth, Annu. Rev. Phys. Chem. **58**, 375 (2007)
58. N. Sato, K. Seki, H. Inokuchi, J. Chem. Soc., Faraday Trans. II **77**, 1621–1633 (1981)
59. P.U. Biedermann, A. Levy, J.J. Stezowski, I. Agranat, Chirality **7**, 199–205 (1995)
60. Z. Zhi-lin et al., J. Phys. D: Appl. Phys. **31**, 32–35 (1998)
61. B. Grünbaum, G.C. Shephard, *Tilings and Patterns*, (Freeman, New York, NY 1987)
62. J. Stavans, Rep. Prog. Phys. **56**, 733 (1993), B. Dubertret, T. Aste, H.M. Ohlenbusch, N. Rivier, Phys. Rev. E **58**, 6368 (1998)
63. F. Graner, Y. Jiang, E. Janiaud, C. Flament, Phys. Rev. E **63**, 011402 (2000)
64. G. Ungarn, X. Zeng, Soft Matter **1**, 95 (2005)
65. A.J. Stone, D.J. Wales, Chem. Phys. Lett. **128**, 501 (1986); P. Simonis, C. Goffaux, P.A. Thiry, L.P. Biro, Ph. Lambin, V. Meunier, Surf. Sci. **511**, 319 (2002), A. Hashimoto, K. Suenaga, A. Gloter, K. Urita, S. Iijima, Nature **430**, 870 (2004)
66. V.H. Crespi, L.X. Benedict, M.L. Cohen, S.G. Louie, Phys. Rev. B **53**, R13303 (1996); M. Deza, P.W. Fowler, M. Shtogrin, K. Vietze, J. Chem. Inf. Comput. Sci. **40**, 1325 (2000)
67. G.R. Desiraju, Nature **412**, 397 (2001)
68. M. Tamres, J. Am. Chem. Soc. **74**, 3375 (1952), M. Nishio, M. Hirota, Y. Umezawa, *The CH/π interaction: evidence, nature, and consequences*, (Wiley, New York, NY 1998); U. Samanta, P. Chakrabarti, J. Chandrasekhar, J. Phys. Chem. A **102**, 8964 (1998), S. Tsuzuki, K. Honda, T. Uchimaru, M. Mikami, K. Tanabe, J. Am. Chem. Soc. **122**, 3746 (2000); M. Nishio, Cryst. Eng. Comm. **6**, 230 (2004)
69. R.D. Young, Physics Today, November 1971, 42
70. J.K. Gimzewski, B. Reihl, J.H. Coombs, R.R. Schlittler, Z. Phys. B Cond. Matter **72**, 497 (1988)
71. X.H. Qiu, G.V. Nazin, W. Ho, Science **299**, 542 (2003)
72. Z.-C. Dong, X.-L.Guo, A.S. Trifonov, P.S. Dorozhkin, K. Miki, K. Kimura, S. Yokoyama, S. Mashiko, Phys. Rev. Lett. **92**, 086801 (2004)
73. R. Berndt, R. Gaisch, J.K. Gimzewski, B. Reihl, R. R. Schlittler, W.-D. Schneider, M. Tschudy, Science **262**, 1425 (1993)
74. Z.-C. Dong, A. Kar, P. Dorozhkin, K. Amemiya, T. Uchihashi, S. Yokoyama, T. Kamikado, S. Mashiko, T. Okamoto, Thin Solid Films **438–439**, 262 (2003)
75. R. Berndt, J.K. Gimzewski, P. Johansson, Phys. Rev. Lett. **67,** 3796 (1991)
76. R. Berndt, J.K. Gimzewski, P. Johansson, Phys. Rev. Lett. **71,** 3493 (1993)
77. W.L. Barnes, J. Mod. Opt. **45,** 661 (1998)
78. R Berndt, J.K. Gimzewski, Ann. Physik **2,** 133 (1993)
79. S.W. Wu, G.V. Nazin, X. Chen, X.H. Qiu, W. Ho, Phys. Rev. Lett. **93,** 236802 (2004)

80. W.C. Hung, C.D. Ho, C.P. Liu, Y.P. Lee, J. Phys. Chem. **100,** 3927 (1996)
81. A. Sassara, G. Zerza, M. Chergui, F. Negri, G. Orlandi, J. Chem. Phys. **107,** 8731 (1997)
82. Y. Wang, J.M. Holden, A.M. Rao, P.C. Eklund, U.D. Venkateswaran, D.L. Eastwood, R.L. Lideberg, G. Dresselhaus, M.S. Dresselhaus, Phys. Rev. B **51,** 4547 (1995)
83. V. Capozzi, G. Casamassima, G.F. Lorusso, A. Minafra, R. Piccolo, T. Trovato, A. Valentini, Solid State Comm. **98,** 853 (1996)
84. W. Guss, J. Feldmann, E.O. Gobel, C. Taliani, H. Mohn, W. Muller, P. Haussler, H.U. Termeer, Phys. Rev. Lett. **72,** 2644 (1994)
85. D.J. van den Heuvel, I.Y. Chan, E.J.J. Groenen, M. Matsushita, J. Schmidt, G. Meijer, Chem. Phys. Lett. **233,** 284 (1995)
86. S.M. Argentine, K.T. Kotz, A.H. Francis, J. Am. Chem. Soc. **117,** 11762 (1995)
87. A. Sassara, G. Zerza, M. Chergui, J. Phys. Chem. **102,** 3072 (1998)
88. J.H. Rice, J.P. Galaup, S. Leach, Chem. Phys. **279,** 23 (2002)
89. S. Kazaoui, N. Minami, Synth. Met. **86,** 2345 (1997)
90. M. Ichida, M. Sakai, T. Yajima, A. Nakamura, H. Shinohara, Chem. Phys. Lett. **271,** 27 (1997)
91. V. Capozzi, G. Perna, Thin Solid Films **515**, 7247 (2007)

Computational Nanomechanics of Quasi-one-dimensional Structures in a Symmetry-Adapted Tight Binding Framework

Traian Dumitrica

Abstract The success of many nanotechnologies depends on our ability to understand and control the mechanics of nano objects, such as nanotubes and nanobelts. Because of the numerous experimental difficulties encountered at this scale, simulation can emerge as a powerful predictive tool. For this, new multiscale simulation methods are needed in which a continuum model emerges from a precise, quantum mechanical description of the atomic scale. Because computing nanomechanical responses requires large systems, computationally affordable but less accurate classical atomistic treatments of the atomic scale are widely adopted and only multiscale classical atomistic-to-continuum bridging is achieved. As a first step towards achieving accurate multiscale models for nano objects, based on a quantum-mechanical description of chemical bonding, here we present an ingenious symmetry-adapted atomistic scheme that performs calculations under helical boundary conditions. The utility of the microscopic method is illustrated with examples discussing the nanomechanical response of carbon nanotubes and thermodynamical stability of silicon nanowires.

1 Introduction

The discovery of quasi one-dimensional organizations of matter with distinct shapes – nanotubes [1, 2], and nanowires [3, 4] – opened a new frontier in science and engineering. Understanding the behavior of these structures is not only of fundamental but also of practical importance, because by capitalizing on the science emerging from the newly accessible size range, engineers can develop technologies that will benefit humankind. The success of these nanotechnologies depends on our ability to understand and control the nanoscale mechanics. As experimental difficulties are numerous on this small scale, computer simulations emerge as a powerful predictive tool with great effects on technological innovation.

T. Dumitrica (✉)
Department of Mechanical Engineering, University of Minnesota, Saint Paul, MN, USA
e-mail: td@me.umn.edu

A. Aldea, V. Bârsan (eds.), *Trends in Nanophysics*, Engineering Materials,
DOI 10.1007/978-3-642-12070-1_2,

It should be noted that atomistic simulation methods are required at the nanoscale. However, because of the structure of interest extent over several micrometers or more in one direction, a full atomistic treatment is prohibitive. This situation demands new simulation methods and the vision is that nanoscale modeling will be achieved through a multiscale approach, where the continuum emerges from a precise, quantum mechanical description of the atomic scale [5].

The first important step to achieve multiscale modeling is designing an atomistic scheme to compute the nanomechanical response with high accuracy. Based on the obtained result, one can further construct a continuum model of the nano objects. The chapter is addressing this first critical step by presenting a *symmetry-adapted microscopic modeling approach* applicable to a large class of quasi one-dimensional nanostructures.

Since computing nanomechanical response requires relatively large systems (at least a few hundred of atoms), computationally affordable but less accurate classical atomistic treatments of the atomic scale are heavily adopted [6–11]. To date, only multiscale classical atomistic-to-continuum bridging is achieved [12–19]. Unfortunately, the inaccuracy of the "parent" atomistic model directly transfers to the continuum one and this severely limits our predictive power. The accurate atomistic tools to study the properties of matter are the methods that explicitly account for the quantum mechanics of the electrons. Originally developed in the context of quantum chemistry and condensed matter physics, these powerful methods combine fundamental quantum-mechanical predictive power with atomic resolution in length and time. These methods include density-functional theory [20–22] and tight binding [23, 24] molecular dynamics.

An accurate atomistic description is highly desirable in computational nanomechanics. In order to deliver quantitative prediction, suited for further engineering use, the accurate quantum-mechanical description of chemical bonding is needed. Furthermore, because the electronic subsystem is treated explicitly, electronic, optical or piezoelectric properties can be also derived. Unfortunately, the size range covered by quantum-mechanical methods (of a few hundred atoms) is the major impediment for using these methods in nanomechanics. (This is in spite of the face that the size limit increases rapidly due to parallel computing.)

This work does not intend to advance nanomechanics by trying to enlarge the current computational limit for number of atoms that can be treated with quantum mechanical accuracy. Instead, within the current computational limits, the main idea is to introduce a substantial simplification in the atomistic computations by making recourse to the helical symmetry of the nano objects.

The concept is illustrated in Fig. 1 for a single walled carbon nanotube structure, which can be viewed as the result of rolling up into a seamless tube a flat sheet of graphite, called graphene. The hexagonal structure of graphene, see Fig. 1a, is described by translating a two-atom cell (orange shaded) along the lattice vectors $\mathbf{T_1}$ and $\mathbf{T_2}$. Rolling into a tube can be performed under various curvatures in different directions. For example, Fig. 1b shows a chiral tube obtained by rolling the graphene along the circumference vector $\mathbf{C}$ having distinct components (6 and 3, respectively) along $\mathbf{T_1}$ and $\mathbf{T_2}$. Through folding, the $\mathbf{T_1}$ and $\mathbf{T_2}$ directions wind up in helixes. To

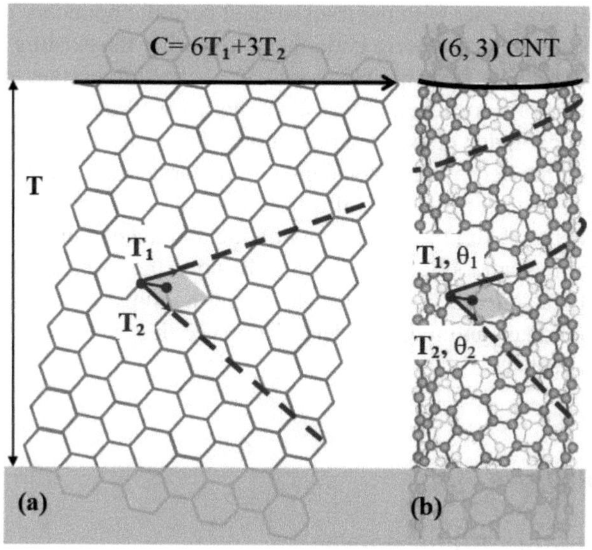

Fig. 1 Translational cell of a (6,3) nanotube in the (**a**) unfolded and (**b**) folded representation

describe the resulting tube structure, one must apply both translations ($\mathbf{T_1}$ and $\mathbf{T_2}$) and rotations (θ_1 and θ_2) around the tube axis to a two-atom cell.

When computing properties of a system with crystalline order, current implementations operating under periodic boundary condition, make recourse to the specific translational symmetry of the structure to simplify the problem to the extent that microscopic calculations are performed on the small translational repeating cells. For example, to obtain bulk properties of graphene (Fig. 1a), a calculation on the two-atom unit cell suffices [25]. However, in a system with helical symmetry the translational periodicity doesn't bring the same advantage. For example, Fig. 1b shows that the translational cell of a chiral nanotube (with axial periodicity T) contains a large number of atoms. Often such translational cells are even too large to allow a quantum treatment. However, calculations would become as convenient as in the flat graphene, if one can take advantage of the helical symmetry and further reformulate the problem with augmented repetition rule to include translations and rotations.

The use of helical symmetry at the nanoscale is very much in its infancy. To date, it has attracted the interest of the research community working on carbon nanotubes [26–28]. Fortunately, the type of simplification portrayed in Fig. 1 actually applies to a large category of quasi one-dimensional nanostructures, recently called objective structures [29, 30].

In the first part, our main objective is to outline the tight binding computational tool able to operate under the new helical boundary condition. We will call symmetry-adapted modeling, the new scheme that operates under helical boundary condition. It is not necessary to create this capability from scratch. Instead, the strategy is to implement the new helical boundary condition into an existing tight binding solver, now able to operate with the translational periodic boundary conditions. We detail the necessary steps that were undertaken to modify the existing computational package *Trocadero* [31]. By using this code we inherited important features such as atomistic models for various chemical elements as well as numerical algorithms (such as dynamics under constant temperature) that can be immediately used in conjunction with the new symmetry-adapted boundary conditions.

In the second part, our goal is to illustrate how the created computational capability can be used to obtain the accurate modeling and the nanomechanical response of carbon nanotubes and silicon nanowires. Besides the immediate interest, these results are of great interest for multiscale modeling since such data is needed to construct and validate continuum models. (However, this aspect is beyond the scope of the present chapter.)

2 Methodology

2.1 Tight Binding Preliminaries

Tight binding is a basic semi-empirical method which offers a satisfying description of the electronic structure and bonding of covalent systems in an intuitive localized picture. The tight-binding method of modelling materials lies between the very accurate, very expensive, ab initio methods and the fast but limited *empirical* methods. When compared with ab initio methods, tight binding is typically two to three orders of magnitude faster, but suffers from a reduction in transferability due to the approximations made; when compared with *empirical* methods, tight binding is two to three orders of magnitude slower, but the quantum mechanical nature of bonding is retained, ensuring that the angular nature of bonding is correctly described far from equilibrium structures. Tight binding is therefore useful for the large number of situations in which quantum mechanical effects are significant, but the system size makes ab initio calculations impractical. More specifically, with relatively modest resources up to about 500 atoms can be comfortably treated with tight binding.

Invoking the Born-Oppenheimer approximation, from the electronic structure one can extract a tight binding potential that can be used to carry out tight binding molecular dynamics simulations.

2.1.1 Formulation of Tight Binding Molecular Dynamics in a One-Dimensional Bravais Lattice

In tight binding the electronic states of a periodic structure are obtained by solving the one-electron Schrödinger equation

$$\left[-\frac{\hbar^2}{2m}\nabla^2 + V(\mathbf{r})\right]|j\rangle = \varepsilon_j|j\rangle, \tag{1}$$

where m is the electron mass, $\hbar$ the Plank's constant, $V(\mathbf{r})$ an effective periodic potential, and $|j\rangle$ and ε_j are the one-electron wavefunction and energy for the state j, respectively. This equation is solved with the ansatz that the single electron states can be represented in terms of atomic orbitals located on each single atom. The number of atomic orbitals n_α is usually taken equal with the number of valence electrons for the atomic species (for example $n_\alpha = 4$ for carbon and silicon). Let N_t (usually a large number) the number of translational cells over which periodic boundary condition are imposed and let N_0 be the number of atoms in each cell. If no explicit recourse to translational symmetry is made, the one-electron eigenfunctions $|j\rangle$ are represented in terms of localized orbitals functions $|\alpha n\rangle$, where α labels the orbital symmetry (s, p_x, p_y, p_z for our case) and n the atomic location:

$$|j\rangle = \sum_{\alpha,n} C_{\alpha n}(j)|\alpha n\rangle, \quad j = 1, ..., N. \tag{2}$$

The number of eigenstates $N = n_\alpha \cdot N_0 \cdot N_t$ equals the total number of valence electrons in the N_t translational cells considered. The expansion coefficients grouped in the vector $\mathbf{C}(j)$ should be obtained from the one-electron generalized $N \times N$ eigenvalue problem

$$\mathbf{H} \cdot \mathbf{C}(j) = \varepsilon_j \mathbf{S} \cdot \mathbf{C}(j), \ j = 1, \ldots, N, \tag{3}$$

which can be easily obtained by substituting Eq. (2) into (1). Equation (3) is known as the matrix form of the Schrödinger equation. In tight binding, the matrix the elements of the Hamiltonian $\mathbf{H}$ matrix – e^0_α diagonal and $t^0_{\alpha',\alpha}$ off-diagonal – and of the overlap $\mathbf{S}$ matrix $s^0_{\alpha',\alpha}$ are not explicitly calculated. They are replaced with a parameter which depends only on the internuclear distance and the symmetry of the orbitals involved. The parameterization is performed by using the available experimental data or by making recourse to more accurate but computationally expensive *first principles* [32, 33] calculations.

The total energy contained in the electronic states writes:

$$E_{\text{band}} = 2\sum_{j=1}^{N} f_j \varepsilon_j, \tag{4}$$

where f is the Fermi distribution function.

2.1.2 Tight Binding Under Periodic Boundary Condition – The Electronic Structure

Of course, in a crystalline structure with N on the order of 10^{23}, the $N \times N$ eigenvalue problem (3) cannot be solved directly. The usual approach is to introduce a significant computational simplification by explicitly accounting for the *translational* symmetry. For simplicity, we are next presenting this approach only for the one-dimensional crystal. (The expansion to the three-dimensional case is trivial.)

Let $\mathbf{T}$ be the lattice periodicity vector and $\{\mathbf{X}_n : n = 1, \ldots, N_0\}$ the atomic positions within the unit cell. The atomic locations in one-dimensional periodic structure can be obtained with

$$\mathbf{X}_{n,\zeta} = \zeta \mathbf{T} + \mathbf{X}_n, \tag{5}$$

where $\zeta = 0, \ldots, N_t - 1$, and $n = 1, \ldots, N_0$. To incorporate the translational symmetry, instead of Eq. (2) the one-electron solutions are represented in terms of the Bloch sums:

$$|\alpha n, k\rangle = \frac{1}{\sqrt{N_t}} \sum_{\zeta} e^{ikT\zeta} |\alpha n, \zeta\rangle, \tag{6}$$

where the wavenumber k takes N_t uniformly spaced values in the interval $-\pi/T \leq k < \pi/T$, with $\delta k = 2\pi/TN_t$. $|\alpha n, \zeta\rangle$ represents the orbital α located on atom n located in the translational cell ζ, with $\zeta = 1, \ldots, N_t$. The advantage is that in the representation (6), the tight binding Hamiltonian and overlap matrix elements between two $\langle \alpha n', k'|$ and $|\alpha n, k\rangle$ sums are vanishing unless $k = k'$. Thus, the eigenvalue problem becomes block-diagonal and it can be solved separately for each N_t block labeled by k and having dimension $(N/N_t)\times(N/N_t)$:

$$\mathbf{H}(k) \cdot \mathbf{C}(j, k) = \varepsilon_j(k)\mathbf{S}(k) \cdot \mathbf{C}(j, k), \quad j = 1, \ldots, N/N_t. \tag{7}$$

In other words, instead of solving the single $N \times N$ eigenvalue problem (3) to obtain the N electronic states, one solves instead N_t eigenvalue problems (7) of size $N/N_t \times N/N_t$ to obtain the same N electronic states, now labeled by j and k.

Formally, in the two-center approximation, the k-dependent elements of Hamiltonian $\mathbf{H}$ and overlap $\mathbf{S}$ matrices write:

$$\langle \alpha' n', k|H|\alpha n, k\rangle = \sum_{\zeta} e^{-ikT\zeta} t^0_{\alpha'\alpha}(\mathbf{X}_{n',\zeta} - \mathbf{X}_n), \tag{8}$$

$$\langle \alpha' n', k|\alpha n, k\rangle = \sum_{\zeta} e^{-ikT\zeta} s^0_{\alpha',\alpha}(\mathbf{X}_{n'\zeta} - \mathbf{X}_n). \tag{9}$$

As before, $\mathbf{X}_n$ represent the atomic coordinates in the initial $\zeta = 0$ cell.

2.1.3 Tight Binding Molecular Dynamics Under Periodic Boundary Condition

In the translational formulation, the total energy contained in the electronic states writes:

$$E_{\text{band}} = 2 \sum_{j=1}^{N/N_t \sum_k^{\delta k(N_t-1)/2} = -\delta k(N_t-1)/2} f_j(k) E_j(k), \tag{10}$$

and depends parametrically on the coordinates of the nuclei. In the framework of the Born-Oppenheimer approximation [34], once the electronic energy is found, it can be directly used to give the motion of the nuclei, which are treated classically. More precisely, to perform molecular dynamics, the forces acting on the N_0 atoms located in the initial cell, are needed. In tight binding molecular dynamics, the force due to the band energy on the atom at $\mathbf{X}_m$ located in the initial cell due to the band energy, $\mathbf{F}_m = -\partial E_{\text{band}}/\partial \mathbf{X}_m$, writes:

$$\mathbf{F}_m = -2 \sum_{j,k} f_j(k) \left(\mathbf{C}^\dagger(j,k) \cdot \frac{\partial \mathbf{H}(k)}{\partial \mathbf{X}_m} \cdot \mathbf{C}(j,k) - \varepsilon_j(k) \mathbf{C}^\dagger(j,k) \cdot \frac{\partial \mathbf{S}(k)}{\partial \mathbf{X}_m} \cdot \mathbf{C}(j,k) \right), \tag{11}$$

which is the Hellmann-Feynman force. Typical derivatives of the tight binding Hamiltonian and overlap matrix elements are:

$$\frac{\partial \langle \alpha' n', \kappa | H | \alpha m, \kappa \rangle}{\partial \mathbf{X}_m} = \frac{1}{N_t} \sum_{\zeta} e^{-i\kappa\zeta} \frac{\partial t^0_{\alpha',\alpha}(\mathbf{X}_{n',(\zeta)} - \mathbf{X}_m)}{\partial \mathbf{X}_m}, \tag{12}$$

$$\frac{\partial \langle \alpha' n', k | \alpha m, k \rangle}{\partial \mathbf{X}_m} = \frac{1}{N_t} \sum_{\zeta} e^{-ik\zeta} \frac{\partial s^0_{\alpha',\alpha}(\mathbf{X}_{n',(\zeta)} - \mathbf{X}_m)}{\partial \mathbf{X}_m}. \tag{13}$$

In tight binding one assumes a form for the Hamiltonian and overlap matrix elements without specifying anything about the atomic orbitals except their symmetry. The values of the matrix elements may be derived approximately or may be fitted to experiment or other theory.

2.2 Detailed Formulation of the Symmetry-Adapted Tight Binding Molecular Dynamics

A major disadvantage of tight binding is still its high computational cost: Computing the electronic structure energy and forces from a tight binding Hamiltonian by direct diagonalization results in a cubic scaling of the computational time with the number of electrons considered $[O(N^3)]$. This limits the system size to about 1000 Si or C atoms. In a Bravais lattice, the number of atoms is small. Thus, the above approach is suitable in simulations of bulk crystals. However, the simplifications due to the periodic boundary condition are insufficient for comprehensive calculations of quasi one-dimensional objects such as nanowires and nanotubes. This is because the number of atoms in the translational unit cells can be very large, leading to a large-size matrix equation for the electronic problem. Note that there is a current research line that attempts to alleviate this notorious dependence of computation time with $O(N)$ methods. However, the errors introduced in the electronic band structure could affect the outcome.

2.2.1 Symmetry-Adapted Tight Binding Molecular Dynamics in the "Angular-Helical" Representation

Figure 1 suggests that the nanotube structures could be described by applying successive commuting helical transformations. However such "helical-helical" representation is not convenient for a quantum mechanical implementation. For this reason, here we introduce the "helical-angular" representation [25].

An infinitely-long CNT exhibits *translational* symmetry. When its unrolled unit cell is projected on the graphene layer, Fig. 2, the circumference turns into the chiral vector $\mathbf{C}_h = l_1\mathbf{a}_1 + l_2\mathbf{a}_2$ of modulus $|\mathbf{C}_h| = a\sqrt{l_1^2 + l_2^2 + l_1 l_2}$. Here a is the lattice constant for the flat honeycomb lattice. The CNT axis aligns to the translational vector $\mathbf{T} = t_1\mathbf{a}_1 + t_2\mathbf{a}_2$, where $t_1 = (2l_2 + l_1)/d_R$ and $t_2 = -(2l_1 + l_2)/d_R$. Here $d_R = \text{gcd}(2l_1 + l_2, 2l_2 + l_1)$, where gcd indicates the greatest common divisor of the two number within the brackets. The length of the translational vector is $|\mathbf{T}| = \sqrt{3}|\mathbf{C}_h|/d_R$. The translational unit cell of the CNT is delineated by $\mathbf{T}$ and $\mathbf{C}_h$. It also follows [25] that the number of carbon pairs inside the translational unit cell writes $N = 2(l_1^2 + l_1 l_2 + l_2^2)/d_R$.

The translational representation of the infinitely long CNT is widely used for computing the elastic response. One disadvantage is that it requires the explicit treatment of all N atoms, and N scales with the CNT's diameter and chirality angle χ. Only a small number of carbon nanotubes (small-diameter and achiral ones) have N computationally affordable for tight binding and density functional theory.

In addition to *translation*, carbon nanotubes exhibit *angular* and *helical* symmetries [25], also conveniently described by vectors in the example of Fig. 2. For *angular* symmetry, note that while $\text{gcd}(t_1, t_2) = 1$ (indicating that $\mathbf{T}$ never spans in the axial direction two identically-oriented carbon pairs), $d = \text{gcd}(l_1, l_2)$ can take any value between 1 and n (supposing that $0 \leq l_2 \leq l_1$). Thus, $\mathbf{C}_h$ spans d

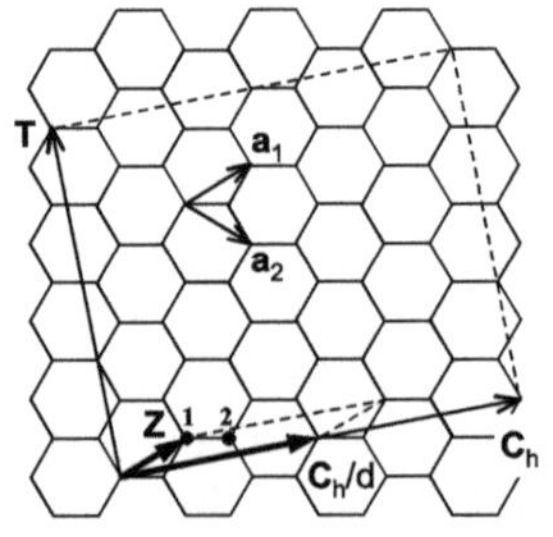

Fig. 2 Unfolded representation of a (4,2) CNT. The big rectangle shows the translational unit cell bounded by the chiral $\mathbf{C}_h$ and translational $\mathbf{T}$ vectors. The small parallelogram shows the "helical-angular" cell bounded by the rotational $\mathbf{C}_h/d$ and screw $\mathbf{Z}$ vectors. It contains atoms 1 and 2. Graphene's lattice vectors $\mathbf{a}_1$ and $\mathbf{a}_2$ are also shown

identically-oriented carbon pairs in the circumferential direction and the rotation vector is given by $\mathbf{C}_h/d$. For *helical* symmetry, we note that among the $N-1$ non-equivalent vectors connecting the graphene lattice points within the translational unit cell, $d-1$ correspond to pure angular rotations, while the remaining $N-d$ correspond to distinct screw operations. Among these vectors, here we are concerned only with vector $\mathbf{Z} = u\mathbf{a}_1 + v\mathbf{a}_2$, which has the smallest components in both axial and circumferential directions [25]. Using simple geometric considerations, it follows that integers u and v must satisfy $l_2u - l_1v = d$ and $0 < t_1v - t_2u < N/d$. More useful, $\mathbf{Z}$ is decomposed along $\mathbf{T}$ and $\mathbf{C}_h$ as [25]:

$$N\mathbf{Z} = D\mathbf{C}_h + d\mathbf{T}, \tag{14}$$

where $D = t_1v - t_2u$. The infinitely-long CNT is alternatively described from the two-atom reduced cell bounded by the vectors $\mathbf{C}_h/d$ and $\mathbf{Z}$. For example, one can see from eq. (14) that N operations of $\mathbf{Z}$ extend over d translational cells. The rest of $(d-1)N$ carbon pairs of the d translational cells are filled by $d-1$ applications of $\mathbf{C}_h/d$.

Moving to the rolled-up geometry, a (l_1, l_2) CNT is described the "angular-helical" as an objective molecular structure with

$$\mathbf{X}_{n,\zeta_1,\zeta_2} = \mathbf{R}_2^{\zeta_2}\mathbf{R}_1^{\zeta_1}\mathbf{X}_n + \zeta_1\mathbf{T}_1, \quad j = 1, N_0. \tag{15}$$

The rotational matrix $\mathbf{R}_2$ corresponds to vector $\mathbf{C}_h/d$ and indicates an angular rotation of angle $\theta_2 = 2\pi/d$. Rotational matrix $\mathbf{R}_1$ of angle $\theta_1 = 2\pi D/N$ and the axial vector $\mathbf{T}_1 = d\mathbf{T}/N$ correspond to the screw vector $\mathbf{Z}$. Index n runs over the two atoms (the molecule) at locations $\mathbf{X}_j$ inside the reduced computational domain. Integers ζ_1 and ζ_2, with $-\infty < \zeta_1 < \infty$ and $0 \le \zeta_2 \le d-1$, label the various replicas of the initial domain.

We will accommodate this representation into the one-electron wavefunction solutions by representing them in terms of symmetry-adapted sums of localized orbitals, suitable for the helical and angular symmetry. We discuss next the steps that will be undertaken to enhance to symmetry-adapted modeling the current periodic boundary condition implementation in the computational package *Trocadero* [31]. We remind the reader that the microscopic model of interest is the nonorthogonal density functional theory-based tight binding model with two-center terms of Porezag et al. [32].

2.2.2 Symmetry-Adapted Tight Binding – Treatment of the Electronic Structure

Consider N_s the number of screw operations (typically ∞) over which the helical boundary condition is imposed and let N_a be the number of θ_2 rotations needed to fill the circumference of the quasi one-dimensional nanostructure. The symmetry-adapted Bloch sums write

$$|\alpha n, l\kappa\rangle = \frac{1}{\sqrt{N_a \cdot N_s}} \sum_{\zeta_1=0}^{N_s-1} \sum_{\zeta_2=0}^{N_a-1} e^{il\theta_2\zeta_2 + i\kappa\zeta_1} |\alpha n, \zeta_1\zeta_2\rangle, \tag{16}$$

where the Bloch factors are the eigenvalues of the commuting rotation and screw operators [27]. Here $l = 0, 1, \ldots, (N_a - 1)$ represents the angular number. To avoid the discomfort of introducing helical distances, $-\pi \leq \kappa < \pi$ represents the helical wavevector already normalized by the helical periodicity. As in Eq. (15), index n runs over the atoms located in the symmetry-adapted computational cell and $|\alpha n, \zeta_1\zeta_2\rangle$ refers to the orbital with symmetry α located on atom n, all in the symmetry-adapted cell indexed by ζ_1 and ζ_2. Lastly, to satisfy the generalized Bloch theorem, the $|\alpha n, \zeta_1\zeta_2\rangle$ orbitals are obtained by applying a $\mathbf{R}^{(\zeta_1,\zeta_2)}$ rotation to the orbitals $|\alpha n\rangle$ located in the (ζ_1, ζ_2) cell that are parallel with those situated in the initial (0, 0) cell. Specifically, for the sp case:

$$\begin{pmatrix} |sn, \zeta_1\zeta_2\rangle \\ |p_x n, \zeta_1\zeta_2\rangle \\ |p_y n, \zeta_1\zeta_2\rangle \\ |p_z n, \zeta_1\zeta_2\rangle \end{pmatrix} = \begin{pmatrix} 1 & 0 & 0 & 0 \\ 0 & \cos(\zeta_1\theta_1 + \zeta_2\theta_2) & -\sin(\zeta_1\theta_1 + \zeta_2\theta_2) & 0 \\ 0 & \sin(\zeta_1\theta_1 + \zeta_2\theta_2) & \cos(\zeta_1\theta_1 + \zeta_2\theta_2) & 0 \\ 0 & 0 & 0 & 1 \end{pmatrix} \cdot \begin{pmatrix} |sn\rangle \\ |p_x n\rangle \\ |p_y n\rangle \\ |p_z n\rangle \end{pmatrix}. \tag{17}$$

Note that the s orbitals are not affected because of their intrinsic symmetry while the p_z ones are invariant as they are oriented along the long axis.

The symmetry-adapted Bloch elements between different angular and helical numbers of the Hamiltonian and overlap matrices vanish. Therefore, the eigenvalue problem for the atoms contained in the $N_s \cdot N_a$ symmetry-adapted domains becomes block-diagonal and it can be solved separately for each block labeled by l and κ:

$$\mathbf{H}(l\kappa) \cdot \mathbf{C}(j, l\kappa) = \varepsilon_j(l\kappa)\mathbf{S}(l\kappa) \cdot \mathbf{C}(j, l\kappa), \quad j = 1, \ldots, N. \tag{18}$$

The elements of eigenvector $\mathbf{C}$ represent the expansion coefficients in the basis (16) of the one-electron wavefunction solutions of the Schrödinger equation. Having [32] the elements of the Hamiltonian matrix – ε_α^0 diagonal and $t_{\alpha',\alpha}^0$ off-diagonal – and of the overlap matrix $s_{\alpha',\alpha}^0$, the elements of $l\kappa$-dependent Hamiltonian $\mathbf{H}$ and overlap $\mathbf{S}$ matrices are:

$$\langle\alpha' n', l\kappa|H|\alpha n, l\kappa\rangle = \varepsilon_\alpha^0 \delta_{nn'}\delta_{\alpha\alpha'} + \sum_{\zeta_2,\zeta_1} e^{-il\theta_2\zeta_2 - i\kappa\zeta_1} t_{\alpha'\alpha}(\mathbf{X}_{n',(\zeta_1,\zeta_2)} - \mathbf{X}_n), \tag{19}$$

$$\langle\alpha' n', l\kappa|\alpha n, l\kappa\rangle = \delta_{nn'}\delta_{\alpha\alpha'} + \sum_{\zeta_2,\zeta_1} e^{-il\theta_2\zeta_2 - i\kappa\zeta_1} s_{\alpha',\alpha}(\mathbf{X}_{n',(\zeta_1,\zeta_2)} - \mathbf{X}_n), \tag{20}$$

where

$$t_{\alpha'\alpha}(\mathbf{X}_{n',(\zeta_1,\zeta_2)} - \mathbf{X}_n) = \sum_{\alpha''=1}^{n_\alpha} t_{\alpha'',\alpha}^0(\mathbf{X}_{n',(\zeta_1,\zeta_2)} - \mathbf{X}_n) R_{\alpha',\alpha''}^{(\zeta_1,\zeta_2)}, \tag{21}$$

$$s_{\alpha',\alpha}(\mathbf{X}_{n',(\zeta_1,\zeta_2)} - \mathbf{X}_n) = \sum_{\alpha''=1}^{n_\alpha} s^0_{\alpha'',\alpha}(\mathbf{X}_{n',(\zeta_1,\zeta_2)} - \mathbf{X}_n) R^{(\zeta_1,\zeta_2)}_{\alpha',\alpha''}. \tag{22}$$

In comparison with the tight binding under periodic boundary conditions [24], in Eq. (18) we obtained a size reduction of the eigenvalue problem at the expense of carrying out diagonalizations at additional $l\kappa$ points. In view of the $O(N^3)$ computational time scaling, the advantage of the symmetry-adapted basis is evident.

2.2.3 Symmetry-Adapted Tight Binding Molecular Dynamics

The total electronic energy contained in the $N_s \cdot N_a$ symmetry-adapted domains writes:

$$E_{\text{band}} = 2 \sum_{j=1}^{n_\alpha N_0} \sum_{\kappa=-\delta\kappa(N_s-1)/2}^{\delta\kappa(N_s-1)/2} \sum_{l=0}^{N_a-1} f_j(l\kappa)\varepsilon_j(l\kappa), \tag{23}$$

where f is the Fermi function and $\delta\kappa = 2\pi/N_s$. Because E_{band} is invariant under the screw and angular rotations as well as under the permutations of atoms, the conditions for carrying out Symmetry-adapted MD from the tight binding potential are fulfilled. To perform MD, the forces acting on the N_0 atoms are needed. The force on the atom at $\mathbf{X}_m$ located due to the band energy, $\mathbf{F}_m = -\partial E_{\text{band}}/\partial \mathbf{X}_m$, writes:

$$\mathbf{F}_m = -2\sum_{j,l\kappa} f_j(l\kappa) \left(\mathbf{C}^\dagger(j,l\kappa) \cdot \frac{\partial \mathbf{H}(l\kappa)}{\partial \mathbf{X}_m} \cdot \mathbf{C}(j,l\kappa) \right.$$
$$\left. - \varepsilon_j(l\kappa)\mathbf{C}^\dagger(j,l\kappa) \cdot \frac{\partial \mathbf{S}(l\kappa)}{\partial \mathbf{X}_m} \cdot \mathbf{C}(j,l\kappa) \right), \tag{24}$$

which is the Hellmann-Feynman force. The above expressions are considered as $N_s \to \infty$ and κ becomes continuous. The derivatives of the Hamiltonian and overlap matrix elements are:

$$\frac{\partial \langle \alpha' n', l\kappa | H | \alpha m, l\kappa \rangle}{\partial \mathbf{X}_m} = \frac{1}{N_a N_s} \sum_{\zeta_2,\zeta_1} e^{-il\theta_2\zeta_2 - i\kappa\zeta_1} \sum_{\alpha''=1}^{n_\alpha} \frac{\partial t^0_{\alpha'',\alpha}(\mathbf{X}_{n',(\zeta_1,\zeta_2)} - \mathbf{X}_m)}{\partial \mathbf{X}_m} R^{(\zeta_1,\zeta_2)}_{\alpha',\alpha''}. \tag{25}$$

$$\frac{\partial \langle \alpha' n', l\kappa | \alpha m, l\kappa \rangle}{\partial \mathbf{X}_m} = \frac{1}{N_a N_s} \sum_{\zeta_2,\zeta_1} e^{-il\theta_2\zeta_2 - i\kappa\zeta_1} \sum_{\alpha''=1}^{n_\alpha} \frac{\partial s^0_{\alpha'',\alpha}(\mathbf{X}_{n',(\zeta_1,\zeta_2)} - \mathbf{X}_m)}{\partial \mathbf{X}_m} R^{(\zeta_1,\zeta_2)}_{\alpha',\alpha''}. \tag{26}$$

Derivatives $\partial t^0_{\alpha'',\alpha}/\partial \mathbf{X}_m$ and $\partial s^0_{\alpha'',\alpha}/\partial \mathbf{X}_m$ are given in [32]. We note that the two-body repulsive part of the tight binding potential [32] does not necessitate specific adjustments for symmetry-adapted molecular dynamics.

3 Applications of Symmetry-Adapted Modeling in Nanomechanics

3.1 Linear- and Nonlinear-Elastic Response of Carbon Nanotubes

3.1.1 Linear-Elastic Response

We now illustrate the applicability of the method in the linear-elastic deformation regime. One could carry out microscopic calculations of relaxed, elongated, and twisted carbon nanotubes described from a two-atom domain with Eq. (15) serving as the objective boundary conditions. The electronic states are modeled with the presented symmetry-adapted scheme coupled with the non-orthogonal two-centre tight binding model of carbon [32], implemented in the package Trocadero [31].

In order to define infinitesimal elastic moduli of a nanotube, we first identified the stress-free equilibrium geometries. We find that the structural parameter predictions given by the roll-up are adequate. The isometric mapping that wraps the graphene sheet into the cylindrical geometry is not precisely a nanotube at equilibrium as the carbon-carbon bond lengths will differ in general from their values in the flat geometry. We identified the equilibrium carbon nanotubes through conjugate gradient potential energy surface scans performed on the two-atom cell under different $|\mathbf{T}_1|$ values.

Figure 3a plots the obtained strain energy (W_0), defined as the energy (measured per atom) of the equilibrium CNT configuration measured with respect to the graphene, as a function of the CNT diameter ($2R$). The characteristic behavior $W_0 = C/2R^2$ is obtained with $C = 4.13$ eVÅ$^2/atom$, in very good agreement with density functional theory data [35]. Note the large difference with the 2.2 eVÅ$^2/atom$ and 1.8 eVÅ$^2/atom$ values given by the first- and second-generation Brenner's potentials, respectively. No χ-dependence was obtained, as can be seen also from the insert showing W_0 versus χ for an equal-radius CNT family. Adopting a surface-without-thickness membrane [13] representation of graphene, the associated bending rigidity is $D_b = C/S_0$, where $S_0 = 2.6$ Å^2 is the surface per atom defined by the CNT at equilibrium.

Axial strain (ε) to an (l_1, l_2) CNT was applied by changing $|\mathbf{T}_1|$ to $(1+\varepsilon)|\mathbf{T}_1|$, with ε in the (-0.5:0.5%) range. The intrinsic θ_1 and θ_2 angles were kept fixed. The obtained size dependence of $Y_s = (1/S_o)d^2W_0/d\varepsilon^2$ is displayed in Fig. 4b. In contrast with other results, Y_s appears insensitive to χ. The axial elasticity of carbon nanotubes appears similar to the in-plane one of graphene apart from effects due the tube curvature. Above $\sim$1.25 nm Y_s is practically constant and takes the 430 GPa nm value of graphene. The agreement with previous tight binding calculations [36], carried out in the translational representation of the carbon nanotubes, confirms the validity of our scheme. For $2R < 1.25$ nm Y_s softens, which is in disagreement for example with the data obtained using the second-generation Brenner's potential.

An axial twist rate (γ) was imposed by varying the intrinsic angle θ_1 to $\theta_1 + \gamma|\mathbf{T}_1|$. Because the symmetry-adapted tight binding scheme doesn't rely on translational symmetry, any γ can be prescribed. By contrast, limited γ choices are

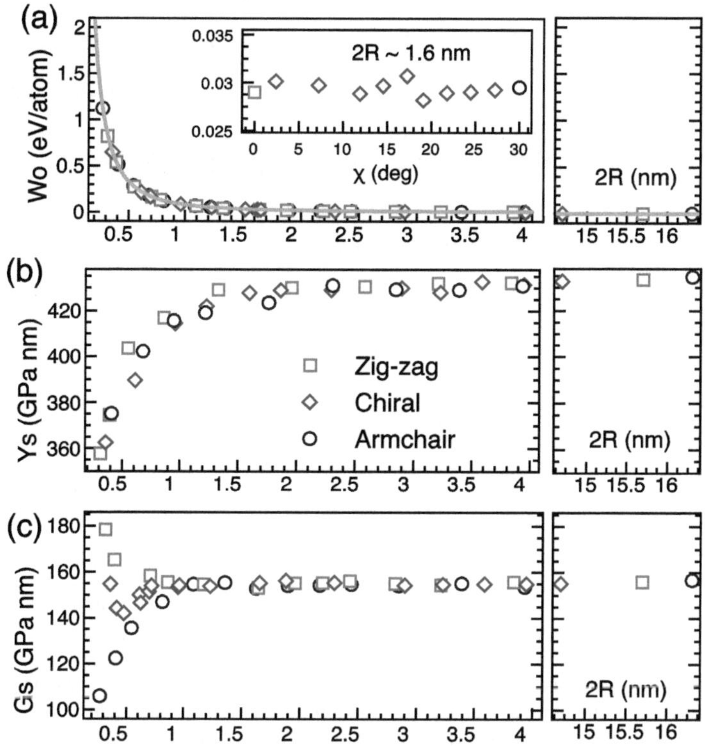

Fig. 3 (**a**) Curvature-strain energy W_0 (**b**) Young's Y_s and (**c**) shear modulus G_s as a function of the CNT diameter $2R$. Insert shows W_0 dependence with χ at a constant diameter

available in the translational formulation, making calculations cumbersome. The resulting shear strain $\varepsilon' = \gamma R$ was varied in the (−0.5%:0.5%) range, while $|\mathbf{T}_1|$ was kept constant. Figure 4c displays the obtained surface shear modulus (G_s), defined as $G_s = (1/S_o)d^2W_0/d\varepsilon'^2$. For $2R > 1.25$ nm, G_s is not sensitive to χ and converges quickly to the 156 GPa nm value of graphene. This value is higher than the 113 GPa nm given by the second-generation Brenner's potential and is in excellent agreement with the 157 GPa nm from density functional theory. For $2R < 1.25$ nm there is a pronounced χ splitting with G_s bounded from above by zig-zag and from below by armchair carbon nanotubes.

The χ-dependence for G_s for $2R < 1.25$ nm unequivocally shows anisotropy. Interestingly, the HiPco-produced carbon nanotubes fall within this size-range. However, our tight binding data for $2R > 1.25$ nm shows that both Y_s and G_s are practically converged to the values of isotropic graphene. Thus, the isotropic CNT model is justified and we have $G_s = Y_s/2(1+\nu_s)$. The resulting Poisson ratio $\nu_s = 0.38$ is lower than the one obtained with the Brenner's potential [13] but still larger than in experiment. For practical reasons it is also useful to give a CNT elastic thin-shell model. Thus, we expressed $D_b = Yh^3/12$ and $Y_s = Yh$. Using our data we obtained $h = 0.8$ Å and $Y = 5.2$ TPa.

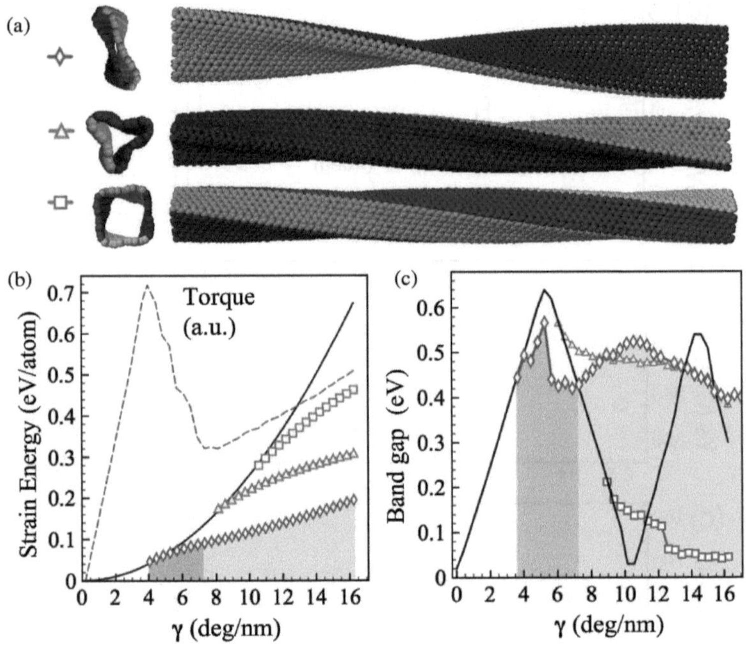

Fig. 4 (**a**) Rippling patterns in a (12,12) CNT under 10°/nm twist. Colors correspond to distinct azimuthal replicas of the objective domain. (**b**) Strain energy and (**c**) band gap vs. twist rate γ for cylinder and rippling modes. Torque (in arbitrary units) for the two-lobe mode is plotted in order to indicate the two stages of rippling, also marked with two gray levels

3.1.2 Nonlinear Elastic Response of Carbon Nanotubes in Torsion

Can the symmeyty-adapted modeling be useful beyond the linear elastic regime? We illustrate now the use of the method for studying torsional rippling deformations. Rippling represents a way of lowering energy by creating an inhomogeneous torsional strain coupled with the development of helicoidal ridges and furrows of positive and negative curvature, respectively. Due to the distributed nature, the long-range inter-wall interactions are also perturbed. For high fidelity modeling, accurate accounting of both covalent and van der Waals forces is needed. In principle, quantum mechanical calculations can provide both the morphological changes and the electronic behavior. There is ample evidence that an explicit quantum mechanica treatment of bonding gives an accurate description of carbon nanotube's mechanics. However, the usual translation-invariant formulation is computationally prohibitive for long-range elastic deformations with helical symmetry. To reduce the computational complexity, empirical classical potentials and continuum idealizations derived from classical potentials are widely used. When compared with quantum mechanical descriptions, these approaches are sometimes inaccurate, including for describing the linear elastic regime. Another obvious disadvantage is that the electronic response is not revealed.

We discuss single-wall carbon nanotubes in torsion under the constraint of constant length. Renouncing the translational symmetry allows to compute in an economical fashion the expected helical [13] rippling morphologies that characterize the initial stages of the nonlinear elastic response. For example, the wave-like modes for (12,12) CNT shown in Fig. 4a, were individually computed with objective MD from three objective cells delineated by the translational vector **T** but different θ_2 values, 180°, 120°, and 90°. Because these cells contain only 24, 18, and 12 atoms, respectively, the tight binding treatment can be comfortably applied. We studied in a decoupled way the rippling morphologies, unambiguously identify the fundamental (lowest-energy) mode, and pinpoint the level of twist beyond which the cylinder shape becomes first unstable, i.e., the bifurcation point.

The importance range of the various modes can be judged from Fig. 4b, which shows the computed strain-energies vs. γ. The curve obtained from a 4 atom objective cell with $\theta_2 = 30°$ (thin line) is a useful reference as the CNT's cylinder shape is preserved. At lower strains this state is stable but when strain exceeds 4°/nm the two-lobe rippling significantly lowers the elastic energy. The perfect structure can also assume the three- and four-lobe buckled states beyond 7°/nm and 9°/nm, respectively, but the two-lobe morphology stays favorable by a large margin. We conjecture that the various wave-like modes shown in Fig. 4a exist as distinct metastable states and twisting doesn't mix them by driving further circumferential rippling.

The evolution of torque (energy derivative with γ), Fig. 4b, further reveals that the two-lobe rippling develops in two stages. The first is a *transient* one and initiates at the bifurcation, where the linear torque variation is interrupted by a rather abrupt drop. The cylinder shape develops ridges and furrows as the torque's rate of change is negative. The minimum distance between furrows decreases until a van der Waals equilibrium distance of 3.4 Å is reached at 7.2°/nm. Here the transient stage is complete and torque's rate of growth switches from negative to positive. In this second stage there is no notable change in shape of the collapsed cross section.

Figure 4c shows the variations in the electronic band gap for the idealized cylinder and rippling modes. Band-gap information is important because of the relationship with the experimentally-relevant CNT resistance. For the idealized case, the band gap is periodically modulated. During one period $\Delta\gamma$, it grows in a linear fashion, reaches a maximum at halfway, then decreases to zero. These features are in agreement with predictions of Yang and Han [37] developed from the band structure of graphene: uniformly twisted carbon nanotubes are metallic only when the metallic wavefunction of graphene is commensurated with the CNT's circumference. Otherwise, they are semiconducting. Due to the linearity of the dispersion relation of graphene near its Fermi level, band gap varies in a linear fashion with a $3t_0R$ slope. R is the CNT's radius and t_0 is the tight binding hopping element. It also follows [37] that $\Delta\gamma = d_{C-C}/R^2$, where d_{C-C} is the C–C bond length. Fitting to our data gives $t_0 = 2.72$ eV and $d_{C-C} = 1.42$ Å, in agreement with the actual values of these parameters.

The generality of the above behavior is demonstrated in Fig. 5, summarizing results from a series of similar calculations performed on armchair carbon

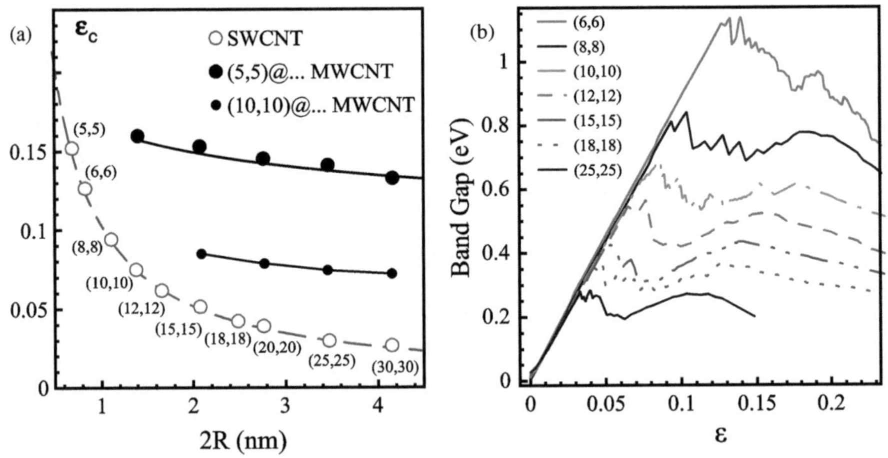

Fig. 5 (**a**) Critical shear strain vs. CNT's diameter for SW (*open circles*) and MW (*full circles*) armchair carbon nanotubes. (**b**) Changes in the band gap with the shear strain for armchair SWcarbon nanotubes

nanotubes. During torsion the ideal CNT wall experiences a shear strain $\varepsilon = \gamma R$. Figure 5a plots the obtained critical shear ε_c, which exhibits a scaling $\varepsilon_c = 0.1([nm]/2R)^{0.99}$. Figure 5b reveals the band gap dependence with ε. In agreement with the idealized model, the linear regime data collapses onto a common line with a $3t_0$ slope. However, beyond ε_c, the idealized model based on the homogeneous strain assumption cannot be used.

3.2 Stability of Polycrystalline and Hexagonal Si Nanowires

In spite of a large body of experimental [38–40] and theoretical [41–49] research, the ground state of quasi one-dimensional silicon structure at the lowest diameters is not yet known. Relying on thermodynamic arguments, one conjectures that in relatively thick nanowires the influence of surfaces is less important and arrangements with bulklike cores are more likely. However, as the diameter is decreased, surfaces are becoming increasingly important in the nanowire energetic balance [45], and quasi one-dimensional organizations with non-cubic core structures but low surface energies are possible.

To describe the ground state Si nanowire structure at sizes of below 10 nm, several candidates have been already envisioned: It was predicted [45] that an enhanced stability can be obtained in polycrystalline achiral Si nanowires constructed with five identical crystalline prisms exposing only low energy (001) Si surfaces. In another recent study [46], hexagonal nanowires with hexagonal cross sections were found to be the most stable. Note that although nanowires with hexagonal cores are prevalent in III-V zinc-blende semiconductors [50], they have been synthesized also in Si [40]. Finally, icosahedral Si quantum dots [51, 52] constructed from tetrahedral blocks were proposed. Because of the low formation energy of the exposed (111) surfaces, these dots are very stable. One-dimensional nanowires assembled from icosahedral dots appeared [48] more favorable than the achiral pentagonal nanowires of [45].

3.2.1 Structures of Polycrystalline and Hexagonal Si Nanowires

We now discus the structure of the three nanowire motifs presented in Figs. 6 and 7. Although all three motifs have translational symmetry, they can be more economically described in the "angular-helical" representation discussed earlier.

Table 1 summarizes the exact values of the domain parameters θ_1 and θ_2 as well as algebraic expressions for the number of atoms N_0 and the number of surface atoms N_{sf} in the Symmetry-adapted domain as a function of the number of atomic layers L. The precise T_1 value is model dependent and will be determined by simulations.

The nanowire shown in Fig. 6a, labeled P, exhibits a five-fold rotational symmetry. The exposed five equivalent (001) surfaces have dimer rows oriented parallel with respect to the nanowire axis. At the center of this nanowire there is a channel of pentagonal rings. The symmetry-adapted computational domain shown in blue (dark gray) is a triangular prism limited by two (111) and one (001) surfaces. The associated angular parameters are $\theta_1 = 0$ and $\theta_2 = 2\pi/5$ while T_1 equals the periodic boundary condition periodicity T. Thus, in comparison with the traditional periodic boundary condition scheme, symmetry-adapted molecular dynamics reduces to 1/5 the number of atoms that need to be accounted for. Note that to form the nanowire, the (111) surfaces of the symmetry-adapted domains are connected through low energy stacking fault defects. Because in the bulk Si the two (111) planes of the

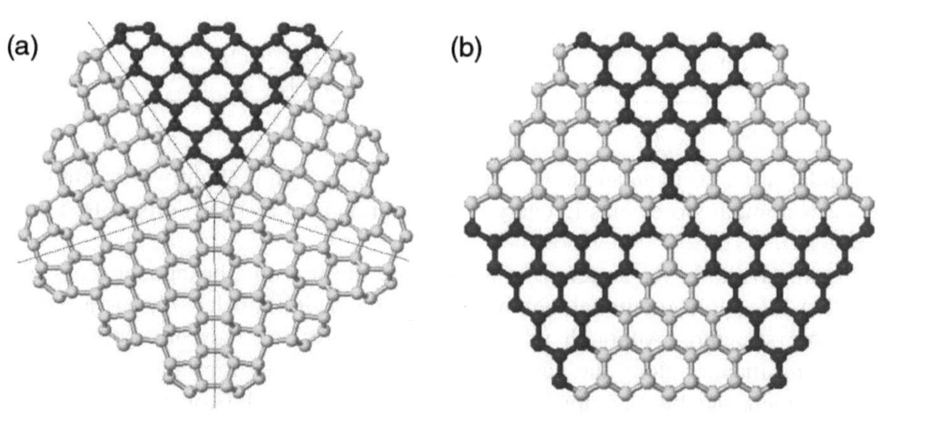

Fig. 6 (**a**) Pentagonal Si nanowire (labeled P). Surface is shown in a reconstruction with *symmetric* dimers, as described with a Tersoff potential. *Dotted lines* indicate the stacking fault planes. The computational "angular-helical" domains are shown in blue atoms. (**b**) Hexagonal Si nanowire (labeled H) with hexagonal cross section. The surface is shown as unrelaxed. Domains with same ζ_1 are shown in same color. For both nanowires only axial views are shown

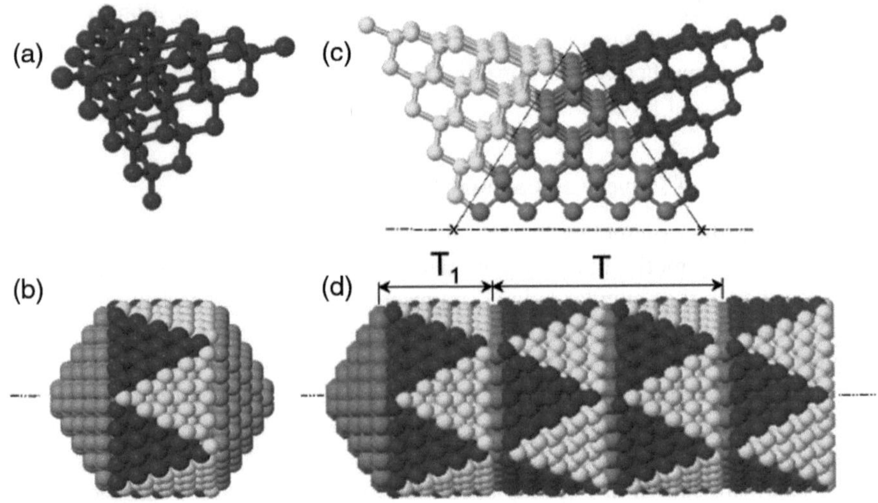

Fig. 7 (**a**) A tetrahedron building block truncated from the bulk Si exposes four (111) surfaces. (**b**) Icosahedral Si dot composed of twenty equivalent tetrahedron building blocks. With a domain composed of three tetrahedron blocks, shown in (**c**), a nanowire (labeled I) with pentagonal cross section, shown in (**d**), can be constructed based on formula describing the nanowire in the "angular-helical" and parameters entered in Table 1 corresponding to $L = 5$. Alternatively, the I nanowire can be constructed from aligned icosahedral Si dots sharing the tetrahedron building blocks shown in gray

Table 1 Comparison of the symmetry-adapted cells for the P, H, and I nanowires. L is the number of layers, N_0 is the total number of atoms, and N_{sf} is the number of surface atoms

Nanowire	P	H	I
θ_1	0	$\pi/3$	$\pi/5$
θ_2	$2\pi/5$	$2\pi/3$	$2\pi/5$
N_0	$2(L-1)(2L-1)$	$2L^2$	$0.5L(L+1)(2L+1)$
N_{sf}	$2(L-1)$	$2L$	$L(L+1)$

triangular prism form an angle $\theta = 2\tan^{-1}(1/\sqrt{2})$, which is slightly different from θ_2, each domain stores elastic energy corresponding to a shear deformation with $\varepsilon = \theta_2 - \theta$.

Figure 6b shows the structure of a nanowire motif with a hexagonal cross section, but three-fold rotational symmetry. This nanowire, labeled H, has a hexagonal core structure. Although its surfaces are equivalent, they are shifted in an alternating manner along the nanowire axis by half of the translational period T. The symmetry-adapted computational domain represents 1/6 of the translational domain. Its parameters are $\theta_1 = \pi/3$, $T_1 = 0.5T$, and $\theta_2 = 2\pi/3$. The non-zero value of θ_1 indicates that the H nanowires is generated by repeated screw and pure rotations.

The last nanowire motif considered, labeled I, also exhibits a five-fold rotational symmetry. The tetrahedron block shown in Fig. 7a is truncated directly from the bulk. It exposes only low-energy (111) surfaces and all faces are equilateral triangles. Twenty such blocks are combined through low-energy stacking faults to form the dot shown in Fig. 7b, having icosahedral I_h symmetry. The combination of three blocks shown in Fig. 7c represents the symmetry-adapted domain for the I nanowire shown in Fig. 7d. The associated angular parameters are $\theta_1 = \pi/5$, $\theta_2 = 2\pi/5$. Thus, with Eq. (15) this nanowire is built from screw and pure rotation isometries.

In the I nanowire one can identify two types of blocks: blocks (shown with yellow/light gray and blue/dark gray) that, like in the I_h dot, expose one facet and blocks (shown in gray) in which all facets are in contact with other blocks. As discussed before [48], the I nanowire can be also thought of as being composed of aligned polycrystalline I_h Si quantum dots that share the tetrahedron blocks shown in gray (i.e., five tetrahedral blocks at each interface). In this respect, the two crosses separated by a distance T_1 in Fig. 7c mark the centers of two virtual I_h dots sharing the gray blocks. This picture helps to comprehend the strain accumulated in the I nanowire, discussed next.

As in the case of the P nanowires, the mismatch resulted when combining the bulk tetrahedron blocks to form the I_h dot and the I nanowire introduces strain. In the high symmetry of the I_h dot the 20 tetrahedron blocks are equivalent and thus the mismatch strain will be equally distributed among them. More specifically, since at each vortex a radial channel of pentagonal rings is formed, each tetrahedron block around it accumulates elastic shear energy corresponding to a strain $\varepsilon = \theta_2 - \theta$. As each tetrahedron is delimited by three radial channels, it follows that the exposed faces are equilateral triangles, as can be also noted from Fig. 7b. The symmetry lowering to the I nanowire superposes additional elastic energy. In an icosahedron the tetrahedron building blocks are not regular (i.e., the exposed faces are equilateral while the internal ones are isosceles triangles). Thus, the (110) planes of the gray blocks of Fig. 7b (to be shared in the I nanowire construction) are not perpendicular on the nanowire long axis. Consequently, the internal blocks will undergo additional elastic deformation, specifically by decreasing the angle made by the two (111) facets that join into the circumferential edge, Fig. 7b, such that the middle transversal (110) plane becomes perpendicular to the nanowire axis. The other tetrahedron type will also be affected by this adjustment and the exposed faces will become isosceles triangles in the one-dimensional structure. From this qualitative picture one can immediately conjecture that the I nanowires will store a larger strain than the P nanowires and the I_h dots.

We finally note that in the I nanowires the periodic boundary condition treatment is unnatural as T is L-dependent and the number of atoms in the cell under periodic boundary condition increases significantly (as $\sim L^3$). For example, for $L = 6$ there are 2730 Si atoms in the cell under periodic boundary condition but only 273 atoms in the symmetry-adapted one.

3.2.2 Structural Optimizations

Which non-bulk nanowire structural motif is in fact more energetically favorable and therefore more likely to be stable? To answer this question we recognize that the accuracy of any microscopic investigation depends critically on the level of theory behind the description of the Si-Si interatomic interactions. Unfortunately, accurate *ab initio* methods formulated in the typical periodical boundary conditions context are computationally demanding and studies employing these methods can be carried out only at the smallest diameters [42, 44, 47]. They are used to complement larger scale microscopic calculations based on empirical classical potentials [53, 54], which are typically assumed to be less accurate only in the smallest size range due to the enhanced role of surfaces.

Consider for instance the description of the surface reconstruction as obtained with the widely used Stillinger-Weber [53] and Tersoff [54] classical potentials, both used in [45, 46, 48] to describe the pentagonal nanowire that exposes (001) facets. The (001) Si surface has been studied with these potentials [55] and a p(2×1) *symmetric* dimer pattern was found to be the most stable. However, with density-functional theory it was found that the *alternating* asymmetric buckling of surface dimers p(2×2) is more favorable. Since the *asymmetric* buckling is not captured, the classical treatment [53, 54] for the thinnest nanowires exposing (001) surfaces will not be accurate.

We performed the structural optimization on the three selected nanowires by the proposed symmetry-adapted tight binding molecular dynamics method, carried out with a 1 fs time step, followed by conjugate gradient energy minimization scans for several T_1 parameter values until the optimal configuration was identified. Computations were carried out on the fundamental domains addressed in Table 1 for all I nanowires and for the H nanowires with odd L. Larger domains, with $2T_1$ for the P nanowires and $2\theta_2$ for the H nanowires with even L, were also used in order to describe the alternating reconstructions of the surfaces. Exploiting symmetry has the potential danger of missing minima with lower symmetry. Our additional calculations under periodic boundary condition carried out for the smallest nanowires (with $L = 2$) showed agreement with the symmetry-adapted data.

Figure 8a presents the optimal surface reconstruction for two P nanowire domains, showing an alternating buckled dimer reconstruction on top. Accounting correctly for the surface reconstruction appeared important especially for the thinnest nanowires where the surface to volume ratio is largest. For example, we obtained a 20 meV/atom energy lowering from the non-alternating buckled to the alternating buckled pattern for the P nanowire with $L = 2$.

The H nanowire surface exposes dimer rows aligned perpendicularly on the nanowire axis. Our symmetry-adapted tight binding molecular dynamics optimization procedure obtained again an alternating buckled pattern of the surface dimers, as presented in Figure 8b. This appears in disagreement with our obtained classical description which does not account for the buckling effect. Regarding the surfaces of the I nanowires, we noted that the characteristic surface buckling of the (111) surface was severely reduced with the increase in diameter. In fact, above $2R = 2.5$ nm the surface was practically built up from flat hexagonal rings. The alternating buckled pattern was obtained on the edge surface dimer rows formed at the interface between tetrahedrons. By contrast, the surface did not appear flattened and the edge dimers did not buckle in the classical treatment.

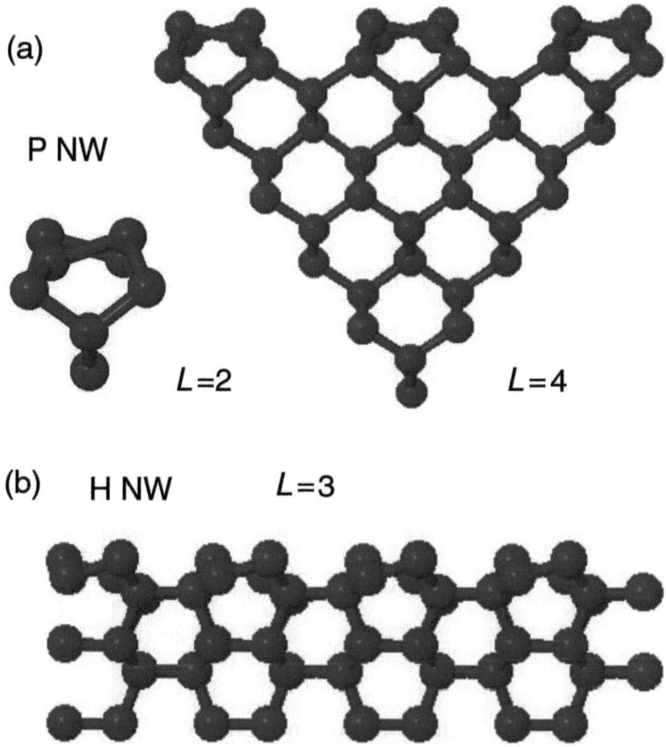

Fig. 8 (**a**) Symmetry-adapted domain (axial view) for the P nanowires with $L = 2$ and $L = 4$. The alternating buckled dimer reconstruction pattern of the surface can be observed. (**b**) Side view of the symmetry-adapted domain for the H nanowires with $L = 3$ of length $4T_1$, showing on top the buckled dimer alternating pattern

In Figure 9 we report the obtained stability data for the three nanowire motifs as described by the two microscopic models. As can be seen from Figure 9a, which plots the nanowires formation energies E (measured with respect to atomic energy values) in the tight binding model as a function of diameter $2R$, the P and H nanowire motifs emerge as the most favorable. (Practically the two curves are overlapping at diameters above 2 nm.) Surprisingly, the I nanowire is favorable only at the very small diameters, below ~2.5 nm, when the I curve intersects the P one. Above this value, this nanowire becomes unfavorable. Interestingly, Figure 9b shows that a very different conclusion can be obtained if one relies on the classical potential data. The formation energy curves for the three nanowires are very close together. In agreement with previous investigations [48], the I wire appears now more favorable than the P one and the intersections between the I and P curves, marked by the down arrow, is delayed until ~9 nm. In agreement also with the previous comparison [46] based on the classical Stillinger-Weber potential, the H nanowire motif appears overall more favorable than the P one. Comparing now the I and H nanowires, we see that below 6 nm in diameter the I nanowire is favored.

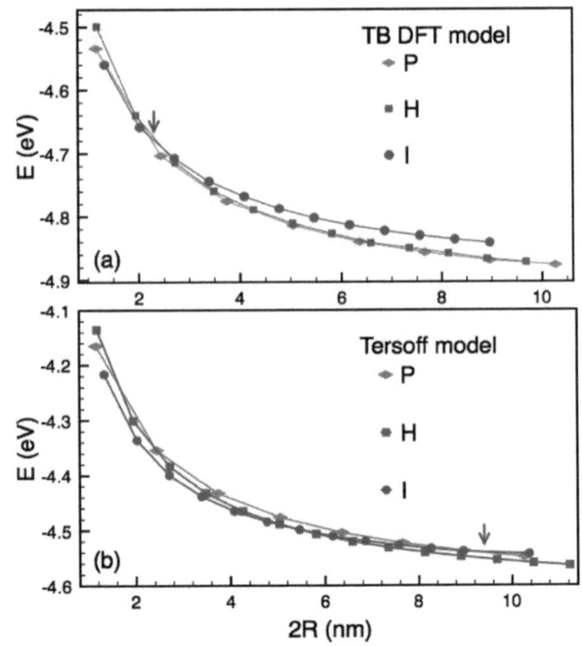

Fig. 9 Size dependence of the cohesive energy for the P, H, and I Si nanowires. The (**a**) density functional theory-based tight binding potential and (**b**) classical Tersoff potential give different energetic orderings. Arrows mark intersections of I and P energy curves

3.2.3 Comparison of the Tight Binding and Classical Potential Simulation Results

To rationalize the differences between Fig. 9a, b we have analyzed in more detail how the two microscopic models are describing each nanowire motif. Figure 10a,b,c plots over a large diameter range the obtained $E - E_{\text{bulk}}$ values for all nanowires as obtained with the tight binding and Tersoff atomistic descriptions. We found it instructive to perform a Wulff decomposition of the obtained cohesive energy data as:

$$E - E_{\text{bulk}} = \delta E_{\text{bulk}} + E_{\text{sf}} + E_{\text{e}}, \tag{27}$$

where E_{bulk} represents the cohesive energy of the crystalline nanowire bulk, i.e., cubic diamond bulk (−4.953 eV and −4.628 eV for tight binding and Tersoff potential, respectively) for the P and I nanowires and the hexagonal bulk Si (−4.943 eV and −4.625 eV for the tight binding and Tersoff potential, respectively) for the H nanowire. In Eq. (13) δE_{bulk} is the bulk energy correction, which captures the elastic strain stored in the nanowire core. The surface E_{sf} and edge E_{e} energies have analytic expressions constructed by taking into account the structural parameters of the nanowires. A Wulff decomposition extrapolation approach was used before to predict formation energies of nanowires at larger sizes [45]. Here, by identifying in the atomistic data the magnitude of the various contributions, we use it to obtain more insight about the differences between the two models.

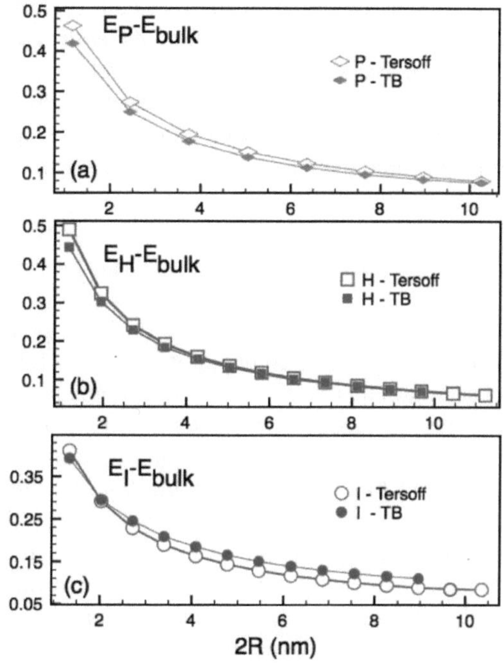

Fig. 10 (**a**) *P*, (**b**) *H*, and (**c**) *I* Si nanowires described with the tight binding (*full markers*) and Tersoff (*open markers*) potential. The zero of energy is taken to be the cohesive energy of the cubic diamond Si in (**a**) and (**c**), and the hexagonal Si in (**b**). The energy is measured in eV per atom

In Fig. 10a we see that overall the tight binding description of the *P* nanowire gives a lower energy. The differences are especially significant at the smaller sizes (44 meV at $L = 2$) and they diminish as the nanowire diameter is increased. Using the structural information entered in Table 1 and recognizing that this nanowire structure has no edge energy penalty, we obtain that the cohesive energy should scale with the number of layers L as

$$E_P(L) - E^c_{\text{bulk}} = \delta E_{\text{bulk}} + \frac{\gamma_P}{2L-1}, \tag{28}$$

where γ_P is the surface energy penalty per surface atom and δE_{bulk} represents the shear elastic energy stored in these structures. Fitting to the atomistic data we obtained γ_P of 1.25 eV/atom with tight binding and 1.39 eV/atom with Tersoff description. For both models we found that δE_{bulk} is small and can be neglected. Thus, the energetic differences noted in Fig. 10a can be attributed mainly to the inability of the Tersoff potential to correctly describe the exposed Si (001) surfaces.

Moving on to the *H* nanowire, Fig. 10b shows similar differences with a classical–tight binding agreement at large diameters. The *H* nanowire does not contain edge and bulk correction terms. Relying on the algebraic form

$$E_H(L) - E^w_{\text{bulk}} = \frac{\gamma_H}{L}, \tag{29}$$

a least-squares fitting of the two data sets obtained γ_H as 0.88 eV/atom in tight binding and 0.97 eV/atom with the classical potential.

Figure 10c shows even larger differences between the two I nanowire descriptions. While at the lowest diameter the tight binding model yields a lower energy, the classical potential-based modeling gives a lower energy for $L \geq 4$ or $2R \geq 2.5$ nm. We fitted the data to the algebraic form

$$E_I(L) - E^c_{\text{bulk}} = \delta E_{\text{bulk}} + \frac{\gamma_I}{L + 0.5} + \frac{12\delta}{(L+1)(2L+1)}, \tag{30}$$

which was constructed based on the structural parameters entered in Table 1 and the observation that there are $6L$ edge atoms in each computational domain. Due to the different surface reconstruction above 2.5 nm in diameter, the first two tight binding data points were not included in the fitting. For δE_{bulk} we obtained 36 meV/atom with tight binding versus 21 meV/atom with classical. Both models confirm the I nanowire stores significant strain energy but due to the errors in reproducing the elastic constants, the strain component is not well represented by the classical model. Further, we found the flattened surface obtained with tight binding has a higher 980 meV/atom cost when compared with the 900 meV/atom obtained classically. Finally, we found that the buckling of the dimers located on the edges lowers the tight binding energy as $\delta = -37$ meV/atom. A δ value of 38 meV/atom obtained from the classical data agrees with our observation that this model does not capture the dimer buckling. As can be seen from Eq. (30), the edge energy component is the least important, and the tight binding description gives a higher energy because the bulk correction and the surface energy count more in the energy balance. We have summarized all the fitted data of the energy components in the Wulff decomposition in Table 2.

Table 2 Summary of energy values obtained from the Wulff-like energy decomposition of the three nanowires

Nanowire		P	H	I
E_{bulk}(eV)	Tight binding	−4.953	−4.943	−4.953
	Tersoff	−4.628	−4.625	−4.628
δE_{bulk}(meV)	Tight binding	0	–	36
	Tersoff	0	–	21
γ (eV)	Tight binding	1.25	0.88	0.98
	Tersoff	1.39	0.97	0.90
δ (meV)	Tight binding	–	–	−37
	Tersoff	–	–	38

4 Conclusion

In conclusion, symmetry adapted tight binding modeling is a versatile and efficient computational technique for nanomechanics. Based on the obtained modeling results, one has the possibility to next construct a continuum, computational efficient model of the nano objects that can be used in engineering applications. Here we proved the abilities of this method to uncover the linear and nonlinear elastic response of carbon nanotubes from a density-functional-based tight binding modeling of the covalent bond. The critical strain beyond which nanotubes behave nonlinearly in torsion, the most favorable rippling morphology, and the twist- and morphology-related changes in fundamental band gap were identified from a rigorous atomistic description.

From the presented investigations on the stability of Si nanowires we see that the use of a quantum-mechanical description of the atomic bonding is critical as the classical potentials [54] introduces significant errors. Moreover, these errors are not systematic: On one hand, for the P and H nanowires the errors are larger at smaller diameters as they are mainly caused by the inability of the classical potential to describe the surface reconstruction. On the other hand, for the I nanowire the errors are large even at large diameters, as they are due to the inability to describe the strain stored in the nanowire core as well as the higher surface energy penalty. The magnitude of these errors are causing the energetic ordering of the nanowire structures not to be identical in the two microscopic descriptions. Hence, different stability conclusions [46, 48] can be reached from the two data sets. We note that the usual approach [45, 46] of complementing the large scale classical calculations with expensive quantum calculations carried out at the smallest diameters (i.e., that assumes that corrections are needed at the smallest diameters only) should be used with care. For example, judging the I nanowire stability by combining the Tersoff data with the affordable first two tight binding data points under periodic boundary condition would be deceiving.

Acknowledgements TD acknowledges NSF CAREER CMMI-0747684 for supporting this research.

References

1. S. Iijima, Helical microtubules of graphitic carbon. Nature, **354** (6348), 56–58 (1991)
2. N.G. Chopra, R.J. Luyken, K. Cherrey, V.H. Crespi, M.L. Cohen, S.G. Louie, A. Zettl, Boron nitride nanotubes. Science, **269** (5226), 966–967 (1995)
3. H. Dai, E.W. Wong, Y.Z. Lu, S. Fan, C.M. Lieber, Synthesis and characterization of carbide nanorods. Nature, **375** (6534), 769–772 (1995)
4. P. Yang, C.M. Lieber, Nanorod-superconductor composites: a pathway to materials with high critical current densities. Science, **273** (5283), 1836–1840 (1996)

5. D. Qian, G.J. Wagner, W.K. Liu, A multiscale projection method for the analysis of carbon nanotubes. Comput. Method Appl. Mechan. Eng. **193** (17–20), 1603–1632 (2004)
6. T. Belytschko, S.P. Xiao, G.C. Schatz, R.S. Ruoff, Atomistic simulations of nanotube fracture. Phys. Rev. B, **65** (23), 235430 (2002)
7. B.I. Yakobson, C.J. Brabec, J. Bernholc, Nanomechanics of carbon tubes: instabilities beyond linear response. Phys. Rev. Lett. **76** (14), 2511–2514 (1996)
8. K.M. Liew, C.H. Wong, X.Q. He, M.J. Tan, S.A. Meguid, Nanomechanics of single and multiwalled carbon nanotubes. Phys. Rev. *B*, **69** (11), 115429 (2004)
9. X. Li, W. Yang, B. Liu, Bending induced rippling and twisting of multiwalled carbon Nanotubes. Phys. Rev. Lett., **98** (20), 205502 (2007)
10. W.K. Liu, E.G. Karpov, S. Zhang, H.S. Park, An introduction to computational nanomechanics and materials. Comput. Meths Appl. Mechan. Eng. **193** (17–20), 1529–1578 (2004)
11. W.W. Gerberich, W.M. Mook, C.R. Perrey, C.B. Carter, M.I. Baskes, R. Mukherjee, A. Gidwani, J. Heberlein, P.H. McMurry, S.L. Girshick, Superhard silicon nanospheres. J. Mechan. Phys. Solids, **51** (6), 979–992 (2003)
12. M. Arroyo, T. Belytschko. An atomistic-based finite deformation membrane for single layer crystalline films. J. Mechan. Phys. Solids, **50** (9), 1941–1977 (2002)
13. M. Arroyo, T. Belytschko, Nonlinear mechanical response and rippling of thick multiwalled carbon nanotubes. Phys. Rev. Lett. **91** (21), 215505 (2003)
14. S. Zhang, S.L. Mielke, R. Khare, D. Troya, R.S. Ruoff, G.C. Schatz, T. Belytschko, Mechanics of defects in carbon nanotubes: atomistic and multiscale simulations. Phys. Rev. *B*, **71** (11), 115403 (2005)
15. A.Y.T. Leung, X. Guo, X.Q. He, H. Jiang, Y. Huang, Postbuckling of carbon nanotubes by atomic-scale finite element. J. Appl. Phys., **99**, 124308 (2006)
16. B. Liu, H. Jiang, Y. Huang, S. Qu, M.F. Yu, K.C. Hwang, Atomic-scale finite element method in multiscale computation with applications to carbon nanotubes. Phys. Rev. *B*, **72** (3), 35435 (2005)
17. P. Zhang, Y. Huang, P.H. Geubelle, P.A. Klein, K.C. Hwang, The elastic modulus of single-wall carbon nanotubes: a continuum analysis incorporating interatomic potentials. Int J Solids Struct., **39** (13–14), 3893–3906 (2002)
18. P. Zhang, H. Jiang, Y. Huang, P.H. Geubelle, K.C. Hwang, An atomistic-based continuum theory for carbon nanotubes: analysis of fracture nucleation. J. Mechan. Phys. Solids, **52** (5), 977–998 (2004)
19. L.V. Zhigilei, C. Wei, D. Srivastava, Mesoscopic model for dynamic simulations of carbon nanotubes. Phys. Rev. *B*, **71** (16), 165417 (2005)
20. R.M. Martin, Electronic structure: basic theory and practical methods. (*Cambridge University Press*, Cambridge, UK, 2005)
21. N.M. Ghoniem, E.P. Busso, N. Kioussis, H. Huang, Multiscale modelling of nanomechanics and micromechanics: an overview. Philos. Mag., **83** (31), 3475–3528 (2003)
22. W. Kohn, L.J. Sham, Self-consistent equations including exchange and correlation effects. Phys. Rev., **140** (4A), A1133–A1138 (1965)
23. J.C. Slater, G.F. Koster, Simplified LCAO method for the periodic potential problem. *Phys. Rev.*, 94 (6), 1498–1524 (1965)
24. C.M. Goringe, D.R. Bowler, E. Hernández, Tight binding modelling of materials. Rep Prog Phys. **60** (12), 1447–1512 (1997)
25. R. Saito, G. Dresselhaus, M.S. Dresselhaus, Physical properties of carbon nanotubes. (Imperial College Press, London, 2005)
26. P.B. Allen, Nanocrystalline Nanowires: III. Electrons. Nano Lett, **7** (5), 1220–1223 (2007)
27. V.N. Popov, An equivalent non-orthogonal tight binding formalism using Bloch sums adapted for screw and translation operations can be found in . *New J. Phys.*, **6** (1), 17 (2004)
28. C.T. White, D.H. Robertson, J.W. Mintmire, Helical and rotational symmetries of nanoscale graphitic tubules. Phys. Rev. *B*, **47** (9), 5485–5488 (1993)
29. R.D. James, Objective structures. J. Mechan. Phys. Solids, **54**: 2354–2390 (2007)
30. T. Dumitrică, R.D. James, Objective molecular dynamics. J. Mechan. Phys. Solids, **55** (10), 2206–2236 (2007)

31. R. Rurali, E. Hernández, Trocadero: a multiple-algorithm multiple-model atomistic simulation program. Comput. Mater. Sci., **28** (2), 85–106 (2003)
32. D. Porezag, T. Frauenheim, T. Köhler, G. Seifert, R. Kaschner, Construction of tight binding-like potentials on the basis of density-functional theory: application to carbon. Phys. Rev. *B*, **51** (19), 12947–12957 (1995)
33. C.H. Xu, C.Z. Wang, C.T. Chan, K.M. Ho, A transferable tight binding potential for carbon. *J. Phys.:* Condens. Matt., **4**, 6047–6054 (1992)
34. M. Born, G. R. Oppenheimer. Zur Quantentheorie der Molekeln. Ann. Physik, **84**, 457 (1927)
35. K.N. Kudin, G.E. Scuseria, B.I. Yakobson, C_2, F, BN, and C nanoshell elasticity from ab initio computations. Phys. Rev. *B*, **64** (23), 235406 (2001)
36. E. Hernández, C. Goze, P. Bernier, A. Rubio, Elastic properties of C and $B_xC_yN_z$ composite nanotubes. Phys. Rev. Lett., **80** (20), 4502–4505 (1998)
37. L. Yang, J. Han, Electronic structure of deformed carbon nanotubes. Phys. Rev. Lett., **85** (1), 154 (2000)
38. Y. Wu, Y. Cui, L. Huynh, C.J. Barrelet, D.C. Bell, C.M. Lieber, Controlled growth and structures of molecular-scale silicon nanowires. Nano Lett, **4** (3), 433–436 (2004)
39. M. Takeguchi, M. Tanaka, H. Yasuda, K. Furuya, Real-time high-resolution transmission electron microscopy observation of the growth process of (001) surfaces on a nanometer-sized Si multiply twinned particle. Surf Sci, **493** (1), 414–419 (2001)
40. A.F. Morral, J. Arbiol, J.D. Prades, A. Cirera, J.R. Morante, Synthesis of silicon nanowires with wurtzite crystalline structure by using standard chemical vapor deposition. Adv. Mater, **19**, 1347–1351 (2007)
41. J. Bai, X.C. Zeng, H. Tanaka, J.Y. Zeng, Metallic single-walled silicon nanotubes. Proc Nat Acad Sci., **101** (9), 2664–2668 (2004)
42. T. Dumitrică, M. Hua, B.I. Yakobson, Endohedral silicon nanotubes as thinnest silicide wires. Phys. Rev. *B*, **70** (24), 241303 (2004)
43. R. Rurali, N. Lorente, Metallic and semimetallic silicon $< 100 >$ nanowires. Phys. Rev. Lett., **94** (2), 26805 (2005)
44. R. Kagimura, R.W. Nunes, H. Chacham, Structures of si and ge nanowires in the subnanometer range. Phys. Rev. Lett., **95** (11), 115502 (2005)
45. Y. Zhao, B.I. Yakobson, What is the ground-state structure of the thinnest si nanowires? Phys. Rev. Lett., **91** (3), 35501 (2003)
46. I. Ponomareva, M. Menon, D. Srivastava, A.N. Andriotis, Structure, stability, and quantum conductivity of small diameter silicon nanowires. Phys. Rev. Lett., **95** (26), 265502 (2005)
47. I. Ponomareva, M. Menon, E. Richter, A.N. Andriotis, Structural stability, electronic properties, and quantum conductivity of small-diameter silicon nanowires. Phys. Rev. *B*, **74** (12), 125311 (2006)
48. K. Nishio, T. Morishita, W. Shinoda, M. Mikami, Molecular dynamics simulations of self-organized polyicosahedral Si nanowire. J. Chem. Physi., **125**, 074712 (2006)
49. J.F. Justo, R.D. Menezes, L.V.C. Assali, Stability and plasticity of silicon nanowires: the role of wire perimeter. Phys. Rev. *B*, **75** (4), 45303 (2007)
50. F. Glas, J.C. Harmand, G. Patriarche, Why does wurtzite form in nanowires of III-V zinc blende semiconductors? Phys. Rev. Lett., **99** (14), 146101 (2007)
51. Y. Zhao, Y.H. Kim, M.H. Du, S.B. Zhang, First-Principles prediction of icosahedral quantum dots for tetravalent semiconductors. Phys. Rev. Lett., **93** (1), 15502 (2004)
52. S.G. Hao, D.-B. Zhang, T. Dumitrică, Effect of small shape changes on the optical response of highly symmetric silicon quantum dots. Phys. Rev. *B*, **76** (8), 81305 (2007)
53. F.H. Stillinger, T.A. Weber, Computer simulation of local order in condensed phases of silicon. Phys. Rev. *B*, **31** (8), 5262–5271 (1985)
54. J. Tersoff, Modeling solid-state chemistry: interatomic potentials for multicomponent systems. Phys. Rev. *B*, **39** (8), 5566–5568 (1989)
55. H. Balamane, T. Halicioglu, W. A. Tiller, Comparative study of silicon empirical interatomic potentials. Phys. Rev. *B*, **46** (4), 2250–2279 (1992)

Part II
Nanowires: Growth and Properties

Thin-Film Metamaterials Called Sculptured Thin Films

Akhlesh Lakhtakia and Joseph B. Geddes III

Abstract Morphology and performance are conjointed attributes of metamaterials, of which sculptured thin films (STFs) are examples. STFs are assemblies of nanowires that can be fabricated from many different materials, typically via physical vapor deposition onto rotating substrates. The curvilinear-nanowire morphology of STFs is determined by the substrate motions during fabrication. The optical properties, especially, can be tailored by varying the morphology of STFs. In many cases prototype devices have been fabricated for various optical, thermal, chemical, and biological applications.

1 Introduction

The emergence of metamaterials at the end of the 20th century heralded a major motivational shift in research on materials and coincided with the ramping up of nanotechnology [1]. Materials researchers began to consider the design of composite materials, called metamaterials, to perform more than one role each in specific environments. Among optics researchers today, the term *metamaterial* is often taken to mean a material with negative refractive index [2], but there is much more to metamaterials than that. In the year 2000, Rodger Walser [3] coined this term for certain types of artificial materials, and later formally defined metamaterials as "macroscopic composites having a manmade, three-dimensional, periodic cellular architecture designed to produce an optimized combination, not available in nature, of two or more responses [emphasis in the original] to specific excitation." We can relax the requirements of periodicity today, though not of cellularity. Indeed, cellularity in morphology engenders multifunctional performance.

Sculptured thin films (STFs) [4] exemplify metamaterials. An STF is an assembly of nanowires typically grown by physical vapor deposition, whose bent and twisted forms are engineered via the growth process. As a result of the flexibility in

A. Lakhtakia (✉)
Department of Engineering Science and Mechanics, Pennsylvania State University, University Park, PA 16802, USA
e-mail: axl4@psu.edu

A. Aldea, V. Bârsan (eds.), *Trends in Nanophysics*, Engineering Materials,
DOI 10.1007/978-3-642-12070-1_3,

Fig. 1 Scanning electron micrograph showing the helical nanowires of a chiral STF. Courtesy: M. W. Horn, The Pennsylvania State University

controlling the evolving nanostructure of the films during fabrication, their performance characteristics can be engineered.

The curvilinear-nanowire morphology of STFs is exemplified by Fig. 1. Note the individual nanowires that make up the STF in this case are helical; they are also nominally identical and parallel. The constituent nanowires of an STF have diameters of $\sim$10–300 nm, and lengths on the order of tens of nanometers to several micrometers. The nanowires are made of clusters 1–3 nm in linear dimension. Therefore, STFs may be classified as nanomaterials. For many applications dependent on optical or infrared radiation, the closely packed nanowires can be treated effectively as equivalent to an anisotropic and nonhomogeneous continuum.

Possibly the first STF was fabricated by Young and Kowal in 1959 [5], who made fluorite films via oblique evaporation onto a rotating substrate. Although the structure of the resulting film was not unambiguously determined, the film did display optical activity. Most likely, the film was structurally chiral, as in Fig. 1. This development was preceded by about 50 years of growing columnar thin films by thermal evaporation [6]. Columnar thin films are assemblies of parallel straight nanowires, and are optically akin to biaxial crystals. The emergence of scanning electron microscopy in the late 1950s and early 1960s led to the confirmation of nanowire morphology. In 1966, a significant event occurred when Nieuwenhuizen and Haanstra showed that the nanowire inclination could be altered within a transitional distance of about 30 nm [7].

Messier and colleagues identified the conditions for the emergence of the nanowire morphology [8]. In 1989, Motohira and Taga showed that the direction of growing nanowires could be altered often within a few nanometers by appropriately switching the direction of the incoming vapor mid-deposition [9]. Lakhtakia and Messier realized that nanowires with almost arbitrary shapes could be fashioned by continuously changing the direction of the incoming vapor during deposition [10]. Robbie and colleagues reported the first STF with unambiguously determined structure [11]. Since that time, hundreds of research papers, a monograph [4], and many book chapters have been written on STFs [12–16].

This article contains a brief description of and the physical phenomenons exhibited by STFs. In many cases, those phenomenons are due directly to the symmetries that the structures of the constituent nanowires break. In Sect. 2, we review the fabrication of STFs. Then in Sect. 3, we present the mathematical description of the optical constitutive relations of STFs. In Sect. 4, we review the optical, thermal, chemical, and biological applications to which STFs either have been put or have been suggested. The advantages and disadvantages of STF technologies compared to other technologies are presented in the same section. We end in Sect. 5 with some remarks on the extension of the STF concept for acoustic applications.

2 Fabrication of STFs

STFs are chiefly fabricated by physical vapor deposition [17]. Several variants exist, of which thermal evaporation is the simplest. A source material is evaporated under high vacuum and the vapor allowed to flow at an oblique incidence angle χ_v onto a substrate, as schematically depicted in Fig. 2. When the temperature of the source material is less than approximately one–third the melting temperature, there is little surface diffusion, and straight nanowires form. The morphology is enabled by the shadowing of nanowires by those in front of them.

The nanowire morphology can be changed during deposition by tilting and rotating the substrate. When the substrate is rotated about a direction perpendicular to the substrate and passing through it, the nanowires become helical. These structures form the basis of chiral STFs, also known as thin-film helicoidal bianisotropic mediums (TFHBMs) in optical parlance. Several scanning electron micrographs of STFs, including TFHBMs, are shown in Fig. 3.

In addition, the substrate can be rotated about an axis in the plane of the substrate. Changing the tilt along either or both of two such orthogonal axes can lead to zigzag,

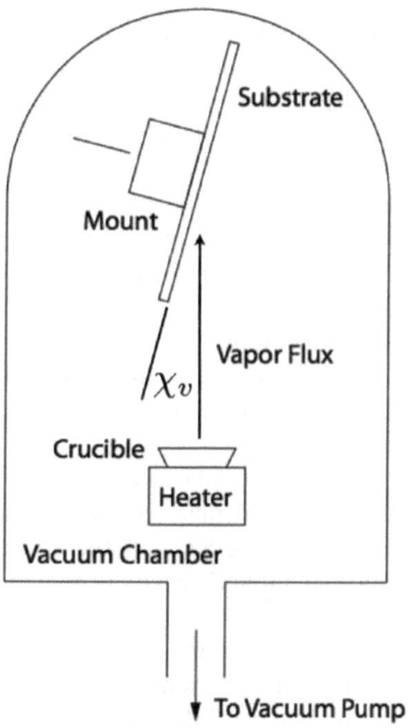

Fig. 2 Schematic for fabrication of STFs by thermal evaporation. The source material is placed in a heated crucible, from which the vapor flux is allowed to fall obliquely on a substrate. A motorized mount allows the orientation of the substrate to be varied with time

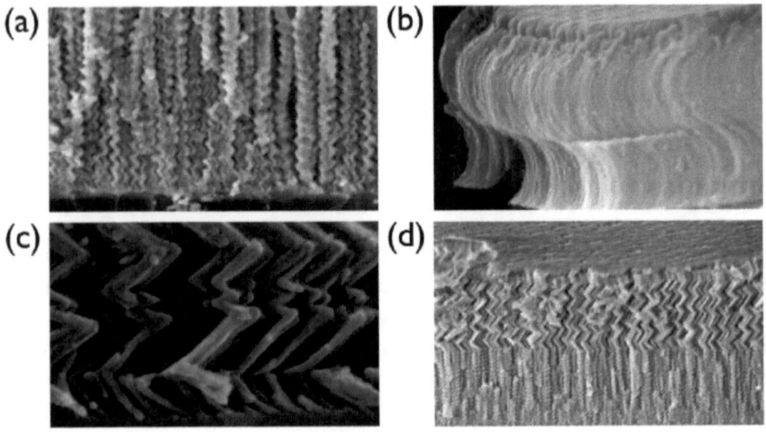

Fig. 3 Examples of (**a**) thin–film helicoidal bianisotropic medium, (**b**) and (**c**) sculptured nematic thin films, and (**d**) hybrid STFs. Courtesy: M. W. Horn and R. Messier, The Pennsylvania State University

c-shaped, and s-shaped nanowires. Such STFs are known as sculptured nematic thin films (SNTFs). Several scanning electron micrographs of SNTFs are shown in Fig. 3.

Hybrid STFs that consist of both TFHBM and SNTF sections grown one after another are also possible. For example, Suzuki and Taga fabricated hybrid STFs with zigzag, helicoidal, and straight nanowire sections [18]. More recently, Park et al. grew a hybrid STF comprising three different chiral sections [19].

Several variations on the basic scheme are possible, including serial bideposition, multideposition, prepatterning of substrates, and hybrid chemical/physical vapor deposition. In serial bideposition, a single source is used to fabricate films with large local birefringence [20, 21]. For example, when growing chiral STFs, the source is manipulated to deposit alternately on either side of the growing nanowires. Multideposition consists of two or more sources being simultaneously evaporated to grow STFs [22]. Hybrid chemical/physical vapor deposition has been used to fabricate polymeric STFs [23, 24], because such materials cannot be heated to produce a collimated vapor.

Inverse structures can be formed by infilling of STF voids via electrodeposition of metal or infiltration of molten polymer, and subsequent removal of the initial STF via etching [33].

STFs can be fabricated from just about any material that can be evaporated. A wide variety of materials, including dielectrics, metals, and polymers have been used to fabricate STFs. Table 1 lists several materials from which STFs have been fabricated. GeSbSe chalcogenide glasses and luminescent Alq_3 have been used most recently [39–42].

The structures into which STFs can be formed has also expanded to include three-dimensional STF architectures. STFs can be deposited on substrates that have been prepatterned via lithography or other processes. The STF nanowires grow on the

Table 1 Materials from which STFs have been fabricated. Note that the list includes oxides, fluorides, metals, semiconductors, and polymers. Materials marked with * indicate inverse structures formed by infilling of STF voids with the indicated material and subsequent removal of the initial STF

Material(s)	Reference(s)	Material(s)	Reference(s)
Iron oxide	[25]	Silicon oxide	[17, 26–28]
Tantalum oxide	[29]	Tin oxide	[27]
Titanium oxide	[21, 29, 30]	Zirconium oxide	[29, 31]
Calcium fluoride	[11, 26]	Magnesium fluoride	[11, 17, 26]
Indium nitride	[32]	Aluminum	[27]
Chromium	[26, 27]	Copper	[26]
Gold	[33]*	Manganese	[26, 34]
Molybdenum	[27]	Nickel	[33]*
Carbon	[35]	Silicon	[36–38]
GeSbSe chalcogenide	[39–41]	Alq3	[42]
Parylene	[23, 24]	Polystyrene	[33]*

raised areas (and, depending on the angle of incidence of the vapor, the side walls) to form three-dimensional structures [27, 28]. Alternatively, the pre–deposition of arrays of small islands, or seeds, of source material can result, after STF deposition, in patterns of single nanowires or clusters of nanowires [43–45].

3 Optical Constitutive Relations of STFs

Consider an STF that occupies the region $0 \leq z \leq L$. We use the cartesian coordinate system (x, y, z) with the triad of unit vectors $\{\underline{u}_x, \underline{u}_y, \underline{u}_z\}$. The frequency-domain constitutive relations of such a dielectric, spatially local STF within this region are as follows:

$$\underline{D}\,(x, y, z, \omega) = \epsilon_o \,\underline{\underline{\epsilon}}_r\,(z, \omega) \cdot \underline{E}\,(x, y, z, \omega) \ , \tag{1}$$

$$\underline{B}\,(x, y, z, \omega) = \mu_o \,\underline{H}\,(x, y, z, \omega) \ , \tag{2}$$

where $\underline{E}$ and $\underline{B}$ are the macroscopic primitive electric and magnetic field phasors; $\underline{D}$ and $\underline{H}$ are the corresponding induction field phasors; ϵ_o and μ_o are the permittivity and permeability of vacuum, respectively; ω is the angular frequency; and $\underline{\underline{\epsilon}}_r$ is the relative permittivity dyadic. The latter can be factored into pieces which depend either only on the structure of the film or the dispersive properties; thus,

$$\underline{\underline{\epsilon}}_r\,(z, \omega) = \underline{\underline{S}}\,(z) \cdot \underline{\underline{\epsilon}}_{ref}\,(\omega) \cdot \underline{\underline{S}}\,(z) \ , \tag{3}$$

where $\underline{\underline{\epsilon}}_{ref}$ is the reference permittivity dyadic, so named to correspond to a reference plane $z = z_o$ where the rotation dyadic $\underline{\underline{S}}$ equals the identity dyadic (i.e., $\underline{\underline{S}}\,(z_o) = \underline{\underline{I}}$).

There are two canonical forms for the rotation dyadic. When

$$\underline{\underline{S}}(z) = \underline{\underline{S}}_y(z) = \underline{u}_y\underline{u}_y + \left(\underline{u}_x\underline{u}_x + \underline{u}_z\underline{u}_z\right)\cos\tau(z) + \left(\underline{u}_z\underline{u}_x - \underline{u}_x\underline{u}_z\right)\sin\tau(z)\ , \quad (4)$$

the STF exhibits an essentially two-dimensional morphology (in the xz plane), and hence is called an SNTF. The angle $\tau(z)$ determines whether the nanowires are chevrons, c-shaped, s-shaped, or of other more complex shapes.

When

$$\underline{\underline{S}}(z) = \underline{\underline{S}}_z(z) = \underline{u}_z\underline{u}_z + \left(\underline{u}_x\underline{u}_x + \underline{u}_y\underline{u}_y\right)\cos\zeta(z) + \left(\underline{u}_y\underline{u}_x - \underline{u}_x\underline{u}_y\right)\sin\zeta(z)\ , \quad (5)$$

the STF exhibits a three-dimensional morphology, and hence is called a TFHBM. Such films are structurally chiral. Often such STFs are formed with $\zeta(z) = \pi z/\Omega_c$ and are therefore periodic, where Ω_c is the structural half–period. These periodically nonhomogeneous STFs are called chiral STFs.

In more complicated STFs, $\underline{\underline{S}}(z)$ can be some combination of $\underline{\underline{S}}_y(z)$ and $\underline{\underline{S}}_z(z)$. In hybrid STFs, both $\underline{\underline{\epsilon}}_{ref}(\omega)$ and $\underline{\underline{S}}(z)$ can have sectionwise variations along the z axis.

A link between the microscopic structure of an STF and its macroscopic constitutive properties is found by local homogenization in a nominal model of STF morphology [46]. This nominal model considers each STF nanowire as being composed of a string of homogenous ellipsoids; the voids are also treated as conglomerates of ellipsoids. Local homogenization of this model, for example with the Bruggeman formalism, yields $\underline{\underline{\epsilon}}_{ref}(\omega)$. An extended model shows how STFs infiltrated with a chiral fluid could exhibit bianisotropic optical properties [47, 48].

4 Applications of STFs

A variety of the optical, electronic, chemical, and biological applications of STFs have been proposed and demonstrated.

4.1 Optical

The area of application for which STFs have been developed the most thoroughly is optics. The nanometer-scale control over the structure of STFs has meant that the optical properties of the films could be controlled on subwavelength scales. The recent development of three-dimensional STF architectures means that simultaneous control at both wavelength and subwavelength scales is possible – as in three-dimensional photonic crystals.

Even within the optics category, the most well–developed area is filter technology, particularly that relying on the circular Bragg phenomenon exhibited by chiral STFs. The chirality and periodicity of the latter can give rise to a stop band along the thickness direction. That is, left/right circularly polarized light over a bandwidth called the Bragg regime is mostly reflected by a sufficiently thick chiral STF that is structurally left/right handed, but right/left circularly polarized light is either transmitted or absorbed. In a more abstract sense, the chiral STF breaks both rotational and translational symmetries, and this gives rise to the circular Bragg phenomenon.

Due to their controllable void fraction, STFs have been suggested as low–permittivity dielectrics to reduce crosstalk in integrated circuits [49]. Also, efforts have begun to integrate active light emission and gain elements into STFs for optoelectronic applications. For example, a layer of optically pumped quantum dots was embedded between two chiral STF layers to create a device which emitted circularly polarized light [50]. Chiral STFs were also used to create a circularly polarized external-cavity diode laser [51]. In addition, the possibility of tunable polymeric chiral STF lasers, whose microstructure can be compressed and released by a piezoelectric element to control the output wavelength, has been analyzed [52]. A similar scheme has been suggested for tunable optical filters [53].

Another route to tunability is to infiltrate the pores of, for example, chiral STFs with liquid crystals. The aciculate molecules of the liquid crystal tend to align with the STF nanowires; this effect has been shown to increase the optical rotary power of the resulting composite films [54]. Moreover, one can apply an electric field to align the liquid crystal molecules along the field, with a concomitant change in optical properties for the composite [55].

A third route to tunability is to fabricate STFs from electro optic materials [22, 56, 57]. Most notably, the chiral STFs exhibiting the Pockels effect can exhibit a Bragg regime in the presence of a low–frequency electric field [58].

A major emerging area in optics is to exploit STFs for plasmonics–based sensing of chemical and biological materials. For example, ultrathin metal STFs have shown a sensitivity that is significantly larger than for bulk films: 113°/RIU for metal STFs in contrast to 79°/RIU for bulk films [59]. Most recently, the propagation of multiple surface–plasmon–polariton waves, of the same frequency but different phase speeds, guided by a planar metal/STF interface has been theoretically predicted [60–63] and experimentally verified [64], with exciting prospects for multianalyte sensing and error–free sensing.

Regardless of the excellence of optical performance characteristics, no STF can be used reliably if it cannot withstand mechanical stimuli within reasonable ranges. Various mechanical properties of STFs, such as hardness, compliances, yield stress, and ultimate strength, need to be modeled and measured in relation to STF morphology and composition. Furthermore, there is no experimental evidence showing how the optical properties change with the application of stress. This is practically virgin territory for research. It is known from indentation experiments that chiral STFs are permanently deformed when applied pressures exceed 30 μPa [65]. This problem could be alleviated by impregnating an STF with a polymer [33, 49] which shall have to be accounted for during optical design; however, that step will require experimentation. Moreover, polymer infiltration would allow STFs to be used as optical pressure sensors [4] as well as further the development of mechanically tunable light emitters and optical filters [52, 53].

4.2 Thermal

Harris et al. deposited alternating layers of solid and porous SNTFs of yttria-stabilized zirconia [31]. They found that the composite structure had a reduced thermal diffusivity compared to bulk films, with the added benefit of providing stress relief from expansion. The particular form of the SNTF structure they used was required to prevent cracking in the films.

4.3 Chemical

A variety of photochemical, electrochemical, and fluid sensor applications of STFs have also begun to be investigated.

Titanium dioxide STFs have been integrated into Grätzel cells for solar-power applications [66]. The photocatalytic properties of titanium dioxide STFs have been found to depend on their morphology [18].

The porous structure of indium nitride SNTFs and TFHBMs was shown to improve their electrochromic properties [32]. In addition, Brett and colleagues measured the electrohemical properties of carbon TFHBMs and chevronic manganese SNTFs [34, 35].

The porosity of STFs means that chemicals can infiltrate their pores and change their optical and electrical properties. For example, by depositing titanium dioxide STFs on coplanar interdigitated electrodes, sensors whose capacitance increases with increasing relative humidity can be made [67]. Both the optical rotation and location of spectral hole within the circular Bragg phenomenon of a chiral STF also respond to the relative humidity of the environment [68, 69]. Shifts in spectrums have also been observed on complete infiltration of the voids in chiral STFs by acetone, methyl alcohol, isopropyl alcohol, and hexamethyldisilizane [66]. This experimental work was anticipated by earlier theoretical studies of both SNTFs and TFHBMs as fluid sensors [70–72]. The susceptibility of the optical properties of STFs to ambient moisture underscores the need for the proper packaging of films not intended for use as sensors.

Although most STFs tend to absorb ambient moisture, they can under certain conditions exhibit superhydrophobic behavior, especially when functionalized with siloxane [73, 74]. Fluorination should also promote superhydrophobicity [75].

4.4 Biological

These chemical applications are quickly being joined by biological ones. Magnesium–alloy STF coatings on stents are being fabricated to be bioabsorbable [66]. Parylene STFs have been fabricated as scaffolds on which, for example, both kidney and fibroblast cells can grow [22, 76]. STFs are particularly suited to serve as scaffolds because their structure effectively mimics that of a tissue's extracellular matrix [22]. The ability to make transparent STFs means that cell cultures grown within them can be imaged, for example with confocal microscopy [22]. The degradation of iron(III) oxide STFs by metal-reducing bacteria has been shown to affect their optical properties, a result that indicates the possibility of using STF-based fiber optic sensors to measure bacterial activity below ground [25].

4.5 Advantages and Disadvantages

STF technology presents several advantages as compared with competing technologies. The fabrication process of STFs is very simple and cheap, especially as compared to conventional semiconductor patterning technologies. The fabrication process affords a high degree of control over STF morphology, and consequently properties. STFs can be made of many different kinds of materials. However, there are disadvantages too. STFs can be fragile, and so must be suitably packaged to prevent damage. Suitable packaging is also needed to prevent absorption of moisture and concomitant change in properties, unless the films are being used as sensors.

Thus STFs are probably most appropriate for applications where a high degree of control over film properties and cost are important, but mechanical stability is of lesser concern.

5 Concluding Remarks

Extension of the STF concept for acoustic purposes requires a different length scale than that for optical purposes: nanometers have to be replaced by micrometers. Very recently, polymer STFs have been made on the microscale by a hybrid vapor deposition process; one such STF is shown in Fig. 4. Such films can be used as acoustic filters, both for polarization and bandwidth [77–81]. These films could also be used

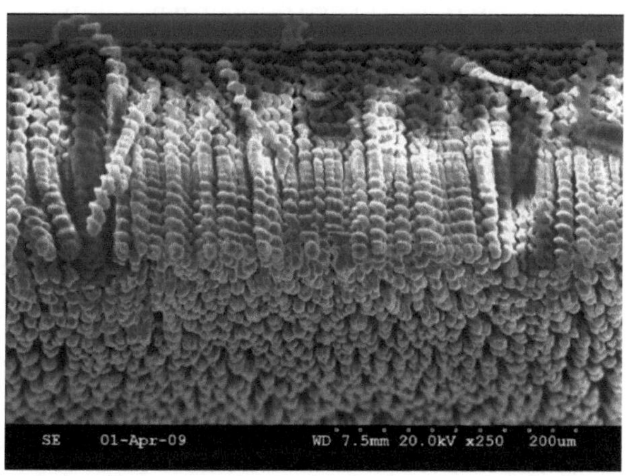

Fig. 4 Scanning electron micrograph of a chiral STF of parylene fabricated for acoustic purposes

as MEMS elements, particularly for sensing. The acoustic properties of these STFs require experimental research coupled with theoretical modeling [82–84].

To conclude, STFs show promise for several optical, electronic, chemical, biological, and acoustic applications. Of these, the optical applications are the most developed. The controllable morphology and wide variety of constituent materials makes these thin-film metamaterials attractive for these and other applications.

Acknowledgements J. B. Geddes III gratefully acknowledges support of a Beckman Postdoctoral Fellowship. A. Lakhtakia thanks the Binder Endowment at Penn State for financial support of his research.

References

1. A. Lakhtakia, T.G. Mackay, Meet the metamaterials. OSA Opt. Photon. News, **18**(1), 32–37 (2007)
2. T.G. Mackay, A. Lakhtakia, Negative refraction, negative phase velocity, and counterposition in bianisotropic materials and metamaterials. Phys. Rev. B, **79**, 235121, (2009)
3. R. Walser, Metamaterials: an introduction, in Introduction to Complex Mediums for Optics and Electromagnetics ed. by W.S. Weiglhofer, A. Lakhtakia, (SPIE Press, Bellingham, WA, USA, 2003) pp. 295–316
4. A. Lakhtakia, R. Messier, *Sculptured Thin Films: Nanoengineered Morphology and Optics*. (SPIE Press, Bellingham, WA, USA, 2005)
5. N.O. Young, J. Kowal, Optically active fluorite films. Nature, **183** 104–105 (1959)
6. R. Messier, The nano-world of thin films. J. Nanophoton., **2** 021995 (2008)
7. J.M. Nieuwenhuizen, H.B. Haanstra, Microfractography of thin films. Phillips Tech. Rev., **27** 87–91 (1966)
8. R. Messier, A.P. Giri, R.A. Roy, Revised structure zone model for thin film physical structure. J. Vac. Sci. Technol. A, 2, 500–503 (1984)
9. T. Motohiro, Y. Taga, Thin film retardation plate by oblique deposition. Appl. Opt., **28**, 2466–2482 (1989)
10. A. Lakhtakia, R. Messier, The key to a thin film HBM: The Motohiro–Taga interface. in *Proceedings of Chiral '94; 3rd International Workshop on Chiral, Bi-isotropic and Bi-anisotropic Media*, ed. by F. Mariotte, J.-P. Parneix (Périgueux, France, 1994) pp 125–130
11. K. Robbie, M.J. Brett, A. Lakhtakia, First thin-film realization of a helicoidal bianisotropic medium. J. Vac. Sci. Technol. A, **13**, 2991–2993, (1995)
12. V.C. Venugopal, A. Lakhtakia, Sculptured thin films: conception, optical properties and applications, in *Electromagnetic Fields in Unconventional Materials and Structures*, ed. by O.N. Singh and A. Lakhtakia (Wiley, New York, NY, USA, 2000) pp 151–216
13. A. Lakhtakia, R. Messier, The past, the present, and the future of sculptured thin films, in *Introduction to Complex Mediums for Optics and Electromagnetics*, ed. by W.S. Weiglhofer, A. Lakhtakia (SPIE Press, Bellingham, WA, USA, 2003) pp 447–478
14. J.B. Geddes III, Towards shaping of pulsed plane waves in the time domain via chiral sculptured thin films, *Frontiers in Optical Technology: Materials and Devices*, ed. by P.K. Choudhury, O.N. Singh, (Nova Science, Hauppauge, NY, USA, 2006) pp 1–21
15. J.A. Polo, Sculptured thin films, in *Micromanufacturing and Nanotechnology*, ed. by N.P. Mahalik, (Springer, Berlin, Germany, 2006) pp 357–381
16. F. Wang, in Optics of slanted chiral STFs, *Frontiers in Surface Nanophotonics: Principles and Applications*, ed. by D.L. Andrews Z. Gaburro, (Springer, New York, NY, USA, 2007) pp 129–167

17. R. Messier, V.C. Venugopal, P.D. Sunal, Origin and evolution of sculptured thin films. J. Vac. Sci. Technol. A, **18**, 1538–1545 (2000)
18. M. Suzuki, Y. Taga, Integrated sculptured thin films. Jap. J. Appl. Phys. Part 2, **40** L358–L359 (2001)
19. Y.J. Park, K.M.A. Sobahan, C.K. Hwangbo, Wideband circular polarization reflector fabricated by glancing angle deposition. Opt. Exp., **16**, 5186–5192 (2008)
20. I.J. Hodgkinson, Q.H. Wu, Serial bideposition of anisotropic thin films with enhanced linear birefringence. Appl. Opt. **38**, 3621–3625 (1999)
21. I.J. Hodgkinson, Q.h. Wu, B. Knight, A. Lakhtakia, K. Robbie, Vacuum deposition of chiral sculptured thin films with high optical activity. Appl. Opt., **39**, 642–649 (2000)
22. A. Lakhtakia, M.C. Demirel, M.W. Horn, J. Xu, Six emerging directions in sculptured-thin-film research. Adv. Solid State Phys. **46**, 295–307 (2008)
23. S. Pursel, M.W. Horn, M.C. Demirel, A. Lakhtakia, Growth of sculptured polymer submicronwire assemblies by vapor deposition. Polymer, **46**, 9544–9548 (2005)
24. M.C. Demirel, S. Boduroglu, M. Cetinkaya, A. Lakhtakia. Spatially organized free–standing poly(p-xylylene) nanowires fabricated by vapor deposition. Langmuir, **23**, 5861–5863 (2007)
25. H. Tan, O.K. Ezekoye, J. van der Schalie, M.W. Horn, A. Lakhtakia, J. Xu, W.D. Burgos, Biological reduction of nanoengineered iron(III) oxide sculptured thin films. Environ. Sci. Technol. **40**, 5490–5495 (2006)
26. K. Robbie, M.J. Brett, A. Lakhtakia, Chiral sculptured thin films. Nature, **384**, 616 (1996)
27. M.W. Horn, M.D. Pickett, R. Messier, A. Lakhtakia, Blending of nanoscale and microscale in uniform large-area sculptured thin-film architectures. Nanotechnology, **15**, 303–310 (2004)
28. M.W. Horn, M.D. Pickett, R. Messier, A. Lakhtakia, Selective growth of sculptured nanowires on microlithographic lattices. J. Vac. Sci. Technol. B, **22**, 3426–3430 (2004)
29. I. Hodgkinson, Q.h. Wu, J. Hazel, Empirical equations for the principal refractive indices and column angle of obliquely deposited films of tantalum oxide, titanium oxide, and zirconium oxide. Appl. Opt. **37**, 2653–2659 (1998)
30. A.C. van Popta, J.C. Sit, M.J. Brett, Optical properties of porous helical thin films. Appl. Opt. **43**, 3632–3639 (2004)
31. K.D. Harris, D. Vick, E.J. Gonzalez, T. Smy, K. Robbie, M.J. Brett, Porous thin films for thermal barrier coatings. Surf. Coat. Technol. **138**, 185–191 (2001)
32. Y. Inoue, A. Yamaguchi, T. Fujihara, J. Yamazaki, O. Takai, Biomimetic improvement of electrochromic properties of indium nitride. J. Electrochem. Soc. **154**, J212–J216 (2007)
33. A.L. Elias, K.D. Harris, M.J. Brett, Fabrication of helically perforated gold, nickel, and polystyrene thin films. J. MEMS, **13** 808–813 (2004)
34. J.N. Broughton, M.J. Brett, Electrochemical capacitance in manganese thin films with chevron microstructure. Electrochem. Solid-State Lett. **5**, A279–A282 (2002)
35. G.K. Kiema, M.J. Brett, Electrochemical characterization of carbon films with porous microstructures. J. Electrochem. Soc. **150** E342–E347 (2003)
36. E. Schubert, J. Fahlteich, B. Rauschenbach, M. Schubert, M. Lorenz, M. Grundmann, G. Wagner, Recrystallization behavior in chiral sculptured thin films from silicon. J. Appl. Phys. **100**, 016107 (2006)
37. D.-X. Ye, T. Karabacak, R.C. Picu, G.-C. Wang, T.-M. Lu, Uniform Si nanostructures grown by oblique angle deposition with substrate swing rotation. Nanotechnology, **16**, 1717–1723 (2005)
38. D.-X. Ye, Z.-P. Yang, A.S.P. Chang, J. Bur, S.Y. Lin, T.-M. Lu, R.Z. Wang, S. John. Experimental realization of a well-controlled 3D silicon spiral photonic crystal. J. Phys. D: Appl. Phys. **40**, 2624–2628 (2007)
39. R.J. Martín-Palma, J.V. Ryan, C.G. Pantano, Surface microstructure of gesbse chalcogenide thin films grown at oblique angle. J. Appl. Phys. **101**, 083513 (2007)
40. R.J. Martín-Palma, J.V. Ryan, C.G. Pantano, Spectral behavior of the optical constants in the visible/near infrared of gesbse chalcogenide thin films grown at glancing angle. J. Vac. Sci. Technol. *A*, **25**, 587–591 (2007)

41. R.J. Martín-Palma, A. Redondo-Cubero, R. Gago, J.V. Ryan, C.G. Pantano. Rutherford backscattering spectrometry characterization of nanoporous chalcogenide thin films grown at oblique angles. J. Anal. At. Spectrom. **23**, 981–984 (2008)
42. P.C.P. Hrudey, K.L. Westra, M.J. Brett, Highly ordered organic Alq_3 chiral luminescent thin films fabricated by glancing–angle deposition. Adv. Mater. **18**, 224–228 (2006)
43. M. Malac, R.F. Egerton, M.J. Brett, B. Dick, Fabrication of submicrometer regular arrays of pillars and helices. J. Vac. Sci. Technol. *B*, **17**, 2671–2674 (1999)
44. B. Dick, M.J. Brett, T. Smy, Controlled growth of periodic pillars by glancing angle deposition. J. Vac. Sci. Technol. *B*, **21**, 23–28 (2003)
45. M.O. Jensen, M.J. Brett Periodically structured glancing angle deposition thin films. IEEE Trans. Nanotechnol. **4**, 269–277 (2005)
46. J.A. Sherwin, A. Lakhtakia, Nominal model for structure–property relations of chiral dielectric sculptured thin films. Math. Comput. Modelling, **34**, 1499–1514 (2001) Corrections: **35** 1355–1363 (2002)
47. J.A. Sherwin, A. Lakhtakia, Nominal model for the optical response of a chiral sculptured thin film infiltrated by an isotropic chiral fluid. Opt. Commun. **214**, 231–245 (2002)
48. J.A. Sherwin, A. Lakhtakia, Nominal model for the optical response of a chiral sculptured thin film infiltrated by an isotropic chiral fluid-oblique incidence. Opt. Commun. **222** 305–329 (2003)
49. V.C. Venugopal, A. Lakhtakia, R. Messier, J.-P. Kucera, Low–permittivity materials using sculptured thin film technology. J. Vac. Sci. Technol. B, **18**, 32–36 (2000)
50. F. Zhang, J. Xu, A. Lakhtakia, S.M. Pursel, M.W. Horn, A. Wang, Circularly polarized emission from colloidal nanocrystal quantum dots confined in cavities formed by chiral mirrors. Appl. Phys. Lett. **91**, 023102 (2007)
51. F. Zhang, J. Xu, A. Lakhtakia, T. Zhu, S.M. Pursel, M.W. Horn, Circular polarization emission from an external cavity diode laser. Appl. Phys. Lett., **92** 111109, 2008.
52. F. Wang, A. Lakhtakia, R. Messier, Towards piezoelectrically tunable chiral sculptured thin film lasers. Sens. Actuat. A: Phys., **102**, 31–35 (2002)
53. F. Wang, A. Lakhtakia, R. Messier, On piezoelectric control of the optical response of sculptured thin films. J. Mod. Opt. **50**, 239–249 (2003)
54. K. Robbie, D.J. Broer, M.J. Brett, Chiral nematic order in liquid crystals imposed by an engineered inorganic nanostructure. Nature, 399, 764–766 (1999)
55. J.C. Sit, D.J. Broer, M.J. Brett, Alignment and switching of nematic liquid crystals embedded in porous chiral thin films. Liq. Cryst. **27**, 387–391 (2000)
56. A. Lakhtakia, J.A. Reyes, Theory of electrically controlled exhibition of circular Bragg phenomenon by an obliquely excited structurally chiral material-Part 1: Axial dc electric field. Optik, **119**, 253–268 (2008)
57. A. Lakhtakia, J.A. Reyes, Theory of electrically controlled exhibition of circular Bragg phenomenon by an obliquely excited structurally chiral material-Part 2: Arbitrary dc electric field. Optik, **119**, 269–275, 2008.
58. J.A. Reyes, A. Lakhtakia. Electrically controlled optical bandgap in a structurally chiral material. Opt. Commun. **259**, 164–173 (2006)
59. A. Shalabaney, A. Lakhtakia, I. Abdulhalim, A. Lahav, C. Patzig, I. Hazek, A. Karabchevky, B. Rauschenbach, F. Zhang, J. Xu, Surface plasmon resonance from metallic columnar thin films. Photon. Nanostruct. Fund. Appl, **7**, 176–185, 2009
60. M.A. Motyka, A. Lakhtakia, Multiple trains of same-color surface plasmon-polaritons guided by the planar interface of a metal and a sculptured nematic thin film. J. Nanophoton. 2, 021910 (2008)
61. M.A. Motyka, A. Lakhtakia, Multiple trains of same-color surface plasmon-polaritons guided by the planar interface of a metal and a sculptured nematic thin film. Part II: Arbitrary incidence. J. Nanophoton. **3**, 033502 (2008)
62. J.A. Polo, Jr., A. Lakhtakia, On the surface plasmon polariton wave at the planar interface of a metal and a chiral sculptured thin film. Proc. R. Soc. Lond. A, **465**, 87–107 (2009)

63. J.A. Polo, Jr., A. Lakhtakia, Energy flux in a surface-plasmon-polariton wave bound to the planar interface of a metal and a structurally chiral material. J. Opt. Soc. Am. A, **26** 1696–1703 (2009)
64. A. Lakhtakia, Y.-J. Jen, C.-F. Lin. Multiple trains of same-color surface plasmon-polaritons guided by the planar interface of a metal and a sculptured nematic thin film. Part III: Experimental evidence. J. Nanophoton. **3**, 033506 (2009)
65. M.W. Seto, K. Robbie, D. Vick, M.J. Brett, L. Kuhn, Mechanical response of thin films with helical microstructures. J. Vac. Sci. Technol. B, **17**, 2172–2177 (1999)
66. S.M. Pursel, M.W. Horn, Prospects for nanowire sculptured-thin-film devices. J. Vac. Sci. Technol. B, **25** 2611–2615 (2007)
67. J.J. Steele, G.A. Fitzpatrick, M.J. Brett, Capacitive humidity sensor with high sensitivity and subsecond response times. IEEE Sensors J., **7** 955–956 (2007)
68. I.J. Hodgkinson, Q.h. Wu, K.M. McGrath, Moisture adsorption effects in biaxial and chiral optical thin film coatings. Proc. SPIE, **3790**, 184–194 (1999)
69. J.J. Steele, A.C. van Popta, M.M. Hawkeye, J.C. Sit, M.J. Brett, Nanostructured gradient index optical filter for high-speed humidity sensing. Sens. Actuat. B: Chem. **120** 213–219 (2006)
70. E. Ertekin, A. Lakhtakia, Sculptured thin film šolc filters for optical sensing of gas concentration. Eur. Phys. J. Appl. Phys. **5**, 45–50 (1999)
71. A. Lakhtakia, Enhancement of optical activity of chiral sculptured thin films by suitable infiltration of void regions. Optik, **112** 145–148, 2001. Corrections: 112:544, 2001.
72. A. Lakhtakia, M.W. McCall, J.A. Sherwin, Q.H. Wu, I.J. Hodgkinson. Sculptured-thin-film spectral holes for optical sensing of fluids. Opt. Commun. **194**, 33–46 (2001)
73. S. Tsoi, E. Fok, J.C. Sit, J.G.C. Veinot, Superhydrophobic, high surface area, 3-D SiO_2 nanostructures through siloxane-based surface functionalization. Langmuir, **20**, 10771–10774 (2004)
74. N. Verplanck, Y. Coffinier, V. Thomy, R. Boukherroub, Wettability switching techniques for superhydrophobic surfaces. Nanoscale Res. Lett. **2**, 577–597 (2007)
75. Y. Zhou, X. Song, M. Yu, B. Wang, H. Yan, Superhydrophobic surfaces prepared by plasma fluorination of lotus-leaf-like amorphous carbon films. Surf. Rev. Lett., **13**, 117–122 (2006)
76. M.C. Demirel, E. So, T.M. Ritty, S.H. Naidu, A. Lakhtakia, Fibroblast cell attachment and growth on nanoengineered sculptured thin films. J. Biomed. Mater. Res., Part B: Appl. Biomater. **81**, 219–223 (2006)
77. A. Lakhtakia, Elastodynamic wave propagation in a continuously twisted structurally chiral medium along the axis of spirality. J. Acoust. Soc. Am., **95**, 597–600 (1994) Corrections: **95**, 3669 (1994)
78. A. Lakhtakia, K. Robbie, M.J. Brett, Spectral Green's function for wave excitation and propagation in a piezoelectric, continuously twisted, structurally chiral medium. J. Acoust. Soc. Am. **101**, 2052–2059 (1997)
79. A. Lakhtakia, M.W. Meredith, Shear axial modes in a PCTSCM. Part IV: Bandstop and notch filters. Sens. Actuat. A: Phys., **73**, 193–200 (1999)
80. A. Lakhtakia, Shear axial modes in a PCTSCM. Part VI: Simpler transmission spectral holes. Sens. Actuat. A: Phys., **87**, 78–80 (2000)
81. R.J. Carey, A. Lakhtakia, Shear axial modes in a PCTSCM. Part VIII: Spectral holes with dissimilar materials but without phase defects. Sens. Actuat. A: Phys., **126**, 382–385 (2006)
82. A. Lakhtakia. Axial loading of a chiral sculptured thin film. Model. Simul. Mater. Sci. Eng. **8**, 677–686 (2000)
83. A. Lakhtakia, Perturbational solution for quasi-axial propagation in a piezoelectric, continuously twisted, structurally chiral medium. Appl. Acoust., **62**, 1019–1023 (2001)
84. A. Lakhtakia, Microscopic model for elastostatic and elastodynamic excitation of chiral sculptured thin films. J. Compos. Mater. **36**, 1277–1298 (2002)

GaN and InN Nanowires: Growth and Optoelectronic Properties

Toma Stoica, Eli Sutter, and Raffaella Calarco

Abstract Self-assembled GaN and InN nanowires (NWs) were synthesized by radio frequency Plasma-Assisted Molecular Beam Epitaxy (PAMBE) without external catalyst. NWs of micrometers length and diameter in the range of 20–200 nm are fabricated using this method under N-rich conditions. Driving mechanisms of the NW nucleation and the growth are discussed. The NWs have been investigated using scanning electron microscopy (SEM), high-resolution transmission electron microscopy (HRTEM), atomic force microscopy (AFM), and photoluminescence (PL). Electric and photoelectric measurements on single wire devices have been performed as well. We establish that the dark, Ultraviolet (UV) photo-current and band-edge absorption tails in GaN NWs are strongly dependent on wire diameter. A model of surface Fermi level pinning and Franz-Keldysh effect in carrier depletion region at wire surface were used to explain the observed behaviors. InN NWs show infrared (IR) photoluminescence strongly dependent on the growth parameters. High electron concentration of 10^{18} – 10^{19} cm^{-3} was evaluated from line shape analysis of PL spectra. The Fermi level pinning at the surface corresponds to a surface accumulation layer. To modify the surface of InN NWs, core-shell InN/GaN NWs were grown. In this paper we focus on the influence of surface effects on the growth and properties of GaN and InN nanowires.

1 Introduction

Among semiconductor nanostructures, GaN and InN nanowires (NWs) are of large interest for future electronic and optoelectronic devices due to the high electron mobility and optical gap of the InGaN alloy which spans a large spectral range from 3.4 eV to less than 0.7 eV. A detailed review of the physical properties of InN and related III-nitrides including electronic, optoelectronic and structural characterizations of films and nanostructures was recently published [1]. When a

T. Stoica (✉)
Institute of Bio- and Nanosystems (IBN-1) and JARA-FIT Jülich-Aachen Research Alliance, Forschungszentrum Jülich, 52425 Jülich, Germany
e-mail: t.stoica@fz-juelich.de

A. Aldea, V. Bârsan (eds.), *Trends in Nanophysics*, Engineering Materials,
DOI 10.1007/978-3-642-12070-1_4,

self-assembling procedure is used, the NWs approach has the potential for low-cost fabrication of future devices, by using a large variety of substrates. There are already many reports on fabrication of optoelectronic devices, chemical and biological sensors using NWs [2–19].

In general two approaches can be used to obtain semiconductor nanostructures and particularly nitrides NWs: (i) top-down method in which a sequence of epitaxial layers is structured usually by e-beam lithography and reactive ion etching (RIE) into columns [20–26]; (ii) bottom-up method in which the structure is grown directly in the form of nanowires by a self-assembling process [27–46]. The first method is limited by the low productivity and resolution of the e-beam patterning, as well as the defects produced during the RIE, but has the advantage of the precise control of position and size of the NWs. The bottom-up method consists of self-assembling of NWs and can efficiently produce a large number of wires with high aspect ratio. The growth occurs with high rate in a preferred crystallographic direction, independent of the use of metal catalyst.

The catalyst nano-droplets act as traps in which the atoms and molecules of the impinging flux are solved and condensed at the solid interface by the well known vapor-liquid-solid (VLS) mechanism [41] of nanowire growth. The use of an external metal catalyst has the disadvantage of a possible undesirable doping of the semiconductor NWs. Therefore a catalyst-free growth of NWs is preferable, and this can be achieved by adjusting the growth parameters as we discuss in the next section.

In this paper, a review of results of our group on the catalyst-free growth of GaN and InN NWs and their properties is presented with an emphasis on surface induced phenomena.

2 Catalyst-Free MBE Growth of GaN and InN NWs: Nucleation and Diffusion-Induced Mechanism

PAMBE catalyst-free grown GaN and InN nanowires with good crystalline quality and strong luminescence can be obtained by varying the growth parameters. In a PAMBE equipment, a radio frequency plasma source for the activated atomic nitrogen flux is used. The metal atoms flux in these experiments is controlled by the temperature of the effusion cell and measured in beam-equivalent pressure (BEP), i.e. the increase of the pressure in the MBE chamber due to Ga or In supply. In general, the higher the deposition temperature, the higher crystalline quality can be obtained. However, there is an upper limit in the growth temperature given in the case of GaN by the Ga desorption and GaN decomposition, which strongly increases above 800°C, and for InN by its decomposition at relatively low temperatures (above 500°C) [29, 45, 46]. The upper-limit of the growth temperature is influenced in some extent by the used nitrogen flux. Additionally, the optimum growth condition for obtaining the desired NW shape of the columns depends also on the nature of the substrate. In the studies presented in this paper, Si(111) substrates were used, the native oxide was removed by in situ annealing at 950°C.

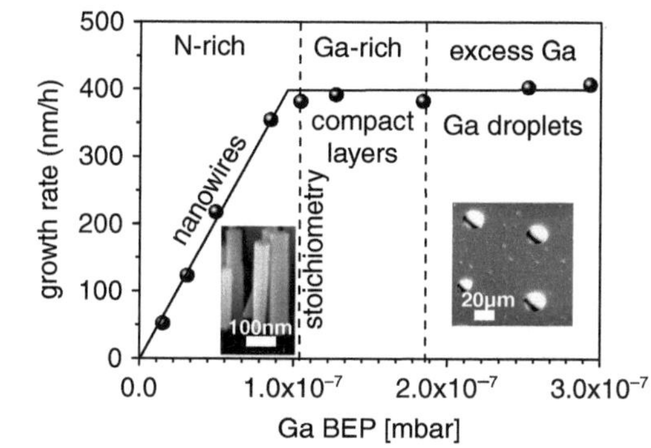

Fig. 1 The growth rate of GaN columnar and compact layers as a function of the Ga flux

The growth rate of GaN (columnar or compact layers grown on a AlN buffer layer) as function of Ga flux for constant growth temperature and nitrogen flux is shown for GaN in Fig. 1. Three growth regimes can be observed. For high Ga flux, compact layers are formed and the nitrogen flux limits the growth rate. The excess Ga is accumulated in droplets, which are observed on the substrate surface. There is an intermediary Ga flux range in which compact layers are formed without massive Ga accumulation on the sample surface. At low Ga flux, in the N-rich regime, the growth is limited by the Ga flux and the columnar growth is observed.

Growth and properties of GaN NWs were reported earlier in the 2000s by studies of E. Calleja et al. [28]. The nucleation and growth of the NWs was explained at that time based on theoretical prediction that the Ga adatom surface mobility is strongly reduced under N-rich conditions [43] and a "Ga-balling" effect takes place. Thus, the growth takes place at the formed Ga droplets by a self-catalytic VLS mechanism [28, 37]. However, Ga droplets formation was neither detected in our investigations and nor was it reported in literature for similar growth conditions. Moreover, we can expect that Ga droplets, if they are formed, are not stable under N-rich conditions and high temperatures, because of the strong desorption and a quick transformation into GaN as shown by our experiments using the droplet nitridation method [44].

From our understanding of the NW catalyst-free growth, GaN crystalline clusters formed on the deposition substrate act as seeds for the NW nucleation. After nucleation, the growth of NWs takes place in a preferable crystalline direction, assisted by adatom diffusion, in addition to the direct impinging flux on nanowire top surface. The lateral growth of the NW cannot be completely suppressed with this growth method, resulting in the increase of NW diameter during growth, as well as in the formation of a shell in case of NW heterostructures. The schematic of this growth process is shown in Fig. 2. To obtain heterostructures NWs, the impinging flux is changed from one type of atoms to another. Typical SEM side- and top-view images for a GaN NW sample are shown in Fig. 3 for an optimized growth condition: deposition temperature Td = 780°C; Ga BEP = 3.0×10^{-8} mbar and 4 sccm N flux with a forward power of the plasma of 500 W. We can see that NWs with fairly uniform diameter along the entire wire and smooth lateral surfaces can be synthesized.

Our model for the nucleation and growth is based on the investigation of a series of GaN NW samples, grown for different times, while all other growth parameters

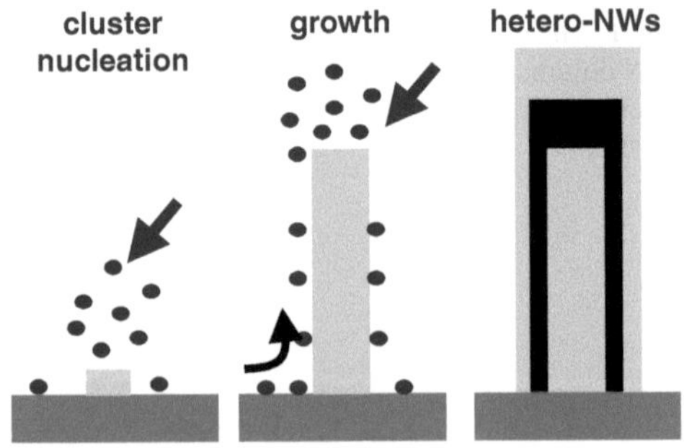

Fig. 2 Schematic of the vapor-solid mechanism in MBE growth process

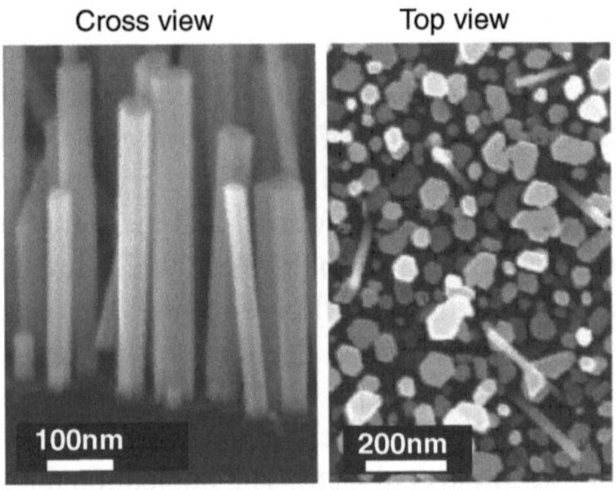

Fig. 3 SEM images of an optimized GaN NW sample

were kept constant. SEM images of the samples were used to perform the statistical evaluation of the NW density and sizes (length and diameter) [36].

The results are shown in Fig. 4 as function of the growth time. It is observed that NW formation needs a long nucleation or incubation time, in which a very low NW density is formed, perhaps as a result of an extrinsic nucleation on defects of the substrate surface. After about 30min of deposition, NWs start to massively nucleate reaching a maximum value at 1h deposition. Finally, at longer deposition time the density decreases due to the coalescence of closer nucleated wires (Fig. 4a). The mean values of the NW diameter and length vary almost linearly with deposition time (Fig. 4b) until the coalescence process dominates the growth kinetics of some wires. If we extrapolate to zero deposition time, a mean value of the nucleation diameter of about 16 nm is obtained. The average axial growth rate evaluated from the data in Fig. 4b is about 3.2 nm/min, and only a 20 times smaller value is obtained for the mean radial growth rate.

To get more insight in the nucleation mechanism a sample grown for 45 min, for which the nucleation process is not finished and well-developed NWs coexist on the same sample with just nucleated ones, was carefully investigated by HRTEM [38]. It was found that along with nucleated NWs, small GaN crystalline clusters are found on top of an interface SiN amorphous layer. This is illustrated by the TEM images in Fig. 5. The interface SiN layer was observed to extend also underneath NWs, with approximately the same thickness of 1 – 2 nm. The nanocrystals

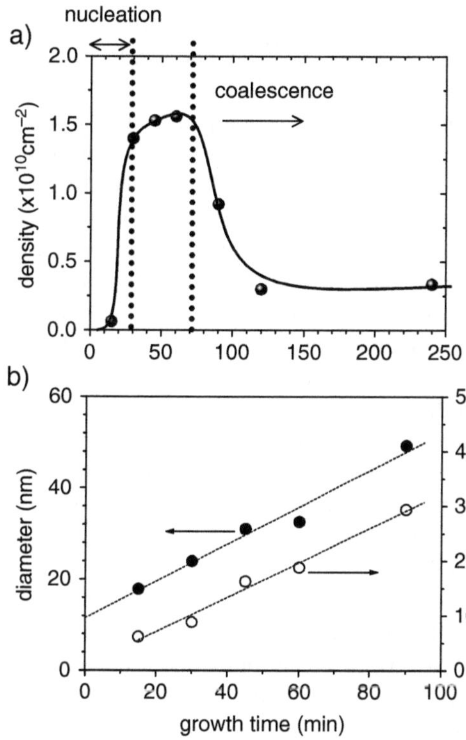

Fig. 4 (**a**) NW density; (**b**) NW length and diameter as a function of the deposition time

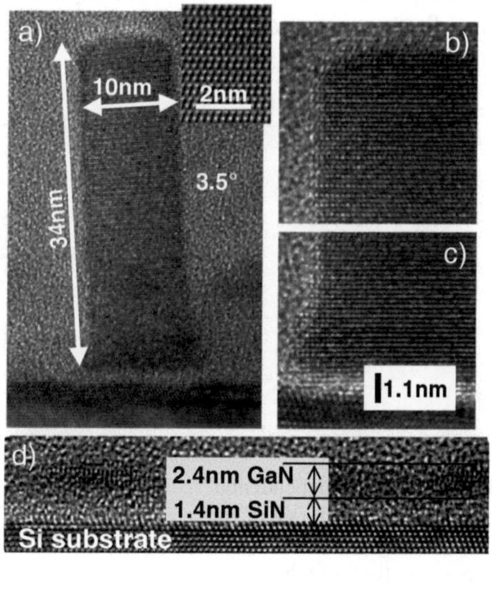

Fig. 5 HRTEM images of Ga NWs grown for 45 min: (**a**) images of a short NW; (**b**) and (**c**) higher magnification of the top and the bottom parts of a NW; (**d**) image of non-uniform GaN nanocrystalline film in-between NWs

observed in-between NWs are of different sizes. The height of nanocrystals varies from 2 nm to 6 nm. Some nanocrystals are larger and have a shape close to a just nucleated NW. The smallest diameter of nanocrystals was measured to be 7 nm. Interestingly, the well-developed long NWs with lengths of about 200 nm, after a statistical analysis of length to diameter correlation, have the lower limit of the nucleation diameter of 7 nm. This suggests that nanocrystals of this diameter are the smallest stable clusters for NW nucleation. In addition it is possible to observe that the radial growth rate is strongly reduced in case of very thin wires.

The HRTEM investigations have shown that the growing axial direction is always parallel to the c-axis. In Fig. 5a, a low magnification TEM image of a NW is shown. The high magnification image in the inset shows the good crystalline quality of the NW. Not all wires grow perpendicular to the substrate, one such NW is shown in Fig. 5a. In general the tilt of the NWs is correlated with some imperfections of the substrate surface. Figures 5b, c show high magnification of the top and the bottom parts of the same NW. This wire is perpendicular to the substrate and the thickness of the interface amorphous layer underneath is about 1.1 nm. Fig. 5d shows some degree of non-homogeneity in the "wetting layer" formed by GaN nanocrystals in-between NWs.

For deposition times longer than 60 min, the nucleation terminates due to lack of space left in-between wires. A coalescence process occurs due to the proximity of some NWs, while many others continue to grow separately. We have investigated the correlation between the length and diameter of NWs for a long deposition of 6 h [35].

The axial growth rate (length/time) evaluated for individual NW using SEM images is shown as function of the diameter in Fig. 6. A dependence of the type 1/diameter was found, i.e. thinner NWs grow faster. This was explained using a simplified model of a diffusion-induced mechanism of the growth kinetics, illustrated by the scheme of the inset of the Fig. 6 (*L* is the wire length, *D* is the diameter). As a first approximation we can write the growth rate of the volume of the NW top

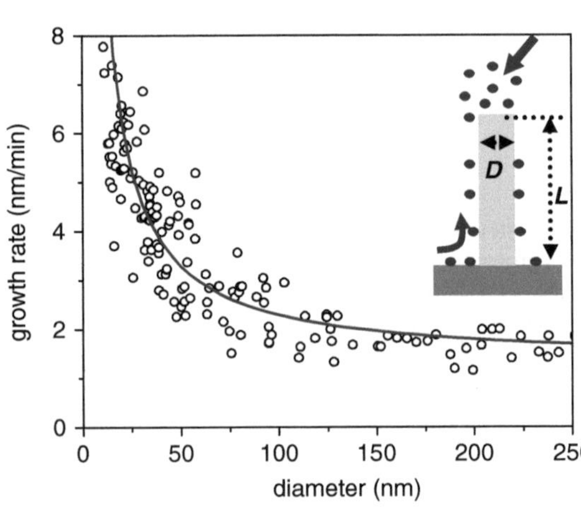

Fig. 6 Diameter dependence of the axial growth rate of the GaN NWs grown for long deposition time of 6 h

part as $(\pi/4)D^2 dL/dt$ and expressed it as a sum of the two contributions: one of the adatoms of the direct impinging flux on top of the NW proportional to the top area $(\pi/4)D^2$; and another term which is due to the adatoms diffusing to the top surface from the substrate area, which should be proportional to the perimeter of the top surface πD. Inserting the factor $\pi/4$ in the proportionality factors the growth rate equation is:

$$D^2 dL/dt = \gamma_T D^2 + \gamma_S D \tag{1}$$

where γ_T and γ_S are proportionality coefficients for direct impinging adatoms on top surface and for adatoms diffusing from lateral sides, respectively. If we suppose a nucleation time t_0 and a nucleation diameter $D = D_0$ with a simple linear time dependence $D = D_0 + \gamma_R(t - t_0)$ we will find:

$$L = \gamma_T(t - t_0) + (\gamma_S/\gamma_R)\ln(D/D_0) \tag{2}$$

which is reduced to the formula

$$L = (\gamma_S/D + \gamma_T) \cdot (t - t_0) \tag{3}$$

for $\gamma_R(t - t_0)/D_0 << 1$ fulfilled for the used growth conditions [35]. Thus, it is obtained the dependence $1/D$ observed experimentally (Fig. 6) as well as the linear time dependence of the data in Fig. 4b. For a given deposition time, Eq. 3 can be simplified to $L/(t - t_0) = C_1(1 + C_2/D)$. By the fit of this dependence to the experimental points in Fig. 6 the value of 80 nm was obtained for C_2. The coefficient C_1 gives the growth rate of thicker wires (large area deposition), while C_2 is a critical diameter for which the diffusion-induced mechanism is significant for the growth rate. It has to be mentioned that the maximum aspect ratio of 250 is obtained for thin NWs with diameter of less than 15 nm.

The correlation of the length and diameter of the wires has been studied using HRTEM measurements of a sample with long growth duration as the previous one, but at which after 6 h deposition of GaN NWs a sequence of AlN/GaN multilayers was deposited on top of the wires, as shown in Fig. 7a. We can see that heterostructure NWs can be grown with a control of the interface at the level of the atomic layer, which supports the vapor-solid mechanism for the NW growth. The Ga droplet self-catalytic VLS mechanism cannot explain such a sharp interface. In Fig. 7a a zoom of the NW side shows the formation of a multilayer GaN/AlN shell. The AlN layers can be used as markers to determine the thickness of GaN of different segments. Many wires have been investigated by HRTEM to obtain again the correlation of the growth rate and wire diameter.

Almost the same $1/D$ dependence of the GaN growth rate was found for the first deposited segment of the heterostructure (Fig. 7b). The fit curve $C_1(1 + C_2/D)$ in Fig. 7b is computed with the same value of the critical diameter $C_2 = 80$ nm and C_1 values as deduce from SEM statistics. The fact that thinner wires grow faster can be seen from the qualitative comparison of two wires with different diameters as

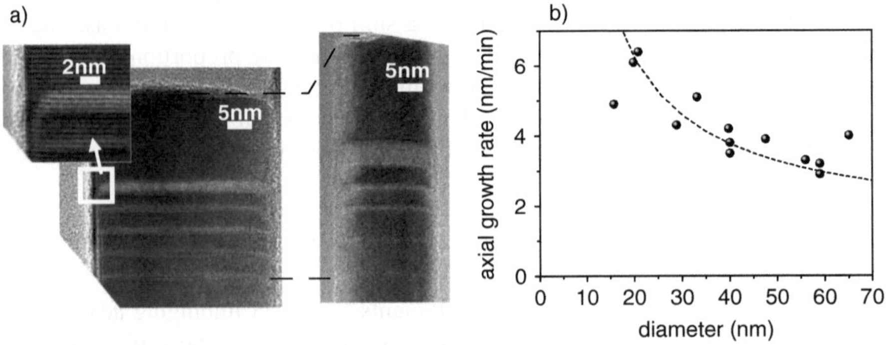

Fig. 7 (**a**) HRTEM images of two wires with different diameters obtained after 6 h GaN deposition followed by the growth of AlN/GaN multilayers. (**b**) axial growth rate of the first deposited 1 min GaN segment of the heterostructure GaN NWs. The deposition sequence was from top: GaN 5 min/AlN 3 min/2x(AlN 1 min/GaN 1 min)/2x(AlN 30 s/GaN 1 min)/GaN 6h/Si-subst

shown in Fig. 7a. However, numerical data show large fluctuations of the growth rate for different GaN segments of the same NW and also between wires with similar diameter. We can explain this observation by two effects: first, for an individual NW the density of neighbors fluctuates and can produce fluctuations of the adatom accumulation for the investigated wire; second, the restart of the GaN growth after an AlN deposition requires to start again with a nucleation process which has a statistical behavior. The fluctuation due to re-nucleation on top of an AlN layer is weak for very thin AlN of a few atomic layers, but becomes stronger when the AlN layer is thicker. In Fig. 8, Z-contrast TEM images of wires with a sequence of AlN/GaN on top demonstrates that not only the thickness of the GaN segments can fluctuate in a NW after thick 5min AlN deposition, but also the diameter can vary when the AlN deposition was 20 min layer. The phenomena related to the NW heterostructure growth need more fundamental studies and optimization to obtain uniform and controllable segments.

The substrate temperature is a very sensitive parameter for InN growth due to its low dissociation temperature. The columnar growth takes place in N-rich conditions. Similar for InN, the thinner NWs seem to grow faster, but statistical investigation of the growth mechanism is more difficult than for GaN, because of larger fluctuations in length and diameter. An optimization process was conducted, taking into account

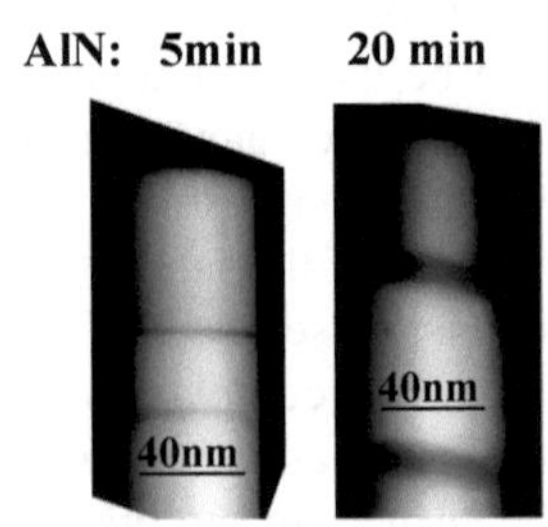

Fig. 8 TEM images of GaN NWs which has on top a heterostructure disk sequence 2x(AlN 5min or 20 min/GaN 5min)

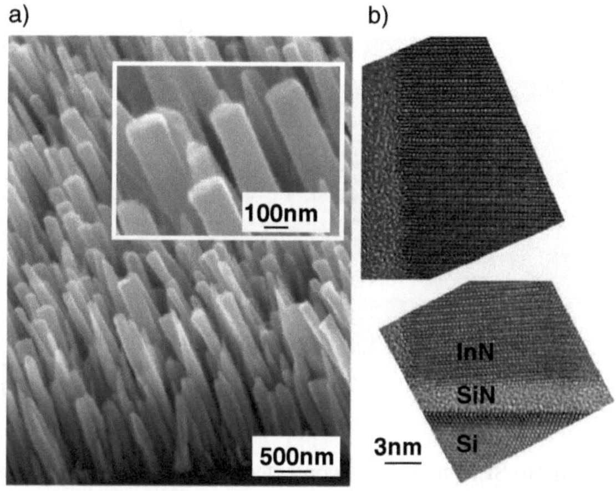

Fig. 9 InN NWs grown at 475°C: (**a**) SEM side view image (the inset shows a zoom of the NWs top); (**b**) HRTEM images of the base and in the middle of the NW

the luminescence properties and NW morphology [45]. In Fig. 9a, a side-view SEM image of InN NW as-grown sample is shown with an inset of a zoom of the top parts of NWs. Deposition temperature Td = 475°C, In BEP = 7.0×10^{-8} mbar, and 4 sccm N flux with forward power of the plasma of 500 W were used as deposition parameters.

HRTEM studies show that InN NWs free of dislocations can be formed. In Fig. 9b, HRTEM images of the bottom part and in the middle of a NW are given, which reveal a good crystalline structure. However, staking fault defects can be frequently observed, especially at the base of the NWs. The lateral surfaces of InN NWs are rougher than the GaN once. Similar to the case of GaN growth, a SiN amorphous thin film ($\sim$ 2 nm thick) is formed as an interface between the NWs and the Si substrate. The composition of this layer was investigated by electron energy loss spectroscopy (EELS) using a TEM spot with diameter of about 2 nm.

In conclusion, we demonstrate that GaN and InN NWs with good crystallinity and wire shape morphology can be grown by PAMBE on Si(111) substrate, via optimization of the growth parameters in N-rich conditions. Prior to the assembly of the NWS we observe formation of a thin amorphous SiN layer as an interface layer for both GaN and InN NWs. The diffusion-induced mechanism can explain the observed 1/diameter linear dependence of the axial growth rate.

3 Surface Influence on Optical and Optoelectronic Properties of GaN and InN NWs

3.1 Optoelectronic Properties of GaN NWs

The as-grown NWs can be easily removed from the original substrate and spread on a different one that is more suitable for the determination of their properties. We have used oxidized c-Si wafers as substrates for fabrication of single NW devices.

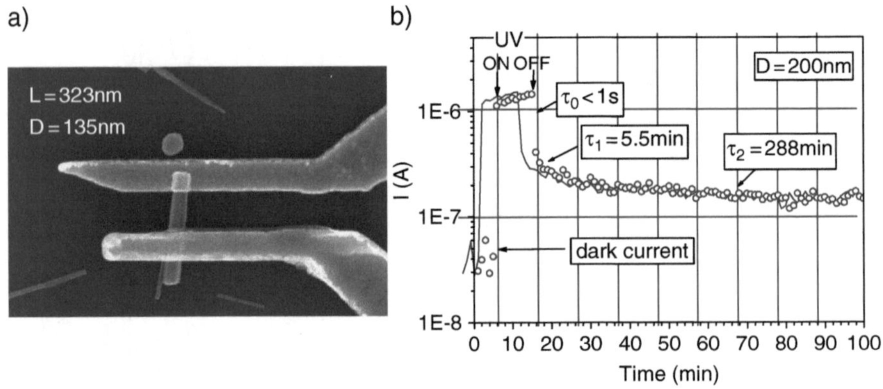

Fig. 10 (**a**) SEM image of a contacted GaN NW; (**b**) time dependence of the current: ON-OFF UV illumination effect and photocurrent relaxation

The wires are located using SEM images and Ti(10 nm)/Au(100 nm) metallic layers are patterned by e-beam lithography to obtain ohmic contacts [17]. In Fig. 10a, the SEM image illustrates a contacted GaN NW.

The photocurrent measurements show long time relaxation behaviors after illumination. In Fig. 10b such a time dependence photocurrent measurements is presented. Under UV integral illumination using a Xe lamp, the current increases about two orders of magnitude relative to the dark current. After turning off the light, at first fast decrease of the current ($\sim$1 s) is measured, after which the current relaxation down to the initial dark values has for some wires the duration of days. Such long-time persistent photocurrent is found for semiconductor structures in which one type of carriers is trapped and has to overcome an energetic barrier to recombine with the opposite carrier type. In Fig. 10b, the evaluated exponential relaxation times are given for different ranges of the curve. The relaxation of the photocurrent is faster for thinner NWs. The persistent photocurrent effect becomes insignificant for wires with diameter smaller than 80 nm.

The relaxation curves for two wires with different diameters are shown in Fig. 11 on a log-log scale. It can be seen that this part of the relaxation can be described by a stretched exponential dependence $\Delta I(t) = \Delta I_0 \exp(-t/\tau)^{\beta}$ in which the stretching exponent β has the value of the order of 0.1 – 0.2, while the values of τ strongly depend on the NW diameter [47]. This relaxation type with $\beta<1$ corresponds to a trapping - detrapping dispersive process of the excited carriers relaxation. We associate these phenomena in GaN NWs to the trapping of holes at the wire surface, as we discuss below.

A strong diameter dependence of the integral photocurrent was found for thinner wires, by measuring single GaN NW devices with different diameter of the NW [17]. The integral illumination photocurrent measured at room temperature under constant UV illumination intensity and applied electric field is shown in Fig. 12 as a function of the wire diameter. For diameters smaller than a critical one of 80 nm, the photocurrent varies 6 orders of magnitude in the short diameter range 40–80 nm.

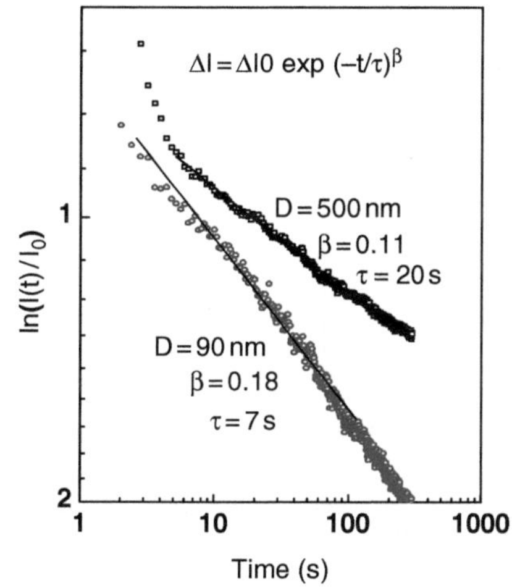

Fig. 11 Persistent photocurrent: current relaxation after UV illumination for two wires with different diameters

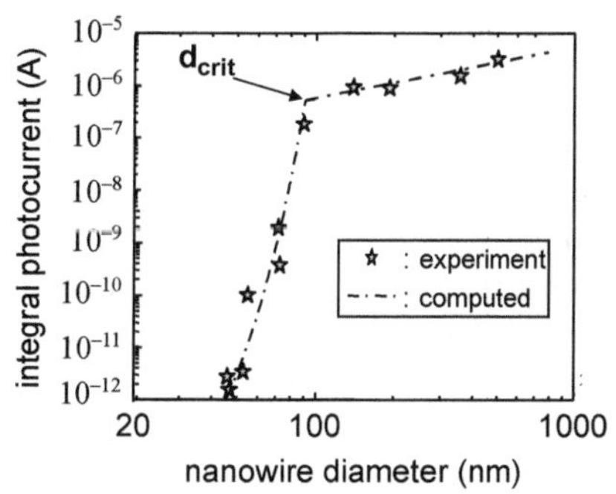

Fig. 12 Diameter dependence of the integral photocurrent: points are experimental data and the dash dot line is a computed curve

For thicker wires, the photocurrent increases almost linearly with the volume of the wire. Similar diameter dependence was found for the dark current as a function of wire diameter [47]. To explain the size dependence of the photocurrent as well as the persistent photocurrent effect, we have elaborated a cylindrical model based on the Fermi level pining at the NW surface. In wide band gap semiconductors, the Fermi pining at the surface creates a depletion region by a transfer of carriers from the bulk to the surface states. Thus, a build-in electric field is created at the surface and consequently band bending occurs.

The formation of the depletion region is schematically shown in Fig. 13a for a cylinder model of n-doped NWs. The depletion width depends on the doping level and on the NW diameter. In our case the NWs have a non-intentional n-doping which has to be taken into account. For thinner wires with diameter d smaller than a critical one d_{crit}, the depletion region is extends to the entire wire volume, while for thicker wires the core remains unperturbed (Fig. 13a). We can understand based on this model the size dependence of the photocurrent: under illumination, the photo-holes drift to the wire surface under the build-in electric field within the depletion

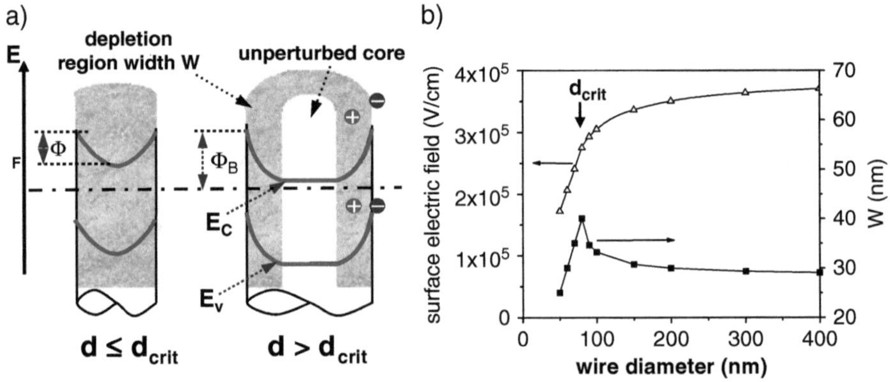

Fig. 13 (**a**) Scheme of the surface depletion region in GaN NWs; (**b**) computed diameter dependence of the surface electric field and the width W of the depletion region

region, while the electrons remain in the NW core and they must overcome the surface potential barrier to recombine with holes.

As illustrated in Fig. 13a, the barrier is reduced for wires with diameter smaller than d_{crit} resulting in strong increase of the surface recombination, and correspondingly rapid decrease of the photocurrent by decreasing the NW diameter (in Fig. 12, diameter smaller than d_{crit}). For diameter larger than d_{crit} the surface potential is constant as well as the surface recombination probability. The separation of holes at surface and electrons in the bulk persists for longer time after illumination in thicker wires, while for thinner wires with a reduced surface barrier the carrier recombination is faster.

To develop a numerical model, we have to solve the Poisson equation in cylindrical coordinates [17, 49]. With the approximation of uniform distribution of n-doping atoms, it is found that for thinner wires, the surface barrier Φ increases with the square of the diameter d, while the electric field at the surface varies linearly with d:

$$d < d_{\text{crit}}; \quad \Phi = \frac{e^2 N_{\text{d}}}{16\varepsilon} d^2 \equiv \Phi_{\text{B}} (d/d_{\text{crit}})^2; \quad \xi(d) = \frac{e N_{\text{d}}}{4\varepsilon} d \tag{4}$$

Here Φ_{B} is the surface potential relative to the unperturbed bulk, ξ is the surface electric field, N_{d} is the density of donors and ε is the dielectric constant. The critical diameter is given by the formula $d_{\text{crit}} = \sqrt{\frac{16\varepsilon}{e^2 N_{\text{d}}} \Phi_{\text{B}}}$. Thus, the critical diameter is strictly correlated to the carrier concentration $n = N_{\text{d}}$ for a given surface potential. For thicker wires we have:

$$d > d_{\text{crit}}; \quad \Phi = \Phi_{\text{B}}; \quad \xi(d) = \frac{e N_{\text{d}}}{4\varepsilon} \left(d/2 - \frac{(d/2 - W)^2}{d/2} \right) \tag{5}$$

and the width of the depletion region W is given by:

$$d > d_{\text{crit}};\quad \Phi_{\text{B}} = \frac{e^2 N_{\text{d}}}{2\varepsilon} W^2 \left(1 + \frac{1}{2W^2}\left(Wd - 3W^2 + 2(d/2 - W)^2 \ln\left(1 - \frac{W}{d/2}\right)\right)\right) \tag{6}$$

The computed diameter dependences of the surface electric field and of the depletion width W are shown in Fig. 13b for a critical diameter of 80 nm.

For dominant surface recombination, the photocurrent has an exponential dependence on the surface potential barrier $\sim \exp(\Phi/kT)$, because of the thermal activation of the electron-hole recombination process described above (kT is the thermal energy). We used this formula for the fit of the experimental data in Fig. 12 and determined donor concentration $N_d = 6.2 \times 10^{17}$ cm^{-3} and the surface potential Φ_B of 0.55 eV [17]. These parameters correspond to a critical diameter of about 80 nm. A mean value of the carrier concentration in GaN NWs was also evaluated using the phonon-plasmon coupling of the longitudinal phonons investigated by Raman scattering [48] in NW ensembles. A good agreement with the value deduced from photocurrent data was found.

The influence of the surface depletion region can also account for the size effect observed in the spectral dependence of the photocurrent [49]. In Fig. 14 the near band edge spectral photocurrent shows a strong diameter dependence of the exponential tail of the sub-band gap absorption. We have explained this size dependence considering the Franz-Keldysh effect [50, 51] induced by the electric field within the surface depletion region.

The Franz-Keldysh sub-band absorption process is shown schematically in Fig. 15a. Due to the band bending at the wire surface, the wave functions of electrons and holes are extended into the forbidden band gap (they tunnel into the gap) and thus optical transitions can take place at smaller energy than the band gap energy.

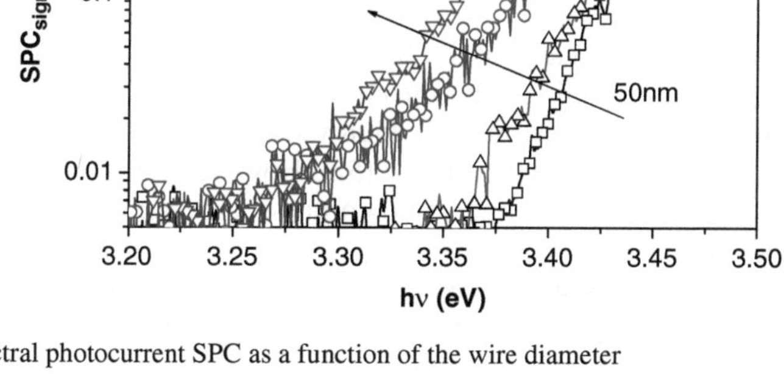

Fig. 14 Spectral photocurrent SPC as a function of the wire diameter

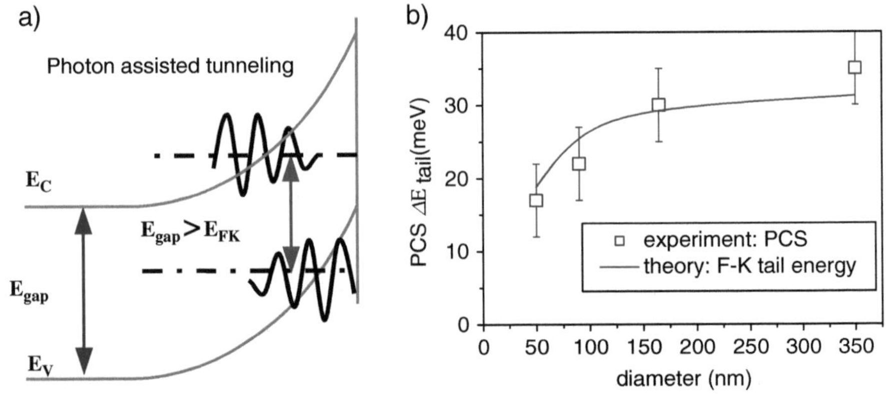

Fig. 15 (**a**) Schematic surface band bending and Franz-Keldysh sub-band absorption effect; (**b**) specific energy of the SPC band tail energy (*points*) and the computed Franz-Keldysh specific energy (*solids line*)

This type of sub-band optical transitions is also called photon assisted tunneling. It causes an exponential absorption tail. The absorption at photon energy $h\nu$ smaller than the gap can be approximately described by an exponential function with specific energy ΔE:

$$\alpha(h\nu) \sim \exp\left(-\left|\frac{h\nu - E_{\mathrm{gap}}}{\Delta E}\right|^{3/2}\right); \quad h\nu < E_{\mathrm{gap}}; \quad \Delta E = \frac{2(eh\xi)^{2/3}}{3(m^*)^{1/3}} \tag{6}$$

where ξ , m^*, e, and h are the electric field, effective mass, electron charge, and Planck constant, respectively. The surface electric field computed in Fig. 13b as a function of the wire diameter can be used for the evaluation of the Franz-Keldysh specific energy ΔE.

The theoretical curve $\Delta E(d)$ is compared in Fig. 15b with the experimental values of the specific energy of the exponential tails of the experimental spectral photocurrent shown in Fig. 14. The same values of the model parameters deduced from the fit of the size dependence of the integral photocurrent of Fig. 12 have been used. We can see in Fig. 15b a good agreement between experiment and the computed curve.

The origin of the space charge at the NWs surface for non-polar lateral surfaces is not clear. Scanning tunneling experiments (STM) on perfect non-polar crystallographic planes of GaN cleaved in ultra-high vacuum have shown that there is no charge transfer from the bulk to the surface [52]. However, fore GaN NWs exposed to air and subjected to the device fabrication process, the lateral NW surface shows a depletion region as discussed above.

Trapping of free carriers at the NWs surface has a strong effect on the I-V characteristics. Similar to the persistent photocurrent phenomena, long-term hysteresis effect can be observed in I-V curves, especially for the case of a dominant space charge limited current effect [17]. The space charge limited current in NWs imply more complex phenomena than for bulk and films, because the charge injected from electrodes can modify not only the NW core, but also the surface potential. The conductance of the NWs depends on the previous changes of the applied voltage, temperature evolution, exposure to different atmospheres and illumination. Trapping effect is also detected in field effect transistor (FET) experiments. Our single NW devices can be used to study the field effect in GaN NWs using the Si substrate as the gate electrode. It was shown that by changing the gate voltage, the wire conductance can be changed by more than 6 orders of magnitude [53]. The trapping at the wire surfaces and/or at the oxide-GaN interface causes hysteresis.

In conclusion, we can say that the electro-optical properties of GaN NWs are consistently explained by the model of the Fermi level pining at the nanowire surface. The surface properties can be modified by changing the environment atmosphere. The strong influence of the surface on the GaN NWs can be successfully used in fabrication of chemical and biological sensors.

3.2 Optoelectronic Properties of InN NWs

The growth of InN NWs encounters difficulties due to the relative low decomposition temperature of InN, as discussed above. In general, the growth at low temperatures may result in high density of local defects. Therefore, it is very important for growth optimization to investigate not only the NW morphology, but also the optical and electronic quality. This motivated us to perform careful photoluminescence (PL) studies [45]. These studies revealed intrinsic properties of the InN NWs [54], which we will also discuss in this section.

In Fig. 16a, low temperature normalized PL spectra are shown for a series of samples grown with different In fluxes. The PL spectra were all acquired at low temperature of 20 K, if not mentioned otherwise. All investigated samples were grown at 475°C and for a duration of 2 h. The peak energy and the broadening of the PL line have a non-monotonic variation with In flux. The lower values of the peak energy and of the full width at half maximum (FWHM) is determined for In-BEP of 1.0×10^{-7} mbar. In Fig. 16b the In flux dependence of the peak intensity and peak energy are shown. The PL intensity varies by more than 2 orders of magnitude with In flux and reaches a maximum at the In-BEP value of 1.0×10^{-7} mbar.

Both, the sharpness of the line and the higher PL efficiency are usually associated with higher quality of the crystalline semiconductor, with low density of nonradiative defect centers, low carrier concentration and low structural disorder. Indeed, the In-flux dependence of the NW morphology discussed above and presented in [45] can be correlated with the PL results. The NWs grown with low In flux are thinner, shorter and denser. Shorter wires show lower PL efficiency due to structural defects at the NW base [45] as well as to non-intentional doping by Si diffusion from the substrate. In addition the presence of an accumulation layer at the NW surface, as discussed below, has strong influence on PL spectra, especially for thinner wires deposited at low In fluxes. The spectra indeed become very broad for lower In flux

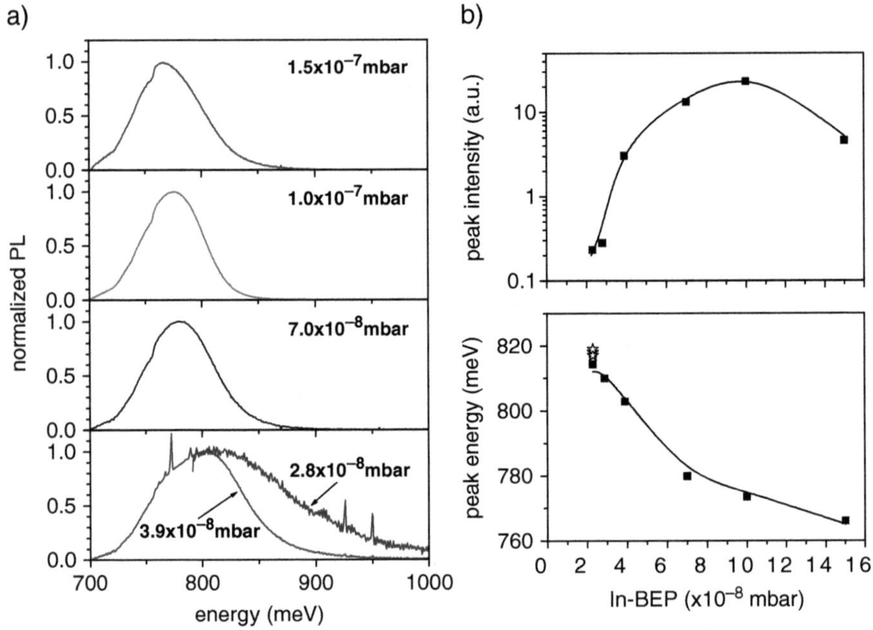

Fig. 16 (**a**) PL spectra for different In fuxes; (**b**) peak intensity (*top*) and energy (*bottom*) of the spectra of figure a, as a function of In flux

(Fig. 16) corresponding to large electron concentration. At high In flux 1.5×10^{-7} mbar, the coalescence of the NWs is very strong resulting in more compact layers. The coalescence regions have high density of structural nonradiative defects, which causes a decrease of the PL intensity as shown in Fig. 16b. We can conclude that there is an optimal value of the In flux for obtaining simultaneously high PL efficiency and good NW morphology.

The growth temperature has a monotonic effect on the PL properties. By increasing the deposition temperature, the PL intensity increases constantly while the FWHM of the PL line decreases. In Fig. 17a low temperature normalized spectra are shown for three different growth temperatures. The intensity and the energy of the PL peak are shown in Fig. 17b as a function of the growth temperature. The PL behavior indicates improved crystalline quality with increased deposition temperature. Indeed, at higher temperatures 525°C large conical InN crystals with clear smooth hexagonal facets on top are formed [45]. These structures have a good crystalline quality, but they might not show NWs properties due to their non-uniform diameter. Therefore the optimal growth temperature occurs at intermediate value ~ 475°C, as a compromise between crystalline quality and NW morphology.

The main broadening of the PL line in InN is due to the high concentration of electrons and their energetic distribution in the degenerated conduction band [54], so called band filling effect. This effect was usually observed for InN material with high non-intentional doping and was first described for low band gap InN

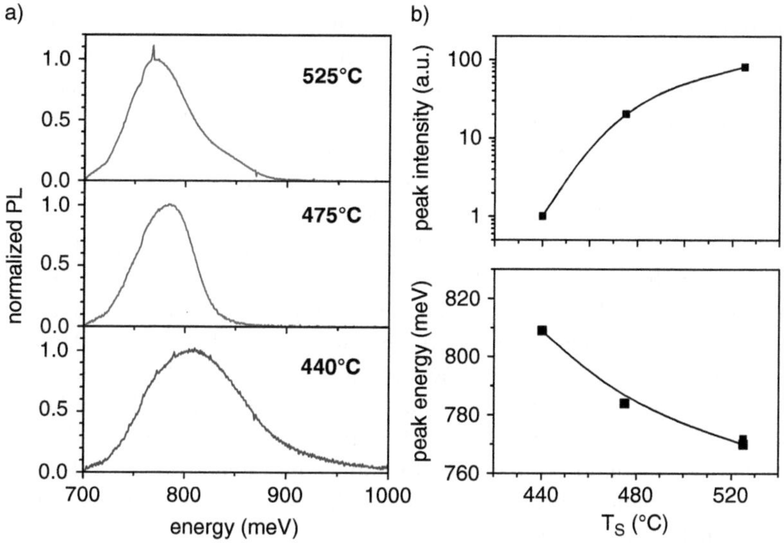

Fig. 17 (**a**) PL spectra for different growth temperatures; (**b**) peak intensity (*top*) and energy (*bottom*) of the spectra of figure a as a function of growth temperature

films in [55]. The recombination model is illustrated in Fig. 18. In Fig. 18a the PL spectra of two different samples are normalized so that the low energy tails are superposed. The positions of the energy gap E_g and the Fermi levels E_F are indicated by arrows. It is evident that the band filling is broader for sample S2 with higher Fermi energy corresponding to higher electron concentration. A schematic of the band-band recombination model used for line shape analysis of the PL spectra is depicted in Fig. 18b. By illumination, the photo-generated holes can recombine radiatively with electrons, leaving in the final state of the recombination process a hole in the Fermi sea of the conduction band which is rapidly annihilated by thermal relaxation. The schematic representation of the accumulation layer in InN NWs shown in Fig. 18c and will be discussed later in this section.

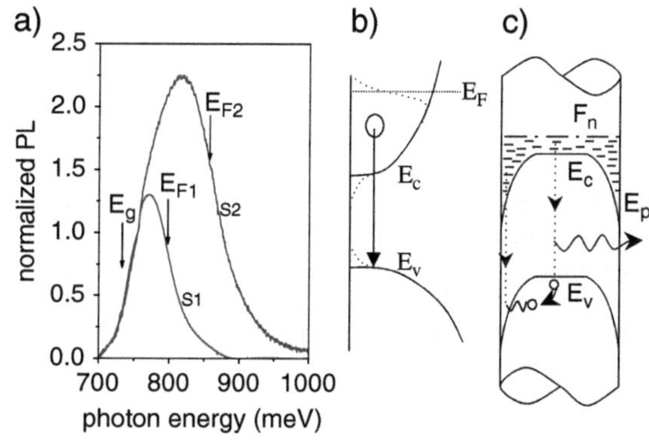

Fig. 18 (**a**) PL spectra of two InN NW samples showing different band filling values; (**b**) schematic diagram of band-band recombination; (**c**) schematic band diagram of the accumulation layer on lateral surface of InN NW and different recombination transitions

Additional broadening of the PL spectra can be caused by different fluctuations between wires and along the wire, such as: fluctuations in the electron concentration (Fermi level) or in the band gap due to doping fluctuations. The Landsberg broadening due to a very short life time of the final state of the electron-recombination process in degenerate semiconductor can also contribute. This was discussed in [54] coming to the conclusion that the electron density fluctuation is the dominant additional contribution to the PL broadening at low temperatures.

With elevating the temperature the PL broadening is increased. The electrons and holes are thermally excited to higher energy levels and thus the high-energy tail of the PL is increased. This can be clearly seen in Fig. 19a where normalized spectra measured at different temperatures are shown on logarithmic scale. The specific energy E_0 of the high energy exponential tail (see Fig. 19a for definition of E_0) is almost constant at low temperatures and increases proportionally to the thermal energy kT at higher temperatures (Fig. 19b). In Fig. 19b, the energy E_0 is shown as a function of temperature for a group of samples. Its value is related at low temperatures to the fluctuations discussed above which have been taken into account in the line shape analysis of the spectra.

To further elucidate the radiative recombination process, the laser power dependence of the spectra was measured. With increasing the power a slight increase of the band filling due to the photo-generated carriers was observed. This results in a blue shift of the PL peak with laser power as shown in Fig. 20 for two measurement tem-

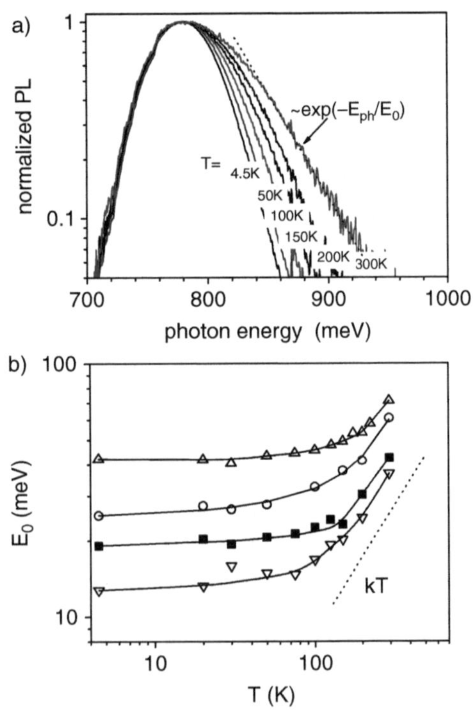

Fig. 19 (**a**) normalized PL spectra acquired at different temperatures to illustrate the increasing of the high energy tail at higher temperatures; (**b**) the temperature dependence of the specific energy E_0 defined in figure a

Fig. 20 Laser power dependence of the PL peak energy at two measurement temperatures

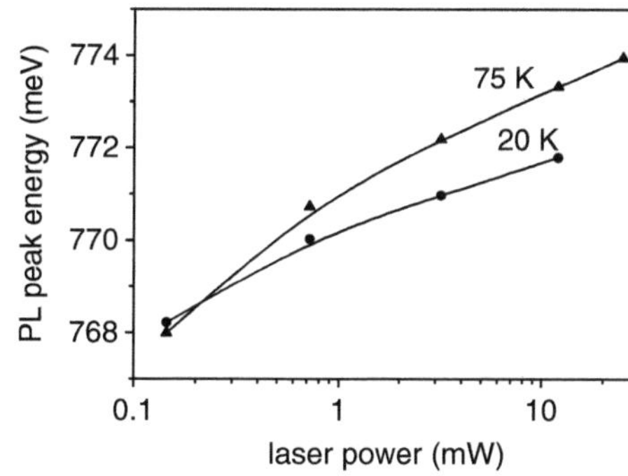

peratures. The shift is nearly linear with the logarithm of the illumination intensity. This was attributed to band filling due to photo holes generated in an exponential tail of the valance band [54]. The occurrence of a small band tail of localized states is quite normal in a degenerate semiconductor, because of local potential fluctuations induced by high electron density as found in InN. The band tails are also depicted in the schematic in Fig. 18b.

The band filling produced by the electrons in thermal equilibrium in the degenerated conduction band, as well as the fluctuations in the Fermi level have been taken into account in the line shape analysis of the PL spectra of different samples. This analysis yielded an estimate of the mean electron density and the energy gap. The variation of the growth parameters lead to a wide range of changes of the PL spectra and thus in the resulting electron density. Although the broadening of the PL due to Fermi level fluctuations can affect the results, the correlation between the PL peak energy and the estimated electron density shown in Fig. 21 is good and can be used to obtain a rough estimate of the electron density, without a complicate fitting procedure. The correlation between the estimated gap values and electron density shows a normal shrinking of the gap with electron density [54].

The PL integral intensity for constant illumination power is shown in Fig. 22 as a function of the estimated electron density n_e. The dependence is much stronger than expected from a local electron-hole recombination process. The strongest car-

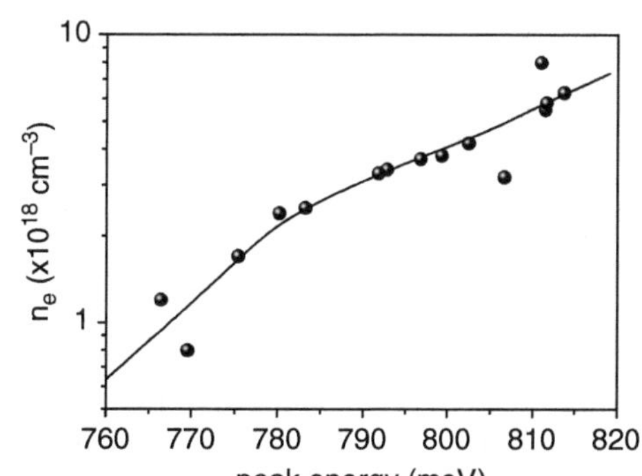

Fig. 21 The estimated mean electron density as function of PL peak energy

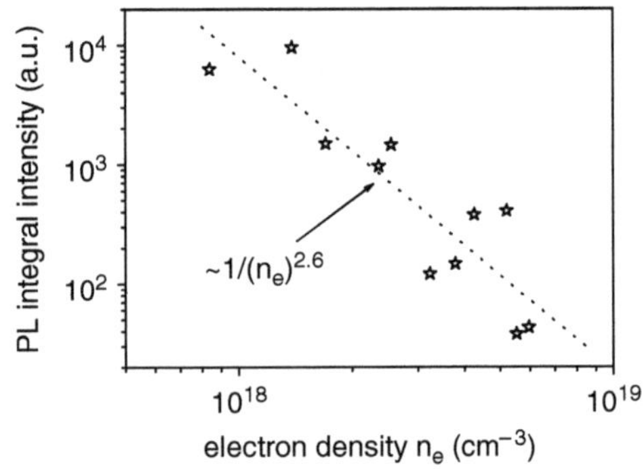

Fig. 22 Correlation of the PL intensity with the estimated mean concentration electron density in InN NWs

rier density dependence is observed in the case of dominant non-radiative Auger recombination for which the recombination probability is proportional to the product pn_e^2 (p is the photo-hole density). With the radiative bimolecular recombination proportional to the product pn_e, we expect that the PL intensity varies as $1/n_e$.

The stronger dependence in Fig. 22 suggests that additional non-radiative surface recombination controls the PL process. This can occur by recombination of holes with the high density electrons in the accumulation layer at the NW surface, as schematically shown in Fig. 18c. The holes are thermally activated and tunnel into the thin accumulation layer where they recombine with electrons of much higher density than that of the wire bulk. As discussed in [54], the temperature dependence of the PL offers also arguments in the favor of a strong influence of the recombination in the surface accumulation layer on the observed PL.

Similar accumulation layer is usually reported for low gap semiconductors such as for example InAs and InSb [56]. The theory predicts accumulation layer for the non-polar surface in InN, with surface potential of 0.6 eV, if the surface is In rich [57]. For a perfect non-polar surface no surface accumulation layer should be formed [57]. The possibility of the formation of an In-rich nonpolar surface in spite of the used N-rich deposition conditions was explained based on the low decomposition temperature and local In-rich on InN NW surface during the growth. Moreover, the surfaces of the investigated InN NWs are not perfect, and atomic steps can be detected by HRTEM. Additionally, the investigated samples are exposed to air and oxidation and adsorption of foreign molecules takes place, changing dramatically the NW surface potential. The accumulation layer was detected in InN for both polar and non-polar surfaces [58].

Quantum computations of the accumulation layer have shown that the distribution of the electron originating from the surface is extends deep in the NW bulk (about tens of nm) at considerable concentration above 10^{17} cm^{-3} [59]. Thus the accumulation layer can influence the PL emission not only by increasing the total non-radiative recombination, but also by contributing to the electrons involved in radiative processes. Then the estimated mean electron density includes these surface induced electrons, especially for thinner wires. The formation of the accumulation layer at the lateral surface of the InN NWs was also evidenced by electrical and magneto-electrical measurements on single NW devices [60–62].

In order to modify the surface potential, core-shell NW heterostructures have been fabricated. InN NWs were grown at the optimal temperature of 475°C. The GaN shell was subsequently deposited for different growth durations. The GaN shell was deposited at the same low growth temperature, to avoid InN decomposition. HRTEM studies show that GaN forms a fairly uniform thin shell with thickness of about 1–2 nm for deposition time shorter than 5 min. For 10 min deposition, the strain accumulated in the GaN shell is relaxed by formation of GaN islands on the lateral surfaces. The islands are also visible in the SEM images of the core-shell structures. The low temperature PL spectra of the core-shell InN/GaN NW samples with the deposition time varied from 1 min up to 10 min are shown in Fig. 23 in comparison with a reference sample of bare InN. The inset shows a side view SEM image of the sample with 10 min GaN shell. When the thickness of the GaN shell increases, the PL spectra show a red shift and a considerable increase of the peak intensity. Significant changes of the PL spectra are observed when the GaN deposition is varied from 0 to 1 min and then to 5 min. The PL spectra show insignificant changes when GaN deposition time increases from 5 to 10 min. Accordingly to the dependence of the peak energy on electron concentration given in Fig. 21, the red shift observed in Fig. 23 for the samples with thicker GaN shells deposited for 5 and 10 min corresponds to two times reduction of the mean electron concentration in comparison to the reference. The PL results can be attributed to a passivation effect of the GaN shell on InN surface states as follows. The GaN shell reduces the surface potential and consequently the accumulation layer. This results in a decrease of the contribution of the accumulation layer to the total nonradiative recombination, as well as to the electrons involved in PL emission. The strain induced in the InN

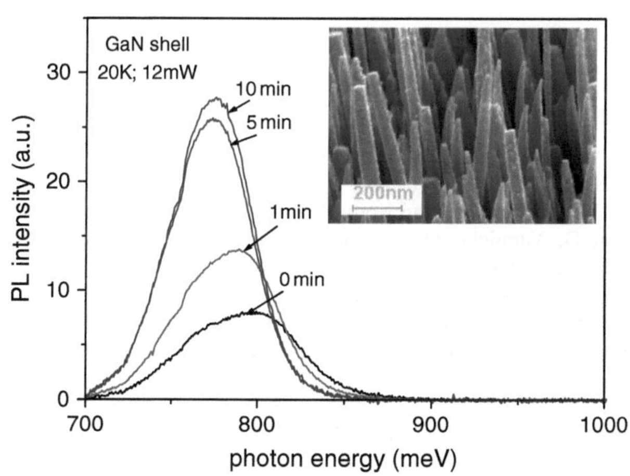

Fig. 23 PL spectra of core-InN/shell-GaN samples for different GaN growth times, in comparison with the reference sample. The inset shows an SEM image of the sample with 10 min GaN

core by the GaN shell should have an insignificant effect because of the relatively small thickness of the shell in comparison with the thickness of the InN core.

We can conclude that the accumulation layer formed on the lateral sides has a strong influence on the electronic properties of InN NWs. The photoluminescence studies have been used successfully in the growth optimization process, as well as for obtaining information about fundamental properties of the InN NWs.

4 Conclusions

High crystalline quality GaN and InN NWs can be formed by catalyst free PAMBE in N-rich conditions. The investigation of the nucleation and growth mechanism using SEM and HRTEM, as well as of the optical and optoelectrical properties of the NWs assisted the optimization of the growth parameters. The NW surface has a strong influence on the growth process and the optoelectronic properties of the NWs. For the investigated NW growth a vapor-solid mechanism was determined. The surface diffusion of adatoms controls the growth rate for thinner wires. The wire diameter dependence of the dark and photocurrent properties of GaN NWs were consistently explained based on surface electron depletion region. The photoluminescence spectra of InN NWs vary in intensity and line shape due to a band-filling of the conduction band. The line shape analysis was used to evaluate the mean electron density in InN samples. The electron accumulation layer formed at the lateral wire surface plays an important role in the optical and optoelectrical properties of the InN NWs.

Acknowledgements The authors thank K. H. Deussen for technical support. This work was financially supported by the German Ministry of Education and Research project "EPHQUAM" and performed under the auspices of the U.S. Department of Energy under contract No. DE-AC02-98CH1-886.

References

1. J. Wu, J. Appl. Phys. **106**, 011101 (2009)
2. E.D. Minot, F. Kelkensberg, M. van Kouwen, J.A. van Dam, L.P. Kouwenhoven, V. Zwiller, M.T. Borgstrom, O. Wunnicke, M.A. Verheijen, E. Bakkers, Nano Lett. **7**, 367 (2007)
3. Y. Huang, X. Duan, C.M. Lieber, Small **1**, 142 (2005)
4. O. Hayden, R. Agarwal, C.M. Lieber, Nat. Mater. **5**, 352 (2006)
5. A.B. Greytak, C.J. Barrelet, Y. Li, C.M. Lieber, Appl. Phys. Lett. **87**, 151103 (2005)
6. J.C. Johnson, H.J. Choi, K.P. Knutsen, R.D. Schaller, P. Yang, R.J. Saykally, Nat. Mater. **1**, 106 (2002)
7. H. Kind, H. Yan, B. Messer, M. Law, P. Yang, Adv. Mater. **14**, 158 (2002)
8. K. Keem, H. Kim, G.T. Kim, J.S. Lee, B. Min, K. Cho, M.Y. Sung, S. Kim, App. Phys. Lett. **84**, 4376 (2004)
9. F. Patolsky, G. Zheng, C.M. Lieber, Nanomedicine **1**, 51 (2006)
10. F. Patolsky, G.F. Zheng, C.M. Lieber, Anal. Chem. **78**, 4260 (2006)
11. Y. Cui, Z. Zhong, D. Wang, W.U. Wang, C.M. Lieber, Nano Lett. **3**, 149 (2003)

12. Y. Huang, X. Duan, Y. Cui, C.M. Lieber, Nano Lett. **2**, 101 (2002)
13. D.M. Bagnall, M. Boreland, Photovoltaic Technol. **36**, 4390 (2008)
14. J. Chen, G. Cheng, E. Stern, M.A. Reed, P. Avouris, Nano Lett. **7**, 2276 (2007)
15. C.M. Lieber, Z.L. Wang, MRS Bull. **32**, 99 (2007)
16. G.M. Jones, B.H. Hu, C.H. Yang, M.J. Yang, Y.B. Lyanda-Geller, Physica E **34**, 612 (2006)
17. R. Calarco, M. Marso, T. Richter, A.I. Aykanat, R. Meijers, A.v.d. Hart, T. Stoica, H. Lüth, Nano Lett. **5**, 981 (2005)
18. J. Wensorra, K.M. Indlekofer, M.I. Lepsa, A. Fo1rster, H. Lüth, Nano Lett. **5**, 2470 (2005)
19. J. Ebbecke, S. Maisch, A. Wixforth, R. Calarco, R. Meijers, M. Marso, H. Lüth, Nanotechnology **19**, 275708 (2008)
20. M.A. Reed, R.J. Randall, R.J. Aggarwal, R.J. Matyi, T.M. Moore, A.E. Wetsel, Phys. Rev. Lett. **60**, 535 (1988)
21. M. Tewordt, L. Martín-Moreno, V.J. Law, M.J. Kelly, R. Newbury, M. Pepper, D.A. Ritchie, J.E.F. Frost, G.A.C. Jones, Phys. Rev. B **46**, 3948 (1992)
22. D.G. Austing, T. Honda, S. Tarucha, Semicond. Sci. Technol. **11**, 388 (1996)
23. A. Randame, G. Faini, H. Launois, Z. Phys. B **85**, 389 (1991)
24. T. Schmidt, M. Tewordt, R.H. Blick, R.J. Haug, D. Pfannkuche, K. von Klitzing, A. Förster, H. Lüth, Phys. Rev. B **51**, 5570 (1995)
25. M. Griebel, K.M. Indlekofer, A. Förster, H. Lüth, J. Appl. Phys. **84**, 6718 (1998)
26. J. Wensorra, M.I. Lepsa, K.M. Indlekofer, A. Förster, P. Jaschinsky, B. Voigtländer, G. Pirug, H. Lüth, Phys. Stat. Sol. (a) **203**, 3559 (2006)
27. V. Mamutin, Tech. Phys. Lett. **25**, 741 (1999)
28. E. Calleja, M.A. Sanchez-Garcia, F.J. Sanchez, F. Calle, F.B. Naranjo, E. Munoz, U. Jahn, K. Ploog, Phys. Rev. B **62**, 16826 (2000)
29. R. Meijers, T. Richter, R. Calarco, T. Stoica, H.P. Bochem, M. Marso, H. Lüth, J. Cryst. Growth **289**, 381 (2006)
30. S. Biswas, S. Kar, T. Ghoshal, V.D. Ashok, S. Chakrabarti, S. Chaudhuri, Mater. Res. Bull. **42**, 428 (2007)
31. H.Y. Chen, H.W. Lin, C.H. Shen, S. Gwo, Appl. Phys. Lett. **89**, 243105 (2006)
32. K.A. Bertness, A. Roshko, L.M. Mansfield, T.E. Harvey, N.A. Sanford, J. Cryst. Growth **300**, 94 (2007)
33. J.F. Fälth, S.K. Davidsson, X.Y. Liu, T.G. Andersson, Thin Solid Films **515**, 603 (2006)
34. L. Cerutti, J. Ristic, S. Fernandez-Garrido, E. Calleja, A. Trampert, K.H. Ploog, S. Lazic, J.M. Calleja, Appl. Phys. Lett. **88**, 213114 (2006)
35. R.K. Debnath, R. Meijers, T. Richter, T. Stoica, R. Calarco, H. Lüth, Appl. Phys. Lett. 90, 123117 (2007)
36. R. Calarco, R.J. Meijers, R.K. Debnath, T. Stoica, E. Sutter, H. Lüth, Nano Lett. **7**, 2248 (2007)
37. A. Trampert, J. Ristic, U. Jahn, E. Calleja, K.H. Ploog, Microsc. Semicond. Mater. Conf., Cambridge Inst. Phys. Conf. Ser. 2003, No.180, p. 167
38. T. Stoica, E. Sutter, R.J. Meijers, R.K. Debnath, R. Calarco, H. Lüth, D. Grützmacher, Small **4**, 751 (2008)
39. Y.L. Chang, F. Li, A. Fatehi Z. Mi, Nanotechnology **20**, 345203 (2009)
40. T. Kuykendall, S. Aloni, I.J. La Plante, T. Mokari1, Int. J. Photoenergy, **2009**, 767951 (2009)
41. R.S. Wagner, W.C. Ellis, Appl. Phys. Lett. **4**, 89 (1964)
42. L. Jensen, M. Björk, S. Jeppesen, A. Persson, B. Ohlsson, L. Samuelson, Nano Lett. **4**, 1961 (2004)
43. T. Zywietz, J. Neugebauer, M. Scheffler, Appl. Phys. Lett. **73**, 487 (1998)
44. R.K. Debnath, T. Stoica, A. Besmehn, K. Jeganathan, E. Sutter, R. Meijers, H. Lüth, R. Calarco, J. Cryst. Growth **311**, 3389 (2009)
45. T. Stoica, R. Meijers, R. Calarco, T. Richter, H. Lüth, J. Cryst. Growth **290**, 241 (2006)
46. R. Calarco, M.Marso, Appl. Phys. A **87**, 499 (2007)
47. L. Polenta, M. Rossi, A. Cavallini, R. Calarco, M. Marso, R. Meijers, T. Richter, T. Stoica, H. Lüth, ACS Nano **2**, 287 (2008)

48. K. Jeganathan, R.K. Debnath, R. Meijers, T. Stoica, R. Calarco, D. Grützmacher, H. Lüth, J. Appl. Phys. **105**, 123707 (2009)
49. A. Cavallini, L. Polenta, M. Rossi, T. Stoica, R. Calarco, R.J. Meijers, T. Richter, H. Lüth, Nano Lett. **7**, 2166 (2007)
50. W. Franz, Naturforschung **13**, 484 (1958)
51. L. Keldysh, Sov. Phys. JETP **7**, 788 (1958)
52. M. Bertelli, P. Löptien, M. Wenderoth, A. Rizzi, R.G. Ulbrich, M.C. Righi, A. Ferretti, L. Martin-Samos, C.M. Bertoni, A. Catellani, Phys. Rev. B **80**, 115324 (2009)
53. S.L. Rumyantsev, M.S. Shur, M.E. Levinshtein, A. Motayed, A.V. Davydov, J. Appl. Phys. **103**, 064501 (2008)
54. T. Stoica, R.J. Meijers, R. Calarco, T. Richter, E. Sutter, H. Lüth, Nano Lett. **6**, 1541 (2006)
55. V. Yu. Davydov, A.A. Klochikhin, Semiconductors **38**, 897 (2004)
56. H. Lüth, *Surface and Interfaces of Solid Materials*, (Springer, Berlin, 1995)
57. C.G. Van de Walle, D. Segev, J. Appl. Phys. **101**, 081704 (2007)
58. P.D.C. King, T.D. Veal, C.F. McConville, F. Fuchs, J. Furthmuller, F. Bechstedt, P. Schley, R. Goldhahn, J. Schoermann, D.J. As, K. Lischka, D. Muto, H. Naoi, Y. Nanishi, H. Lu, W.J. Schaff, Appl. Phys. Lett. **91**, 092101 (2007)
59. L. Wang, D. Wang, P.M. Asbeck, Solid-State Electron **50**, 1732 (2006)
60. C. Blömers, T. Schäpers, T. Richter, R. Calarco, H. Lüth, M. Marso, Appl. Phys. Lett. **92**, 132101 (2008)
61. C. Blömers, T. Schäpers, T. Richter, R. Calarco, H. Lüth, M. Marso, Phys. Rev. B **77**, 201301 (2008)
62. T. Richter, C. Blömers, H. Lüth, R. Calarco, M. Indlekofer, M. Marso, T. Schäpers, Nano Lett. **8**, 2834 (2008)

Part III
Transport Phenomena in Nanostructures

Electronic and Thermal Sequential Transport in Metallic and Superconducting Two-Junction Arrays

T. Kühn and G.S. Paraoanu

Abstract The description of transport phenomena in devices consisting of arrays of tunnel junctions, and the experimental confirmation of these predictions is one of the great successes of mesoscopic physics. The aim of this paper is to give a self-consistent review of sequential transport processes in such devices, based on the so-called "orthodox" model. We calculate numerically the current-voltage (I–V) curves, the conductance versus bias voltage (G–V) curves, and the associated thermal transport in symmetric and asymmetric two-junction arrays such as Coulomb-blockade thermometers (CBTs), superconducting-insulator-normal-insulator-superconducting (SINIS) structures, and superconducting single-electron transistors (SETs). We investigate the behavior of these systems at the singularity-matching bias points, the dependence of microrefrigeration effects on the charging energy of the island, and the effect of a finite superconducting gap on Coulomb-blockade thermometry.

1 Introduction

Quasiparticle transport processes across metallic junctions play a fundamental role in the functioning of many devices used nowadays in mesoscopic physics. One such device is the single electron transistor (SET), invented and fabricated almost two decades ago [1], which has found remarkable applications as ultrasensitive charge detector [2] and as amplifier operating at the quantum limit [3]. In the emerging field of quantum computing the superconducting SET has been proposed and used as a quantum bit [4]. With advancements in lithography techniques, this device can be fully suspended [5], thus providing a new avenue for nanoelectromechanics. Similar devices are currently used (in practice arrays with several junctions turn out to provide a larger signal-to-noise ratio) as Coulomb blockade (CBT) primary thermometers [6]. Also, superconducting double-junction systems with appropriate

G.S. Paraoanu (✉)
Low Temperature Laboratory, School of Science and Technology, Aalto University,
FI-00076 AALTO, Finland
e-mail: paraoanu@cc.hut.fi

A. Aldea, V. Bârsan (eds.), *Trends in Nanophysics*, Engineering Materials,
DOI 10.1007/978-3-642-12070-1_5, © Springer-Verlag Berlin Heidelberg 2010

bias can be operated as microcoolers [7]. The functioning of these three classes of devices is based on the interplay between two out of the three relevant energy scales: the superconducting gap, the charging energy, and the temperature.

For example, in the case of microcoolers, the temperature and the gap are finite, and the charging energy is typically zero. A natural question to raise is then what happens if the charging energy is no longer negligible, for example if one wishes to miniaturize further these devices. In contradistinction, for CBTs the charging energy and the temperature are important, and the superconducting gap is a nuisance. A solution is to suppress the gap by using external magnetic fields, an idea which makes these temperature sensors more bulky and risky to use near magnetic-field sensitive components. Therefore, understanding the corrections introduced by the superconducting gap could provide an interesting alternative route, although, with present technology, the level of control required of the gap value could be very difficult to achieve. Finally, electrometers and superconducting SETs are operated at low temperatures, with the charging energy and the gap being dominant. However, large charging energies are not always easy to obtain for some materials due to technological limitations, while achieving effective very low electronic temperatures is limited by various nonequilibrium processes.

In this article we present a unified treatment of these three devices by solving the transport problem in the most general case, when all three energy scales are present. Our goal is to give an eye guidance for the experimentalist working in the field, showing what are the main characteristics visible in the I–Vs and G–Vs, resulting from sequential tunneling. Both the electrical and the thermal transport are calculated in the framework of a generalized so-called "orthodox" theory, which includes the superconducting gap. For completeness, we offer a self-consistent review of this theory, which has become nowadays the standard model for describing sequential quasiparticle transport processes in these devices. We ignore Josephson effects which, for the purpose of this analysis just add certain well-known features at certain values of the bias (the Josephson supercurrent at zero bias and the Josephson-quasiparticle (JQP) peak at 2Δ for symmetric SETs) to the characteristic. Higher-order tunneling phenomena such as Andreev reflections and cotunneling are neglected as well (the former is visible only for junctions with high transparencies, which is not our case, the latter is a small, second-order transport effect across the island). Also the effect of the environment is not discussed here, since it depends on the specifics of the sample (for example the coupling capacitances of the electrodes). But even stripped down to the essentials of sequential tunneling, the physics of these devices is complex enough to justify a careful analysis, as we attempt below. We present in detail the theory and the numerical methods that can be used to characterize these devices. Also, a lot of interest exists nowadays in the study of hybrid structures – in which the electrodes and the island are made from different materials, thus having different superconducting gaps. Various technologies have been developed to fabricate single-electron transistors with superconducting materials other than Al, most notably with Nb which has a much larger gap [8]. We extend our analysis also to such hybrid SETs.

This paper is organized in the following way: besides the introductory part, there are two main sections, Sect. 2 and Sect. 3, followed by closing remarks in Sect. 4. Sect. 2 discusses tunneling of electrons from one superconductor into another through an insulating barrier. Here we show how to calculate the currents that flow through the junction by using the Fermi golden rule (Sect. 2.1), and we give a comprehensive presentation of the numerical methods used to calculate the integrals over the density of states and Fermi functions which appear naturally in this theory (Sect. 2.2). In Sect. 3 we use these results to describe transport phenomena in two-junction devices. We assume a quasi-equilibrium situation (electron distribution described by a Fermi function with a certain effective temperature) for both the island and the leads. We start with a presentation of the master equation in Sect. 3.1. A first application of this formalism is described in Sect. 3.2, where we analyze the effect of a finite superconducting gap for the functioning of Coulomb-blockade thermometers. Then we discuss the effect of a nonzero charging energy for superconductor-insulator-normal metal-insulator-superconductor (SINIS) structures (Sect. 3.3) – which normally have a negligible charging energy. After that, in Sect. 3.4, we consider the case of superconducting single-electron transistors with either identical materials in the leads and island (symmetric SET) or with an island with a gap different from that of the electrodes (asymmetric SET). The effect of the gate in such transistors is further investigated in Sect. 3.5. We end this section with a calculation of expected cooling effects in such structures (Sect. 3.6). We delegate some of the calculations of integrals in Appendix 1, and of expansions in Appendix 2.

2 Transport in Single Tunnel Junctions

Standard lithographic techniques enable the fabrication of junctions with almost any geometrical characteristics down to the linewidth of the respective system. When using scanning-electron microscopes, the typical linewidths obtained are of the order of 20 nm and even below. More advanced methods, such as focused ion beam lithography, have started to be used recently. The metals are evaporated in an ultra-high vacuum evaporator, using the shadow evaporation technique. This technique consists of evaporation through a mask realized in the resist, at different angles, so that some lines would overlap in a certain region, thus defining the area of the junction. In between two evaporations, the sample is oxidized (typically aluminum oxide), a process which results in the formation of the insulating barrier.

In this section, we assume that the area of the junction thus formed is large enough so that capacitive effects can be neglected (Coulomb-blockade effects are discussed in the next section). This allows us to concentrate on tunneling effects only. The only energy scales relevant here will then be the temperature and the tunneling matrix element.

2.1 General Theory of Tunneling

Consider the case of two conductors (leads) that are connected through a thin insulating barrier. The setup is illustrated in Fig. 1. This creates effectively a potential

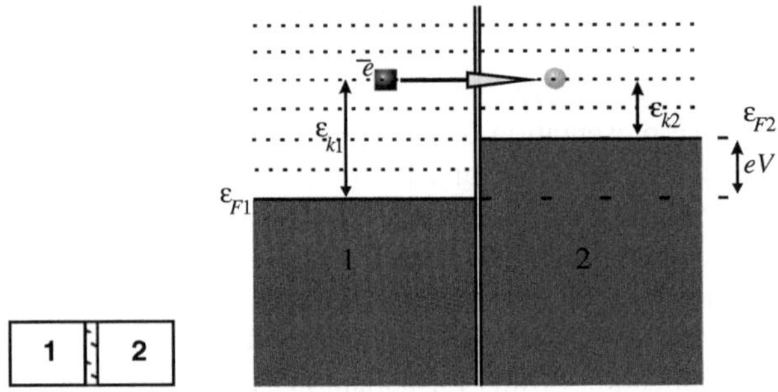

Fig. 1 Schematic of a tunnel junction. In the left figure, the leads are the clear areas marked with 1 and 2, while the insulating barrier is depicted by the shaded area. A schematic of the energy levels is shown for the case of metallic electrodes

barrier between the "left" and "right" electrodes: however electrons can still tunnel from one side to another *via* the tunneling effect. The total many-body Hamiltonian of the system comprises then two parts: one is the Hamiltonian corresponding to the left and right electrodes (which can be either the free-electron Hamiltonian in the case of metals, or the BCS Hamiltonian in the case of superconductors), the other is the tunneling Hamiltonian. This Hamiltonian was first introduced by Bardeen [9]:

$$H_T = \sum_{k_1,k_2,\sigma} T_{k_1,k_2} c^{\dagger}_{k_2\sigma} c_{k_1\sigma} + \text{h.c.}, \tag{1}$$

where k_1 and k_2 are the momentum indices corresponding to the left and right electrodes, and σ is the spin index (conserved under tunneling); $c^{\dagger}_{k_{1,2}\sigma}$ and $c_{k_{1,2}\sigma}$ are creation and annihilation operators for fermions. It is then straightforward to calculate the probability per unit time (the tunneling rate) that an electron residing in the left electrode would tunnel from a state $\{k_1, \sigma\}$ (of energy $\epsilon_{k_1} = \hbar^2 k_1^2/2m - \epsilon_{\text{F}1}$, where $\epsilon_{\text{F}1}$ is the Fermi level of the left electrode) to a state $\{k_2, \sigma\}$ (of energy $\epsilon_{k_2} = \hbar^2 k_2^2/2m - \epsilon_{\text{F}2}$, where $\epsilon_{\text{F}2}$ is the Fermi energy of the right electrode), by a direct application of the Fermi Golden rule,

$$P_{1\to 2}(k_1, k_2) = \frac{2\pi}{\hbar} |T_{k_1,k_2}|^2 \rho_{k_2} f_{k_1} (1 - f_{k_2}), \tag{2}$$

where $f_{k_{1,2}}$ are the Fermi functions corresponding to the states $k_{1,2}$, ρ_{k_2} is the density of states (without the spin degrees of freedom) at k_2, and δE is the energy difference between the states $\{k_2, \sigma\}$ and $\{k_1, \sigma\}$, $\delta E_{1\to 2} = \epsilon_{k_2} - \epsilon_{k_1}$. The energy difference $\delta E_{1\to 2}$ does not depend on the state k_1 or k_2 since it is due to electrical potentials such as an external battery or to the charging of the island (as we will see below in the case of single-electron transistor) which produce a difference in the Fermi energies of the two metals, $\delta E_{1\to 2} = \epsilon_{\text{F}_1} - \epsilon_{\text{F}_2}$, meaning that $\epsilon_{k_1} + \epsilon_{\text{F}1} = \epsilon_{k_2} + \epsilon_{\text{F}2}$, and thus ensuring that the total energy is conserved under tunneling.

Replacing now the sum over states with a continuous integral, we can find the total tunneling rate from left to right by integrating $P_{1\to 2}$, weighted with the density of states ρ_1,

$$\begin{aligned}\Gamma_{1\to 2}(\delta E_{1\to 2}) =& 2\int d\epsilon_1 \rho_1(\epsilon_1) P_{1\to 2}(\epsilon_1, \epsilon_2 = \epsilon_1 + \delta E_{1\to 2}) \\ =& 2\int d\epsilon_1 \int d\epsilon_2 \rho_1(\epsilon_1) P_{1\to 2}(\epsilon_1, \epsilon_2)\delta(\epsilon_2 - \epsilon_1 - \delta E_{1\to 2}).\end{aligned} \quad (3)$$

The last form of the equation, using the Dirac delta function (in no danger of confusion with the δ of $\delta E_{1\to 2}$), reflects indeed the conservation of energy: electrons of energy $\epsilon_1 + \epsilon_{F1}$ can get only to states $\epsilon_2 + \epsilon_{F2}$, if they acquire an energy gain of $\delta E_{1\to 2}$ with respect to their Fermi levels. Inelastic processes are allowed in a more general formulation, which includes the degrees of freedom of the environment, in which case δ is replaced by the so-called P-function [11]. Also, the factor of 2 appearing in front of the integral comes from summation over the spin index. The rate of tunneling from the right electrode to the left one, $\Gamma_{2\to 1}(\delta E_{2\to 1})$, is calculated in a similar way. The difference between these quantities represents the net current of charges flowing through the junction, therefore the net electrical current is

$$\begin{aligned}I_{1\to 2} &= -e\left[\Gamma_{1\to 2}(\delta E_{1\to 2}) - \Gamma_{2\to 1}(\delta E_{2\to 1})\right] \\ &= -e\left[\Gamma_{1\to 2}(\delta E_{1\to 2}) - \Gamma_{2\to 1}(-\delta E_{1\to 2})\right]\end{aligned} \quad (4)$$

We then get, using the expressions Eq. (3) and Eq. (2),

$$I_{1\to 2} = -\frac{4\pi e}{\hbar}\int d\epsilon_1 d\epsilon_2 |T(\epsilon_1, \epsilon_2)|^2 \rho(\epsilon_1)\rho(\epsilon_2)[f(\epsilon_1) - f(\epsilon_2)]\delta(\epsilon_2 - \epsilon_1 - \delta E_{1\to 2}), \quad (5)$$

where we conveniently change all the momentum indices to continuous-energies functions. To make progress with Eq. (5), a few assumptions are needed. First of all, already embedded in the theory above is the idea that tunneling does not significantly change the Fermi distribution function. Nonequilibrium processes however can play an important role, requiring a more involved treatment using Green's functions [10]. Next, the Fermi energy in most metals is of the order of a few eVs; this is large when compared to the typical bias voltages at which such junctions are studied. Then, in the case of metals, the density of states can be considered constant around the corresponding Fermi energies $\rho(\epsilon_1) \approx \rho(0)$, and $\rho(\epsilon_2) \approx \rho(0)$. For superconducting electrodes this assumption does not hold: the density of states is divergent around the gap, and, as we will see below, this has significant effects. It is however possible (see below) to isolate the divergent part of the density of states as an adimensional quantity which, when multiplied with the normal-metal density of states around the Fermi level would give an approximate but still correct superconducting density of states. Finally, the tunneling matrix element T can also be taken as energy-independent, $T(\epsilon_1, \epsilon_2) = T$. With these assumptions, the integral over energies in Eq. (3) can be performed (see [11, 12] and Appendix 1), using the generally useful result

$$\int d\epsilon f(\epsilon)[1 - f(\epsilon - E)] = \frac{E}{\exp(E/k_B T) - 1}, \tag{6}$$

to give for the tunneling rates

$$\Gamma_{1\to 2}(\delta E_{1\to 2}) = -\frac{4\pi}{\hbar}|T|^2 \rho(\epsilon_{\mathrm{F}1})\rho(\epsilon_{\mathrm{F}2}) \frac{\delta E_{1\to 2}}{\exp(-\delta E_{1\to 2}/k_B T) - 1}. \tag{7}$$

For example, at zero temperature we find $\Gamma_{1\to 2}(\delta E_{1\to 2}) = (4\pi/\hbar)|T|^2 \rho(\epsilon_{\mathrm{F}1})\rho(\epsilon_{\mathrm{F}2})$ $\delta E_{1\to 2}$ for $\delta E_{1\to 2} > 0$ and $\Gamma_{1\to 2} = 0$ for $\delta E_{1\to 2} \leq 0$. The value of the current does not depend however on temperature. From Eq. (7) and Eq. (4) we find a linear dependence in δE for the current

$$I_{1\to 2} = -\frac{4\pi e}{\hbar}\rho(\epsilon_{\mathrm{F}1})\rho(\epsilon_{\mathrm{F}2})|T|^2 \delta E_{1\to 2} \tag{8}$$

If $\delta E_{1\to 2} = -eV$, as it is in the case of a voltage-biased single junction (see Fig. 1), we immediately recognize Ohm's law from Eq. (8), $I_{1\to 2}(V) = R_{12}^{-1} V$; with the quanta of resistance defined as $R_K = h/e^2 = 2\pi\hbar/e^2 = 25.8\ \mathrm{k}\Omega$ we get

$$\frac{1}{R_{12}} = \frac{8\pi^2}{R_K}\rho(\epsilon_{\mathrm{F}1})\rho(\epsilon_{\mathrm{F}2})|T|^2. \tag{9}$$

The theory presented above looks very reasonable until one realizes – *horribile dictu*! – that in fact the result Eq. (9) predicts a rather absurd scaling of electrical resistance with the dimensions of the sample. Let us look at the densities of state. The three-dimensional density of states per volume (with spin states excluded) can be calculated using the standard text-book expression

$$\mathcal{G}^{(3D)}(E) = \frac{1}{4\pi^2}\left(\frac{2m}{\hbar^2}\right)\sqrt{E}. \tag{10}$$

The Fermi energy for metals is of the order of a few eV: for example, $\epsilon_{\mathrm{F}}^{(\mathrm{Au})} = 5.53$ eV, $\epsilon_{\mathrm{F}}^{(\mathrm{Ag})} = 5.94$ eV, $\epsilon_{\mathrm{F}}^{(\mathrm{Al})} = 11.63$ eV, $\epsilon_{\mathrm{F}}^{(\mathrm{Cu})} = 7$ eV. At $\epsilon_{\mathrm{F}} = 10$ eV we obtain for example $\mathcal{G}^{(3D)}(10eV) = 0.75 \times 10^{57} J^{-1}\mathrm{m}^{-3}$; the Fermi wavelength at $\epsilon_{\mathrm{F}} = 10$ eV is $\lambda_{\mathrm{F}} = h/\sqrt{2m\epsilon_{\mathrm{F}}} = 0.37$ nm.

The problem now is that in order to find $\rho(\epsilon_{\mathrm{F}})$ we have to multiply the density of states per volume $\mathcal{G}^{3D}$ with the volume of the electrodes. This is of course absurd, as the inverse of resistance should be linear in the junction's surface and not depend on the volume of the electrodes. The solution to this conundrum stems from the idea that the summation in Eq. (3) has to be done under the restriction that transversal momentum is conserved (specular transmission): thus the correct densities of states that appear further have to be densities of states at fixed transversal momentum [13]. Also the assumption of energy-independence of the tunneling matrix element does not necessarily hold true for most practical junctions. A complete theory of tunneling that takes into consideration all these special features has been developed over the years [14]. However, with the trick of absorbing the density of states into the tunneling resistance, the theory described above is still useful, describing correctly the voltage dependence of the current. As we will see below, the theory can be extended to the case of superconductors. Similar tunneling theories can be constructed for scanning tunneling microscopy, where the densities of state refer to the tip of the instrument and to the surface of the sample [15]. Also, the same structure has been successfully applied to systems such as trapped atoms, where the tunneling can be realized by using RF fields coupled to internal hyperfine states of the atoms [16]. In this case, δE is given by the detuning of the RF field with respect to the atomic frequency, and the tunneling matrix element is truly constant (given by the coupling amplitude of the field with the atoms). In this situation, the current of particles between two hyperfine states depends indeed on the volume of the samples, as it should be, since the field penetrates the sample completely and tunneling between the internal states occurs not only at some interface but in the whole volume. This case also has its specifics however: the momentum (and not only the energy) is conserved.

In the case of superconductors, the theory proceeds as above, but the density of states used has to be replaced by that given by the BCS theory. It turns out that this density of states consists of a factor which is the same as for metals, which gets multiplied by a divergent part near the gap. It is then natural to introduce a normalized density of states $\mathcal{N}$, which would take the value 1 for metals,

$$\mathcal{N}_{\text{normal}}(\epsilon) = 1 , \tag{11}$$

and will be a function

$$\mathcal{N}_{\text{BCS}}(\epsilon) = \frac{|\epsilon|}{\sqrt{\epsilon^2 - \Delta^2}} \Theta(|\epsilon| - \Delta) , \tag{12}$$

for BCS superconductors (Δ is the superconducting gap and Θ is the Heavyside function).

Summarizing all the discussion above, we can write the tunneling probability, including both the metallic and superconducting case, as

$$\Gamma_{1\to 2}(\delta\epsilon) = \frac{1}{e^2 R_{12}} \int_{-\infty}^{\infty} \mathrm{d}\epsilon_1 \int_{-\infty}^{\infty} \mathrm{d}\epsilon_2 \, \mathcal{N}_1(\epsilon_1)\mathcal{N}_2(\epsilon_2) f_1(\epsilon_1)[1 - f_2(\epsilon_2)]\delta(\epsilon_2 - \epsilon_1 - \delta E_{1\to 2}) \tag{13a}$$

$$= \frac{1}{e^2 R_{12}} \int_{-\infty}^{\infty} \mathrm{d}\epsilon_1 \, \mathcal{N}_1(\epsilon_1)\mathcal{N}_2(\epsilon_1 + \delta E_{1\to 2}) f_1(\epsilon_1)[1 - f_2(\epsilon_1 + \delta E_{1\to 2})] , \tag{13b}$$

with $\mathcal{N}_{1,2}$ being the normalized density of states of the electrodes, as introduced above. Furthermore, given the tunneling probabilities of Eq. (13), the current through a single superconducting tunnel junction without charging effects is given as

$$I_{1\to 2} = -e\left[\Gamma_{1\to 2}(\delta E_{1\to 2}) - \Gamma_{2\to 1}(\delta E_{2\to 1})\right] = -e\left[\Gamma_{1\to 2}(\delta E) - \Gamma_{2\to 1}(-\delta E_{1\to 2})\right] . \tag{14}$$

2.2 Numerical Methods

For the discussion in this subsection we use the shorthand notation $\delta E = \delta E_{1\to 2}$. Now Eq. (13) can be solved analytically for some limiting cases, as we have already seen for the case of normal metal electrodes, or can be expressed as a combination of special functions. For example, in the case of a SIN junction, the calculation at zero temperature is straightforward. From Eq. (13b) we get

$$\Gamma_{1\to 2} = \frac{1}{e^2 R_{12}} \sqrt{(\delta E)^2 - \Delta_1^2}\,\Theta(\delta E - \Delta_1), \tag{15}$$

which can be seen for example in Fig. 2 bottom left (the additional spike appearing there at $\delta E = 0$ is a singularity-matching peak discussed below; it shows up because the temperature is finite). In this figure, $I = I_{1\to 2}$ and $\delta E = \delta E_{1\to 2} = -eV$.

In practice however, in the general case of non-identical superconductors and finite temperature, it has to be solved numerically, and, due to the singularities in the BCS density of states, the numerical treatment of Eq. (13) is somewhat challenging (see Fig. 2). We therefore want to discuss the numerical problems arising from these divergencies and possible ways to solve them in a bit more detail. Inserting Eq. (12) into Eq. (13b), one can easily see that the expression for the electron tunneling probability splits into four separate integrals,

$$\Gamma_{1\to 2} = \frac{1}{e^2 R_{12}} \left(\Upsilon_1 + \Upsilon_2 + \Upsilon_3 + \Upsilon_4\right) , \tag{16}$$

where the integrals Υ_n are given by

$$\Upsilon_1 = \int\limits_{-\infty}^{\min(-\Delta_1, -\Delta_2 - \delta E)} \mathrm{d}\epsilon \;\; g(\epsilon) \tag{17a}$$

$$\Upsilon_2 = -\Theta(-\Delta_1 - \Delta_2 + \delta E) \int\limits_{\Delta_2 - \delta E}^{-\Delta_1} \mathrm{d}\epsilon \;\; g(\epsilon) \tag{17b}$$

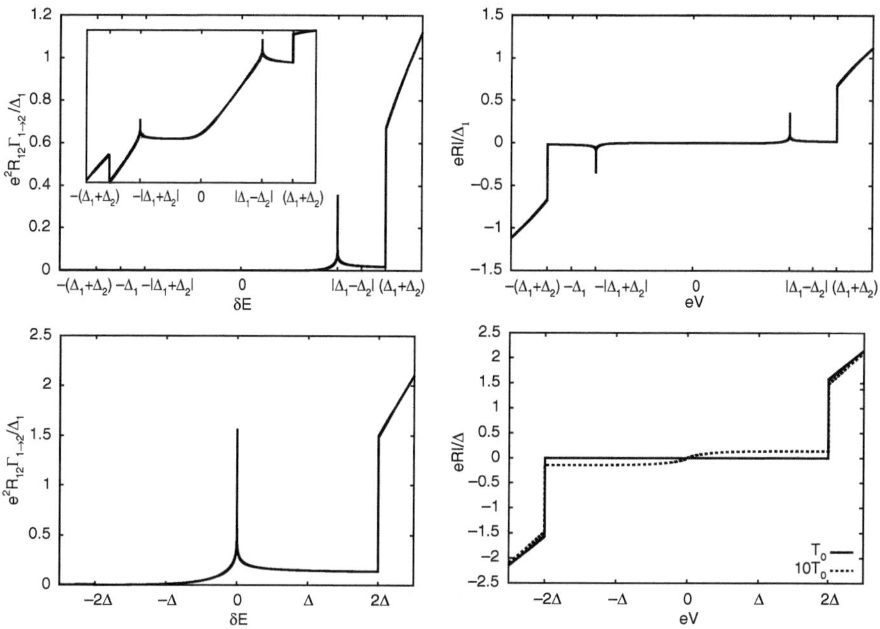

Fig. 2 *Top*: Tunneling rate (*left*) as a function of energy difference, and I–V (*right*) through an asymmetric superconducting tunnel junction with $\Delta_2 = 0.2\Delta_1$. The inset in the tunnel rate plot shows the same data with a logarithmic scale on the y-axis, for better emphasis of all features. Bottom: Tunneling rate (*left*) and I–V (*right*) through a symmetric superconducting tunnel junction. In the I–V we show the current for two different temperatures, with $k_B T_0 = 3.92 \times 10^{-2}\Delta_1$

$$\Upsilon_3 = -\Theta(-\Delta_1 - \Delta_2 - \delta E) \int_{\Delta_1}^{-\Delta_2-\delta E} \mathrm{d}\epsilon\ g(\epsilon) \tag{17c}$$

$$\Upsilon_4 = \int_{\max(\Delta_1, \Delta_2-\delta E)}^{\infty} \mathrm{d}\epsilon\ g(\epsilon)\,. \tag{17d}$$

The limits of these integrals are set by the Θ-functions in the BCS density of states. The function $g(\epsilon)$ is then

$$g(\epsilon) = \frac{\epsilon(\epsilon + \delta E)}{\sqrt{\epsilon^2 - \Delta_1^2}\sqrt{(\epsilon + \delta E)^2 - \Delta_2^2}} f_1(\epsilon)[1 - f_2(\epsilon + \delta E)]\,. \tag{18}$$

Due to the singularities in the BCS density of states, the function $g(\epsilon)$ is singular in each of the finite limits of the integrals Υ_n.

In most cases these singularities can be handled easily enough with the help of a simple integral transformation. As the function $g(\epsilon)$ behaves qualitatively like

$h(x) * x/\sqrt{x^2 - c^2}$, where $h(x)$ is smooth on the interval of integration, we can substitute $x = \sqrt{s^2 + c^2}$ and we are left with an integral which is numerically easily handled:

$$\int_c^d \mathrm{d}x \, \frac{x h(x)}{\sqrt{x^2 - c^2}} = \int_0^{\sqrt{d^2-c^2}} \mathrm{d}s \, h\left(\sqrt{s^2 + c^2}\right) . \tag{19}$$

However, for certain values of δE, the singularities of the densities of states of the first and second lead will coincide, leading to two logarithmic divergences known as singularity matching peaks at $\delta E = \pm(\Delta_1 + \Delta_2)$ and two finite steps at the quasiparticle threshold $\delta E = \pm|\Delta_1 - \Delta_2|$.

Singularity-matching peaks: We will first discuss the singularity matching peaks, which are described by the integrals Υ_1 and Υ_4. When δE is close to $\pm(\Delta_1 + \Delta_2)$, the integration results can become very inaccurate, and the transformation done in Eq. (19) will not help to solve the problem. We can however separate the singularity from the integral by partial integration, leaving us with a finite and easily integrable rest term. We note that the general form of integrals Υ_1 and Υ_4 is

$$\Upsilon(c) = \int_0^\infty \mathrm{d}x \, \frac{h(x)}{\sqrt{x(x+c)}} . \tag{20}$$

As $h(x)$ is proportional to the Fermi distribution, $h(x)$ and all its derivatives exponentially tend to zero for large values of x. Furthermore, $h(x)$ is analytic on the interval $[0, \infty)$. Using partial integration, we obtain

$$\Upsilon(c) = \ln\left(\frac{1}{c}\right) h(x=0) - 2 \int_0^\infty \mathrm{d}x \, \ln\left(\sqrt{x} + \sqrt{x+c}\right) h'(x) , \tag{21}$$

with the big advantage that the singularity itself is described by an explicit expression rather than an integral. The integral in the second term is finite and easily computed. The major nuisance here is to calculate the derivative of the function $h(x)$.

For completeness, it is also possible to derive a complete expansion of $\Upsilon(c)$ with the help of repeated partial integration,

$$\Upsilon(c) = \ln\left(\frac{1}{c}\right) h(x=0) - \sum_{n=1}^{\infty} \frac{(2n)!}{4^n (n!)^3} c^n \ln(c) h^{(n)}(x=0) . \tag{22}$$

Here we denote by $h^{(n)}(x)$ the nth derivative of $h(x)$ with respect to x (for a derivation, see Appendix 2).

Using the recipe of Eq. (21) on the integrals Υ_1 and Υ_4, we obtain

$$\Upsilon_1 = S_1 - 2\int_{\max(\Delta_1,\Delta_2+\delta E)}^{\infty} d\epsilon \ln\left[\sqrt{\epsilon-\Delta_1}+\sqrt{\epsilon-\delta E-\Delta_2}\right] \times \frac{d}{d\epsilon}\left[\frac{\epsilon(\epsilon-\delta E)}{\sqrt{(\epsilon+\Delta_1)(\epsilon-\delta E+\Delta_2)}}[1-f_1(\epsilon)]f_2(\epsilon-\delta E)\right] \tag{23a}$$

$$\Upsilon_4 = S_4 - 2\int_{\max(\Delta_1,\Delta_2-\delta E)}^{\infty} d\epsilon \ln\left[\sqrt{\epsilon-\Delta_1}+\sqrt{\epsilon+\delta E-\Delta_2}\right] \times \frac{d}{d\epsilon}\left[\frac{\epsilon(\epsilon+\delta E)}{\sqrt{(\epsilon+\Delta_1)(\epsilon+\delta E+\Delta_2)}}f_1(\epsilon)[1-f_2(\epsilon+\delta E)]\right], \tag{23b}$$

where the terms S_1 and S_4 describe the singular points, with different expressions if the singularity is approached from above or below,

$$S_1(\delta E > \Delta_1-\Delta_2) = -\frac{\Delta_2(\Delta_2+\delta E)f_1(-\Delta_2-\delta E)f_2(\Delta_2)}{\sqrt{2\Delta_2(\Delta_1+\Delta_2+\delta E)}}\ln\left(-\Delta_1+\Delta_2+\delta E\right) \tag{24a}$$

$$S_1(\delta E < \Delta_1-\Delta_2) = -\frac{\Delta_1(\Delta_1-\delta E)f_1(-\Delta_1)f_2(\Delta_1-\delta E)}{\sqrt{2\Delta_1(\Delta_1+\Delta_2-\delta E)}}\ln\left(\Delta_1-\Delta_2-\delta E\right) \tag{24b}$$

$$S_4(\delta E > \Delta_2-\Delta_1) = -\frac{\Delta_1(\Delta_1+\delta E)f_1(\Delta_1)f_2(-\Delta_1-\delta E)}{\sqrt{2\Delta_1(\Delta_2+\Delta_1+\delta E)}}\ln\left(-\Delta_2+\Delta_1+\delta E\right) \tag{24c}$$

$$S_4(\delta E < \Delta_2-\Delta_1) = -\frac{\Delta_2(\Delta_2-\delta E)f_1(\Delta_2-\delta E)f_2(-\Delta_2)}{\sqrt{2\Delta_2(\Delta_2+\Delta_1-\delta E)}}\ln\left(\Delta_2-\Delta_1-\delta E\right). \tag{24d}$$

In the case of a symmetric junction, where $\Delta_1 = \Delta_2 = \Delta$, only one singularity matching peak appears in the tunneling rates (13). In this case the singularity matching peak disappears in the current through the junction. With the derived expressions we can easily understand this, as the expressions S_1 and S_4 cancel out when inserted into the expression for the current (14). In this special case the current can therefore be computed using only the respective second terms in Eqs. (23).

Quasiparticle threshold: The quasiparticle threshold is described by the integrals Υ_2 and Υ_3 in Eqs. (17). We can find an explicit expression for the height of these jumps with the help of a partial integration and taking the limit $\delta E \to \pm(\Delta_1+\Delta_2)$. Both Υ_2 and Υ_3 are of the general form

$$\Upsilon(c) = \int_{-c}^{c} \mathrm{d}x \, \frac{h(x)}{\sqrt{c^2 - x^2}}\,, \tag{25}$$

where both $h(x)$ and $h'(x)$ are analytic functions on the interval $[-c, c]$. Then we can find the solution of $\Upsilon(c)$ in the limit $c \to 0$ by partial integration,

$$\begin{aligned} \Upsilon_c &= \frac{\pi}{2}\,[h(c) + h(-c)] - \int_{-c}^{c} \mathrm{d}x \,\arctan\left[\frac{x}{\sqrt{c^2 - x^2}}\right] h'(x) \\ &\longrightarrow \pi h(0) + 0\,, \qquad \text{for} \quad c \to 0\,. \end{aligned} \tag{26}$$

Using this method, we find that the height of the step at the quasiparticle threshold voltage is

$$\frac{\pi}{2e^2 R_{12}} \sqrt{\Delta_1 \Delta_2}\, f_1(-\Delta_1) f_2(-\Delta_2) \tag{27a}$$

for $\delta E = -(\Delta_1 + \Delta_2)$, and

$$-\frac{\pi}{2e^2 R_{12}} \sqrt{\Delta_1 \Delta_2}\, f_1(+\Delta_1) f_2(+\Delta_2) \tag{27b}$$

for $\delta E = +(\Delta_1 + \Delta_2)$.

We note here that at zero temperature the height of the first step is $(\pi/2e^2 R_{12})\sqrt{\Delta_1 \Delta_2}$ and the second one is zero (due to the fact that Fermi functions at positive energies are zero). The first peak is always more prominent and positive, while the second threshold is observable only at finite temperatures and it appears as a negative step (see the inset of Fig. 2).

Dynes parametrization: As one can see from the above discussion, it is rather cumbersome to treat superconducting tunnel junctions in the BCS picture. The expressions for the tunneling probabilities of Eq. (13) will also appear in the treatment of superconducting SETs, where the strong features discussed in this chapter will cause further numerical difficulties, especially when computing conductances. It is therefore convenient to somewhat smoothen the jumps and singularities of the tunneling probabilities.

One method to achieve this is to replace the δ-distribution in Eq. (13a) by a Gaussian function of finite width. (This is not only a mathematical trick but it corresponds, in the so-called $P(E)$ theory [11] to the case of a high impedance environment at finite temperature). One is then left to solve a two-dimensional integral with non-trivial integration limits.

A more convenient and at the same time physically more justifiable solution to the problem is to introduce a broadening parameter into the density of states (12),

$$\mathcal{N}_{\text{Dynes}} = \left| \Re \left\{ \frac{E + i\eta}{\sqrt{(E + i\eta)^2 + \Delta^2}} \right\} \right| . \tag{28}$$

The parameter η is called Dynes parameter [17] and it accounts for the finite life-time of quasiparticles in the superconductor. The symbol $\Re$ stands for the real part of the expression in curly brackets. This broadening of the density of states leads to nonzero currents flowing through the tunnel junction at voltages smaller than Δ/e (say for a NIS junction), which are often seen in experiments. These sub-gap currents are believed to be due to additional states formed inside the gap, allowing for the opening of new conduction channels. The origin of these currents is still not clarified in the literature. One should realize that these subgap currents are distinct from the currents due solely to the existence of a finite temperature (which causes subgap excitations) in a superconductor. Such temperature effects are already included in the formulation presented above.

It has been demonstrated [18] that these additional currents due to states formed inside the gap are Coulomb-blockaded as well (as it could be expected physically). Moreover, they can be even used for Coulomb-blockade thermometry. It should be noted also that an intrinsic dependence of the gap on temperature and magnetic field exists, as given by standard BCS theory. By analyzing the change in the zero-bias conductance when either temperature or magnetic field is increased, it is possible to determine the critical temperature and the critical magnetic field of the island [19]. A detailed description of another fitting procedure which uses the Dynes form for the density of states is also given in e.g. [24]. Through replacing the density of states in Eq. (12) with Eq. (28) the tunneling probabilities in Eq. (13) become finite and continuous for all values of δE and integration takes place on the whole real axis. The integral (13b) is still not easy to solve with high precision, but compared to the procedure that follows when using Eq. (12), the numerical treatment becomes certainly more simple. In the following sections we will exclusively use Eq. (28) as density of states, with a small but finite value for η.

3 Two-Junction Devices

Two-junction arrays are very useful devices for a variety of applications. They are denoted usually as a heterostructure - for example, SINIS, NININ, *etc.*, showing the phase of the electrons (metallic or superconducting) at the operating temperature. If a gate is capacitively coupled with the middle electrode (called island) by a coupling capacitance C_g, then the device is usually referred to as a single electron transistor (SET). If some or all of the leads in the SET are superconducting, the device will have characteristics that stem altogether from tunneling, charging effects, and superconducting gap. The fabrication of such devices is possible due to modern nanolithography techniques, as mentioned in the previous section, using which one can easily fabricate small junctions and small metallic or superconducting grains, resulting in non-negligible charging energies. For example, the area S of the junctions obtained by shadow evaporation techniques can be of the order of 100 nm^2 or even less. The insulating layer is usually of the order of d =1 nm (usually 0.5–5 nm) in thickness, and the dielectric constant ϵ_r of the oxide is of the order of 10. Using the formula for the capacitance $C = \epsilon_0 \epsilon_r S/d$, we can readily obtain capacitances of the order of $C \approx 10^{-15} F$ (femtofarads). The energy scale corresponding to this capacitance is $e^2/2C \approx 100\ \mu V$ (or a temperature of the order of 1K). Thus, charging effects can become observable at temperatures that can be achieved by standard cryogenic devices, such as dilution refrigerators. Effort has been put in recent times in reducing this capacitance even further, so that charging effects would be visible at even higher temperatures (preferably at room temperature [20]). This is where novel fabrication methods, based on advances in nanoscience, are essential.

3.1 Master Equation

The Coulomb blockade phenomena discussed here are essentially classical. The tunnel resistance is assumed to have a value much larger than the quanta of resistance $R_K = h/e^2$. This makes the number of electrons on an island a well-defined classical quantity, albeit a discret one. The Coulomb blockade effects then stem from the fact that the biasing potential can be varied continuosly while the potential given by the product between the charge and the capacitance can change only in steps. This gives rise to steps in the current and Coulomb blockade oscillations in the conductance. Except for tunneling, this phenomenon is indeed a classical one: the energy level separation of a small metallic island is still much smaller than the thermal energy $k_B T$, therefore the spectrum can be treated as continuous. This should be contrasted to the case of tunneling in semiconductor nanostructures, where the spectrum is discrete (see [21]).

In the case of single electron transistors, the relevant charging energy is given by the effective capacitance of the island to the ground: this is denoted in the following, using an already standard notation, by C_Σ, expressed as $C_\Sigma = C_L + C_R + C_g$. Here C_L is the capacitance of the left junction, C_R is the capacitance of the right junction, and C_g is the capacitance between the island and the gate electrode. The associated charging energy of the island, corresponding to one electron of extra charge, is

$$E_C = \frac{e^2}{2C_\Sigma} \tag{29}$$

The gate capacitance C_g is usually neglected or it can formally be distributed between the junction capacitances, leaving $C_\Sigma \approx C_L + C_R$. The relevant electrostatic energy of the island with n excess electrons and gated with the potential V_g turns out to be

$$E_{\mathrm{ch}}(n, Q_g) = \frac{1}{2C_\Sigma}\left(ne + q_g\right)^2$$
$$\equiv E_C\left(n + n_g\right)^2 ,$$

where $n_g \equiv q_g/e$ and $q_g \equiv C_g V_g + C_{\mathrm{R}} V_{\mathrm{R}} + C_{\mathrm{L}} V_{\mathrm{L}}$. We want now to calculate the energy changes $\delta E_{\mathrm{L}\to\mathrm{I}}(n)$, $\delta E_{\mathrm{I}\to\mathrm{L}}(n)$, $\delta E_{\mathrm{R}\to\mathrm{I}}(n)$, and $\delta E_{\mathrm{I}\to\mathrm{R}}(n)$, (including the work done by the sources) associated with electrons tunneling onto and off the island). The general idea is that, when such processes occur, n is increased (or decreased), respectively, by one, and the energy difference between the electron states before and after the tunneling event will be given by $E_{ch}(n+1) - E_{ch}(n)$ (or $E_{ch}(n) - E_{ch}(n-1)$) plus the work done by the voltage source (see Fig. 3).

We give now a detailed derivation of these results. Kirchhoff's laws for the circuit presented in Eq. (3) can be written as:

$$-ne = Q_{\mathrm{L}} + Q_{\mathrm{R}} + Q_g, \tag{30a}$$
$$\frac{Q_{\mathrm{L}}}{C_{\mathrm{L}}} - \frac{Q_{\mathrm{R}}}{C_{\mathrm{R}}} = V_{\mathrm{L}} - V_{\mathrm{R}}, \tag{30b}$$
$$\frac{Q_{\mathrm{L}}}{C_{\mathrm{L}}} - \frac{Q_g}{C_g} = V_{\mathrm{L}} - V_g, \tag{30c}$$
$$\frac{Q_{\mathrm{R}}}{C_{\mathrm{R}}} - \frac{Q_g}{C_g} = V_{\mathrm{R}} - V_g, \tag{30d}$$

where Q_{L}, Q_{R} and Q_g are the charges corresponding to capacitors C_{L}, C_{R} and C_g respectively. Combining these equations, we find

$$Q_g(n) = -\frac{C_g}{C_\Sigma}\left[C_{\mathrm{R}} V_{\mathrm{R}} + C_{\mathrm{L}} V_{\mathrm{L}} - V_g(C_{\mathrm{L}} + C_{\mathrm{R}}) + ne\right] \tag{31a}$$
$$Q_{\mathrm{L}}(n) = -\frac{C_{\mathrm{L}}}{C_\Sigma}\left[C_{\mathrm{R}} V_{\mathrm{R}} + C_g V_g - V_{\mathrm{L}}(C_g + C_{\mathrm{R}}) + ne\right] \tag{31b}$$
$$Q_{\mathrm{R}}(n) = -\frac{C_{\mathrm{R}}}{C_\Sigma}\left[C_{\mathrm{L}} V_{\mathrm{L}} + C_g V_g - V_{\mathrm{R}}(C_g + C_{\mathrm{L}}) + ne\right] \tag{31c}$$

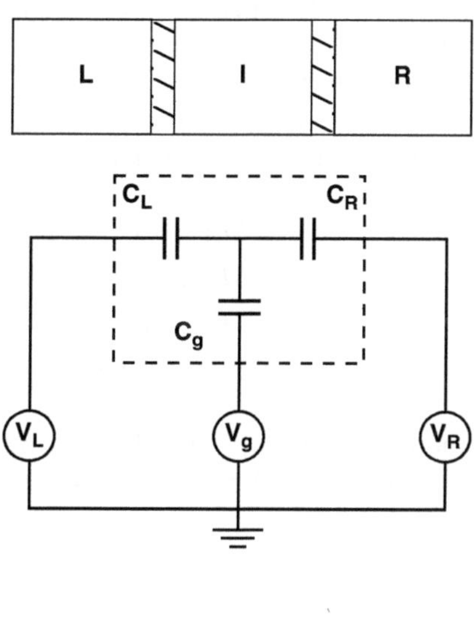

Fig. 3 Schematic of an SET. The top picture shows the left (L) and right (R) leads in contact with the island (I) via a tunneling junction, each. The bottom picture shows the electrical equivalent circuit, where V_{L} and V_{R} are the voltages applied to the left and right lead, respectively, while V_g is the gate voltage, that effectively alters the island charge

The total electrostatic energy of the system stored in the capacitors is given by

$$E^{(\mathrm{el})}(n) = \frac{Q_{\mathrm{L}}^2}{2C_{\mathrm{L}}} + \frac{Q_{\mathrm{R}}^2}{2C_{\mathrm{R}}} + \frac{Q_g^2}{2C_g} \tag{32}$$

Using the expressions Eqs. (31) one can verify immediately an interesting property of $E^{(el)}(n)$, namely that it does not contain any terms linear in n,

$$E^{(\mathrm{el})}(n) = \frac{(ne)^2}{2C_\Sigma} + \text{terms depending only on voltages.} \tag{33}$$

Consider now a process by which one electron tunnels from the left electrode to the island, accompanied by a re-arrangement of the charges so that electrostatic equilibrium is reached corresponding to a state with $n+1$ extra electrons on the island. The work done by the sources is then given by

$$\begin{aligned} W_{\mathrm{L}\to\mathrm{I}} = &\left[Q_{\mathrm{L}}(n+1) - Q_{\mathrm{L}}(n) - e\right] V_{\mathrm{L}} + V_{\mathrm{R}}\left[Q_{\mathrm{R}}(n+1) - Q_{\mathrm{R}}(n)\right] \\ &+ V_g\left[Q_g(n+1) - Q_g(n)\right]. \end{aligned} \tag{34}$$

Let us look a bit in slow motion at what is in fact happening here: as the electron tunnels through the left island, to ensure the neutrality of the conductor connecting the source V_{L} to the left island, another electron must be pulled through the source, which requires the energy $-eV_{\mathrm{L}}$. As the electron arrives on the island, the system is no longer in the electrostatic equilibrium ensured by Eqs. (30) and, to reach the new equilibrium with $n+1$ electrons on the island, the charges $Q_{\mathrm{L}}(n+1) - Q_{\mathrm{L}}(n)$, $Q_{\mathrm{R}}(n+1) - Q_{\mathrm{R}}(n)$, and $Q_g(n+1) - Q_g(n)$ are being transferred through the corresponding sources. The total energy change during this process, which includes the work done by the sources, is

$$\delta E_{\mathrm{L}\to\mathrm{I}} = W_{\mathrm{L}\to\mathrm{I}} + E^{(\mathrm{el})}(n) - E^{(\mathrm{el})}(n+1). \tag{35}$$

In other words, when an electron tunnels from a state k_{L} of the left electrode to a state q_I of the island, the energy is conserved (due to the Dirac delta appearing in the transition rate) in the following way: $\epsilon_{q_I} = \epsilon_{k_{\mathrm{L}}} + \delta E_{\mathrm{L}\to\mathrm{I}}$, or $\epsilon_{q_I} + E^{(\mathrm{el})}(n+1) = \epsilon_{k_{\mathrm{L}}} + W_{\mathrm{L}\to\mathrm{I}} + E^{(\mathrm{el})}(n)$. There are altogether four different possible tunneling events, two increasing and two decreasing the island charge:

$$\delta E_{\mathrm{L}\to\mathrm{I}}(n) = -eV_L - 2E_C\left(n + n_g + 1/2\right), \tag{36a}$$
$$\delta E_{\mathrm{I}\to\mathrm{L}}(n) = +eV_L + 2E_C\left(n + n_g - 1/2\right), \tag{36b}$$
$$\delta E_{\mathrm{I}\to\mathrm{R}}(n) = +eV_R + 2E_C\left(n + n_g - 1/2\right), \tag{36c}$$
$$\delta E_{\mathrm{R}\to\mathrm{I}}(n) = -eV_R - 2E_C\left(n + n_g + 1/2\right). \tag{36d}$$

Another handy way to get these changes in energy is to introduce the free energy of the system as the Legendre transform of the electrostatic energy Eq. (32)

$$E_{\text{ch}}(n) = E^{(\text{el})}(n) - Q_{\text{L}}(n)V_{\text{L}} - Q_{\text{R}}(n)V_{\text{R}} - Q_g(n)V_g. \tag{37}$$

Then, using Eqs. (31) we get (up to a constant)

$$E_{\text{ch}}(n) = \frac{1}{2C_\Sigma}\left(ne + q_g\right)^2, \tag{38}$$

and the expressions Eqs. (36) can be put in a physically more transparent form

$$\delta E_{\text{L}\to\text{I}}(n) = -eV_L + E_{\text{ch}}(n) - E_{\text{ch}}(n+1)\,, \tag{39a}$$
$$\delta E_{\text{I}\to\text{L}}(n) = +eV_L + E_{\text{ch}}(n) - E_{\text{ch}}(n-1)\,, \tag{39b}$$
$$\delta E_{\text{I}\to\text{R}}(n) = +eV_R + E_{\text{ch}}(n) - E_{\text{ch}}(n-1)\,, \tag{39c}$$
$$\delta E_{\text{R}\to\text{I}}(n) = -eV_R + E_{\text{ch}}(n) - E_{\text{ch}}(n+1)\,. \tag{39d}$$

We see that $E_{\text{ch}}(n)$ has the meaning of an effective charging energy of the island, including the work done by the sources.

The probability to tunnel onto or off the island through either the left or right junction can be obtained by inserting the respective energy difference of Eq. (36) into Eq. (13). We define the probabilities for the four different tunneling events as function of the excess charge n on the island as

$$\Gamma_{\text{L}\to\text{I}}(n) \equiv \Gamma[\delta E_{\text{L}\to\text{I}}(n)] \tag{40a}$$
$$\Gamma_{\text{I}\to\text{L}}(n) \equiv \Gamma[\delta E_{\text{I}\to\text{L}}(n)] \tag{40b}$$
$$\Gamma_{\text{I}\to\text{R}}(n) \equiv \Gamma[\delta E_{\text{I}\to\text{R}}(n)] \tag{40c}$$
$$\Gamma_{\text{R}\to\text{I}}(n) \equiv \Gamma[\delta E_{\text{R}\to\text{I}}(n)]\,. \tag{40d}$$

To continue our analysis, we now notice that, due to the fact that the number of electrons on the island is a well-defined number, we can talk about classical states of the system and index them with the excess number of electrons on the island, n. There will then be transitions between these states, namely, with the notations in Eqs. (40), the probabilities of the island charge number n being raised or lowered by one become $\Gamma_\uparrow$ and $\Gamma_\downarrow$ respectively ,

$$\Gamma_\downarrow(n) = \Gamma_{I\to L}(n) + \Gamma_{I\to R}(n) \tag{41a}$$
$$\Gamma_\uparrow(n) = \Gamma_{\text{L}\to\text{I}}(n) + \Gamma_{\text{R}\to\text{I}}(n)\,. \tag{41b}$$

We now want to calculate the rate at which the probability $p(n;t)$ that the system is in state n changes. There are three contributions to this rate:

- If the system is already in the state n, it can make transitions to the state $n+1$ or to the state $n-1$, with transition rates $\Gamma_\uparrow(n)$ and $\Gamma_\downarrow(n)$;
- The system can get to the state n from state $n-1$ with transition rate $\Gamma_\uparrow(n-1)$;
- The system can get to the state n from state $n+1$ with transition rate $\Gamma_\downarrow(n+1)$.

This results in a master equation for $p(n; t)$,

$$\frac{dp(n;t)}{dt} = -[\Gamma_\uparrow(n)+\Gamma_\downarrow(n)]p(n;t)+\Gamma_\downarrow(n+1)p(n+1;t)+\Gamma_\uparrow(n-1)p(n-1;t). \tag{42}$$

This equation describes a classical Markovian process: indeed, embedded in this treatment is the idea that electrons do not have any memory while tunneling. The history of the electrons being at previous times in other states is erased and therefore there is no time-dependence in the transition rates Eqs. (41), which depend only on the "present" state n.

We now want to characterize the stationary state, defined by the condition $dp(n;t)/dt = 0$. From Eq. (42) we can verify immediately that the stationary state is given by a time-independent $p(n)$ satisfying

$$p(n)\Gamma_\uparrow(n) = p(n+1)\Gamma_\downarrow(n+1), \tag{43a}$$

which is equivalent to

$$p(n)\Gamma_\downarrow(n) = p(n-1)\Gamma_\uparrow(n-1), \tag{43b}$$

Using now the normalization condition for probabilities,

$$\sum_{n=-\infty}^{\infty} p(n) = 1, \tag{44}$$

we find

$$p(n) = \frac{\left[\prod_{i=-\infty}^{n-1} \Gamma_\uparrow(i)\right]\left[\prod_{i=n+1}^{\infty} \Gamma_\downarrow(i)\right]}{\sum_{m=-\infty}^{\infty}\left(\left[\prod_{i=-\infty}^{m-1} \Gamma_\uparrow(i)\right]\left[\prod_{i=m+1}^{\infty} \Gamma_\downarrow(i)\right]\right)}. \tag{45}$$

From Eqs. (43) another useful property results immediately,

$$\sum_{n=-\infty}^{\infty} p(n)\Gamma_\uparrow(n) = \sum_{n=-\infty}^{\infty} p(n)\Gamma_\downarrow(n) \tag{46}$$

Using now Eq. (46) and the definitions in Eqs. (41) it is easy to check that there is no charge accumulation on the island, that is, the currents flowing through the left junction and the right junction are equal,

$$I = I_{\mathrm{L}\to\mathrm{I}} = I_{I\to L} = -e \sum_{n=-\infty}^{\infty} p(n)\,[\Gamma_{\mathrm{L}\to\mathrm{I}}(n) - \Gamma_{I\to L}(n)] \tag{47}$$
$$= -e \sum_{n=-\infty}^{\infty} p(n)\,[\Gamma_{I\to R}(n) - \Gamma_{\mathrm{R}\to\mathrm{I}}(n)]$$

This formula is used in the following to calculate the currents and the conductivities.

We point out here that the simplifications leading to Eqs. (44–47) are valid only under the assumption of steady-state conditions. In cases in which biasing is done by RF fields - such as turnstile operation or RF cooling [22] - this assumption could break and the treatment above is insufficient.

3.2 Coulomb Blockade Thermometry

A very useful application of double junction structures with normal-metal electrodes (NININ) is for temperature measurements. Consider such a structure, with charging energy of the island denoted as before by $E_C = e^2/2C_\Sigma$ and at temperature (assuming equilibrium) T. Clear Coulomb-blockade effects can be seen if the temperature is much smaller than the charging energy: the current is zero for voltages lower then e/C_Σ, as it takes that much energy to add an electron on the island. Plateaus are then formed at values of the bias voltage corresponding to multiples of e/C (see Fig. 4 left). The conductance curves then show so-called Coulomb-blockade oscillations (see Fig. 4) right). Gate voltages can be also included in this analysis, in which case it can be shown that, in the gate voltage – bias voltage space the stable states (corresponding to different numbers of electrons on the island) form a diamond pattern [23].

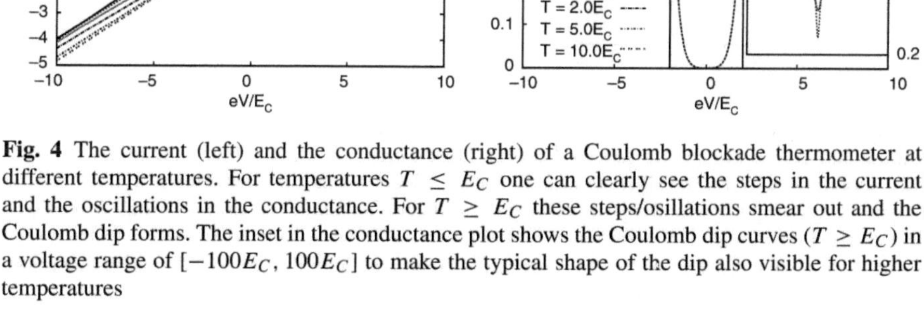

Fig. 4 The current (left) and the conductance (right) of a Coulomb blockade thermometer at different temperatures. For temperatures $T \leq E_C$ one can clearly see the steps in the current and the oscillations in the conductance. For $T \geq E_C$ these steps/osillations smear out and the Coulomb dip forms. The inset in the conductance plot shows the Coulomb dip curves ($T \geq E_C$) in a voltage range of $[-100E_C, 100E_C]$ to make the typical shape of the dip also visible for higher temperatures

We now examine the effect of temperature: intuitively, as the temperature is increased, one expects that the features visible at low temperatures will be smeared out and eventually washed away. Surprisingly however, if the temperature is larger than the charging energy, one clear feature survives in the conductance, namely a dip at low bias voltages. It can be shown [6], within the orthodox theory presented above, that, in the limit $k_B T \gg E_C$, the full width of this Coulomb blockade dip (measured half-way between the dip in the conductance at zero bias and the plateau at $V \to \pm\infty$) is given by

$$\frac{eV_{1/2}}{2k_B T} \approx 5.439, \tag{48}$$

and the relative change of conductance at zero bias voltage is

$$\frac{G(V \to \pm\infty) - G(V = 0)}{G(V \to \pm\infty)} = \frac{E_C}{3K_B T}. \tag{49}$$

It is straigthforward to show that $G(V \to \pm\infty) = (2R)^{-1}$, meaning that at large bias voltage the charging energy of the island does not have any effect and the conductance of the NININ array becomes that of two junctions of resistance R in series. Eq. (48) enables the use of such structures as primary thermometers: from a simple measurement of conductance it is possible to extract $V_{1/2}$ without any need for calibration. Eq. (49) can be used to determine E_C.

What happens now if the electrodes become superconducting? In Fig. 5 left we present the behavior of the Coulomb blockade dip for a SINIS structure as the gap of the electrodes is increased. One can see how the superconducting gap is introducing additional features into the conductance curve which, as the gap is gradually

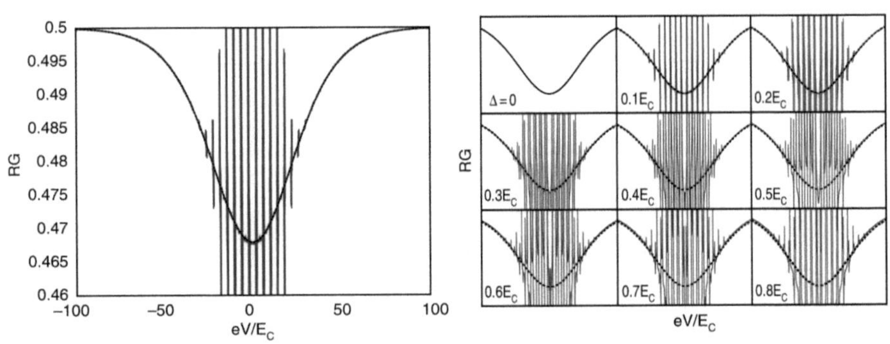

Fig. 5 *Left* The conductance dip in a superconducting CBT at a temperature of $T = 5E_C$, with a finite gap $\Delta_I = \Delta_L = 0.1\ E_C$ (*red color*). For comparison we overlayed the graph with the corresponding curve for the all-metallic case (*black line*). The CB dip is still clearly visible, but the features due to superconductivity are clearly imprinted on the curve, according to Eq. (51). *Right*: The conductance dip in a superconducting CBT, with a growing superconducting gap from 0 to 0.8 E_C. Here we can see how the dip is gradually smeared out by the features that arise due to the superconducting nature of the island and leads

increased, finally distort its shape so much that the CBT dip is very soon not distinguishable anymore. The origin of these features is described in detail in Sect. 3.4.

3.3 SINIS

A superconductor-insulator-normal metal-insulator-superconducting (SINIS) device can be seen as an SET without gate, with superconducting leads and a normal metal island. SINIS are very useful devices, used as local coolers [7], secondary thermometers [24], and electron pumps [25]. Usually the charging energy in a SINIS is so small that it can be neglected. However, in some applications, for instance as thermometers, the SINIS dimensions may become small enough to make charging effects become observable.

Figure 6 shows the I–V-curves and conductances of a SINIS structure at different operating temperatures for both zero and finite charging energy. Without charging energy, the SINIS reproduces the I–V curve of a single NIS junction, with all features being at twice the voltage, as we expect due to the fact that only half of

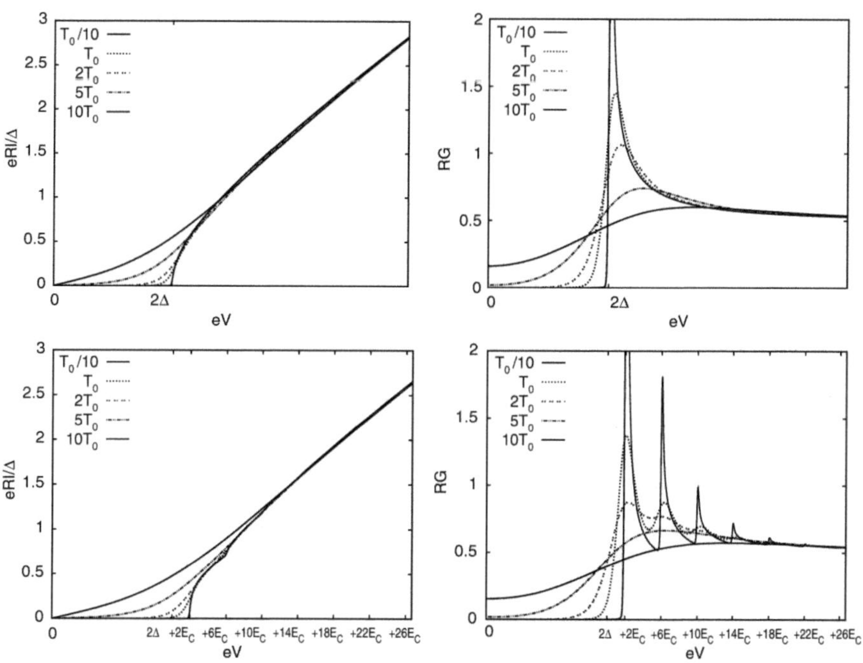

Fig. 6 *Top*: Current (*left*) and conductance (*right*) of a SINIS structure as function of applied voltage at different operating temperatures. The charging energy of the system is set to zero. Bottom: Current (*left*) and conductance (*right*) through a SINIS structure with finite charging energy ($E_C = 0.15\ \Delta$). The offset of the quasiparticle threshold is $2E_C$ and the periodicity of its repetition is 4 E_C. In all plots the temperature is $k_B T_0 = 3.92 \times 10^{-2} \Delta$

the voltage applied to the SINIS drops over a single junction. With a finite charging energy, the quasiparticle threshold is pushed up from $V = 2\Delta/e$ to $(2\Delta + 2E_C)/e$, and is repeated at $4E_C$ intervals with decreasing amplitude. This is visible especially well in the conductance plots in Fig. 6. At operating temperatures that are comparable to or larger than the charging energy, all features are smeared out. In Fig. 6 (for the case that $\Delta = 220\,\mu$ eV and thus $E_C = 33\,\mu$ eV), the curve with still barely visible oscillations conductance would correspond to a temperature of $2k_B T_0 \approx 17\mu$ eV, while in the curve at the next higher temperature of $5\,k_B T_0 \approx 43\,\mu$ eV the oscillations are smeared out entirely.

3.4 Superconducting SETs

When also the island becomes superconducting, we differentiate between two cases, the symmetric SSET, where the gaps (denoted by Δ) of both leads and island are identical, and the asymmetric SSET, where the gap in the island differs from the gaps in the leads (denoted by Δ_{I} and Δ_{L}, respectively). For simplicity of the further discussion, we introduce two new notations, namely the sum of the lead and island gaps,

$$\Delta_\Sigma \equiv \Delta_{\mathrm{L}} + \Delta_{\mathrm{I}}\,, \tag{50a}$$

and the difference between the gaps,

$$\Delta_\delta \equiv |\Delta_{\mathrm{L}} - \Delta_{\mathrm{I}}|\,. \tag{50b}$$

Figure 7 shows the I–V characteristics of an asymmetric SET with $\Delta_{\mathrm{I}} = 0.2\Delta_{\mathrm{L}}$ and $E_C = 0$ for different temperatures. Compared to the I–V of a SINIS struc-

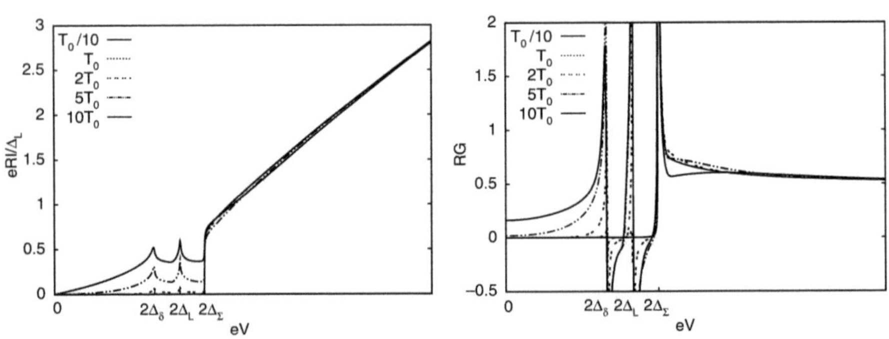

Fig. 7 Current (*left*) and conductance (*right*) of an asymmetric SET as function of applied voltage at different operating temperatures. The charging energy of the system is set to zero, $k_B T_0 = 3.92 \times 10^{-2} \Delta_{\mathrm{L}}$, and $\Delta_{\mathrm{I}} = 0.15\Delta_{\mathrm{L}}$, where the subscripts I and L denote island and lead, respectively

ture, we can see the appearance of a new feature, the singularity-matching peak, which apears at voltages $V^{(sm)} = \pm 2\Delta_\delta/e$. As mentioned above, these peaks appear because of the overlap of two infinite density of states appearing under the integral giving the tunneling rates. Singularity-matching peaks in superconducting single-electron transistors have been investigated theoretically previously in [26] and first observed experimentally by [27] and [28]. In NbAlNb structures subgap features at $2(\Delta_{\mathrm{Nb}} - \Delta_{\mathrm{Al}})$ have been observed in [29]. Also in this figure we see the usual feature due to quasiparticle threshold at $V = \pm 2\Delta_\Sigma/e$. Finally, a third peak appears at $V = 2\Delta_{\mathrm{L}}/e$, where the upper and lower singularity in the density of states of the left and right lead, respectively, are aligned. The right plot in Fig. 7 shows the conductance for the same values. The conductance plot is perhaps not as descriptive as the one of a SINIS, but it still serves two purposes: (1) it emphasizes very well all sharp features and thus makes it easy to determine their position and, (2) it allows to easily distinguish between singularity matching peak- and quasiparticle threshold-like features, as the latter only produces an upward spike, while the former produces a double spike, with the second spike going downwards. This is especially useful when discussing SSETs with charging energy, where one can find a large amount of weak features.

At finite charging energies, we see a similar repetition of features as one sees in SINIS I–Vs with $E_C > 0$. These features are a combination of the usual coulomb blockade features in normal metal SETs and a series of peaks and steps. The origin of the peaks are the singularity matching peaks in the tunneling rates (13). Their positions are easily calculated by equating the energy differences of Eq. (36) with the positions of the singularity matching peaks $\pm\Delta_\delta$, and solving for the applied voltage.

To get a feeling of how these features enter the I–V-curve, we first look at the special case of two identical junctions, where $C_L = C_R = C_\Sigma/2$. In this setup, we have that $V_L = -V_R = V/2$. For further simplification, we also set the gate charge equal to zero, *i.e.* $n_g = 0$. If we now for instance look at the series of peaks that correspond to $\delta E_{L\to I}$, we find $\pm\Delta_\delta = eV^{(\mathrm{sm})}/2 - 2E_C(n + 1/2)$, and therefore $eV^{(\mathrm{sm})} = \pm 2\Delta_\delta + 2E_C(2n + 1)$.

Altogether we find that the bias voltages at which peaks occur (at $n_g = 0$) are

$$V_n^{(\mathrm{sm})} = \pm\frac{2\Delta_\delta}{e} \pm \frac{2E_C}{e}(2n + 1)\,. \tag{51}$$

Similarly, the I–V shows a series of positive and negative steps. These steps originate from the quasiparticle threshold in the tunneling rates (13). We can find them in a similar manner as the peaks, by equating Eqs. (36) with $\pm\Delta_\Sigma$ and find

$$V_n^{(\mathrm{qp})} = \pm\frac{2\Delta_\Sigma}{e} \pm \frac{2E_C}{e}(2n + 1)\,. \tag{52}$$

Equation (52) corresponds to positive steps for $+\Delta_\Sigma$, and to negative steps for $-\Delta_\Sigma$. The negative steps however, which are already small in the tunneling rates, are difficult to observe in practice.

Figures 8, 9 and 10 show the I–V and conductance of SSETs with a finite charging energy of $E_C = 0.15\Delta_L$, with ratios between island and lead gaps of $\Delta_\mathrm{I}/\Delta_\mathrm{L} = 1, 0.5$ and 0.2, respectively. In all figures one can see the distribution of peaks and steps as described in Eqs. (51) and (52), with growing complexity. In the I–V curves, especially in Fig. 8, the features are difficult to distinguish. Therefore the conductance plots become a powerful tool in identifying the position of peaks and steps, and in distinguishing between them (single and double spikes in conductance correspond to steps and peaks, respectively, in current).

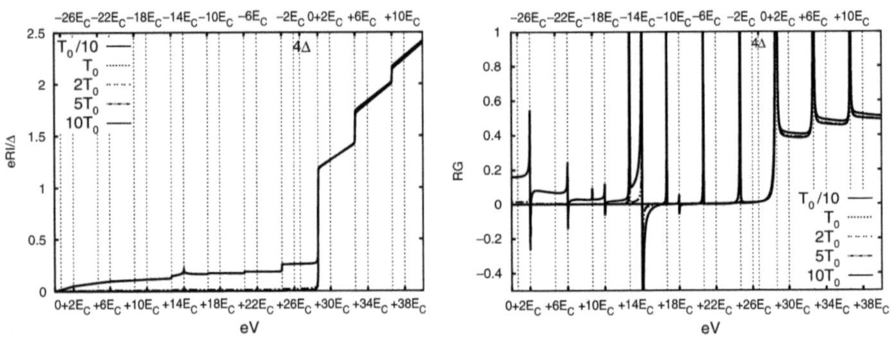

Fig. 8 Current (*left*) and conductance (*right*) of a symmetric SET as function of applied voltage at different operating temperatures. The gaps are chosen such that $\Delta_\mathrm{I} = \Delta_\mathrm{L} \equiv \Delta$, and the charging energy of the system is set to $E_C = 0.15\Delta$. As before, $k_B T_0 = 3.92 \times 10^{-2}\Delta$. In both plots the top most scale denotes offsets from the value of $eV = 4\Delta$

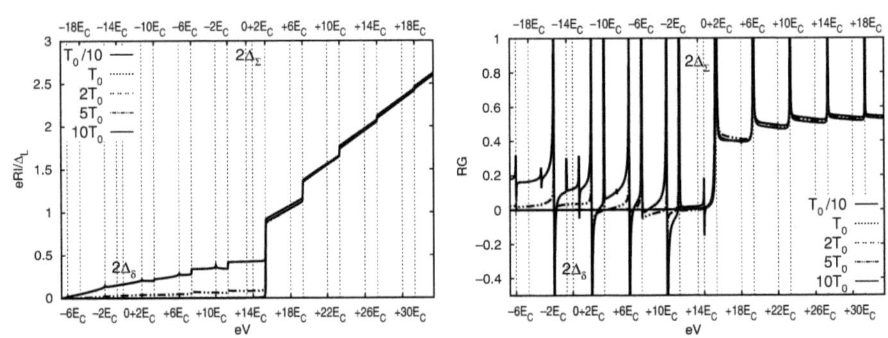

Fig. 9 Current (*left*) and conductance (*right*) of an asymmetric SET as function of applied voltage at different operating temperatures. The gaps are chosen such that $\Delta_\mathrm{I} = 0.5\Delta_\mathrm{L}$, and the charging energy of the system is set to $E_C = 0.15\Delta$. As before, $k_B T_0 = 3.92 \times 10^{-2}\Delta_\mathrm{L}$. In both plots the top most scale denotes offsets from the value of $eV = 2\Delta_\Sigma$, while the bottom most scale denotes offsets from the value of $eV = 2\Delta_\delta$. The three unmarked peaks correspond to $-2\Delta_\delta + 10E_C$, $14E_C$ and $18E_C$, respectively

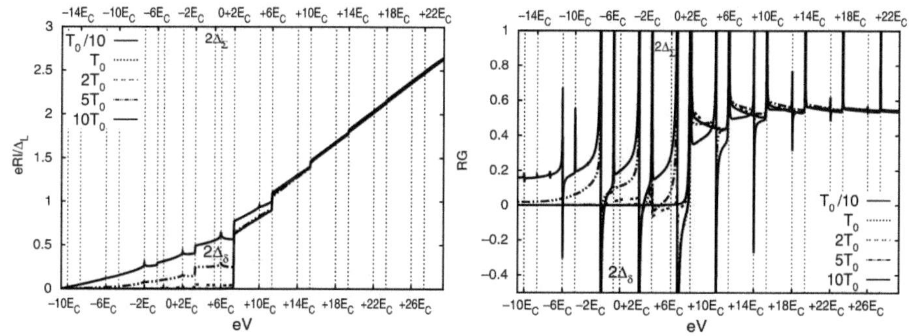

Fig. 10 Current (*left*) and conductance (*right*) of an asymmetric SET as function of applied voltage at different operating temperatures. The gaps are chosen such that $\Delta_{\rm I} = 0.2\Delta_{\rm L}$, the charging energy of the system is set to $E_C = 0.15\Delta$, and $k_B T_0 = 3.92 \times 10^{-2}\Delta_{\rm L}$. In both plots the top most scale denotes offsets from the value of $eV = 2\Delta_\Sigma$, while the bottom most scale denotes offsets from the value of $eV = 2\delta_\Delta$. The unmarke peaks correspond to $-2\Delta_\delta + 14E_C$

3.5 Effect of the Gate

To study how the gate charge affects the IV-curves of an SSET, it is best to do some simplifications again, to somewhat limit the amount of confusing details. In Fig. 11 we have reproduced the first figure of [27], where the IV of a symmetric SSET was studied. The figure shows a set of IV curves through the SSET as function of bias voltage for different values of the gate charge. As one can see, also here appears a series of peaks and steps, but their position is shifted as the gate charge is altered. Again the position of the features is easily calculated by equating the energy differences (for the most general case, use (39)) to $\pm\Delta_\delta$ for peaks, and $\pm\Delta_\Sigma$ for steps. As in this case $\Delta_\delta = 0$, the equation for the position of the peaks is somewhat simplified, as here the derived voltages will not depend on the superconducting gap. As in [27], the positions of the peaks in the system at hand (where possibly $C_{\rm L} \neq C_{\rm R}$) are given by

$$V^{\rm sm}_{\{\rm L,R\},n} = \frac{eC_{\{\rm L,R\}}}{C_{\rm L}C_{\rm R}}\left[\frac{1}{2} \pm \left(n + n'_g\right)\right]. \tag{53}$$

Here we defined $n'_g \equiv n_g + Q_0/e$, with en_g the gate charge and Q_0 the background charge of the island. Accordingly, the positions of the positive and negative steps are given by

$$V^{\rm qp}_{\{\rm L,R\},n} = \frac{eC_{\{\rm L,R\}}}{C_{\rm L}C_{\rm R}}\left[\pm\frac{2\Delta C_\Sigma}{e^2} + \frac{1}{2} \pm \left(n + n'_g\right)\right]. \tag{54}$$

Here the positive steps are obtained for the + sign in from of the fraction $2\Delta C_\Sigma/e^2$, and the negative steps are obtained for a − sign in front of the same fraction. The negative steps can be observed only at higher temperatures – see the comments on

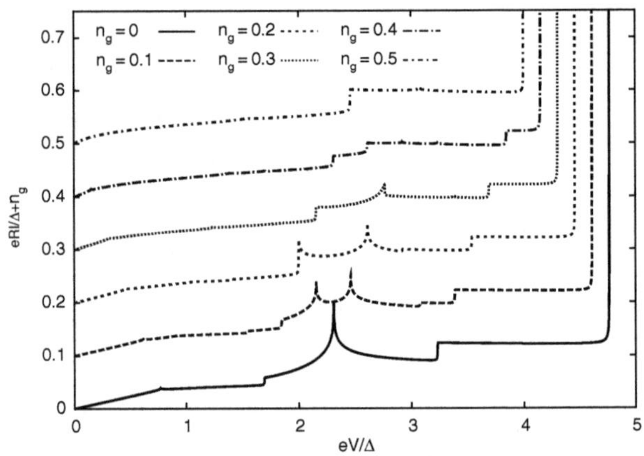

Fig. 11 The effect of the gate charge. Here $\Delta = 2.6E_C$ and $T = 0.8E_C$. The curves are offset by the respective gate charge for better distinguishability

the height of the quasiparticle threshold, Eq. (27) and (27); only the positive steps are visible in Fig. (11).

It is easy to verify that Eqs. (51) and (52) are another case of the expressions derived above Eqs. (53–54); these sets of equations become identical when using the same assumptions (identical junctions and $n'_g = 0$), and remembering that here $\Delta_\delta = 0$ and $\Delta_I = \Delta_L = \Delta_R = \Delta$.

3.6 Cooling

Tunneling of electrons through junctions results in energy being transferred between the electrodes, as the higher energetic electrons are extracted or pumped from or into one of the electrodes. In the case of two-junction structures such as SETs, SINIS, *etc.*, this results in the heating and cooling of the middle electrode (the island). This is due to the fact that, although the currents through the two junctions are the same, as it should be due to the conservation of charge (the charge of the island has to remain constant) the energy transfer (heat) does not need to satisfy such a conservation law.

It is straightforward to calculate the energy transferred per unit time out of the island (cooling power) as a function of the applied voltage,

$$P_{\mathrm{I}}(V) = \sum_{n=-\infty}^{\infty} p(n)\left[P_{\mathrm{I}\to\mathrm{L}}(n) - P_{\mathrm{L}\to\mathrm{I}}(n) + P_{\mathrm{I}\to\mathrm{R}}(n) - P_{\mathrm{R}\to\mathrm{I}}(n)\right] . \tag{55}$$

Here the powers $P_{\mathrm{L}\to\mathrm{I}}(n)$, $P_{\mathrm{I}\to\mathrm{L}}(n)$, $P_{\mathrm{R}\to\mathrm{I}}(n)$, and $P_{\mathrm{I}\to\mathrm{R}}(n)$ correspond to energy being transferred onto/off the island by tunneling into the left and right electrodes, at a given number of excess electrons n on the island. They can be calculated by multiplying the energy carried by each tunneling electron to the probability of tunneling per unit time (given by the density of states and Fermi factors), and then summing over energy states (see also Eqs. (13a), (13b)),

$$P_{\mathrm{I}\to\mathrm{X}}(n) = \frac{1}{e^2 R_{IX}} \int_{-\infty}^{\infty} \mathrm{d}E \;\; E N_{\mathrm{I}}(E) N_{\mathrm{X}}[E + \delta E_{\mathrm{I}\to\mathrm{X}}(n)] f_{\mathrm{I}}(E) \\ \left[1 - f_{\mathrm{X}}(E + \delta E_{\mathrm{I}\to\mathrm{X}}(n))\right] \tag{56a}$$

$$P_{\mathrm{X}\to\mathrm{I}}(n) = \frac{1}{e^2 R_{XI}} \int_{-\infty}^{\infty} \mathrm{d}E \;\; [E + \delta E_{\mathrm{X}\to\mathrm{I}}(n)]\, N_{\mathrm{I}}\,[E + \delta E_{\mathrm{X}\to\mathrm{I}}(n)]\, N_{\mathrm{X}}(E) \\ \left[1 - f_{\mathrm{I}}(E + \delta E_{\mathrm{X}\to\mathrm{I}}(n))\right] f_{\mathrm{X}}(E)\,, \tag{56b}$$

where X = L, R.

In Fig. 12 we present a comparison between the cooling powers on the island Eq.(55) for SINIS (upper plots) and superconducting SET structures (lower plots). We observe that cooling exists only in an interval of bias voltages not much

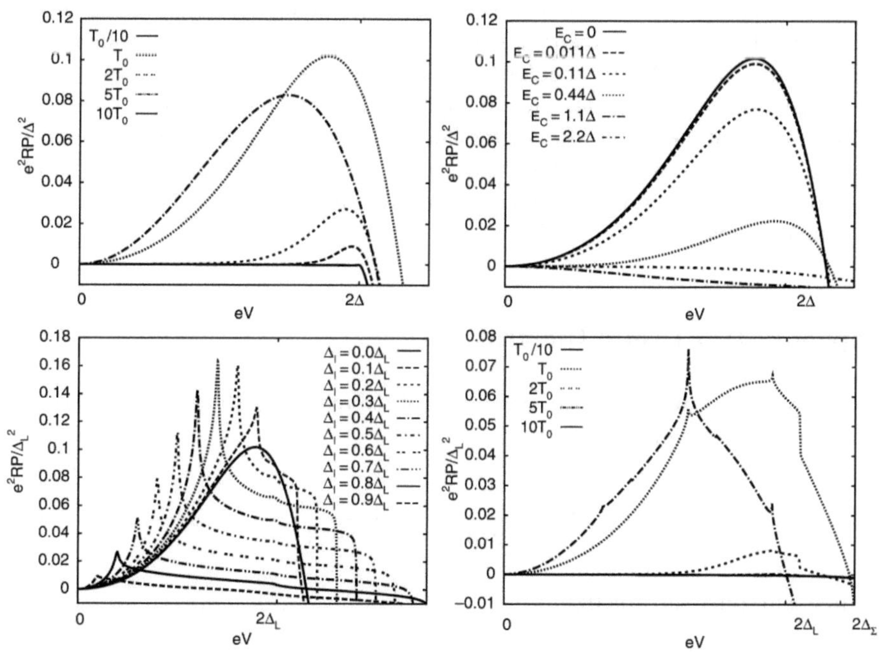

Fig. 12 Cooling powers in different double-island devices. Top left: SINIS structure with $E_C = 0$ at different temperatures. Top right: SINIS at fixed temperature $5T_0$ with different charging energies. Bottom left: SET with $5T_0$ and $E_C = 0$ for different gap sizes $\Delta_{\mathrm{I}} < \Delta_{\mathrm{L}}$ of the island. Bottom right: SET with $\Delta_{\mathrm{I}} = 0.2\Delta_{\mathrm{L}}$ and $E_C = 0.15\Delta_{\mathrm{L}}$ at different temperatures. In all plots $k_B T_0 = 3.92 \times 10^{-2}\Delta_{\mathrm{L}}$

larger than the quasiparticle threshold at $2\Delta/e$ (we take identical superconducting leads with gap Δ). As seen from the left upper plot, as the temperature is lowered the process becomes much less efficient. The general effect of the charging energy is that it is detrimental to cooling (right plot, upper part). In the case of superconducting SET's (lower plots, Fig. 12) the existence of singularity-matching peaks produces relatively sharp spikes in the cooling power. Essentially, the island is cooled down by BCS quasiparticles [30]. The left figure shows these features for the case of negligible charging energy (a SISIS structure with a relatively large middle electrode having a different gap than the leads). An interesting feature is also that the range of voltages over which cooling occurs is extended, due to the fact that now the quasiparticle threshold is at values $2\Delta_{\mathrm{L}}+2\Delta_{\mathrm{I}}$. Finally, cooling in SET structures with finite charging energy is shown in the lower-right plot of Fig. 12.

4 Conclusions

We have presented the theory of tunneling in metallic and superconducting two-junction arrays such as single electron transistors, together with a number of applications. All three energy scales, the charging energy, the superconducting gap, and the temperature, are considered and their role is thoroughly discussed. For example, we examined how a finite superconducting gap affects the Coulomb-blockade based thermometry, the effect of singularity-matching peaks in the current-voltage and conductance-voltage characteristics of superconducting single-electron transistors, and we looked at the effect of charging energy in cooling devices. With the development of the field of nanotechnology, such devices could emerge as very useful tools for high-precision measurements of nano-structured materials and objects at low temperatures.

Appendix 1: A Useful Integral

For completeness, we give a full derivation of the result Eq. (6) (see e.g. [11]). The Fermi functions are given by

$$f(\epsilon) = \frac{1}{\exp(-\epsilon/k_B T) + 1}. \tag{57}$$

We now introduce the function

$$g(x) = \int_{-\infty}^{\infty} \epsilon\left[f(\epsilon) - f(\epsilon + x)\right] \tag{58}$$

Obviously $g(0) = 0$. The first derivative of g gives

$$\frac{dg(x)}{dx} = -\int_{-\infty}^{\infty} d(\epsilon + x)\frac{df(\epsilon + x)}{d(\epsilon + x)} = -f(\infty) + f(-\infty) = 1. \tag{59}$$

This means that $g(x) = x$, so we have the result

$$\int_{-\infty}^{\infty} d\epsilon \left[f(\epsilon) - f(\epsilon + x)\right] = x\,. \tag{60}$$

Now, integrals over expressions of the type $f(E)[1 - f(E + x)]$, which appear in the theory of tunneling due to Pauli exclusion principle, can be solved by noticing that

$$f(\epsilon)[1 - f(\epsilon + x)] = \frac{f(\epsilon) - f(\epsilon + x)}{1 - \exp(-\beta x)}, \tag{61}$$

where $\beta = 1/k_B T$. Now, using Eq. (60) we get immediately

$$\int_{-\infty}^{\infty} d\epsilon f(\epsilon)[1 - f(\epsilon + x)] = \frac{x}{1 - \exp(-\beta x)}. \tag{62}$$

Appendix 2: Derivation of Some Integral Expansions Used

Generic procedure: In the following we want to derive the expansion of Eq. (22). To get a feeling for the procedure, we will first consider a somewhat easier and more common integral, namely

$$\Upsilon(c) = \int_0^{\infty} \mathrm{d}x \, \frac{1}{x + c} h(x)\,, \tag{63}$$

which we assume to not be solvable analytically. For the function $h(x)$ we assume that it is well behaved for all $x \in [0, \infty)$ and that $h(x)$ behaves as $\propto \exp(-x)$ for large x. However, $\Upsilon(c)$ is logarithmically divergent for $c = 0$. We can separate $\Upsilon(c)$ into a divergent and an non-divergent part by performing a partial integration,

$$\Upsilon(c) = \ln(1/c)h(0) - \int_0^{\infty} \mathrm{d}x \, \ln(x + c)h'(x)\,. \tag{64}$$

Here the first term is an explicit expression that describes the divergence around $c = 0$ and the second term is a finite integral that needs to be computed numerically (we use the notation $h'(x) \equiv dh(x)/dx$). Unfortunately, the function over the integral in the second term of Eq. (64) has no Taylor expansion in $x + c \to 0$, making any further analysis of the integral difficult. It is however possible to find an expansion by doing repeated partial integration. If we define the function $F_n(x)$ such that

$$\frac{d^n}{dx^x} F_n(x) = \ln(x)\,, \tag{65}$$

it is easy to show that

$$F_n(x) = \frac{1}{n!} x^n [\ln(x) - H_n]\,, \tag{66}$$

where $H_n = \sum_{k=1}^{n} 1/k$ is the harmonic number (we could actually still add an arbitrary polynomial of nth order to F_n, but due to the repeated partial integrations all its coefficients cancel out and we therefore set them equal zero). These functions have the properties $F_0(x) = \ln(x)$, $F_n(x = 0) = 0$, $(\forall)\ n > 0$ and $F_{n\to\infty}(x) \to 0$, $(\forall)\ x > 0$. With this we can write $\Upsilon(c)$ as

$$\Upsilon(c) = \ln(1/c)h(0) - \sum_{n=1}^{\infty} (-1)^n \frac{1}{n!} c^n [\ln(c) - H_n] h^{(n)}(0)\,, \tag{67}$$

where by $h^{(n)}(x)$ we denote the nth derivative of $h(x)$, $d^n h(x)/dx^n$.

Application: In Sect. 2, we have encountered slightly more complicated integrals, which were of the form

$$\Upsilon(c) = \int_0^{\infty} \mathrm{d}x\ \frac{1}{\sqrt{x(x+c)}} h(x)\,, \tag{68}$$

with a logarithmic singularity in $c = 0$. For this reason we would like to expand $\Upsilon'(c)$ after the recipe that was used when expanding the previous integral, Eq. (63). The only complication we encounter is that we need to find a function $G_n(x)$ with $dG_0(x)/dx = 1/\sqrt{x(x+a)}$ and

$$\frac{d^n}{dx^n} G_n(x) = G_0(x)\,. \tag{69}$$

It is easy to verify that the function G_0 is given by

$$G_0(x) = 2\ln(\sqrt{x} + \sqrt{x+c}). \tag{70}$$

The general function G_n can be found with the trial solution

$$G_n(x) = P_n(x)\ln(\sqrt{x} + \sqrt{x+c}) + Q_n(x)\sqrt{x(x+c)}\,, \tag{71}$$

where $P_n(x)$ and $Q_n(x)$ are polynomials. By inserting Eq. (71) into Eq. (69), we find that P_n and Q_n must fulfill the relations

$$\frac{dP_n(x)}{dx} = P_{n-1}(x)\,, \tag{72}$$

and

$$P_n(x) + (2x+c)Q_n(x) + 2x(x+c)\frac{dQ_n(x)}{dx} = 2x(x+c)Q_{n-1}(x)\,. \tag{73}$$

These relations can finally shown to be fulfilled if

$$P_n(x) = \sum_{k=0}^{n} \frac{1}{k!}\alpha_{n-k}x^k \tag{74a}$$

$$Q_n(x) = \sum_{k=0}^{n-1} \beta_k^n x^k\,, \tag{74b}$$

where

$$\alpha_n = \frac{2}{4^n}\frac{(2n)!}{(n!)^3}c^n \tag{75a}$$

$$\beta_k^n = \sum_{l=0}^{k} \left(\frac{-1}{c}\right)^{l+1} \frac{4^l(l!)^2}{(2l+1)!(k-l)!}\alpha_{n-k+l}\,. \tag{75b}$$

Although these polynomials look somewhat complicated, only the constant term in $P_n(x)$ will enter the expansion of $\Upsilon(c)$,

$$\begin{aligned}\Upsilon(c) &= -\sum_{n=0}^{\infty} G_n(0)h^{(n)}(0)\\ &= -\sum_{n=0}^{\infty} \alpha_n \ln(\sqrt{c})h^{(n)}(0)\\ &= \ln\left(\frac{1}{c}\right)h(0) - \sum_{n=1}^{\infty} \frac{(2n)!}{4^n(n!)^3}c^n \ln(c)h^{(n)}(0),\end{aligned} \tag{76}$$

where the first term of Eq. (76) contains the convergence for $c \to 0$.

The convergence of expression (76) is somewhat tricky to show as it depends on the values of the function $h(x)$ and its derivatives at $x = 0$. With the ratio test of convergence for infinite series we find, writing Eq. (76) as $\sum_n a_n$, that

$$\begin{aligned} \left|\frac{a_{n+1}}{a_n}\right| &= \frac{(2n+1)(2n+2)}{4(n+1)^3} c \frac{h^{(n+1)}(0)}{h^{(n)}(0)} \\ &\to \frac{c}{n}\left(\frac{d}{dx}\ln(h^{(n)}(x))\right)_{x=0} \quad \text{for} \quad n\to\infty\,, \end{aligned} \tag{77}$$

which is smaller than one if $(d/dx)\ln(h^{(n)}(x))|_{x=0}$ increases slower than n. For example, if $h(x) = \exp(-ax)$, we would have $|a_{n+1}/a_n| \to ac/n$, which tends to zero for large n, and thus the series would be convergent.

Acknowledgements We would like to thank J. P. Pekola and I. Maasilta for useful comments on the manuscript. T. K. would like to acknowledge financial support from the Emil Aaltonen foundation. The contribution of G.S.P. was supported by the Academy of Finland (Acad. Res. Fellowship 00857, and projects 129896, 118122, and 135135).

References

1. D.V. Averin, K.K. Likharev, J. Low Temp. Phys. **62**, 345 (1986); T.A. Fulton, G.J. Dolan, Phys. Rev. Lett. **59**, 109 (1987)
2. R.J. Schoelkopf, P. Wahlgren, A.A. Kozhevnikov, P. Delsing, D.E. Prober, Science **280**, 1238 (1998); M.A. Sillanpä; L. Roschier, P.J. Hakonen, Phys. Rev. Lett. **93**, 066805 (2004)
3. M.H. Devoret, R.J. Schoelkopf, Nature **406**, 1039 (2000)
4. Y. Nakamura, Y.A. Pashkin, J.S. Tsai, Nature, **398**, 786 (1999); Yu. Makhlin, G. Schön, A. Shnirman, Rev. Mod. Phys. **73**, 357 (2001); D. Vion et. al., Science **296**, 886 (2002); Y. Nakamura, Yu. A. Paskin, T. Yamamoto, J.S. Tsai, Phys. Rev. Lett. **88**, 047901 (2002); Yu. A. Paskin et. al., Nature **421**, 823 (2003); T. Yamamoto et. al., Nature **425**, 941 (2003); T. Duty, D. Gunnarsson, K. Bladh, P. Delsing, Phys. Rev. B **69** 132504 (2004); J.Q. You F. Nori, Physics Today **58**, 42 (2005); G.S. Paraoanu, Phys. Rev. B **74**, 140504(R) (2006). G.S. Paraoanu, Phys. Rev. Lett. **97**, 180406 (2006); J. Li, K. Chalapat, G.S. Paraoanu, Phys. Rev. B **78**, 064503 (2008); J. Li G.S. Paraoanu, New J. Phys. **11**, 113020 (2009)
5. G.S. Paraoanu, A.M. Halvari, Appl. Phys. Lett. **86**, 093101 (2005); T.F. Li et. al., Appl. Phys. Lett. **91**, 033107 (2007)
6. J.P. Pekola, K.P. Hirvi, J.P. Kauppinen, M.A. Paalanen, Phys. Rev. Lett. **73**, 2903 (1994); K.P. Hirvi, J.P. Kauppinen, A.N. Korotkov, M.A. Paalanen, J.P. Pekola, Appl. Phys. Lett. **67**, 2096 (1995); J.P. Kauppinen and J.P. Pekola, Phys. Rev. Lett. **77**, 3889 (1996); Sh. Farhangfar, K.P. Hirvi, J.P. Kauppinen, J.P. Pekola, J.J. Toppari, D.V. Averin, A.N. Korotkov, J. Low Temp. Phys. **108**, 191 (1997); J.P. Pekola, J.J. Toppari, J.P. Kauppinen, K.M. Kinnunen, A.J. Manninen, A.G.M. Jansen, J. Appl. Phys. **83**, 5582 (1998); J.P. Kauppinen, K.T. Loberg, A.J. Manninen, J.P. Pekola, R.V. Voutilainen, Rev. Sci. Instrum. **69**, 4166 (1998)
7. M.M. Leivo, J.P. Pekola, D.V. Averin, Appl. Phys. Lett. **68**, 1996 (1996); J.P. Pekola, F. Giazotto, O.-P. Saira, Phys. Rev. Lett. **98**, 037201 (2007); O.-P. Saira, M. Meschke, F. Giazotto, A.M. Savin, M. Mottonen, J.P. Pekola, Phys. Rev. Lett. 99, 027203 (2007)
8. A.B. Zorin, S.V. Lotkhov, H. Zangerle, J. Niemeyer, J. Appl. Phys. **88**, 2665 (2000); R. Dolata, H. Scherer, A.B. Zorin, V.A. Krupenin, J. Niemyer, Appl. Phys. Lett. **80** 2776 (2002); N. Kim et. al., Physica B **329–333**, 1519 (2003); G.S. Paraoanu, A. Halvari, Rev. Adv. Mater. Sci. **5**, 265 (2003); M. Watanabe, Y. Nakamura, J.-S. Tsai, Appl. Phys. Lett. **84** 410 (2004); A.M. Savin et. al., Appl. Phys. Lett. **91**, 063512 (2007)
9. J. Bardeen, Phys. Rev. Lett. **6** 57 (1961)
10. F. Giazotto, T.T. Heikkilä, A. Luukanen, A.M. Savin, J. Pekola, Rev. Mod. Phys. **78**, 217 (2006)

11. G.-L. Ingold, Yu.V. Nazarov, "Charge tunneling rates in ultrasmall junctions", in *Single Charge Tunneling*, edited by H. Grabert, M.H. Devoret (Plenum, New York, NY 1992), pp. 21–108
12. G. Schön, "Single-electron tunneling", in T. Dittrich, P. Hänggi, G. Ingold, B. Kramer, G. Schön, W. Zwerger, Quantum Transport and Dissipation (VCH Verlag 1997), Chapter 3.
13. W.A. Harrison, Phys. Rev. **123**, 85 (1961).
14. J.G. Simons, J. Appl. Phys. **34**, 1793 (1963); T.E. Hartman, J. Appl. Phys. **35**, 3283 (1964); W.F. Brinkman, R.C. Dynes, J.M. Rowell, J. Appl. Phys. **41**, 1915 (1970);
15. W.A. Hofer, A.S. Foster, A.S. Shluger Rev. Mod. Phys. **75**, 1287 (2003)
16. P. Törmä, P. Zoller, Phys. Rev. Lett. **85**, 487 (2000); Gh.-S. Paraoanu, M. Rodriguez, P. Törmä, J. Phys. B. **34**, 4763 (2001); Gh.-S. Paraoanu, M. Rodriguez, P. Törmä, Phys. Rev. A **66**, 041603 (2002); J. Kinnunen, M. Rodriguez, P. Törmä, Science **305**, 1131 (2004).
17. R.C. Dynes et. al., Phys. Rev. Lett. **53**, 2437 (1984); M. Kunchur et. al., Phys. Rev. B **36**, 4062 (1987); J.P. Pekola et. al., Phys. Rev. Lett. **92**, 056804 (2004)
18. J.J. Toppari, T. Kühn, A. P. Halvari, J. Kinnunen, M. Leskinen, G.S. Paraoanu, Phys. Rev. B **76**, 172505 (2007)
19. J.J. Toppari, T. Kühn, A.M. Halvari, G.S. Paraoanu, J. Phys.: Conf. Ser. **150**, 022088 (2009)
20. Y.T. Tan, T. Kamiya, Z.A.K. Durrani, H. Ahmed, J. Appl. Phys. 94, 633 (2003); K. Luo, D.-H. Chae1, Z. Yao, Nanotechnol. **18** 465203 (2007)
21. H. van Houten, C.W.J. Beenakker, A.A.M. Staring, "Coulomb-blockade oscillations in semiconductor nanostructures", in *Single Charge Tunneling*, ed by H. Grabert, M.H. Devoret (Plenum, New York, 1992), pp. 167–216
22. S. Kafanov, A. Kemppinen, Yu. A. Pashkin, M. Meschke, J.S. Tsai, J.P. Pekola , Phys. Rev. Lett. **103**, 120801 (2009)
23. D.V. Averin and K.K. Likharev, "Single electronics: a correlated transfer of single electrons and Cooper pairs in systems of small tunnel junctions", in B.L. Altshuler. P.A. Lee, R.A. Webb, *Mesoscopic phenomena in solids*, (Elsevier Science, New York, NY, 1991)
24. P.J. Koppinen, T. Kühn, I.J. Maasilta, J. Low. Temp. Phys. **154**, 179 (2009)
25. J.P. Pekola, J.J. Vartiainen, M. Möttönen, O.-P. Saira, M. Meschke, D.V. Averin, Nature Phys. **4**, 120 (2008)
26. A.N. Korotkov, Appl. Phys. Lett. **69** 2593 (1996)
27. Y. Nakamura, A.N. Korotkov, C.D. Chen, J.S. Tsai, Phys. Reb. B **56**, 5116 (1997)
28. A.J. Manninen, Yu. A. Pashkin, A.N. Korotkov, J.P. Pekola, Europhys. Lett. **39**, 305 (1997)
29. M.G. Blamire, E.C.G. Kirk, J.E. Evetts, T.M. Klapwijk, Phys. Rev. Lett. **66**, 220 (1991)
30. A.J. Manninen, J.K. Suoknuuti, M.M. Leivo, J.P. Pekola, Apl. Phys. Lett. **74** 3020 (1999)

Ballistic Transistors: From Planar to Cylindrical Nanowire Transistors

G.A. Nemnes, U. Wulf, L. Ion, and S. Antohe

Abstract Planar nanoscale transistors and cylindrical nanowire transistors are analyzed in the framework of coherent transport. The multi-terminal devices are modeled in the Landauer-Büttiker formalism, which is efficiently implemented using the R-matrix approach. For the planar case, the linear regime reveals Fabry-Perot type resonances in the source-drain conductance, while in the nonlinear regime a quantum mechanism for the drain current saturation is demonstrated. The advantages of new geometries, like the cylindrical nanowire transistors, are discussed in the context of self-consistent calculations.

1 Introduction

During the last decades the device scaling has been the major driving force of electronics. The Moore's law predicts that the number of transistors per chip doubles every 2 years. In industry, the technology node – as given by major companies like Intel and AMD – is currently moving from 45 to 32 nm for the leading-edge chip products. The major advantages of scaling down the transistor sizes include the almost exponential increase in processing speed and storage capacity. Computing performance per unit cost has increased accordingly. Still, despite these achievements, some new scaling challenges have appeared and need to be addressed, like increased leakage currents, doping fluctuations, proper transistor electrostatics in such small volumes.

Generally, at these system sizes, classical drift-diffusion models have limited success and quantum transport is expected to play an important role [1–4]. Therefore there is a great interest in the current research to develop appropriate models for nanoscale field-effect transistors (nanoFETs), able to give realistic pictures

G.A. Nemnes (✉)
University of Bucharest, Faculty of Physics, "Materials and Devices for Electronics and Optoelectronics" Research Center, P.O. Box MG-11, 077125 Magurele-Ilfov, Romania
e-mail: nemnes@solid.fizica.unibuc.ro

A. Aldea, V. Bârsan (eds.), *Trends in Nanophysics*, Engineering Materials,
DOI 10.1007/978-3-642-12070-1_6, © Springer-Verlag Berlin Heidelberg 2010

for the new designs. In the literature there are many geometries using a wide range of materials, like carbon nanotube field-effect transistors (CNFET) [5, 6], double gate field-effect-transistor (DGFET) [7], silicon-on-insulator metal-oxide-semiconductor field-effect-transistor (SOI MOSFET) [8], silicon nanowire transistor (SNWT) [9, 10] and, in the last few years, cylindrical nanowire transistors have become a reality [11, 12].

Amongst the proposed theoretical methods that have been used for the description of such devices one can mention the non-equilibrium Green's function formalism [14], recursive scattering matrices [15], a 2-D model for ballistic MOSFETs [16, 17], the WKB-approximation [18, 19], Monte-Carlo simulation [20] the Wigner function- [21], and the Master equation approach [22]. A limiting case of quantum transport at low temperatures is the ballistic regime, where the impurity interaction can be completely neglected and the transport is governed by coherent wavefunctions which extend through the entire device [23–26]. The Landauer-Buettiker formalism [1, 2, 27] has proved to be one of the most successful tool for describing ballistic transport, assuming the calculation of currents under different bias conditions, using transmission coefficients between the contacts. The latter are determined by solving the scattering problem associated with the structure potential, which is generally obtained by carrying out self-consistent calculations, so that the Coulomb interaction is taken into account.

An efficient way to obtain the transmission coefficients and the wavefunctions in the open quantum system is the R-matrix approach, which was first developed in the field of nuclear physics [28] and subsequently introduced in the semiconductor physics as well [29]. The main advantage from numerical point of view is that the total computational cost is divided into a first part, independent of the total energy, which is finding the solution of the Wigner-Eisenbud eigenvalue problem. In a second part, the transmissions and wavefunctions are calculated with a much smaller effort for each energy.

This review is organized as follows. In Sect. 2, the Landauer-Büttiker formalism is described for a multi-terminal device. Next, in Sect. 3, the advantages of the R-matrix approach are pointed out. The numerical results are presented and discussed for a planar nanoscale device in Sect. 4, in both linear and nonlinear regime for the source-drain bias. In the end, self-consistent potentials are calculated for a cylindrical nanowire transistor in Sect. 5 and comments are made upon the quantum screening and the linear regime conductance.

2 Landauer-Büttiker Formalism for a Multi-terminal Device

A general multi-terminal system is sketched in Fig. 1 [30]. The system consists of N macroscopic contact reservoirs R_s, $s = 1 \dots N$, a central quantum system (QS), and N connecting leads L_s. One of the reservoirs ($s = g$) is grounded and $N - 1$ biases U_{sg} are applied on the other terminals.

Following the arguments given by Büttiker [27], next to the reservoirs a relatively wide part Ω_s of the connecting lead L_s is assumed to be invariant along its length,

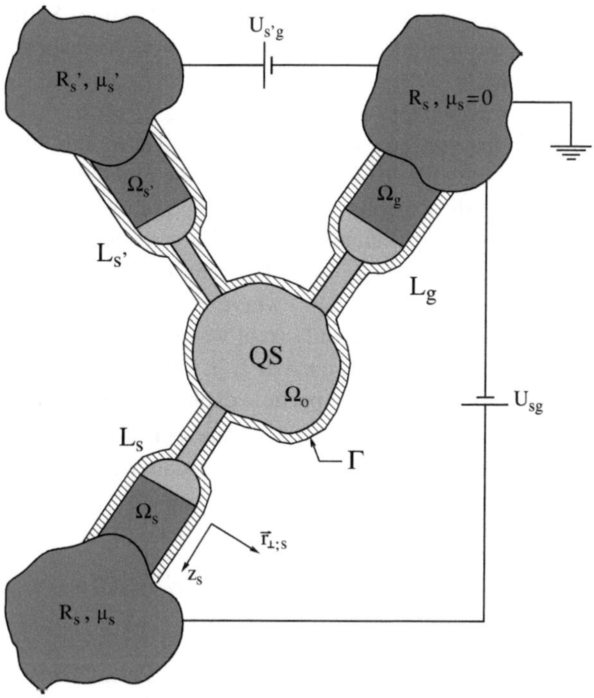

Fig. 1 Multi-terminal device in the Landauer-Büttiker formalism: the quantum system and the narrow part of the leads (Ω_0, *light gray*), the wide portion of the leads with separable Hamiltonian (Ω_s, *dark gray*) and the reservoirs (R_s, *dark gray*)

in which the Hamiltonian is separable. As illustrated in Fig. 1 in each Ω_s we define a local orthogonal coordinate system $(z_s, \boldsymbol{r}_{\perp;s})$ with z_s along the lead and pointing towards the reservoir. Furthermore, we assume that upon applying the biases U_{sg} the potential energy W in Ω_s varies by the same amount as in R_s and the confinement potential $w_s(\boldsymbol{r}_{\perp;s})$ remains the same: $W(\boldsymbol{r}; [U]) = w_s(\boldsymbol{r}_{\perp;s}) - eU_{sg}$, where $[U]$ constitutes the $N-1$-tupel of the applied biases.

The incoming particles (electrons, holes) from each of the semi-infinite leads are scattered on the device potential, obeying the stationary Schrödinger equation

$$\mathcal{H}\Psi(\boldsymbol{r}; E) = E\Psi(\boldsymbol{r}; E), \tag{1}$$

with the one-particle Hamiltonian

$$\mathcal{H} = -\frac{\hbar^2}{2m^*} \triangle + W(\boldsymbol{r}; [U]) \tag{2}$$

and scattering boundary conditions. Assuming the leads are translationally invariant, one can write the solution of Eq. (1) inside the lead Ω_s as:

$$\Psi_s(\boldsymbol{r} \in \Omega_s, E) = \sum_i \Psi_\nu^{in} \exp(-ik_\nu z_s)\Phi_\nu(\boldsymbol{r}_{\perp;s}) + \sum_i \Psi_\nu^{out} \exp(ik_\nu z_s)\Phi_\nu(\boldsymbol{r}_{\perp;s}), \tag{3}$$

where Ψ_{ν}^{in} and Ψ_{ν}^{out} are complex coefficients. Setting $\Psi_{\nu'}^{in} = 1/\sqrt{2\pi}\delta_{\nu\nu'}$, one obtains the scattering boundary condition corresponding to an incoming particle on channel ν.

The wavevectors along the leads are $k_{\nu} = \hbar^{-1}\sqrt{2m^*(E - E_{\perp}^{\nu} + eU_{sg})}$, where $E_{\perp}^{\nu}$ are the energies of the perpendicular modes Φ_{ν}. The composite index $\nu = (s, i)$ denotes the lead index (s) and the channel index (i). The latter are solutions of the two-dimensional Schrödinger equation in the perpendicular direction, $\boldsymbol{r}_{\perp;s}$. The complex coefficients Ψ_{ν}^{in}, $\Psi_{\nu}^{\mathrm{out}}$ are related by the scattering S-matrix, $\Psi^{\mathrm{out}} = \mathsf{S}\Psi^{\mathrm{in}}$. The total transmission between two leads is determined from the S-matrix: $T_{ss'}(E) = \sum_{i,i'} |\tilde{\mathsf{S}}_{\nu\nu'}(E)|^2$, where $\tilde{\mathsf{S}} = \mathsf{k}^{1/2}\mathsf{S}\mathsf{k}^{-1/2}$ and the summations are done over the open channels only (real wavevectors k_{ν}).

To calculate the total current I_s in terminal (i.e. lead) s we sum the current contributions coming from the particle exchange with all other terminals s', $I_s = \sum_{s'} I_s^{s'}$. Each of these contributions is composed of two parts, $I_s^{s'} = I_{s\to s'} - I_{s'\to s}$, where $I_{s\to s'}$ is the current provided by the scattering states outgoing from terminal s and transmitted into terminal s'. For $I_{s\to s'}$ we have

$$I_{s\to s'}([U]) = \frac{2e}{h}\int_{-\infty}^{\infty} dE\, f_{\mathrm{FD}}(E + eU_{sg})T_{ss'}(E; [U]), \tag{4}$$

where $f_{FD}(E) = (1 + e^{E/k_BT})^{-1}$ is the Fermi-Dirac distribution function.

The calculation of S-matrix is done efficiently for each energy using the R-matrix approach, as it will be shown in the next section.

3 The R-Matrix Approach

The R-matrix representation of the S-matrix can be expressed in the compact form [29, 25]:

$$\mathsf{S} = -\frac{1 + \frac{\mathrm{i}}{m^*}\mathsf{Rk}}{1 - \frac{\mathrm{i}}{m^*}\mathsf{Rk}}, \tag{5}$$

where m^* is the effective mass, $(\mathsf{k})_{\nu,\nu'} = k_{\nu}\delta_{\nu\nu'}$ and the R-matrix is given by

$$(\mathsf{R})_{\nu\nu'}(E) = -\frac{\hbar^2}{2}\sum_{l=0}^{\infty}\frac{(\chi_l)_{\nu}(\chi_l^*)_{\nu'}}{E - \epsilon_l}, \tag{6}$$

with

$$(\chi_l)_{\nu} = \int_{\Gamma_s} d\Gamma_s\, \Phi_{\nu}(\boldsymbol{r}_{\perp;s})\, \chi_l(\boldsymbol{r} \in \Gamma_s). \tag{7}$$

In the above equations, $\chi_l(\boldsymbol{r} \in \Omega_0)$ and ϵ_l are the Wigner-Eisenbud (WE) functions and energies. These are the eigenfunctions and eigenvalues of the Hamiltonian with new and fixed boundary conditions on Γ_s, namely

$$\left[\frac{\partial \chi_l}{\partial z_s}\right]_{\Gamma_s} = 0. \tag{8}$$

In the context of resonant transport, with narrow peaks in transmission, one has to find the S-matrix for a large number of energies. The numerical advantage of the R-matrix formalism becomes clear in Eqs. (5), (6), and (7). Once the Wigner-Eisenbud eigensystem (χ_l, ϵ_l) is determined, the overlap integrals $(\chi_l)_\nu$ can be calculated. This first step is the most time consuming and it is independent on the total energy E. Then, in a second step, the R-matrix is easily determined from Eq. (6) by a simple summation for each energy. The S-matrix is obtained using Eq. (5), where finding the inverse of the $[1 - i/m^* \mathsf{Rk}]$ matrix represents a negligible effort (compared to first step) if the number of involved channels is moderate.

4 Planar Nanoscale Transistor

4.1 Planar Transistor Model

The planar nanoscale transistor depicted in Fig. 2 [30] is a down-sized conventional n-type FET with well-known building blocks: source/drain contacts (I,VII), shallow junctions extensions (II,VI), source/drain controlled regions of the electron channel

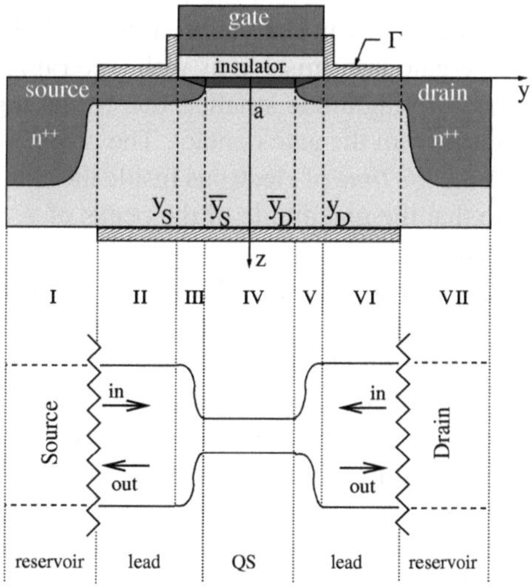

Fig. 2 Planar transistor model. The main features, source/drain contacts, shallow junctions extensions and the inversion channel have each a corespondent in the Landauer-Büttiker picture. The gate was omitted in the lower sketch for better clarity

(III,V) and the inversion channel (IV). Each has a corresponding element in the Landauer-Büttiker picture, as it is indicated in the lower part of the figure: the bulk parts of the source- (I), drain- (VII), and gate contact correspond to the reflectionless and ideally absorbing reservoirs R_s in Fig. 1. The shallow junction extensions (II,VI) have a constant typical width a and we assume a particle-in-a-box-type approximation for the potential. They correspond to the wide portions Ω_s of the lead, $s = \mathrm{S, D}$, with a separable Hamiltonian. Next, the shallow junction extensions narrow into the so called source- and drain controlled regions of the electron channel (II,V), where the potential depends on both coordinates (y, z). The inversion channel (IV) of length $L_y = y_\mathrm{D} - y_\mathrm{S}$ corresponds to the quantum system QS. For the gate, the lead Ω_G is the part of the contact which is located directly above the insulator barrier lying still within the coherence length. The regions III, IV, V together with the insulator make up the scattering region Ω_0. The system is assumed to be invariant in the x-direction within the transistor width L_x. See Refs. [25, 26] for more details.

4.2 Linear Regime

The planar transistor structure described in the previous section has a typical *wide-narrow-wide* geometry: the shallow junction extensions have a relatively wide rectangular potential, while the inversion channel introduces a triangular narrow constriction (see Fig. 3a [30]). The two-dimensional source-drain transmission T^0_SD vs. the in-plane energy E_{yz} is similar to the transmission of a point contact [31]. Yet, the mismatch of the wavefunctions at the interfaces between the electron channel and the contacts leads to oscillations which are superimposed to the step characteristic (see Fig. 3b). These are Fabry-Perot type resonances, which depend on the channel length L_y as $j^2(\hbar^2/2m^*)(\pi/L_y)^2$, with $j \geq 1$. The peaks correspond to standing waves in the electron channel. This type of phenomena is visible in the source-gate transmission as well (see Fig. 3c). The magnitude of the peaks is a few orders of magnitude smaller, due to the high insulator barrier which separates the channel from the gate contact. The probability of decaying into gate is proportional with the *life time* of electrons inside the channel. The latter decreases with increasing j, so that the magnitude of the peaks of T^0_SG decreases for higher order Fabry-Perot resonances.

By varying the gate bias U_GS in linear regime source-drain, the resonances become visible in the source-drain conductance g_DS. In Fig. 4 [30] is depicted g_DS for different temperatures $T = 0, 10, 20\,K$. As the temperature is increased, the slope of the subtreshold (OFF-state regime) characteristic decreases, while the sharpness of the Fabry-Perot resonances (in the ON-state regime) is reduced. Moreover, introducing a smooth transition between the leads and the channel, which is closer to a more realistic device, the effective channel length is decreased, but the sign-changing of the differential conductance is still kept.

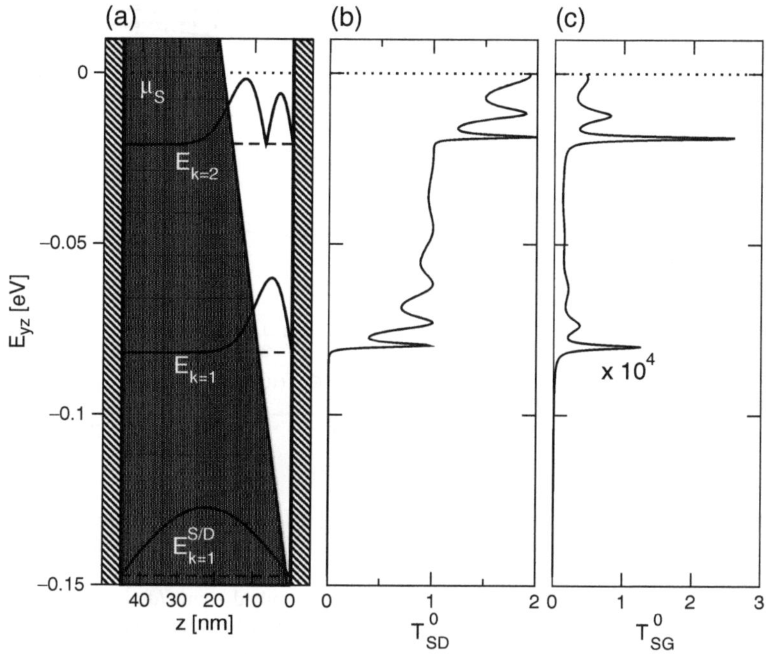

Fig. 3 (**a**) Particle-in-a-box potential inside the shallow junction extension (*dashed borders*) of width $a = 45$ nm and triangular potential inside the inversion channel (*shaded area*) for $U_{DS} = 0V$ and $U_{GS} = 0.3V$. Also shown are sketches of the lowest energy levels and the corresponding localized wavefunctions (*absolute value*). In (**b**) source-drain and (**c**) source-gate transmission for $\bar{y}_S = y_S$ and $\bar{y}_D = y_D$

4.3 Nonlinear Regime

In the classical drift-diffusion model the saturation voltage is given by the pinch-off voltage of the electron channel. In Fig. 5 [32] a completely different saturation mechanism in a ballistic quantum-transistor is demonstrated [26]. It is found that at least in the case of the narrow transistor, where only one subbands participate to the transport, the transmission in the full model can well be approximated by the transmission in the one-dimensional effective model. The two-dimensional *wide-narrow-wide* geometry can be successfully replaced by a one-dimensional effective potential. More precisely, at zero source-drain bias, the 1D effective potential is a rectangular barrier, of length equal the channel length (L_y) and height $V_0 = E_\perp^{\text{chan}} - E_\perp^{S,1}$, i.e. the difference between the position of the subband in the channel and the first perpendicular mode in source (see Fig. 5a). In the ON-state ($E_\perp^{\text{chan}} < \mu_S$, set by the applied gate voltage), for a finite positive source-drain bias, the chemical potential in drain μ_D travels to lower energies and at a certain point, it goes below $E_\perp^{\text{chan}}$. The current, calculated according to Eq. (4) in the energy window $[\mu_S, \mu_D]$, has two components: the first one corresponds to a transmission close to unity, above

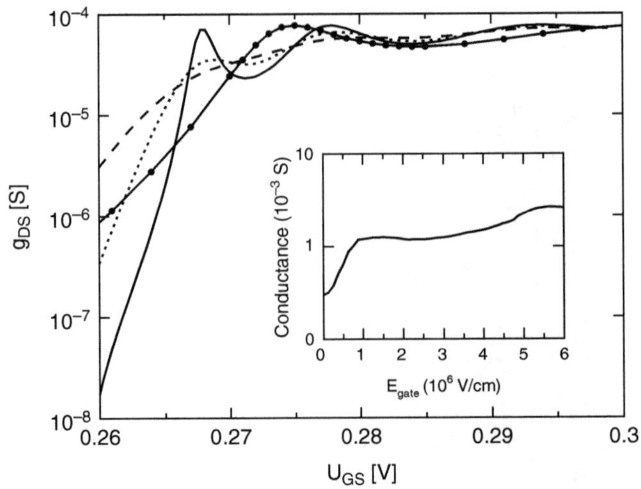

Fig. 4 Source-drain conductance g_{DS} vs. gate bias U_{GS}, for different temperatures, for the sample with $L_x = 10$ nm, at a small source-drain bias $U_{DS} = 0.1\,mV$. The temperatures considered are $T = 1\,K$ (*solid lines*), $T = 10\,K$ (*dotted lines*) and $T = 20\,K$ (*dashed lines*). The *dotted line* shows the result for the same sample, but with a smoothed source/drain-channel transition ($\bar{y}_S = -L_y/2$ and $\bar{y}_D = L_y/2$), so that the effective channel length is reduced by almost a half. The inset contains in logarithmic scale experimental results taken from Fig. 3 in Ref. [33] corresponding to a channel length of 9.1 nm, and a sample width of 10 μm

the barrier in the interval $[\mu_S, E_\perp^{chan}]$; the second component corresponds to the tunneling current and is much smaller compared to the first one. Increasing further the U_{DS} bias, the drain current saturates, as the growing contribution from the tunneling current is still relatively small.

Figure 6 [32] shows the drain current saturation, for a narrow transistor with a width $L_x = 10\,nm$ and channel length $L_y = 50\,nm$. The dots represent the

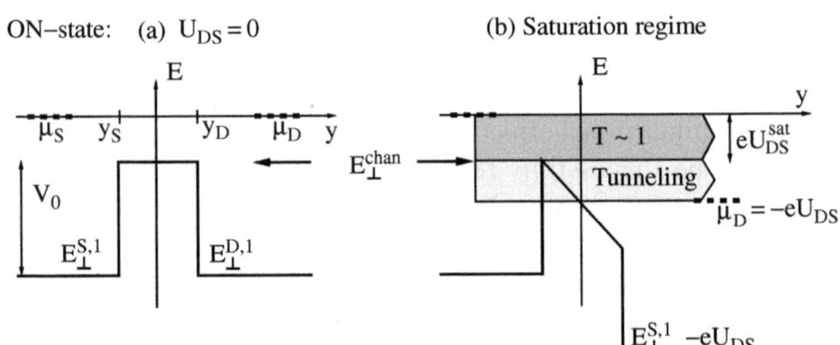

Fig. 5 Quantum saturation mechanism. (**a**) For $U_{DS} = 0$ the *wide-narrow-wide* geometry can be described by an effective one-dimensional potential, i.e. a barrier of height $V_0 = E_\perp^{chan} - E_\perp^{S,1}$ and length L_y. (**b**) Under applied source-drain bias, in the saturation regime, the current has two contributions, as indicated by the shaded areas

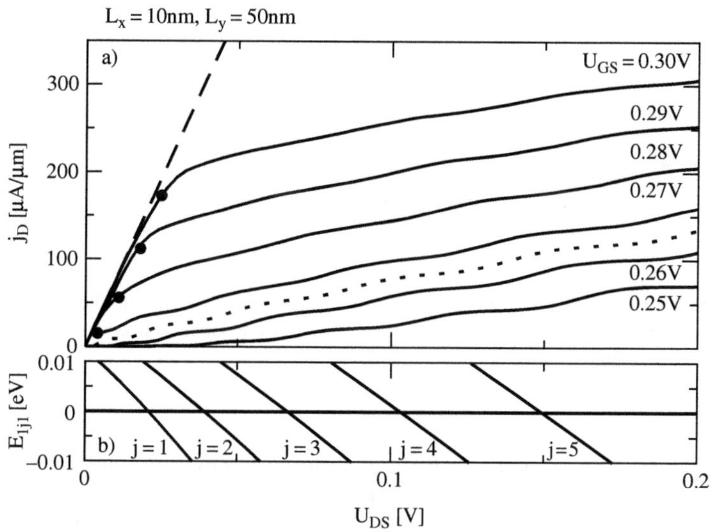

Fig. 6 Drain current for several gate biases $U_{GS} = 0.25 - 0.30\,V$. The dotted line characteristic separates ON- and OFF- states regimes. The dots mark the saturation points according to Eq. (9). In the lower plot are represented the eigenenergies of the closed system, at $U_{GS} = 0.25V$, $E_{\perp}^{\text{chan}} + E_j$ which concur with the Fowler-Nordheim type oscillations visible in the drain characteristics above

saturation points, which are determined by:

$$eU_{DS} = |E_{\perp}^{\text{chan}}|. \tag{9}$$

The almost linear characteristic (dotted line) separates ON- and OFF- states regimes. At lower U_{GS}, we have $E_{\perp}^{\text{chan}} > \mu_S$ and the device behaves as a tunneling transistor. The oscillations visible especially in the OFF-state regime are Fowler-Nordheim type resonances, which correspond to a sloped barrier.

5 Cylindrical Nanowire Transistor

5.1 CNT Model

New concepts and recent technological developments made possible the design of cylindrical nanowire transistors [11–13]. The one-dimensional character of such devices confers certain advantages over their planar counterparts, especially with respect to scaling down the dimensions: increased drive current through the entire body of the nanowire, better electrostatic control using wrap-around gates and improved sub-threshold regime. Following the Landauer-Büttiker picture described in the Sect. 4.1, the n++ regions play the role of the semi-infinite leads, while the small region centered on the intrinsic layer represents the scattering region.

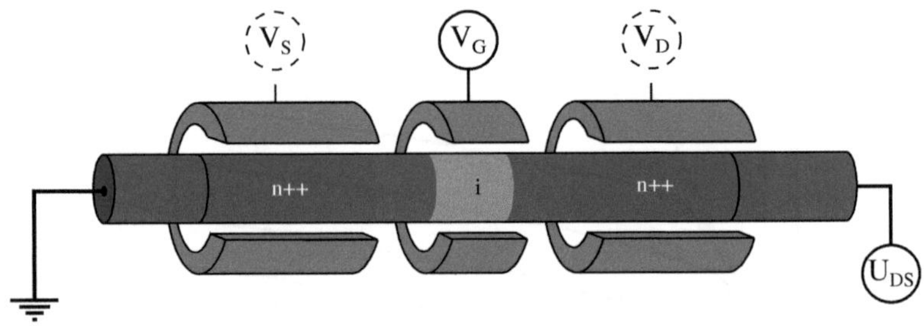

Fig. 7 Cylindrical nanowire transistor with wrap all-around gates. The middle gate controls the intrinsic region. Source- and drain gates can be optionally introduced in order to reduce the field effect of the middle gate

Figure 7 illustrates schematically a cylindrical nanowire transistor. The highly doped n++ semiconductor nanowire of radius R includes a small intrinsic layer of length L_i, which has a much lower free electron concentration. In this way, the potential in the intrinsic region can be efficiently controlled by the middle gate, ensuring the transistor functionality. The other two gates may be introduced in order to reduce the field penetration into the source- and drain contacts.

Assuming rotational symmetry, the solution of the three-dimensional Schrödinger Eq. (1) decouples:

$$\Psi_m(r, z, \theta; E) = \frac{e^{im\theta}}{\sqrt{2\pi}} \psi_m(r, z; E), \tag{10}$$

where $m = 0, \pm 1, \pm 2, \ldots$ is the magnetic quantum number. Each index m defines a two-dimensional scattering problem in (r, z) coordinates. Using the separation of variables method, one obtains:

$$\mathcal{H}_{rz} \psi_m(r, z; E) = E \psi_m(r, z; E), \tag{11}$$

with

$$\mathcal{H}_{rz} = -\frac{\hbar^2}{2m^*} \left(\frac{\partial^2}{\partial r^2} + \frac{1}{r} \frac{\partial}{\partial r} - \frac{m^2}{r^2} + \frac{\partial^2}{\partial z^2} \right) + W(r, z). \tag{12}$$

Furthermore, we assume the asymptotic lead potential is given by the potential of an infinite, uniformly doped nanowire, which will be discussed in the next section.

5.2 Uniformly Doped Nanowire

We start by considering an infinite ZnO nanowire, with cylindrical cross-section, surrounded by an insulating layer. The semiconductor is highly n++ doped and therefore degenerate. In the case of ZnO, native point defects like oxygen vacancies and zinc interstitials usually amount to the external doping. Therefore we will further refer to an effective doping concentration N_{d}, which includes all these contributions.

Due to the rotational symmetry, the potential inside the wire is found using the Hartree approximation, solving the coupled system of Poisson and the one-dimensional Schrödinger equation in the radial direction. The latter can be written as:

$$\mathcal{H}_r \psi_{nm}(r; E_{\perp}^{nm}) = E_{\perp}^{nm} \psi_{nm}(r; E_{\perp}^{nm}), \tag{13}$$

with

$$\mathcal{H}_r = -\frac{\hbar^2}{2m^*}\left(\frac{\partial^2}{\partial r^2} + \frac{1}{r}\frac{\partial}{\partial r} - \frac{m^2}{r^2}\right) + W(r), \tag{14}$$

where $m^* = 0.24m_0$ is the ZnO effective mass, $m = 0, \pm 1, \pm 2, \ldots$ is the magnetic quantum number and $W(r)$ denotes the Coulomb interaction. In the degenerate semiconductor the impurity bands merge with the conduction band, leading to small ionization energies. Therefore we may assume that all donors of the degenerate semiconductor are ionized. This leads to the following relation to be fulfilled self-consistently:

$$N_{\mathrm{d}} = \frac{2}{\pi R^2} \int_0^{E_{\mathrm{F}}} dE \sum_{n,m} g_{1\mathrm{D}}(E - E_{\perp}^{nm}), \tag{15}$$

with

$$g_{1\mathrm{D}}(E) = \frac{m^{*1/2}}{\sqrt{2}\pi\hbar} \frac{\Theta(E - E_{\perp}^{nm})}{\sqrt{E - E_{\perp}^{nm}}}, \tag{16}$$

where Θ is the Heaviside distribution and $g_{1\mathrm{D}}(E)$ is the one-dimensional density of states. The quantities which emerge from the self-consistent scheme are the Fermi energy E_{F} and the energies of the perpendicular modes $E_{\perp}^{\mathrm{nm}}$.

Figure 8 shows the radial self-consistent potential $W_{\mathrm{sc}}(r)$ and electron density $n_{\mathrm{sc}}(r)$ (inset). The mobile charge is confined inside the nanowire due to the hard-wall boundary condition at $r{=}R$, creating a built-in potential. As the nanowire radius is increased, more and more perpendicular modes are occupied below E_{F} and the electron density develops a plateau which matches the positive background. Moreover, the Fermi energy for $R = 20\,nm$ is $E_{\mathrm{F}} = 0.501\,eV$, i.e. $\approx 0.33\,eV$ measured

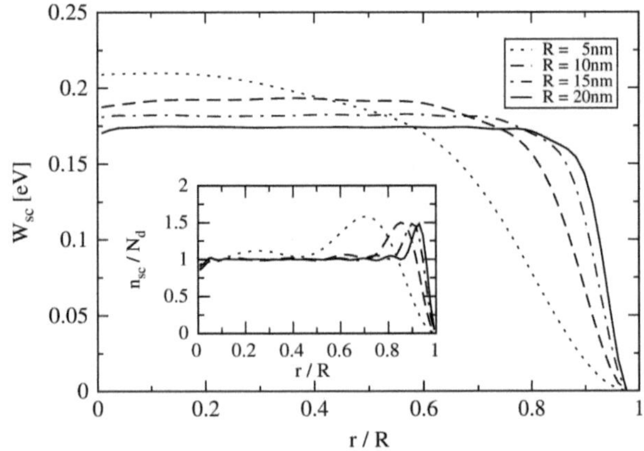

Fig. 8 Radial self-consistent potential $W_{sc}(r)$ and electron density $n_{sc}(r)$ (*inset*) of an uniformly doped nanowire, with $N_d = 10^{20}\,\mathrm{cm}^{-3}$. The Fermi energies are, in increasing nanowire radius, $E_F = 0.542, 0.520, 0.509, 0.501\,eV$

from the potential plateau $\approx 0.17\,eV$ (see Fig. 8), which becomes very close to the Fermi energy of a similar three-dimensional system $E_F^{3D} = \hbar^2(3\pi^2 N_d)^{2/3}/(2m^*) = 0.327\,eV$.

5.3 Doped Nanowire with Intrinsic Region

Introducing an intrinsic layer in the uniformly n++ doped nanowire, a potential barrier appears and, correspondingly, a decrease in the free electron density, as one can see from Fig. 9. The mobile charge density is constructed in the Hartree approximation, in the limit of low temperatures:

$$\rho(r,z) = e \int_0^{E_F} dE \sum_{s,n,m} g_{1D}(E - E_\perp^{nm}) \times |\Psi_{nm}^{(s)}(r,z;E)|^2, \qquad (17)$$

where $\Psi_{nm}^{(s)}(r,z;E)$ are solutions to Eq. (11), $E_\perp^{nm}$ are the perpendicular mode energies determined by Eq. (13) and $s = S, D$ i.e. source, drain. All two-dimensional problems corresponding to each quantum number m are solved, before the charge

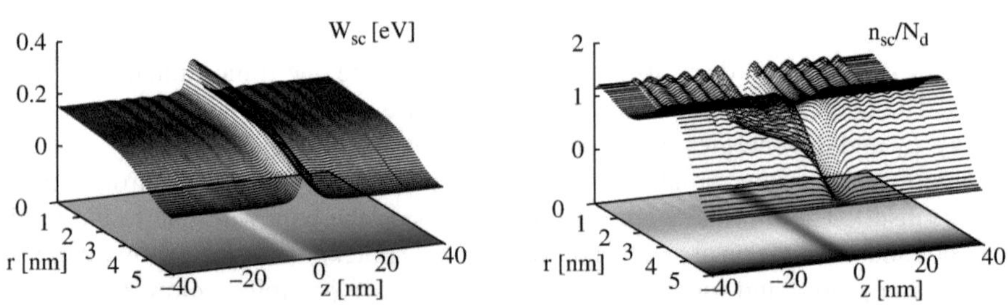

Fig. 9 Self-consistent potential $W_{sc}(r,z)$ (*upper*) and electron concentration $n_{sc}(r,z)$ (*lower*) of an infinite, uniformly n++ doped nanowire ($N_d = 0.5 \times 10^{20} cm^{-3}$), with an intrinsic region $L_i = 4\,\mathrm{nm}$

density in Eq. (17) can be constructed. Therefore, the difficulty of this self-consistent calculation lays in-between the two- and the three-dimensional case.

5.4 CNT: Linear Regime Characteristics

Adding wrap-all-around gates to the structure described in the previous section (see Fig. 7), the transistor functionality is enabled. Due to the short intrinsic region, the height of the barrier is relatively small and the source-drain transmission is relatively high (~ order of 1) even if there is no applied gate bias V_G, i.e. the transistor is said to be normally ON-state. Applying a negative potential on the middle gate, the barrier is enhanced and the electron density inside the intrinsic region decreases (see Fig. 10). Lowering V_G, the potential energy of the nanowire along the radius changes the bending near the gate, as one can see better from the Fig. 11. Thus, finite

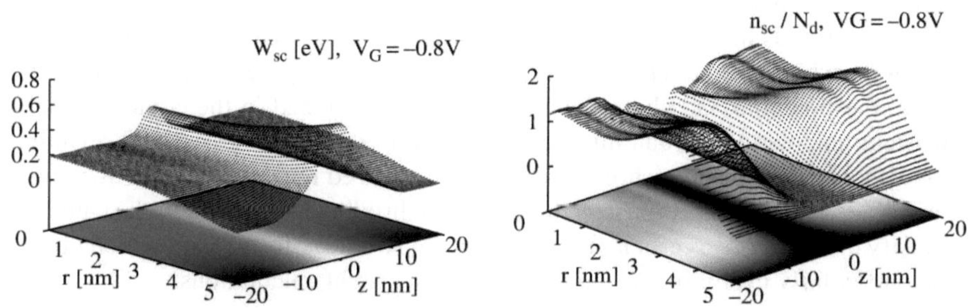

Fig. 10 Self-consistent potential and electron concentration in the considered CNT under gate bias $V_G = -0.8V$. The gate has a radius $R_G = 6\,\text{nm}$ and a length equal to L_i

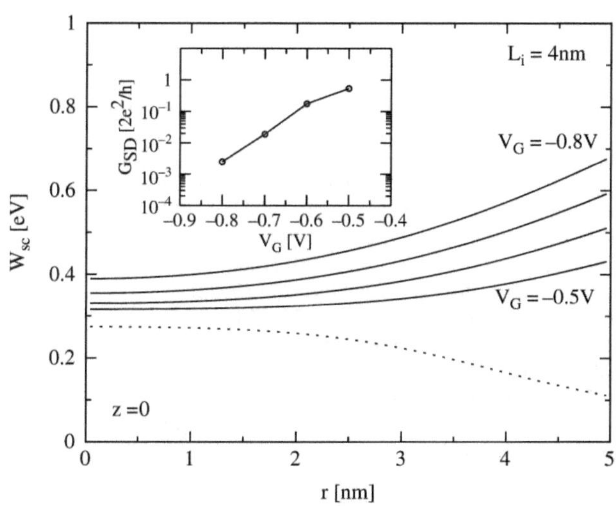

Fig. 11 Radial potential energy in cross section (at $z = 0$), for different gate biases (*solid*). The dotted line represents similar data for the same system without gate. The inset shows the conductance vs. middle gate bias in the low temperature limit

quantum screening effects are becoming visible: inside the nanowire body ($r \approx 0$) there are much smaller potential energy variations than at the outer regions ($r \approx R$) near the gate. Even though the intrinsic region is quite small ($L_i = 4\,\mathrm{nm}$), the conductance drops a few orders of magnitude at the lowest V_G (see inset of Fig. 11), which demonstrates a reasonable subtreshold characteristic and a relatively good gate control.

6 Summary

Important features of planar nanoscale transistors have been reviewed in ballistic limit, using the Landauer-Büttiker picture. In linear regime, the wide-narrow-wide geometry of the shallow junction extensions and the inversion channel produces Fabry-Perot type resonances, which are visible in both source-drain and source-gate transmission. In nonlinear regime, the quantum mechanism of drain current saturation was presented.

The new trends have brought up novel geometries, like the cylindrical nanowire transistors, which are better candidates for aggressively scaled devices. In such a system, the potential and charge distribution were analyzed in the context of self-consistent calculations. Quantum charge screening can be observed and quantitatively described. The subtreshold characteristics proved to posses a good switching behavior, despite the short intrinsic layer. The Schrödinger equation with scattering boundary conditions was efficiently solved in the R-matrix formalism, which proves to be particularly useful in the context of self-consistent calculations.

References

1. D.K. Ferry, S.M. Goodnick, *Transport in nanostructures* (Cambridge University Press, UK, 1999)
2. S. Datta, *Electronic Transport in Mesoscopic Systems* (Cambridge University Press, UK, 1995)
3. C. Weisbuch, B. Vinter, *Quantum semiconductor structures* (Academic, New York, NY, 1991)
4. W.C.J. Magnus, W.J. Schoenmaker, *Quantum Transport in Sub-Micron Devices*, Springer Series in Solid-State Sciences (Springer, Berlin, 2002) p. 137
5. R. Martel et al., Appl. Phys. Lett. 73, 2447 (1998)
6. J. Appenzeller et al., Phys. Rev. Lett. 92, 226802 (2004)
7. B. Yu, Tech. Dig. -Int. Electron Devices Meet. **251** (2002)
8. Y-.K. Choi et al., IEEE Electron Device Lett. **21**, 254 (2000)
9. I. Shorubalko et al., Appl Phys. Lett. **83**, 2369 (2003)
10. S.H. Son et al., J. Appl. Phys. **96**, 704 (2004)
11. K.H. Cho et al., Appl. Phys. Lett. **92**, 052102 (2008)
12. T. Bryllert, L.-E. Wernersson, T.L., L. Samuelson, Nanotechnology **17**, S227 (2006)
13. G.A. Nemnes, L. Ion, S. Antohe, J. Appl. Phys. **106**, 113714 (2009)
14. R. Venugopal et al., J. Appl. Phys. **95**, 292 (2004)
15. M.J. Gilbert et al., J. Appl. Phys. **95**, 7954 (2004)
16. J. Guo et al., Tech. Dig. -Int. Electron Devices Meet., 711 (2002)
17. A. Rahman et al., IEEE Trans. Electron Devices 50, 1853 (2003)

18. K. Natori, J. Appl. Phys. **76**, 4879 (1994)
19. K. Natori, IEICE Trans. Electron. **E84-C**, 1029 (2001)
20. J. Lusakowski et al., J. Appl. Phys. **101**, 114511 (2007)
21. M.D. Croitoru et al., J. Appl. Phys. **93**, 1230 (2003)
22. M. Fischetti et al., Tech. Dig. -Int. Electron Devices Meet. **467**, (2003)
23. A. Svizhenko et al., J. Appl. Phys. **91**, 2343 (2002)
24. R. Venugopal et al., J. Appl. Phys. **92**, 3730 (2002)
25. G.A. Nemnes et al., J. Appl. Phys. **96**, 596 (2004)
26. G.A. Nemnes et al., J. Appl. Phys. **98**, 084308 (2005)
27. M. Buettiker, Phys. Rev. Lett. **57**, 1761 (1986)
28. E.P. Wigner, L. Eisenbud, Phys. Rev. **72**, 29 (1947)
29. L. Smrcka, Superlattices and Microstructures **8**, 221 (1990)
30. G.A. Nemnes, U. Wulf, P.N. Racec, J. Appl. Phys., **96**, (1) 596 (2004) Reprinted with permission from, Copyright 2004, American Institute of Physics.
31. B.J. van Wees et al., Phys. Rev. Lett. **60**, 848 (1988)
32. G.A. Nemnes, U. Wulf, P.N. Racec, J. Appl. Phys., **98**, (8) 084308, (2005) Reprinted with permission from Copyright 2005, American Institute of Physics.
33. A. Hartstein, Appl. Phys. Lett. **59**, 2028 (1991)

R-matrix Formalism for Electron Scattering in Two Dimensions with Applications to Nanostructures with Quantum Dots

P.N. Racec, E.R. Racec, and H. Neidhardt

Abstract We investigate the scattering phenomena in two dimensions produced by a general finite-range nonseparable potential. This situation can appear either in a Cartesian geometry or in a heterostructure with cylindrical symmetry. Increasing the dimensionality of the scattering problem, new processes as the scattering between conducting channels and the scattering from conducting to evanescent channels are allowed. For certain values of the energy called resonance energy, the transmission through the scattering region changes dramatically in comparison with an one-dimensional problem. If the potential has an attractive character, even the evanescent channels can be seen as dips of the total transmission. The multi-channel current scattering matrix is determined using its representation in terms of the R-matrix. The resonant transmission peaks are characterized quantitatively through the poles of the current scattering matrix. Detailed maps of the localization probability density sustain the physical interpretation of the resonances. Our formalism is applied to a quantum dot in a two-dimensional electron gas and to a conical quantum dot embedded inside a cylindrical nanowire.

1 Introduction

There is a permanent requirement of shrinking the semiconductor devices in integrated circuits [1]. As feature sizes shrink into the nanometer scale regime, the device behavior becomes increasingly complicated since new physical phenomena at short distances occur and limitations in material properties are reached. In order to keep the good characteristics for transistors, new transistor architectures were developed progressively in the last decade.

P.N. Racec (✉)
Weierstraß-Institut für Angewandte Analysis und Stochastik, Mohrenstr. 39 10117 Berlin, Germany; National Institute of Materials Physics, PO Box MG-7, 077125 Bucharest Magurele, Romania
e-mail: racec@wias-berlin.de

A. Aldea, V. Bârsan (eds.), *Trends in Nanophysics*, Engineering Materials,
DOI 10.1007/978-3-642-12070-1_7, © Springer-Verlag Berlin Heidelberg 2010

Nowadays, there are developments of planar MOSFETs (metal oxide semiconductor field effect transistors) [2] as well as of gate-all-around (GAA) MOSFETs [3–5]. Both systems are also strongly related to more fundamental research structures developed in the last years, like in-plane-gate transistors [6], single-electron transistors [7], silicon-on-insulator planar double-gate transistors [8], non-planar double-gate FinFETs [9], non-planar trigate transistors [10], nanowire-based field-effect transistors (FET) [11], nanowire resonant tunneling diodes [12, 13], nanowire lasers [14], or nanowire qubits [15], whose maturity has still to be proven for industrial applications. Their structural complexity has also progressively increased, allowing for double-barrier structures [12, 13], or multiple core-shell layers [14, 16]. The material composition includes mainly III-V materials GaAs/AlGasAs [6, 7, 13], InAs/InP [12], GaN/InGaN [14], but also group IV materials Si [4, 17], and Si/Ge [11, 15], predominant in the industry.

The transport phenomena in these mesoscopic devices go beyond the semi-classical limit, and a quantum mechanical description of the current and charge densities [18] is necessary. The most appropriate method for analyzing semiconductor devices with an active region in the nanometer scale and which are almost open (i.e. showing a strong coupling between the active region and contacts) is the scattering theory.

This paper is focused on systems for which the scattering process is a two-dimensional one. Such systems can appear either in a Cartesian geometry, like for devices tailored in a two-dimensional electron gas (2DEG) [6, 7], or in a cylindrical geometry, like for nanowire-based devices [3–5, 12, 13]. In these systems there is a strong confinement of the motion in one direction, called *transversal direction*, while the transport occurs in the other direction, called *longitudinal direction*. The scattering problem is a two-dimensional one because the scattering potential is non-separable, and also the incoming electrons can choose different energy channels for transport, which are mixed due to the scattering.

We present in this work a general method, valid within the effective mass approximation, for solving the two-dimensional (2D) Schrödinger equation with scattering boundary conditions. Its solutions are found using the scattering theory and the R-matrix formalism [19–29]. This method is a semi-analytical one, and it gives the scattering functions in each point inside and outside the scattering area and for each energy as a function of the solutions of the Wigner-Eisenbud problem. As known from the nuclear physics, the Wigner-Eisenbud problem is the eigenvalue problem of the Hamilton operator for the closed counterpart of the considered open quantum system [30]. The R-matrix formalism is not only numerically very efficient, but it is also suitable for higher dimensional nanostructures with complex geometry [20–25] and general nonseparable scattering potential [31, 47]. It can also deal with more than two terminals [26, 29].

Using the scattering functions we analyze further the transport properties of the open quantum structures, especially the conductance. Besides the low dimensionality of such systems, the open character is also an essential feature which controls the transport phenomena through the structure. When the quantum system becomes open, its eigenstates yield resonance states which do not have an infinite life time anymore and which are not strictly localized inside the quantum system. In this paper we identify the signature of the resonance in each conductance peak and study the influence of the nonseparable character of the potential on the resonances and on the conductance through the system.

An interesting effect in a multi-channel scattering problem is that as soon as the potential is not separable anymore, the channels get mixed. If furthermore the scattering potential is attractive, then it leads to unusual scattering properties, like resonant dips in the transmission coefficient just below the next channel minimum energy. As it was shown analytically for a δ scattering potential [32] and later on for a finite-range scattering potential [33, 34] the dips are due to the quasi-bound-states splitting off from a higher evanescent channel. So that, evanescent channels can not be neglected when analyzing scattering in two- or three-dimensional quantum systems. These findings were recently confirmed numerically for a Gaussian-type scatterer [35] and also for a quantum dot or a quantum ring [36] embedded inside nanowires tailored in a two-dimensional electron gas (2DEG), or inside cylindrical nanowires [31]. The high resolution maps for the "near field" scattering wave functions presented in Refs. [31, 35, 36] show explicitly increased localization probability around the scatterer for energies of the quasi-bound states, in agreement with the resonant reflection or resonant back-scattering interpretation of these dips [32, 33]. The Cartesian and cylindrical geometries present different "selection rules" for the intersubband transmission [31].

2 Model

The electronic states in mesoscopic systems are easily described within the effective mass approximation whose validity requires that the envelope function $\Psi(E, \vec{r})$ must be slowly varying over dimensions comparable to the unit cell of the crystal [37].

In the spherical effective mass approximation, the envelope function associated to the energy E_{3D} satisfies a Schrödinger-type equation

$$\left[-\frac{\hbar^2}{2m^*}\Delta + V(\vec{r})\right]\Psi(\vec{r}) = E_{3D}\Psi(\vec{r}). \tag{1}$$

The so-called *scattering potential* $V(\vec{r})$ contains the information about the confinement in the transversal direction, and, inside the allowed area, it is a sum of the heterojunction conduction band discontinuities, the electrostatic potential due to the ionized donors and acceptors, the self-consistent Hartree and exchange potentials due to free carriers, and external potentials. We use the symbol m^* to denote the effective mass of the electrons, while m denotes the magnetic quantum number.

For systems tailored in the 2DEG, the growth direction is chosen the z-direction, while the plane of the 2DEG is (x, y). The wave function in the z-direction, $\xi(z)$, is taken as known (the simplest form is provided by Fang and Howard, [38]) so that the three-dimensional (3D) wave function can be written as

$$\Psi(E_{3D};\vec{r}) = \xi(z)\psi(x,y). \tag{2}$$

The total energy

$$E_{3D} = E_{2DEG} + E, \tag{3}$$

where E_{2DEG} is the energy of the 2DEG level, and E is the energy associated with the motion in the plane of the 2DEG. The 3D Schrödinger type equation reduces to a two-dimensional Schrödinger equation [47]

$$\left[-\frac{\hbar^2}{2m^*}\left(\frac{\partial^2}{\partial x^2}+\frac{\partial^2}{\partial y^2}\right)+V(x,y)\right]\psi(x,y) = E\,\psi(x,y),$$
$$x \in (-\infty,\infty),\, y \in [-d_y, d_y]. \tag{4}$$

In the transversal direction the electron motion is limited at the interval $[-d_y, d_y]$ by a confining potential that we have considered as infinite. The nonseparable potential $V(x, y)$ varies strongly with the position only inside a small domain ($|x| \le d_x$, $|y| \le d_y$) which is usually called *scattering region* and is quasi-constant outside this domain.

For cylindrical nanowires, the azimuthal symmetry suggests to use cylindrical coordinates, with z axis along the nanowire [31]. As long as there are not split gates on the surface of the nanowire, the potential energy $V(\vec{r})$ is rotational invariant

$$V(\vec{r}) = V(r,z) \tag{5}$$

and nonseparable inside the scattering region. A scattering potential which does not explicitly depend on the azimuthal angle θ imposes the eigenfunctions of the orbital angular momentum operator L_z as solutions of Eq. (1)

$$\Psi_m(E_{3D}; r,\theta,z) = \zeta_m(\theta)\psi_m(r,z), \tag{6}$$

where

$$\zeta_m(\theta) = \frac{e^{im\theta}}{\sqrt{2\pi}}, \tag{7}$$

and $m = 0, \pm 1, \pm 2, \ldots$ is the magnetic quantum number. This is an integer number due to the requirement that the function $e^{im\theta}$ should be single-valued. The functions $\psi_m(r, z)$ are determined from the equation

$$\left[-\frac{\hbar^2}{2m^*}\left(\frac{\partial^2}{\partial r^2}+\frac{1}{r}\frac{\partial}{\partial r}-\frac{m^2}{r^2}+\frac{\partial^2}{\partial z^2}\right)+V(r,z)\right]\psi_m(r,z)=E\psi_m(r,z),$$
$$r\in[0,R], z\in(-\infty,\infty), \quad (8)$$

where E denotes here the kinetic energy associated with the 3D motion of the electron inside the nanowire, $E = E_{3D}$. We have also considered an infinite potential outside the nanowire. In such a way, every magnetic quantum number m defines a *two-dimensional (2D) scattering problem*. Furthermore, these 2D scattering problems can be solved separately if the scattering potential is rotational invariant. How many of these problems have to be solved, depends on the specific physical quantity which has to be computed.

2.1 Scattering Problem for Two Dimensions

We consider the following Schrödinger type equation in two dimensions, $(x_\parallel, x_\perp)$, denoting generically the longitudinal and the transversal direction, respectively,

$$\left[-\frac{\hbar^2}{2m^*}\Delta_{x_\parallel,x_\perp}+V(x_\parallel,x_\perp)\right]\psi(E;x_\parallel,x_\perp)=E\psi(E;x_\parallel,x_\perp),$$
$$x_\perp\in\Omega, x_\parallel\in(-\infty,\infty). \quad (9)$$

One could consider here different effective masses in the longitudinal and transversal directions and also, for a layered heterostructure, a position-dependent effective mass. These effects can be incorporated in the formalism, but within this paper we neglect them for the simplicity of the exposure.

For the Cartesian geometry [47], in comparison with Eq. (4), we have

$$\begin{aligned} x_\parallel &= x, x_\perp = y, \Omega=[-d_y,d_y],\\ \Delta_{x_\parallel,x_\perp} &= \frac{\partial^2}{\partial x^2}+\frac{\partial^2}{\partial y^2},\\ E &= E_{3D}-E_{2DEG}, \end{aligned} \quad (10)$$

while for the cylindrical geometry [31], in comparison with Eq. (8), we have

$$\begin{aligned} x_\parallel &= z, x_\perp = r, \Omega=[0,R],\\ \Delta_{x_\parallel,x_\perp} &= \frac{\partial^2}{\partial r^2}+\frac{1}{r}\frac{\partial}{\partial r}-\frac{m^2}{r^2}+\frac{\partial^2}{\partial z^2},\\ E &= E_{3D}, \end{aligned} \quad (11)$$

which depends on the magnetic quantum number m.

The analogy between the both geometries appears more evident considering for the cylindrical geometry the unitary transformation $U : L^2([0, R] \times \mathbb{R}, rdrdz) \rightarrow L^2([0, R] \times \mathbb{R}, drdz)$, with $Uf(r, z) = g(r, z) = \sqrt{r} f(r, z)$. The inverse transformation is $U^\dagger : L^2([0, R] \times \mathbb{R}, drdz) \rightarrow L^2([0, R] \times \mathbb{R}, rdrdz)$, with $U^\dagger g(r, z) = f(r, z) = (1/\sqrt{r}) g(r, z)$. In such a way, the Schrödinger operator becomes

$$\begin{aligned} \tilde{H}_{r,z} &= U H_{r,z} U^\dagger = U \left[-\frac{\hbar^2}{2m^*} \Delta_{z,r} + V(z, r) \right] U^\dagger \\ &= -\frac{\hbar^2}{2m^*} \left[\frac{\partial^2}{\partial r^2} - \frac{m^2 - 1/4}{r^2} + \frac{\partial^2}{\partial z^2} \right] + V(z, r). \end{aligned} \tag{12}$$

In turn, the term which contains $1/r^2$ and which is specific for the cylindrical geometry plays the role of a potential, and the Laplace operator appears as known for two dimensions in Cartesian coordinates. The current scattering matrix and the localization probability distribution density of an electron in a scattering state remain unchanged under this unitary transformation.

Due to the localized character of the scattering potential, it is appropriate to solve Eq. (9) within the scattering theory. The potential energy which appears in Eq. (9) has generally two components:

$$V(x_\parallel, x_\perp) = V_\perp(x_\perp) + V_{scatt}(x_\parallel, x_\perp). \tag{13}$$

The first one, $V_\perp(x_\perp)$, describes the lateral confinement of the electrons and is translation invariant along the parallel direction $x_\parallel$. We consider a hard wall potential

$$V_\perp(x_\perp) = \begin{cases} 0, & x_\perp \in \Omega \\ \infty, & x_\perp \notin \Omega \end{cases}, \tag{14}$$

which defines a quantum wire for the Cartesian geometry and a cylindrical nanowire for the second considered geometry. A parabolic wall like in Ref. [35, 36] may also be considered.

The scattering potential energy inside the nanowire, $V_{scatt}(x_\parallel, x_\perp)$, has generally a nonseparable character in a domain of finite-range and is constant outside this domain. We consider here the nonseparable potential localized within the area $\Omega \times [-d_\parallel, d_\parallel]$, see Fig. 1,

$$V_{scatt}(x_\parallel, x_\perp) = \begin{cases} V_1, & x_\perp \in \Omega, x_\parallel < -d_\parallel \\ W(x_\parallel, x_\perp), & x_\perp \in \Omega, -d_\parallel \leq x_\parallel \leq d_\parallel \\ V_2, & x_\perp \in \Omega, x_\parallel > d_\parallel \end{cases}. \tag{15}$$

There are not material definitions for the interfaces $x_\parallel = \pm d_\parallel$. Usually, they are chosen inside the highly doped regions of the heterostructure characterized by a

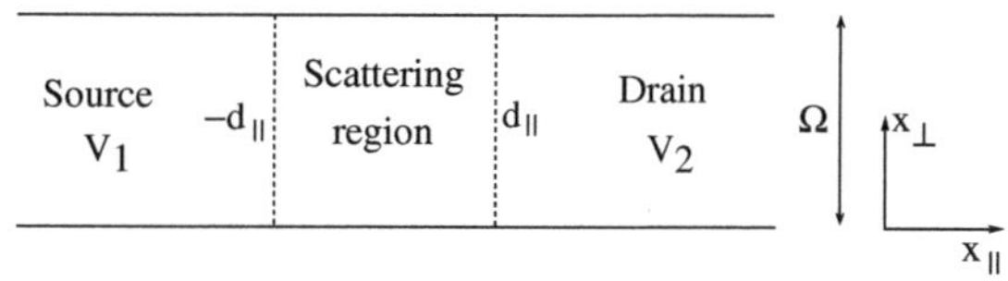

Fig. 1 The generic geometry of the 2D scattering problem

slowly variation of the potential in the longitudinal direction, practically by a constant potential. These regions play the role of the source and drain contacts.

2.2 Scattering States

In the asymptotic regions, $|x_\parallel| > d_\parallel$ i.e. source and drain contacts, the potential energy is separable in the transversal (i.e. confinement) and the longitudinal (i.e. transport) direction, i.e. $V(x_\parallel, x_\perp) = V_\perp(x_\perp) + V_s$, $s = 1, 2$, and Eq. (9) can be directly solved using the separation of variables method

$$\psi(E; x_\parallel, x_\perp) = \phi(x_\perp)\varphi(x_\parallel). \tag{16}$$

The function $\phi(x_\perp)$ satisfies the transversal equation

$$\left[-\frac{\hbar^2}{2m^*}\Delta_{x_\perp} + V_\perp(x_\perp)\right]\phi(x_\perp) = E_\perp\phi(x_\perp), \qquad x_\perp \in \Omega, \tag{17}$$

where

$$\Delta_{x_\perp} = \begin{cases} \frac{d^2}{dy^2}, & \text{Cartesian geometry} \\ \frac{d^2}{dr^2} + \frac{1}{r}\frac{d}{dr} - \frac{m^2}{r^2}, & \text{cylindrical geometry} \end{cases}. \tag{18}$$

The hard wall confinement potential requires Dirichlet boundary condition at the boundaries $\partial\Omega$ of the interval Ω, $\phi(\partial\Omega) = 0$. As a remark, for the cylindrical geometry, the boundary $r = 0$ is an artificial one introduced in order to use the cylindrical symmetry. At this boundary it is sufficient that $\phi(x_\perp)$ remains finite.

Due to the electron confinement in the transversal direction $x_\perp$ the solutions of Eq. (17) define the *transversal modes*, $\phi_n(x_\perp)$, with the corresponding transversal energies $E_{\perp n}$, $n \geq 1$. The eigenfunctions $\phi_n(x_\perp)$ depend on the geometry (Cartesian or cylindrical) and on the confinement potential. In the case of a hard wall confinement, the transversal modes are given for the Cartesian geometry by sine functions [47], while for the cylindrical geometry they are expressed in terms of the Bessel functions of the first kind [31]. The transversal modes form an orthonormal and complete system of functions.

The function $\varphi(x_\parallel)$ satisfies the one-dimensional Schrödinger type equation called longitudinal equation

$$\left[-\frac{\hbar^2}{2m^*}\frac{d^2}{dx_\parallel^2}+V_s\right]\varphi(x_\parallel)=(E-E_\perp)\varphi(x_\parallel),\quad x_\parallel\in(-\infty,-d_\parallel)\cup(d_\parallel,\infty),\tag{19}$$

where $s = 1$ stays for the source contact ($x_\parallel < -d_\parallel$) and $s = 2$ for the drain contact ($x_\parallel > d_\parallel$). In the case of different effective masses in transversal and longitudinal direction, one can use the corresponding effective mass in each of the above equations.

Every transversal mode together with the associated motion on the transport direction defines a *scattering channel* on each side of the scattering area. The scattering channels are indexed by (sn), $n \geq 1$, $s = 1, 2$ for each E. In contradistinction to the Cartesian geometry, in the case of a cylindrical geometry there is a set of 2D scattering problems, indexed by the magnetic quantum number m, that have to be solved. Consequently the scattering channels should be also indexed by m. For simplicity we omit the index m in this section, but we keep in mind that we solve here a 2D scattering problem and obtain the scattering functions for a fixed value of m.

If the total energy E and the lateral eigenenergy $E_{\perp n}$ are fixed, there are at most two linearly independent solutions of Eq. (19). In the asymptotic region they are given as a linear combination of exponential functions

$$\varphi_{sn}(x_\parallel)=\begin{cases}A_s\mathrm{e}^{\mathrm{i}k_{1n}x_\parallel}+B_s\mathrm{e}^{-\mathrm{i}k_{1n}x_\parallel}, & x_\parallel<-d_\parallel\\ C_s\mathrm{e}^{\mathrm{i}k_{2n}x_\parallel}+D_s\mathrm{e}^{-\mathrm{i}k_{2n}x_\parallel}, & x_\parallel>d_\parallel\end{cases}\tag{20}$$

where A_s, B_s, C_s and D_s are complex coefficients depending on n and E for each value of $s = 1, 2$. The wave vector is defined for each scattering channel (sn) as

$$k_{sn}(E)=k_0\sqrt{(E-E_{\perp n}-V_s)/u_0},\tag{21}$$

where $k_0 = \pi/2d_\parallel$ and $u_0 = \hbar^2k_0^2/2m^*$. In the case of the conducting or open channels

$$E-E_{\perp n}-V_s\geq 0,\tag{22}$$

k_{sn} are positive real numbers and correspond to propagating plane-waves. For the evanescent or closed channels

$$E-E_{\perp n}-V_s<0,\tag{23}$$

k_{sn} are given from the first branch of the complex square root function, $k_{sn} = \mathrm{i}|k_{sn}|$, and describe exponentially decaying functions away from the scattering region. Thus, the number of the conducting channels, $N_s(E)$, $s = 1, 2$, is a function of energy, and for a fixed energy E this is the largest value of n, which satisfies the inequality (22) for given values of s.

Each conducting channel corresponds to one degree of freedom for the electron motion through the system and, consequently, there exists only one independent solution of Eq. (9) for a fixed channel (sn) associated with the energy E,

$\psi_n^{(s)}(E; x_\parallel, x_\perp)$. For describing further the transport phenomena in the frame of the scattering theory it is convenient to consider this solution as a *scattering state*, i.e. as a sum of an incoming component on the channel (sn) and a linear combination of outgoing components on each scattering channel. One can write the scattering wave functions in a compact form [47]

$$\psi_n^{(s)}(E; x_\parallel, x_\perp) = \frac{\theta(N_s(E) - n)}{\sqrt{2\pi}} \times \begin{cases} \delta_{s1} e^{ik_{1n}(x_\parallel + d_\parallel)} \phi_n(x_\perp) + \sum_{n'=1}^{\infty} S_{1n',sn}(E) e^{-ik_{1n'}(x_\parallel + d_\parallel)} \phi_{n'}(x_\perp), & x_\parallel < -d_\parallel \\ \delta_{s2} e^{-ik_{2n'}(x_\parallel - d_\parallel)} \phi_n(x_\perp) + \sum_{n'=1}^{\infty} S_{2n',sn}(E) e^{ik_{2n'}(x_\parallel - d_\parallel)} \phi_{n'}(x_\perp), & x_\parallel > d_\parallel \end{cases} \tag{24}$$

The step function θ in the above expressions, with $\theta(x \geq 0) = 1$ and $\theta(x < 0) = 0$, assures that the scattering functions are defined only for the conducting channels. Writing explicitly the position of the interfaces $\pm d_\parallel$ at the exponent has advantages for the analytical treatment of the scattering problem [22, 39]. As it is discussed in Refs. [26, 31], it is necessary to consider the sum until infinity in the second term of the above expression, in order to keep the mathematical completeness of the transversal channels.

The physical interpretation of the expressions (24) is that, due to the nonseparable character of the scattering potential, a plane-wave incident onto the scattering domain is reflected on every channel – open or closed for transport – on the same side of the system and transmitted on every channel - open or closed for transport – on the other side. The reflection and transmission amplitudes are described by the complex coefficients $S_{sn',sn}$ and $S_{s'n',sn}$ with $s \neq s'$, respectively, and all of them should be nonzero. These coefficients define a matrix with $N_1(E) + N_2(E)$ infinite columns. For an elegant solution of the scattering problem we extend $S(E)$ to an infinite square matrix and set at zero the matrix elements without physical meaning, $S_{s'n',sn}(E) = 0$, $n > N_s(E)$, $s = 1, 2$. In this way we define the *wave transmission matrix* or *wave-function amplitudes matrix* [32]. It is also called *generalized scattering matrix* [40]. This is not the well-known scattering matrix (current transmission matrix) whose unitarity reflects the current conservation. The generalized scattering matrix is a non-unitary matrix, which has the advantage that it allows for a description of the scattering processes not only in the asymptotic region but also inside the scattering area.

The three-dimensional scattering states, solutions of Eq. (1) can be now written as

$$\Psi_n^{(s)}(E; x_\parallel, x_\perp, x_3) = \omega(x_3) \psi_n^{(s)}(E; x_\parallel, x_\perp), \tag{25}$$

where $\omega(x_3)$ stays for $\xi(z)$ in the case of the Cartesian geometry and for $\zeta_m(\theta)$ in the case of the cylindrical geometry. Being eigenfunctions of an open system, the scattering states are ortho-normalized in the general sense [27]

$$\int_{\Omega} d\Omega \int_{-\infty}^{\infty} dx_{\parallel}\ \psi_n^{(s)}(E; x_{\parallel}, x_{\perp})\psi_{n'}^{(s')}(E'; x_{\parallel}, x_{\perp})^* = \delta_{ss'}\delta_{nn'}\frac{\delta(E-E')}{g_{sn}(E)}, \quad (26)$$

where $g_{sn}(E) = m^*/[\hbar^2 k_{sn}(E)]$ is the 1D density of states. We have to mention that for the Cartesian coordinates [47], the measures are $d\Omega = dy$, $dx_{\parallel} = dx$, while for the cylindrical geometry [31] they are $d\Omega = rdr$, $dx_{\parallel} = dz$.

2.3 R-matrix Formalism For Two Dimensions

The scattering functions inside the scattering region are determined using the R-matrix formalism, i.e. they are expressed in terms of the eigenfunctions corresponding to the closed counterpart of the scattering problem [19–26, 29]. In our opinion this is a more appropriate method than the common mode space approach which implies the expansion of the scattering functions inside the scattering area in the basis of the transversal modes $\phi_n(x_{\perp})$. As it is shown in Ref. [34],[41] the mode space approach has limitations for structures with abrupt changes in the potential or sudden spatial variations in the widths of the wire; it breaks even down for coupling operators that are not scalar potentials, like in the case of an external magnetic field. In the R-matrix formalism the used basis contains all the information about the scattering potential, and this type of difficulties can not appear.

Thus, the scattering functions inside the scattering region are given as

$$\psi_n^{(s)}(E; x_{\parallel}, x_{\perp}) = \sum_{l=1}^{\infty} a_{ln}^{(s)}(E)\chi_l(x_{\parallel}, x_{\perp}), \quad (27)$$

with $x_{\perp} \in \Omega$ and $x_{\parallel} \in [-d_{\parallel}, d_{\parallel}]$.

The so-called *Wigner-Eisenbud functions*, $\chi_l(x_{\parallel}, x_{\perp})$, firstly used in the nuclear physics [30, 42], satisfy the same equation as $\psi_n^{(s)}(x_{\parallel}, x_{\perp})$, Eq. (9), but with different boundary conditions in the transport direction. Since the scattering function $\psi_n^{(s)}(x_{\parallel}, x_{\perp})$ satisfies energy dependent boundary conditions derived from Eq. (24) due to the continuity of the scattering function and its derivative at $x_{\parallel} = \pm d_{\parallel}$, the Wigner-Eisenbud function $\chi_l(x_{\parallel}, x_{\perp})$ has to satisfy Neumann boundary conditions at the interfaces between the scattering region and leads

$$\left.\frac{\partial \chi_l}{\partial x_{\parallel}}\right|_{x_{\parallel}=\pm d_{\parallel}} = 0, \quad l \geq 1. \quad (28)$$

The hard wall confinement potential requires Dirichlet boundary condition at $\partial\Omega$ also for the Wigner-Eisenbud functions, $\chi_l(\partial\Omega, x_{\parallel}) = 0$. As already mentioned for the scattering states, for the cylindrical geometry it is sufficient that the Wigner-Eisenbud function remains finite at $r = 0$. The functions χ_l, $l \geq 1$, build a basis which verifies the orthogonality relation

$$\int_{\Omega} d\Omega \int_{-d_{\|}}^{d_{\|}} dx_{\|}\ \chi_l(x_\perp, x_\|)\chi_{l'}(x_\perp, x_\|) = \delta_{ll'} \tag{29}$$

and the closure relation

$$\sum_{l=1}^{\infty} \chi_l(x_\perp, x_\|)\chi_l(x'_\perp, x'_\|) = \delta(x_\perp - x'_\perp)\delta(x_\| - x'_\|). \tag{30}$$

Note that for the cylindrical geometry $\delta(x_\perp - x'_\perp)$ in the relation (30) means $\delta(r - r')/r$. The corresponding eigenenergies to χ_l are denoted by E_l and are called *Wigner-Eisenbud energies*. Since the Wigner-Eisenbud problem is defined on a closed volume with self-adjoint boundary conditions, the eigenfunctions χ_l and the eigenenergies E_l can be chosen as real quantities. The Wigner-Eisenbud problem is, thus, the closed counterpart of the scattering problem.

In the case of the one-dimensional system without spherical symmetry, it was recently proven mathematically rigorous that the R-matrix formalism allows for a proper expansion of the scattering matrix on the real energy axis [28]. In this section we present an extension of the R-matrix formalism for 2D scattering problem.

To calculate the expansion coefficients $a_{ln}^{(s)}(E)$ we multiply Eq. (9) by $\chi_l(x_\|, x_\perp)$ and the equation satisfied by the Wigner-Eisenbud functions by $\psi_n^{(s)}(E; x_\|, x_\perp)$. The difference between the resulting equations is integrated over $\Omega \times [-d_\|, d_\|]$, with the corresponding measures, and one obtains on the right-hand side the coefficient $a_{ln}^{(s)}(E)$. After using the Green's theorem and the boundary conditions one finds $a_{ln}^{(s)}(E)$ and feeds in it into Eq. (27). So, *the scattering functions inside the scattering region* ($x_\| \in [-d_\|, d_\|],\ x_\perp \in \Omega$) are obtained in terms of their derivatives at the edges of this domain,

$$\psi_n^{(s)}(E; x_\|, x_\perp) = \frac{1}{k_0}\int_{\Omega} dx'_\perp \left[R(E; -d_\|, x'_\perp, x_\|, x_\perp)\left.\frac{\partial \psi_n^{(s)}(E; x'_\|, x'_\perp)}{\partial x'_\|}\right|_{x'_\|=-d_\|} - R(E; d_\|, x'_\perp, x_\|, x_\perp)\left.\frac{\partial \psi_n^{(s)}(E; x'_\|, x'_\perp)}{\partial x'_\|}\right|_{x'_\|=d_\|}\right], \tag{31}$$

where the R-function is defined as

$$R(E; x_\|, x_\perp, x'_\|, x'_\perp,) \equiv \frac{u_0}{k_0}\sum_{l=1}^{\infty}\frac{\chi_l(x_\|, x_\perp)\chi_l(x'_\|, x'_\perp)}{E - E_l}. \tag{32}$$

The functions $\partial\psi_n^{(s)}/\partial x_\parallel$ at $x_\parallel = \pm d_\parallel$ are calculated from the asymptotic form (24) based on the continuity conditions for the derivatives of the scattering functions on the interfaces between the scattering region and leads.

With these results the scattering functions inside the scattering domain are expressed in terms of the wave transmission matrix **S**

$$\vec{\Psi}(E; x_\parallel, x_\perp) = \frac{i}{\sqrt{2\pi}} \mathbf{\Theta}(E)[\mathbf{1} - \mathbf{S}^T(E)]\mathbf{K}(E)\vec{R}(E; x_\parallel, x_\perp), \tag{33}$$

where the component (sn) of the vector $\vec{\Psi}$ is the scattering function $\psi_n^{(s)}(E; x_\parallel, x_\perp)$, $n \geq 1$, $s = 1, 2$ and $\mathbf{S}^T$ denotes the matrix transpose. The diagonal matrix **K** has on its diagonal the wave vectors (21) of each scattering channel

$$\mathbf{K}_{sn,s'n'}(E) = \frac{k_{sn}(E)}{k_0}\delta_{nn'}\delta_{ss'}, \tag{34}$$

$n, n' \geq 1$, $s, s' = 1, 2$, and the vector $\vec{R}(E; x_\parallel, x_\perp)$ is given as

$$\vec{R}(E; x_\parallel, x_\perp) = \frac{u_0}{\sqrt{k_0}} \sum_{l=1}^{\infty} \frac{\chi_l(x_\parallel, x_\perp)\vec{\chi}_l}{E - E_l}, \tag{35}$$

where $\vec{\chi}_l$ has the components

$$(\vec{\chi}_l)_{sn} = \frac{1}{\sqrt{k_0}} \int_\Omega \chi_l(x_\perp, (-1)^s d_\parallel)\phi_n(x_\perp)d\Omega, \tag{36}$$

$n \geq 1$, $s = 1, 2$. The diagonal Θ-matrix, $\mathbf{\Theta}_{sn,s'n'}(E) = \theta(N_s(E) - n)\,\delta_{ss'}\,\delta_{nn'}$, $n \geq 1$, $s = 1, 2$, assures non-zero values only for the scattering functions corresponding to the conducting channels.

Using further the continuity of the scattering functions on the surface of the scattering area and expanding $\vec{R}(E; \pm d_\parallel, x_\perp)$ in the basis $\{\phi_n(x_\perp)\}_{n\geq 1}$ we find the relation between the matrixes **S** and **R**

$$\mathbf{S}(E) = \left[\mathbf{1} - 2\left(\mathbf{1} + i\mathbf{R}(E)\mathbf{K}(E)\right)^{-1}\right]\mathbf{\Theta}(E), \tag{37}$$

with the R-matrix given by means of a dyadic product

$$\mathbf{R}(E) = u_0 \sum_{l=1}^{\infty} \frac{\vec{\chi}_l\,\vec{\chi}_l^T}{E - E_l}. \tag{38}$$

According to the above relation, **R** is an infinite-dimensional symmetrical real matrix and its elements defined by Eq. (38) are dimensionless. The above form allows for a very efficient numerical implementation for computing the R-matrix.

The expression (37) of the S-matrix in terms of the R-matrix is the key relation for solving 2D scattering problems using only the eigenfunctions and the eigenenergies of the closed quantum system. They contain the full information about the scattering potential and carry it over to the R-matrix. The matrix **K** characterizes the contacts and can be constructed using only the information about the potential in these regions. On the base of Eq. (37), the wave transmission matrix is calculated and after that the scattering functions in each point of the system are obtained using Eqs. (24) and (33). Further on, each transport property of the open quantum system can be derived from the scattering function in terms of the scattering matrix.

2.4 Reflection and Transmission Coefficients

Using the density current operator

$$\vec{j}(\vec{r}) = \frac{\hbar}{2im^*}\Big(\Psi(\vec{r})\nabla\Psi(\vec{r})^* - \Psi(\vec{r})^*\nabla\Psi(\vec{r})\Big), \tag{39}$$

one can define, as usually, the transmission and reflection probabilities [43]. Here $\Psi(\vec{r})^*$ denotes the complex conjugate of the scattering wave function (25).

The transversal component of the density current $j_\perp(x_\perp, x_\parallel, x_3)$ is zero in leads, because $\phi_n(x_\perp)$ are real functions. The component x_3 of the incident density current is also zero, either due to the confinement in the third direction, like in Cartesian geometry [47], or due to the symmetry reasons like for the cylindrical geometry [31]. What remains is the longitudinal component of the particle density current $j_\parallel(x_\perp, x_\parallel, x_3)$, which provides after the integration over the cross section of the lead with the corresponding measure, $d\Omega$, the very well-known relations for the transmission and reflection probabilities. The probability for an electron incident from the source, $s = 1$, on the channel n to be reflected back into the source on the channel n' is

$$R^{(1)}_{nn'} = \frac{k_{1n'}}{k_{1n}}|S^T_{1n,1n'}|^2, \tag{40}$$

and the probability to be transmitted into the drain, $s = 2$, on the channel n' is

$$T^{(1)}_{nn'} = \frac{k_{2n'}}{k_{1n}}|S^T_{1n,2n'}|^2. \tag{41}$$

The reflection and transmission probabilities for the evanescent (closed) channels are zero. The total transmission and reflection coefficients for an electron incident from reservoir $s = 1$ are defined as

$$T^{(1)} = \sum_{n,n'} T^{(1)}_{nn'}, \quad R^{(1)} = \sum_{n,n'} R^{(1)}_{nn'}. \tag{42}$$

More detailed properties of the many-channel tunneling and reflection probabilities are given in [43], but note that our indexes are interchanged with respect to the definitions used there.

2.5 Current Scattering Matrix

Further, we define the energy dependent *current scattering matrix* as

$$\tilde{\mathbf{S}}(E) = \mathbf{K}^{1/2}(E)\boldsymbol{\Theta}(E)\mathbf{S}(E)\mathbf{K}^{-1/2}(E), \tag{43}$$

so that its elements give directly the reflection and transmission probabilities

$$\begin{aligned} |\tilde{S}_{1n',1n}(E)|^2 &= R^{(1)}_{nn'}(E), & |\tilde{S}_{2n',2n}(E)|^2 &= R^{(2)}_{nn'}(E), \\ |\tilde{S}_{2n',1n}(E)|^2 &= T^{(1)}_{nn'}(E), & |\tilde{S}_{1n',2n}(E)|^2 &= T^{(2)}_{nn'}(E). \end{aligned} \tag{44}$$

The diagonal Θ-matrix assures that the matrix elements of $\tilde{\mathbf{S}}$ are nonzero only for the conducting channels, for which the transmitted flux is nonzero. Using the R-matrix representation of $\mathbf{S}$, Eq. (37), we find from the above relation

$$\tilde{\mathbf{S}}(E) = \boldsymbol{\Theta}(E)\left[\mathbf{1} - 2(\mathbf{1} + \mathrm{i}\boldsymbol{\Omega}(E))^{-1}\right]\boldsymbol{\Theta}(E), \tag{45}$$

with the infinite dimensional matrix $\boldsymbol{\Omega}$

$$\boldsymbol{\Omega}(E) = \mathbf{K}^{1/2}(E)\mathbf{R}(E)\mathbf{K}^{1/2}(E) = u_0 \sum_{l=1}^{\infty} \frac{\vec{\alpha}_l\,\vec{\alpha}_l^T}{E - E_l} \tag{46}$$

and the column vector

$$\vec{\alpha}_l(E) = \mathbf{K}^{1/2}(E)\,\vec{\chi}_l, \tag{47}$$

with $l \geq 1$.

Further, we express the total tunneling coefficient in terms of the current transmission matrix,

$$T(E) = \mathrm{Tr}[\boldsymbol{\sigma}(E)\boldsymbol{\sigma}^{\dagger}(E)], \tag{48}$$

where $\boldsymbol{\sigma}$ denotes the part of $\tilde{\mathbf{S}}$ which contains the transmission, $\boldsymbol{\sigma}_{nn'}(E) = \tilde{S}_{2n',1n}(E)$, $n = \overline{1, N_1(E)}$ and $n' = \overline{1, N_2(E)}$.

According to the definition (46), the matrix $\boldsymbol{\Omega}$ is a symmetrical one, $\boldsymbol{\Omega} = \boldsymbol{\Omega}^T$, and from Eq. (45) it follows that $\tilde{\mathbf{S}}$ also has this property, $\tilde{\mathbf{S}} = \tilde{\mathbf{S}}^T$. On this basis, one can demonstrate that the tunneling coefficient characterizes one pair of open channels irrespective of the origin of the incident flux $T^{(1)}_{nn'} = \left|\tilde{S}_{2n',1n}\right|^2 = \left|\tilde{S}_{1n,2n'}\right|^2 = T^{(2)}_{n'n}$.

This is a well-known property of the transmission through a scattering system and it shows that the current scattering matrix used here is properly defined. The restriction of $\tilde{S}$-matrix to the open channels is the well known current scattering matrix [20, 22, 23], commonly used in the Landauer-Büttiker formalism. For a given energy E this is a $(N_1+N_2)\times(N_1+N_2)$ matrix which has to satisfy the unitarity condition, according to the flux conservation.

In the numerical computations, the matrixes $\mathbf{S}$, $\mathbf{R}$, $\boldsymbol{\Omega}$, $\tilde{\mathbf{S}}$ and $\boldsymbol{\Theta}$ have the dimension $2N\times 2N$, and the vectors $\vec{\chi}_l$, $\vec{\alpha}_l(E)$ have $2N$ components, where N is the number of scattering channels (open and closed) taken numerically into account. The number of the Wigner-Eisenbud functions and energies computed numerically establishes the maximum value for the index l.

2.6 Resonances

The relation (45) is the starting point for a resonance theory of the transmission through a structure with a scattering region [22, 47]. The singularities of the current scattering matrix $\tilde{\mathbf{S}}$ which satisfy the equation

$$\det\left[\mathbf{1}+\mathrm{i}\boldsymbol{\Omega}(E)\right]=0 \tag{49}$$

are usually classified as bound states and resonances. The bound states are characterized by real negative energies while the resonance energies, $\bar{E}_{0l}=E_{0l}-\mathrm{i}\Gamma_l/2$, $l\geq 1$, lie in the complex energy plane below the real positive axis according to the causality requirements [44]. The scattering matrix $\tilde{\mathbf{S}}$ and consequently the total transmission $T(E)$ are defined only for energies in the continuum spectrum (E real positive) of the scattering problem and they are analytical functions over the whole domain. Although they have no singularities in the definition domain, their energy dependence is determined by the resonances, especially by those ones which lie in the vicinity of the real axis. In the resonance domain, i.e. inside a circle of radius Γ_l around $\bar{E}_{0l}$, the elements of the current scattering matrix $\tilde{\mathbf{S}}$ vary strongly with the energy. In turn, $T(E)$ has also an important variation for the real energies included in the resonance domain. Thus the resonances appear usually as peaks in the tunneling coefficient and can be directly seen in the transport properties of the structure. While in the case of a 1D scattering potential the peaks are light asymmetric maxima [22], for a 2D scattering potential the peak shapes cover all ranges of the Fano lines, from asymmetric maxima through "S-type" Fano lines up to antiresonances. These profiles have been already seen experimentally, for example, in the conductance of a single-electron transistor [7]. In the next section we demonstrate that the two-dimensional character of the scattering potential and the strong coupling of the quantum system to the contacts allow for the transmission profiles which are far from Breit-Wigner lines.

The representation of the $\tilde{S}$-matrix in terms of $\boldsymbol{\Omega}$, Eq. (45) allows for an efficient numerical procedure to determine its poles and the resonances. When the quantum system, for example a quantum dot, is coupled to the contacts, it becomes open, and the real eigenenergies of the closed problem, E_l, migrate in the lower part of the complex energy plane, becoming resonant energies, $\bar{E}_{0l} = E_{0l} - i\Gamma_l/2, l \geq 1$. On the base of this correspondence, we fix an energy E_λ of the isolated dot and determine the resonance energy $\bar{E}_{0\lambda}$ as a solution of Eq. (49) in the complex energy plane. The matrix $\mathbf{\Omega}$ contains contributions from all Wigner-Eisenbud functions and energies, i.e. χ_l and E_l, and from all scattering channels, i.e. all matrix elements of $\mathbf{K}$. Thus, the resonance energy $\bar{E}_{0\lambda}$ can strongly differ from E_λ, and only in the case of a very low coupling of the dot to the contacts the eigenenergies of the isolated system, E_λ, can properly approximate the real part of the resonance energy.

The resonance theory presented above is general and can be applied to a variety of structures with a 2D scattering potential, regardless if the geometry of the system is Cartesian or cylindrical. The information about the geometry is contained only in the Wigner-Eisenbud functions and energies. After solving the eigenvalue problem of the closed counterpart of the scattering system one can construct the scattering matrix and analyze it without bearing in mind the geometry of the system.

The expression (45) of the $\tilde{S}$-matrix shows that all matrix elements of $\tilde{\mathbf{S}}$ are singular at the resonance energy. That means that all transmission coefficients $T_{nn'}$ between different scattering channels have a similar dependence on energy around a resonance, and it is enough to analyze the total transmission which is a sum of them in order to characterize the resonance.

3 Model Systems

Further we analyze the total tunneling coefficient $T(E)$ for a large energy interval in the case of a quantum dot isolated inside of a 2DEG (*Cartesian geometry*) and in the case of a conical quantum dot in a cylindrical nanowire (*cylindrical geometry*). The transmission peaks are directly connected to the resonances and they have different profiles depending on the coupling strength between the quantum system and contacts, but also between resonances.

3.1 Quantum Dot in Two-Dimensional Electron Gas

We consider here a quite simple dot, a square dot, isolated inside a quantum wire by the constant barriers V_0 as seen in Fig. 2. The smaller barriers V_{b1} and V_{b2} characterize the coupling between the quantum dot and contacts, and the strength of this coupling can be varied individually. Although our model allows for an arbitrary form of the potential, we have chosen this square dot in order to compare the scattering functions at the resonant energies with the eigenfunctions of an isolated dot $(V_0, V_{b1}, V_{b2} \to \infty)$.

For the numerical calculations we have set $d_x = d_y = 50$ nm, and the width of all barrier 20 nm. Thus the region where the electrons are localized is about 60×60 nm. The barrier which isolates the quantum dot inside the quantum wire

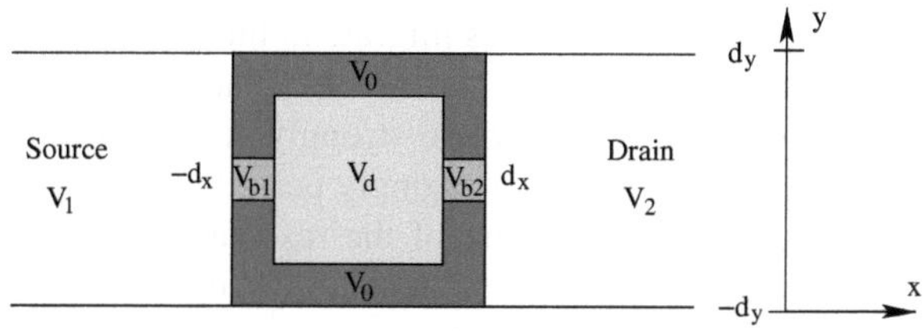

Fig. 2 Potential energy in the 2D quantum wire: constant potential energy in the source and drain contacts, $V_1 \simeq V_2$, and position dependent potential energy in the dot-region. The quantum dot is isolated inside the quantum wire by the barrier with the height V_0. The coupling between dot and contacts is set by the potential energy in the aperture regions, V_{b1} and V_{b2}. The electrons inside the dot experience the potential energy V_d

has been taken as $V_0 = 0.4$ eV, and the potential energy in the aperture regions $V_{b1} = V_{b2} = 0.005$ eV. In the source and drain contacts the potential energy has been considered as the energy reference, $V_1 = V_2 = 0$ eV. The Fermi energy of the electrons has been taken as $E_F = 12$ meV. For this quantum dot, there are four open channels, $N_1(E_F) = N_2(E_F) = 4$, and the closed channels do not have a significant contribution to the conductance.

The quantum dot described above is strongly coupled to the source and drain contacts because the potential energy in the aperture regions lies under the Fermi energy. For explaining the transport phenomena through the dot it is necessary to take properly into account the open character of the system and to analyze the transport properties in terms of the resonances. The eigenstates which characterize the closed counterpart of the open dot have an infinite life time and can not explain the broaden peaks which are experimentally measured in the conductance of an open quantum dot [7].

The resonance energies of the considered dot, solutions of Eq. (49), are presented in Fig. 3. For comparison, the Wigner-Eisenbud energies, i.e. the eigenenergies of the corresponding closed dot are also given. Due to the coupling of the quantum

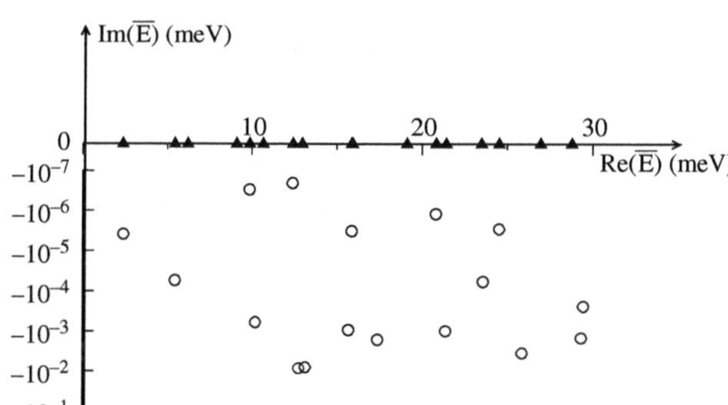

Fig. 3 Resonance energies $\bar{E}_{0l}$ (empty circles) of the open quantum dot given in Fig. 2 and the real eigenenergies (filled triangles) of the closed counterpart of the considered quantum dot

dot to the contacts, the resonance energies migrate in the lower part of the complex energy plane and have different widths. There are very narrow resonances associated with the modes of the dot, which are not so strongly coupled to the contacts, and broad resonances, which describe modes strongly perturbed by the interaction with the reservoirs. For a better understanding of the resonance modes we will examine the localization probability distribution density of the electrons for the energies given by the real part of the resonance energy $\bar{E}_{0l}$.

The potential energy felt by the electrons inside the dot is V_d. This energy can be modified continuously by varying the voltage of a plunger gate [20–23], and the conductance through the dot is measured as a function of V_d. In the linear regime experiments, i.e. small source drain biases, and for very low temperatures the conductance G can be directly connected to the total tunneling coefficient at the Fermi energy [7, 22, 47],

$$G(V_d) = \frac{2e^2}{h} T(E_F; V_d). \tag{50}$$

In the above relation the potential energy V_d appears as a parameter in the expression of the total tunneling coefficient. A new value of V_d means a new scattering potential and a new scattering matrix. But there is no analytic dependence of T on V_d, so that the scattering matrix and after that the conductance have been numerically computed for each value of V_d. In Figs. 4 and 5 the conductance is plotted as a function of $E_F - V_d$. This fact has the great advantage that the position of the maxima in conductance are given with respect to the bottom of the quantum well, and in this way a direct comparison with an infinite quantum well is possible. There are narrow and broad peaks in the conductance and in order to understand why they have different profiles, we have also plotted the electron probability distribution density $P_n(x, y) = |\psi_n^{(1)}(E; x, y)|^2$ for $E = E_F$ and $V_d = V_{0l}$, for the eight peaks considered here, $l = \overline{1, 8}$. In principle, the potential energy V_{0l} is associated with the maximum of the conductance peak, but the conductance curve shows also a "S-type" Fano line, and a rigorous method to fix V_{0l} is necessary.

Using the R-matrix representation of the S-matrix, Eq. (45), we can provide an approximative relation for $T(E_F, V_d)$ around a resonance

$$T(E_F, V_d) = T(E_F, V_{0l} + \delta V) \simeq T(E_F - \delta V, V_{0l}), \tag{51}$$

where V_{0l} is the value of V_d for which the real part of the resonance energy ($\bar{E}_{0l} = E_{0l} - i\Gamma_l/2$) matches the Fermi energy, $E_{0l} = E_F$, and δV is a small variation with values in the interval $(-\Gamma_l, \Gamma_l)$. For a detailed discussion of this approach see Ref. [22], Appendix A. The expression (51) of the total tunneling coefficient allows for a direct connection to the resonances. We can simultaneously plot the conductance as a function of $E_F - \delta V$, $\delta V \in (-\Gamma_l, \Gamma_l)$ and the resonances with the real energies in the interval $(E_F - \Gamma_l, E_F + \Gamma_l)$. These plots are given in Figs. 4 and 5. The dashed vertical lines correspond in each picture to $\delta V = 0$, i.e. $V_d = V_{0l}$ in the plots in the middle part and $E = E_F$ in the plots in the

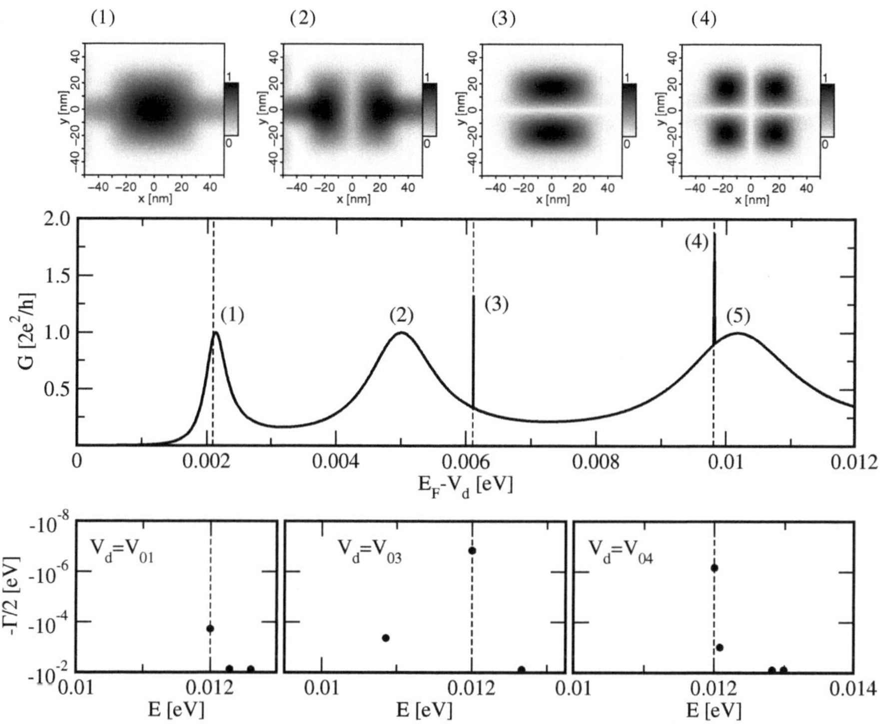

Fig. 4 (*Middle part*) Conductance (solid line) as a function of the potential energy in the dot region, V_d. The vertical dashed lines give the position of the potential energy V_{0l}, $l = 1, 3, 4$. (*Upper part*) The space dependence of the electron probability distribution density $P_n(x, y)/P_{max}$, $P_n(x, y) = |\psi_n^{(s)}(E; x, y)|^2$, $P_{max} = \max[P_n(x, y)]$, $n = 1$ or $n = 2$, for $E = E_F$ and $V_d = V_{0l}$, $l = 1, 2, 3, 4$. For the peaks $l = 1, 2$ we have considered the channel number $n = 1$, while for the peaks $l = 3, 4$, $n = 2$. (*Lower part*) Resonance energies in the complex energy plane with the real part around E_F. The vertical dashed lines correspond to the Fermi energy

lower part. From the simultaneous analysis of these graphics it is evident that we can associate each peak in the conductance with a resonance l. At the resonance energy the electrons show a strong localization in the dot region as, seen in the upper part of the figures. A narrow peak corresponds to a resonance energy with a very small imaginary part and to a resonance state that is almost decoupled from the contacts, i.e. the electron probability distribution density is nearly zero in the aperture regions. These are the resonances denoted by (3), (4), (6) and (8) in Figs. 4 and 5. The other peaks are broaden and they correspond to resonances with a larger imaginary part and to states which are strongly coupled to the contacts. Generally, the peaks for which the resonance states have a maximum at $y = 0$ couple strongly to the contacts and become broaden. These are the peaks (1), (2), (5) and (7) in Figs. 4 and 5. But there is an exception which does not depend on the parameters of the system: the peak denoted by (6) in Fig. 5. Although the probability distribution density has a maximum in the central region of the dot, this state is strongly localized

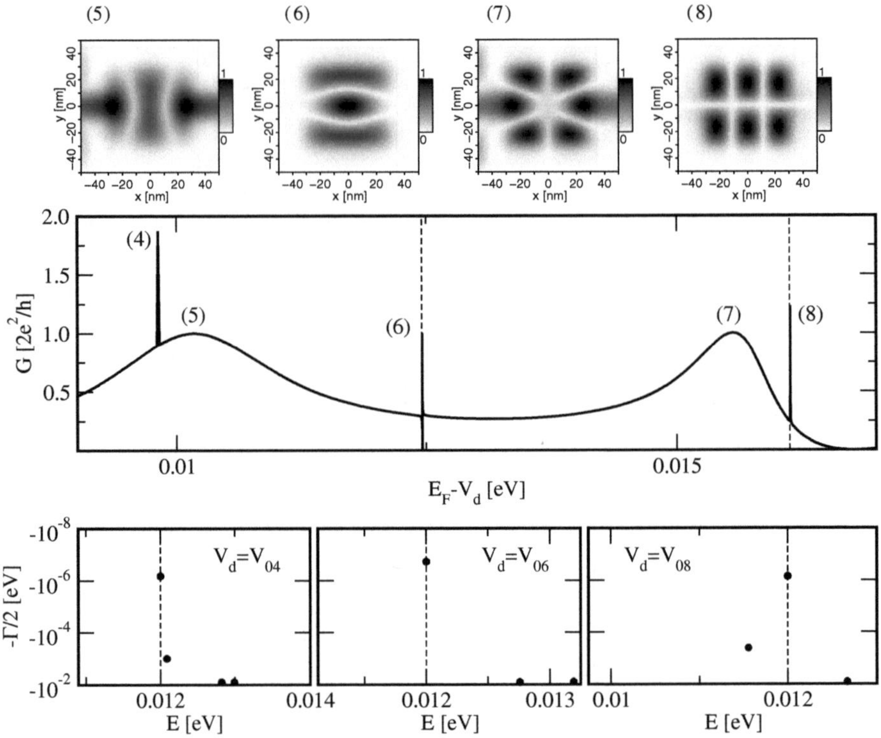

Fig. 5 (*Middle part*) Conductance (solid line) as a function of the potential energy in the dot region, V_d. The vertical dashed lines give the position of the potential energy V_{0l}, $l = 6, 8$. (*Upper part*) The space dependence of the electron probability distribution density $P_n(x, y)/P_{max}$, $P_n(x, y) = |\psi_n^{(s)}(E; x, y)|^2$, $P_{max} = max[P_n(x, y)]$ for $E = E_F$ and $V_d = V_{0l}$, $l = 5, 6, 7, 8$. For the peaks $l = 5, 6, 7$ we have considered the channel number $n = 1$, while for the peaks $l = 8$, $n = 2$. (Lower part) Resonance energies in the complex energy plane with the real part around E_F. The vertical dashed lines correspond to the Fermi energy

inside the dot, and the corresponding peak is very narrow. But this peak is not an asymmetric maximum anymore, it has a "S-type" Fano line shape. This behavior can be explained only taking into account the interaction between resonances. The state which corresponds to the peak (5) has three maxima in the x-direction and one maximum in the y-direction, while the peak (6) has one maximum on x- and three maxima on y-direction. They are states with the same symmetry in the both directions and they influence each other. As a result, there are two hybrid modes, one of them very strongly coupled to the contacts and the other one almost isolated. This interaction between resonant states with the same symmetry in the lateral direction is a general phenomenon which has at the origin the scattering between different energy channels due to the nonseparable character of the scattering potential. For systems with an effective 1D scattering potential the interaction between resonances is weak and the strong asymmetric Fano line shapes ("S-type" or antiresonance) do not appear. Also the hybrid modes do not exist in those cases.

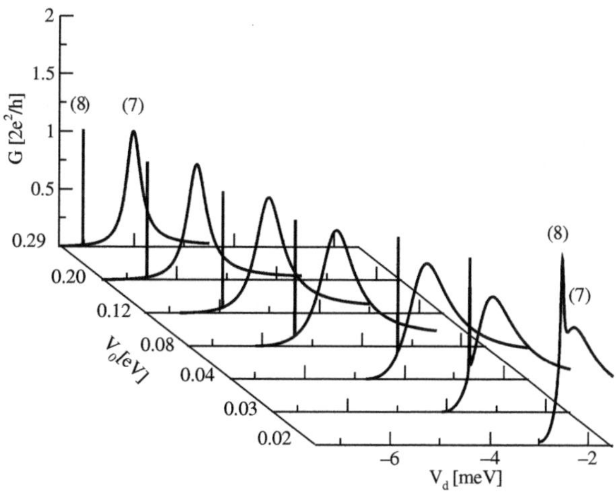

Fig. 6 Conductance as a function of the potential energy in the dot region, V_d, for different values of the lateral confinement potential, V_0

Besides the potential energy in the dot region V_d, the confinement potential of the dot V_0 can be also modified by the top gates. Decreasing the strength of the confinement, the coupling of the resonance states to the contacts increases, and the conductance peaks become broader. We present in Fig. 6 the evolution of the conductance peaks (7) and (8) when the potential V_0 decreases. The width of the two peaks increases, and the maximum of each peak migrates to higher energies. But the shift in energy is different. The state (7) which is strongly coupled to the contacts in the aperture regions is much less influenced by the variation of V_0. The state which corresponds to the peak (8), with nearly zero probability distribution density in the region of the apertures, can have a coupling to the contacts only in the case of a rather transparent confinement barrier V_0. This explains the significant broadening of the line shape and also the larger shift of the peak energy.

3.2 Conical Quantum Dot Inside a Cylindrical Nanowire

We consider a conical quantum dot, embedded in an infinite cylindrical nanowire with the same radius, as is sketched in Fig. 7a. The parameters considered are the height of the dot $h = 5$nm, the radius of the nanowire $R = 5$nm and the effective mass $m^* = 0.19m_0$. We set in our computations $d_{\|} = d_z = 16$nm and the total number of channels (open and closed) $N = 8$. In our calculations, the results do not change if more channels are added.

Depending on the band-offsets between the dot material and the host material, the potential produced by the dot can be repulsive, yielding a quantum barrier, or attractive, yielding a quantum well. We consider here that the dot yields an attractive potential $V(z, r)$, represented in Fig. 7b by a quantum well of depth $W_b = -0.125$eV.

The total tunneling coefficient $T^{(1)}$ versus the incident energy E is plotted in Fig. 8 for different magnetic quantum numbers m. The transmission increases with

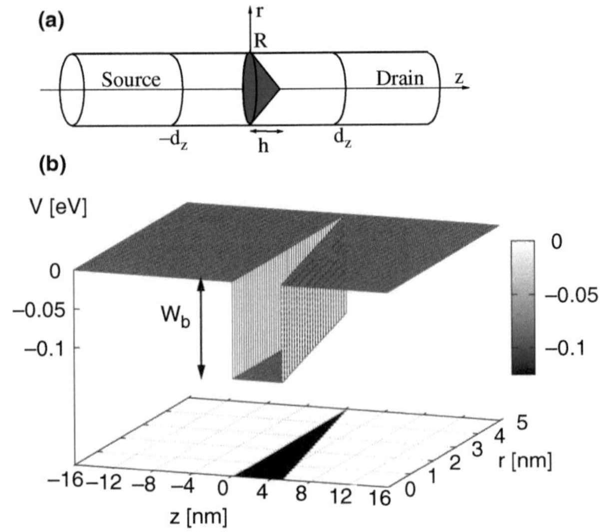

Fig. 7 (**a**) Sketch of a conical quantum dot embedded into a nanowire with the same radius. The dot yields an attractive potential $V(z,r)$, represented in (**b**) by a quantum well of depth $W_b = -0.125\text{eV}$

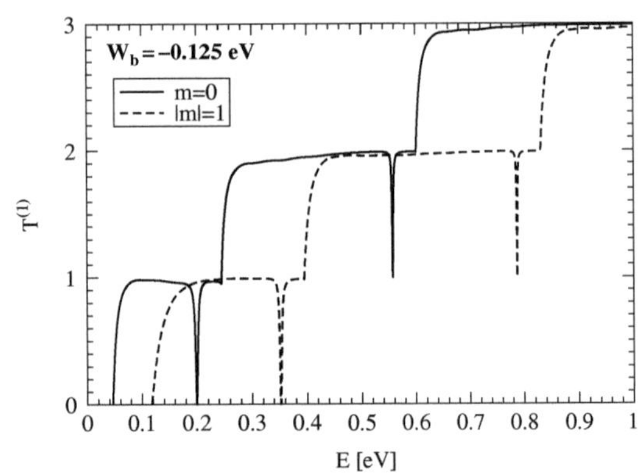

Fig. 8 Total tunneling coefficient as a function of incident energy E for the scattering potential represented in Fig. 7 (b), for different magnetic quantum numbers m (continuous line for $m = 0$, dashed line for $|m| = 1$) and for the well depth $W_b = -0.125\text{eV}$

a unity, every time a new channel $E_{\perp,n}^{(m)}$ becomes available for transport, i.e. becomes open. The length of the plateaus is given by the difference between two successive transversal mode energies, and this length increases with m. Due to the presence of the quantum well, deviations appear from the step-like transmission. Just before a new channel gets open, there is a dip, i.e. a sharp drop, in the tunneling coefficient. These dips are owing to modification of the tunneling coefficient due to the evanescent (closed) channels [32]. This is a multichannel effect that was also put in evidence in Cartesian coordinates for quantum wires tailored in a two-dimensional electron gas [32–36].

The dips can be understood considering the simple couple-mode model [32–34, 31]. For a dot surrounded by the host material, the scattering potential $V(z,r)$ is not anymore separable, so that the scattering mixes the channels [31–34]. As soon as the scattering potential is attractive, the diagonal coupling matrix element

$$V_{nn}(z) = \int_0^R \phi_n(r)V(z,r)\phi_n(r)rdr < 0 \qquad (52)$$

acts for every channel n as an effective one-dimensional (1D) attractive potential [33], which always allows for at least one bound state [45, 46] below the threshold of the continuous spectrum. By mixing the channels, this bound state becomes a *quasi-bound state* or resonance, i.e. with complex energy, whose real part gets embedded into continuous spectrum of the lower channel and the imaginary part describes the width of the resonance. These resonances can be seen now as dips in the tunneling coefficient. The energy difference between the position of the dips and the next subband minima $E_{\perp,n}^{(m)}$ gives the quasi-bound state energy. The positions of the dips, i.e. the quasi-bound state energy, depend on the channel number n and on the magnetic quantum number m and, of course, on the detailed system parameters. In Cartesian coordinates the specific symmetry of the channels (odd and even) do not allow for dips in the first plateau [35]. In the cylindrical geometry this symmetry is broken, so that we obtain a dip in front of every plateau. Our numerical method allows for a high energy resolution in computing the tunneling coefficient, so that we were able to find the dips also in front of the higher-order plateaus.

Further insight about the quasi-bound states of the evanescent channels can be gained looking at the wave functions, whose square absolute value $|\psi_n^{(s)}(E;z,r)|^2$ gives the *localization probability density*.

The R-matrix formalism allows us to produce high resolution maps of the wave functions inside the scattering region, see Eq. (33). In Figs. 9a, b the localization probability distribution density is represented in arbitrary units, for an electron incident from source ($s = 1$) and with a total energy corresponding to the dips in Fig. 8. The total energy E and the channel n, on which the electron is incident, are specified at every plot. Let discuss Fig. 9a. The total energy $E = 0.199eV$ is less than the energy of the second transversal mode, $E_{\perp,2}^{(0)} = 0.244eV$, so that only the first channel is open. Thus the incident wave from the source contact is node-less in r-direction. But, as it can be seen in Fig. 9a, the scattering wave function inside the scattering region has a node in the r-direction, i.e. position in r where the wave

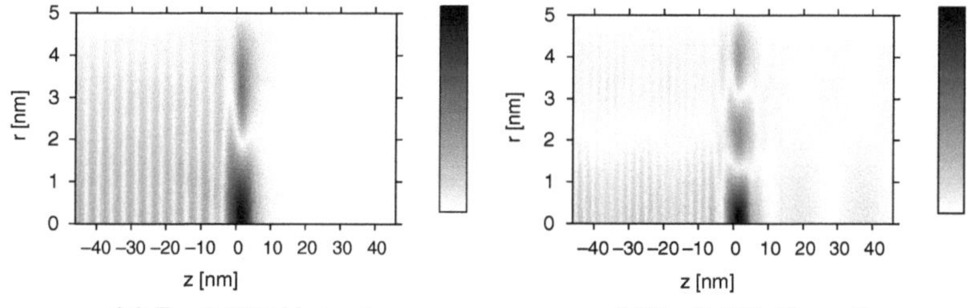

Fig. 9 Localization probability distribution density, $|\psi_n^{(1)}(E;z,r)|^2$, for an electron with $m = 0$, incident from the reservoir $s = 1$ into the channel n and with the total energy E. The energies are the dips in Fig. 8

function is zero. This means that the wave function corresponds to the quasi-bound state splitting off from the second transversal mode, which is an evanescent one. The quasi-bound state is reachable now in a scattering formulation due to channel mixing. The scattering wave function has a pronounced peak around the scattering potential, i.e. $z \in [0, 5]$nm, and decreases exponentially to the left and to the right. On the left side of the scattering potential one observes the interference pattern produced by the incident and the reflected waves, while on the right side there exists only the transmitted wave.

The scattering wave function considered in Fig. 9b has the energy less than the third transversal channel, $E_{\perp,3}^{(0)} = 0.6006eV$, so that the incident part of the scattering state on the second mode $n = 2$ has one node in r-direction. But the scattering function shows inside the scattering region two nodes in the r-direction, so it corresponds to a quasi-bound state splitting off from the above evanescent channel, the third one.

One gets similar pictures for all m-values, with the difference that for $m \neq 0$ the wave functions are zero for $r = 0$. In Fig. 9a, b one can observe that the transmitted part of the scattering wave function is zero, in agreement with the resonant backscattering specific to the quasi-bound states of the evanescent channels [32, 33]. Increasing the strength of the attractive potential one can see more dips [35, 31] in the tunneling coefficient. Another systems embedded inside the cylindrical nanowire, like a cylindrical dot, a quantum ring or a double barrier heterostructure, which also show a similar behavior, were studied in Ref. [31].

We have analyzed until now only the classical allowed energy domain, with a continuous, double degenerated spectrum. The wave functions of the electrons for these energies are extended states, presented here as scattering states. The classically forbidden spectrum contains the bound states or the localized states. The R-matrix formalism can provide also these states, as long as the boundary points $\pm d_z$ are far enough from the quantum dot, so that the bound states fulfill the Neumann boundary condition (28). In such a way, the energies of the bound states are the negative Wigner-Eisenbud energies and the wave functions for the bound state

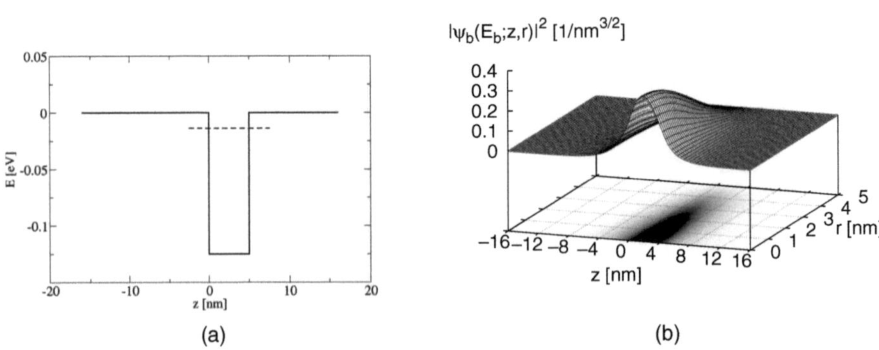

Fig. 10 The bound state for a conical quantum dot inside the nanowire, as in Fig. 7. (a) Value of the bound energy $E_b = -0.013$eV (dashed line). The potential energy along $r = 0$ is represented by the solid line. (b) The absolute value square of the wave function corresponding to the bound state, $|\psi_b(E_b; z, r)|^2$

are the corresponding Wigner-Eisenbud functions. For the conical dot presented here, there is only a bound state represented by the lowest Wigner-Eisenbud energy, $E_b = E_1 < 0$ and $\psi_b(E_b; z, r) = \chi_1(z, r)$. In Fig. 10a we present the energy value E_b of the bound state together with a cut of the potential energy along $r = 0$. The corresponding wave function, namely the absolute value square, is represented in Fig. 10b. One can see that the bound state is mainly localized inside the quantum dot.

4 Summary and Discussion

We have presented a general theory for computing the scattering matrix and the scattering wave functions for a general finite-range extended scattering potential in two dimensions. The theory is based on the R-matrix formalism, which allows a semi-analytical treatment of the scattering problem, yielding in such a way a powerful and efficient numerical method.

This formalism was applied to a quantum dot defined inside a two-dimensional electron gas, as well as to a conical quantum dot embedded inside a cylindrical nanowire.

It is pointed out the role of the evanescent channels, which for a nonseparable attractive scattering potential in a multi-channel nanowire produces resonant dips in the tunneling coefficient. Furthermore, the cylindrical symmetry does not yield the same "selection rules" for tunneling coefficient as the Cartesian symmetry.

It is also presented a general resonance theory, which shows that the two-dimensional character of the scattering potential and the strong coupling of the quantum system to the contacts allow for the transmission profiles which ranges from asymmetric Fano line shapes, through "S-type" Fano lines until antiresonances.

Detailed maps of localization probability distribution density sustain the physical interpretation of the resonances (dips and peaks) found in the studied heterostructures.

Acknowledgements It is a pleasure for us to acknowledge the fruitful discussions with Klaus Gärtner, Vidar Gudmundsson, Andrei Manolescu and Gheorghe Nenciu. One of us (P.N.R.) also acknowledge partial support from German Research Foundation through SFB 787 and from the Romanian Ministry of Education and Research through the Program PNCDI2, Contract number 515/2008.

References

1. Semiconductor Industry Association, International Technology Roadmap for Semiconductors, http://www.itrs.net
2. B. Doyle, R. Arghavani, D. Barlage, S. Datta, M. Doczy, J. Kavalieros, A. Murty, R. Chau, Intel Tech J **6**, 42 (2002)
3. T. Bryllert, L.E. Wernersson, T. Loewgren, L. Samuelson, Nanotechnology **17**, S227 (2006)
4. K.H. Yeo, et al., Tech. Dig. - Int. Electron Devices Meet. 539 (2006)
5. K.H. Cho, K.H. Yeo, Y.Y. Yeoh, S.D. Suk, M. Li, J.M. Lee, M.S. Kim, D.W. Kim, D. Park, B.H. Hong, Y.C. Jung, S.W. Hwang, Appl. Phys. Lett. **92**, 052102 (2008)
6. A.D. Wieck, K. Ploog, Appl. Phys. Lett. **56**, 928 (1990)

7. J. Göres, D. Goldhaber-Gordon, S. Heemeyer, M.A. Kastner, H. Shtrikman, D. Mahalu, U. Meirav, Phys. Rev. B **62**, 2188 (2000)
8. H.S. Wong, K. Chan, Y. Taur, IEDM Tech. Dig. 427–430 (1997)
9. X. Huang, W.C. Lee, C. Kuo, D. Hisamoto, L. Chang, J. Kedzierski, E. Anderson, H. Takeuchi, Y.K. Choi, K. Asano, V. Subramanian, T.J. King, J. Bokor, C. Hu, IEDM Tech. Dig. 67–70 (1999)
10. B. Doyle, S. Datta, M. Doczy, S. Hareland, B. Jin, J. Kavalieros, T. Linton, A. Murthy, R. Rios, R. Chau, IEEE Electron Dev. Lett. **24**, 263 (2003)
11. J. Xiang, W. Lu, Y. Hu, Y. Hu, H. Yan, C.M. Lieber, Nature **441**, 489 (2006)
12. M.T. Bjork, B.J. Ohlsson, C. Thelander, A. Persson, K. Deppert, L. Wallenberg, L. Samuelson, Appl. Phys. Lett. **81**, 4458 (2002)
13. J. Wensorra, K.M. Indlekofer, M.I. Lepsa, A. Forster, H. Lüth, Nano Lett. **5**, 2470 (2005)
14. F. Qian, Y. Li, S.G. Caronak, H.G. Park, Y. Dong, Y. Ding, Z.L. Wang, C.M. Lieber, Nat Mater. **7**, 701 (2008)
15. Y. Hu, H.O.H. Churchill, D.J. Reilly, J. Xiang, C.M. Lieber, C.M, Marcus, Nat. Nanotechnol. **2**, 622 (2007)
16. B. Tian, X. Zheng, T.J. Kempa, Y. Fang, N. Yu, G. Yu, J. Huang, C.M. Lieber, Nature **449**, 885 (2007)
17. K.H. Cho, et al., Tech. Dig. Int. Electron Devices Meet. 543 (2006)
18. D. Vasileska, D. Mamaluy, H.R. Khan, K. Ravela, S.M. Goodnick, J. Comput. Theor. Nanosci. **5**, 999 (2008)
19. L. Smrčka, Superlatt, and Microstruct. **8**, 221 (1990)
20. U. Wulf, J. Kučera, P.N. Racec, E. Sigmund, Phys. Rev. B **58**, 16209 (1998)
21. E. Onac, J. Kučera, U, Wulf, Phys. Rev. B **63**, 85319 (2001)
22. E.R. Racec, U. Wulf, Phys. Rev. B **64**, 115318 (2001)
23. P.N. Racec, E.R. Racec, U. Wulf, Phys. Rev. B **65**, 193314 (2002)
24. G.A. Nemnes, U. Wulf, P.N. Racec, J. Appl. Phys. **96**, 596 (2004)
25. G.A. Nemnes, U. Wulf, P.N. Racec, J. Appl. Phys. **98**, 084308 (2005)
26. T. Jayasekera, M.A. Morrison, K. Mullen, Phys. Rev. B **74**, 235308 (2006)
27. U. Wulf, P.N. Racec, E.R. Racec, Phys. Rev. B **75**, 075320 (2007)
28. J. Behrndt, H. Neidhardt, E.R. Racec, P.N. Racec, U. Wulf, J. Differ. Equ. **244**, 2545 (2008)
29. G. Mil'nikov, N. Mori, Y. Kamakura, Phys. Rev. B **79**, 235337 (2009)
30. E.P. Wigner, L. Eisenbud, Phys. Rev. **72**, 29 (1947)
31. P.N. Racec, E.R. Racec, H. Neidhardt, Phys. Rev. B **79**, 155305 (2009)
32. P.F. Bagwell, Phys. Rev. B **41**, 10354 (1990)
33. S.A. Gurvitz, Y.B. Levinson, Phys. Rev. B **47**, 10578 (1993)
34. J.U. Nöckel, A.D. Stone, Phys. Rev. B **50**, 17415 (1994). DOI 10.1103/PhysRevB.50.17415
35. J.H. Bardarson, I. Magnusdottir, G. Gudmundsdottir, C.S. Tang, A. Manolescu, V. Gudmundsson, Phys. Rev. B **70**, 245308 (2004)
36. V. Gudmundsson, Y.Y. Lin, C.S. Tang, V. Moldoveanu, J.H. Bardarson, A. Manolescu, Phys. Rev. B **71**, 235302 (2005)
37. G. Bastard, J.A. Brum, R. Ferreira, *Solid State Physics*, vol. 44, eds. H. Ehrenreich, D. Turnbull (Academic Press, Boston, MA, 1991), p. 229
38. F.F. Fang, W.E. Howard, Phys. Rev. Lett. **16**, 797 (1966)
39. M. Baro, H.C. Kaiser, H. Neidhardt, J. Rehberg, Rev. Math. Phys. **16**, 281 (2004)
40. H. Schanz, U. Smilansky, Chaos Solitons & Fractals **5**, 1289 (1995)
41. M. Luisier, A. Schenk, W. Fichtner, J. Appl. Phys. **100**, 043713 (2006)
42. A.M. Lane, R.G. Thomas, Rev. Mod. Phys. **30**, 257 (1958)
43. M. Büttiker, Y. Imry, R. Landauer, S. Pinhas, Phys. Rev. B **31**, 6207 (1985)
44. A. Bohm, *Quantum Mechanics* (Springer, New York, NY, 1993)
45. B. Simon, Ann. Physics **97**, 279 (1976)
46. M. Klaus, Ann. Physics **108**, 288 (1977)
47. E.R. Racec, U. Wulf, P.N. Racec, arXiv.1004.0472 [condmat.mes-hall]

Fractional Charge (and Statistics) in Luttinger Liquids

Jon Magne Leinaas

Abstract Charge fractionalization is the phenomenon where quasi-particle excitations in a many-particle system appear with non-integer values relative to the fundamental charge unit. Examples of such systems are known from field theoretical models and from physical realizations. Recently there has been an interest in charge fractionalization in one-dimensional systems described by Luttinger liquid theory. These are gapless systems and that gives rise to the question whether non-integer charges should be regarded as sharp in the same meaning as in a gapped system. In this talk I first give an introduction to charge fractionalization as an effect in gapped systems and discuss next in what sense the charge fractionalization effect is found in gapless Luttinger liquids. The talk is based on a recent paper with Mats Horsdal and Thors Hans Hansson (Phys. Rev B 80: 1153 27, 2009).

1 Introduction

There are certain physical variables that appear, at the fundamental level, always with values that are multiples of a basic unit. A particular example is the electric charge which is quantized in multiples of the electron charge e. This discretization of electric charge is generally believed to have a quantum mechanical origin, even if there at this point is no completely general argument for this to be the case. In spite of the discreteness of electric charge at the fundamental level, *charge fractionalization* is a phenomenon that may occur in certain, sufficiently large many-particle systems. The fractionalization is then based on a redefinition of the charge operator, so that contributions from the ground state (vacuum state) of the many-particle system is suppressed and therefore only local contributions are included which measure deviations from their background values. Such a suppression of the background is

J.M. Leinaas (✉)
Department of Physics, University of Oslo, N-0316 Oslo, Norway
e-mail: j.m.leinaas@fys.uio.no

This is a written version of a talk given at the Joint IFIN-HH,ICTP, IAEA Workshop on Trends in Nanoscience: Theory, Experiment, Technology, in Sibiu, Romania, August 23-30, 2009.

A. Aldea, V. Bârsan (eds.), *Trends in Nanophysics,* Engineering Materials,
DOI 10.1007/978-3-642-12070-1_8, © Springer-Verlag Berlin Heidelberg 2010

often the most sensible way to define physical variables in a low-energy description of the system. Under certain conditions the elementary, particle-like excitations of the a many-particle system, usually referred to as quasiparticles, will then appear with charges that have non-integer values when measured in terms of the original charge unit.

There exist several examples of how charge fractionalization may occur, both in theoretical models and in physical realizations. In particular, two cases of charge fractionalization in condensed matter systems have caught much attention. One is the case of half-integer charges in certain one-dimensional crystals characterized by a Peierls instability. In a field theoretical description of this system, the fractional charges are associated with soliton excitations [1–4]. The second case has to do with quasi-particle excitations in the two-dimensional electron gas of the (fractional) quantum Hall effect. In this case the excitations appear with fractional charges as well as fractional statistics, and the values they take are determined by the ground state, with different rational values for each plateau of the Hall conductivity [5–7]. In addition there are more recent suggestions that charge fractionalization may occur in two-dimensional graphene-like structures [8, 9].

There are certain features that seem to link these cases of charge fractionalization. The fractionalization can thus be viewed as a vacuum polarization effect, where the background defined by the many-particle system "dresses" the quasiparticle with a non-integral number of the original particles, and the fact that there is in these systems a new, fractional unit of charge, reflects a certain property of the ground state which in the field theoretic description takes the form of a topological condition imposed on the fields. Thus, the presence of fractional quantum numbers may seem, quite generally, to be related to the presence of some kind of *topological order* in the system [10, 11], and this connection is one of the reasons that there have been much focus on charge fractionalization effects in many-particle systems. It is of interest to note that the physical dimensionality of the system may be important for such topological effects to take place, and the cases discussed so far are mainly restricted to systems in one and two dimensions.

Another feature that is shared by these examples of charge fractionalization is that the charged excitations are separated from the ground state by an energy gap. The gap suppresses the low frequency contributions to the background fluctuations which therefore may be filtered out in the appropriated "smoothened" definition of the low energy variables. This is the basis for the new fractional units of charge to appear as quantum mechanically sharp in the low energy description.

Recently there has been an increased interest in the question of charge fractionalization in one-dimensional systems described by Luttinger liquid theory [12–18]. This theory is assumed to apply under quite general conditions to many-particle systems in one dimension, and to be relevant to quasi-one-dimensional electron systems in quantum wires and carbon nano-tubes. The interest has been focussed not only on the theoretical analysis of the fractionalization effect in such systems, but also on how to demonstrate the effect experimentally. In fact charge fractionalization has been reported to have been seen in an experiment with electrons in a quantum wire [16]. However, one should note that the actual experiment shows this only in an average sense, and not in the form of individual, sharply defined fractional charges. In addition to this experiment there has quite recently been suggested another way to demonstrate the effect, by use of edge currents in a quantum Hall sample [17].

The possibility of charge fractionalization in Luttinger liquids raises some questions which I have recently discussed in a paper with Mats Horsdal and Hans Hansson [18], and which gives the background for the present talk. These questions arise since the fractionalization effect in Luttinger liquids does not fit into the general picture outlined above. Thus, the system is gapless, and this makes it less clear in what sense fractional charges may be quantum mechanically sharp. There also seems to be no topological constraint on the possible values that such charges may take. Instead various fractional values may appear depending on the initial conditions which give rise to the fractionalization.

The intention here is to discuss some of these questions. In the first part I will discuss charge fractionalization in the more conventional sense, by way of three examples. In the second part I outline how fractional charges may appear in a Luttinger liquid and discuss in what sense these can be viewed as sharp fractional values rather than being fractional in the average sense. The sensitivity of the charge values to different initial conditions are demonstrated by way of two examples, and finally in the summary I discuss the significance this has for our understanding of the effect. In the talk I avoid many details and refer to [18] for a more complete discussion.

2 Examples of Charge Fractionalization

To illustrate the charge fractionalization effect I will briefly discuss three examples, first the classic example provided by the theoretical model of Jackiw and Rebbi [1].

2.1 The Jackiw-Rebbi Soliton

This fractionally charged state appears in a one-dimensional field theory with a scalar field $\phi(x,t)$ and a two-component Dirac field $\psi(x,t)$. In a semi-classical description the ϕ-field is treated as a classical field, while the Dirac field is treated as a quantum field in the background of $\phi(x)$. The field $\phi(x)$ satisfies a non-linear field equation, and this equation has a soliton solution which interpolates between two degenerate energy minima. In the background of the soliton the Dirac field has a localized zero energy solution, and the Dirac vacuum in this background is therefore degenerate, with two solutions corresponding to whether the zero energy state is occupied or not.

The interesting point is that the effect of the Dirac field is to "dress" the soliton so it appears with half-integer charge, either $+e/2$ or $-e/2$, with e as the fermion charge. The soliton of the full theory, which can be interpreted as a physical particle of the theory, then has properties inherited from both the original fields ϕ and ψ, and the charge of this particle is not identical to that of the fundamental Dirac fermion.

A specific realization of the model is provided by the following Lagrangian for the ϕ-field,

$$\mathcal{L}_\phi = \frac{1}{2}\partial_\mu\phi\,\partial^\mu\phi - \frac{1}{2}(\lambda/\kappa)^2(\kappa^2 - \phi^2)^2 \tag{1}$$

here written in relativistic form, with λ and κ as positive constants. The classical ground states correspond to the two constant fields $\phi_\pm = \pm\kappa$, and any finite energy field has to approach one of these values asymptotically for $x \to \pm\infty$. A stationary soliton localized at the origin $x = 0$ is described by the following solution of the field equation

$$\phi_s(x) = \kappa \tanh \lambda x \tag{2}$$

It satisfies the boundary conditions $\phi_s(x) \to \pm\kappa$ for $x \to \pm\infty$. There is also a corresponding anti-soliton given by $\phi_{\bar{s}}(x) = -\kappa \tanh \lambda x$, and there are other solutions where the solitons are shifted along the x-axis and where they are moving with constant velocity (Fig. 1).

The Dirac field of the one-dimensional theory is described by a two component function $\psi(x)$, which is in the simplest case coupled linearly to the ϕ-field. The Hamiltonian has then the form

$$H = \alpha c p + \beta m c^2 \phi(x)/\kappa \tag{3}$$

with α and β as anti-commuting 2×2 Dirac matrices, which can be represented by Pauli matrices, for example as $\alpha = \sigma_x$ and $\beta = \sigma_y$. When the scalar field is in its ground state, either as $\phi = +\kappa$ or $\phi = -\kappa$, the Dirac Hamiltonian describes a free Dirac theory of fermions with mass m and the vacuum state corresponds to a filled Dirac sea, where all negative energy states are occupied. However, in the background of the soliton the mass term tends to zero at the center of the soliton where the scalar field vanishes, and in the neighborhood of this point the localized zero energy state is found. It has the explicit form

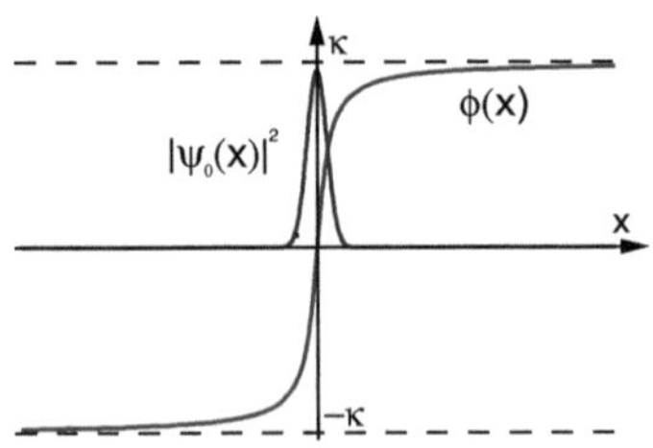

Fig. 1 The soliton solution of the Jackiw-Rebbi model. The curve passing through the origin shows the form of the scalar field $\phi_s(x)$ of the soliton, and the other one shows the absolute square of the zero mode $\psi_0(x)$ of the Dirac field, which appears on the background of the scalar soliton field

$$\psi(x) = N \exp\left(-\frac{mc}{\hbar\kappa}\int_0^x dy\phi_s(y)\right)\chi = N \exp\left[-\frac{mc}{\hbar\lambda}\ln(\cosh\lambda x)\right]\chi \qquad (4)$$

with N as a normalization factor and χ as a two-component spinor which satisfies $\alpha\beta\chi = -i\chi$. For the Dirac vacuum in the background of the soliton field ϕ_s this implies the two-fold degeneracy where all negative energy states are filled, but where the zero-energy state (4) may either be occupied or empty. The integrated charges of these two states can be evaluated and are found to be $\pm e/2$, and these are interpreted as the charges of the solitons dressed by the Dirac vacuum.

The charge values $\pm e/2$ of the solitons are determined by topological properties of the fields, and are therefore insensitive to local changes in the background field ϕ. What is important is the asymptotic behavior of the field, with interpolation between the two minima. A further analysis of the soliton solutions shows that the half-integer charges are quantum mechanically sharp [19, 20].

2.2 The Dressed Flux Tube

In this case we consider the effect of a thin magnetic flux tube that penetrates a two-dimensional plane with Dirac fermions. The fermions are considered as spin polarized and are described by a two-component Dirac field. If the flux tube is assumed to be point-like in the plane, the stationary solutions of the Dirac equation can be found and charge and current densities of the vacuum state in the background of the flux tube can be determined [21–23]. Also in this case there is a polarization of the Dirac vacuum due to the presence of the magnetic flux, and this dresses the flux with charge and creates a vacuum current that encircles the flux tube. These effects are schematically pictured in Fig. 2.

An interesting point is that the total charge attracted to the flux tube is directly proportional to the magnetic flux ϕ,

$$q = -\frac{1}{2}\frac{\phi}{\phi_0}e \qquad (5)$$

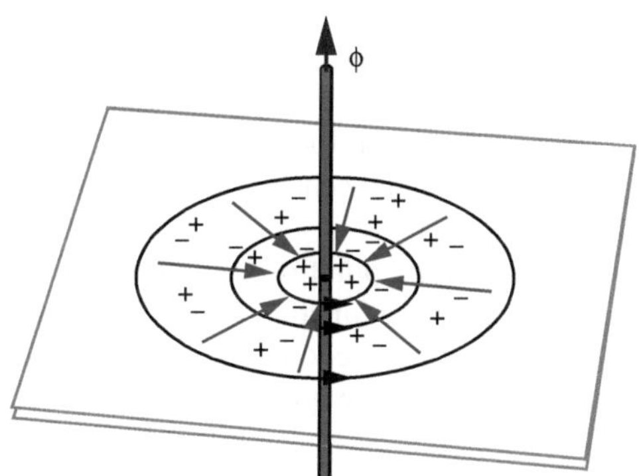

Fig. 2 The dressed flux tube in the two-dimensional Dirac vacuum. The figure shows schematically the effect of the magnetic flux on the vacuum fluctuations, due to the presence of the Aharnov-Bohm field of the flux tube. A surplus charge is attracted to the flux tube and a vacuum current is created that encircles the tube

as long as $|\phi| < \phi_0$, with $\phi_0 = h/e$ as the flux quantum and e as the charge of the Dirac fermions. For larger values of ϕ the charge is a piecewise linear function of ϕ, in a periodic fashion. This linear relation between q and ϕ reflects again a topological property of the charge fractionalization. Thus, the integrated charge is determined by the boundary condition satisfied by the Dirac fields, which in turn depends only on the total magnetic flux inside the boundary. Therefore details of the local form of the magnetic flux are unimportant for the value of the integrated vacuum charge that is attracted to the flux tube [4].

2.3 The Laughlin Quasiparticle

In the quantum Hall effect, where a two-dimensional electron gas is subject to a strong perpendicular magnetic field, the elementary charged excitations for a plateau state with fractional filling are known to carry a fraction of a unit charge and to be anyons rather than bosons or fermions. This was first discussed by R.B. Laughlin [5] and the fractional charge carried by the quasiparticles was later verified experimentally in tunneling experiments [24–26].

The simplest case of such a charge fractionalization effect appear in the Laughlin state at 1/3 filling of the lowest Landau level. A way to understand the appearance of the fractional charged excitation is to imagine the two-dimensional electron gas pierced with a thin magnetic flux tube in a similar way as outlined in the previous example. The effect of this additional magnetic flux is to create a vortex in the electron gas which repels electron charge from the flux tube and creates a current encircling the tube [5]. In this case the charge and current created by the flux tube can be viewed as a polarization effect of the "sea" of electrons in the ground state, in much the same way as the polarization of the Dirac vacuum in the previous example. Figure 3 shows a schematic picture of the situation.

When the magnetic flux is increased to be exactly one flux unit, then the non-local effect that the flux tube has on the electrons will disappear. It becomes in this sense invisible to the electrons and the ground state returns to the original ground state of the 1/3 state without the additional flux of the tube. However, if the flux is adiabatically turned up, it will leave a footprint behind. A hole in the electron density is created even if the direct influence of the flux on the electrons disappears, and this hole represents a true quasiparticle excitation of the two-dimensional electron. The use of the imaginary flux tube is simply a theoretical construction to derive the form of the quasihole, but for the physical system the fractionally charged excitation can instead be created by changing for example the strength of the background magnetic field.

An interesting point of Laughlin's analysis is the suggestion of a simple many-electron trial wave function for the 1/3 ground state and for the quasi-hole excitation. With z_i as complex coordinates for the electrons in the plane (measured in units of magnetic length), the quasi-hole wave function has the form

$$\psi(z_1, z_2, ..., z_N) = \mathcal{N} \prod_i (z_i - z_0)\, \psi_0(z_1, z_2, ..., z_N) \tag{6}$$

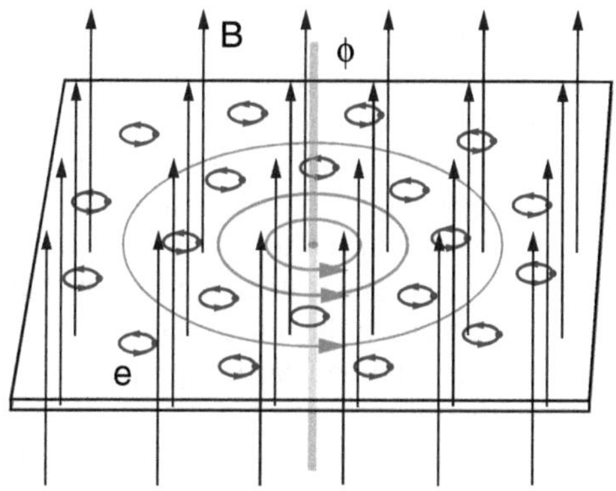

Fig. 3 Creation of a fractionally charged quasiparticle in the two-dimensional electron gas of a quantum Hall system. A strong magnetic field **B** (illustrated by the *vertical arrows*) penetrates the plane of the electrons and trap them in circular orbits in the lowest Landau level. The electrons form an incompressible fluid for certain fractional fillings. The figure illustrates the effect of piercing the fluid with a magnetic flux tube (*in-plane concentric circles*). When the flux ϕ is turned up a charged hole is created, and when a full flux quantum is reached a fractionally charged quasiparticle has been produced

with $\mathcal{N}$ as a normalization factor, z_0 as the quasihole coordinate and $\psi_0(z_1, z_2, ..., z_N)$ as the N-electron ground state,

$$\psi_0(z_1, z_2, ..., z_N) = \mathcal{N}' \prod_{i<j} (z_i - z_j)^3 \exp\left(-\frac{1}{4}\sum_k |z_k|^2\right) \quad (7)$$

For this trial function the properties of the quasihole such as fractional charge and fractional statistics can readily be checked [7]. Numerical studies have also demonstrated the quantum mechanical sharpness of these fractional charges [27].

As already pointed out, there are several features these three examples of charge fractionalization outlined above have in common. The first one is that the local concentration of a fraction of a unit charge can be viewed as a polarization effect of the medium in which the excitation is created. In the first two examples it is a vacuum polarization effect, whereas in the third example it is a polarization effect of the two-dimensional electron gas. In all three cases this polarization can be viewed as due to the presence of a background field, which is in the first case, based on a semi-classical treatment, and in the third example, on the use of a fictitious magnetic flux. The second point is that the fractionalization effect in these cases may be viewed as having a topological origin. In the first case the topology is related to the twist in the scalar soliton field when it interpolates between the two minima, and this affects the positive and negative energy states of the Dirac field in an asymmetric way. In the two other cases the topological effect can be ascribed to the asymptotic behavior of the vector potential introduced by the magnetic field of the flux tube. The topology gives a robustness to the charge fractionalization effect, and makes the values of the fractional charges independent of local details [4].

A third point to mention is that the fractional charges in the cases we have discussed are to be regarded as sharp in the quantum mechanical sense. This has been explicitly demonstrated by evaluating the variance of the local charge in a region that includes a fractionally charged excitations. The sharpness of the charge can be linked to the presence of a gap in the energy spectrum, where in the first two cases the gap is determined by the mass of the Dirac fermions, and in the last example it is given as the energy gap from the ground state to the lowest charged excitation. This sharpness of the charge is important for the interpretation of the fractionally charged excitations as quasiparticles with modified particle properties relative to those of the fundamental particles of the theory.

The question of sharp fractional charges in Luttinger liquids is less clear, precisely because of the lack of a gap in the spectrum. The idea is to discuss this question, with reference to the results of the paper [18], but first I would like to discuss in general terms the meaning of notion of sharp fractional charge, when this is applied to excitations in a many-particle system.

3 Fractional Charge Versus Fractional Probability

For a single particle the question of sharpness of an observable can readily be answered. If the statistical variance of the observable vanishes, which means that the quantum state is an eigenstate of the observable, we refer to the observable as sharp, otherwise it is not. This can be illustrated by comparing two situations, where in the first one (case A) the wave function of a particle with charge e is divided in two parts, with equal probability for the particle to reside inside one of the two boxes. In the other situation (case B) the charge itself is split in two, where half the charge, $e/2$, is located in each of the boxes.

Let us consider a charge operator Q_1 which measures the charge only in box 1. In both situations the expectation value of the operator is $e/2$, but the variance is different. We have for the two cases

$$\text{case A}: \quad (\Delta Q_1)^2 = e^2/4\,, \qquad \text{case B}: \quad (\Delta Q_1)^2 = 0 \tag{8}$$

Clearly, in the first case it is the probability rather than the charge that is divided, so that the charge is either 0 or e, both with probability 1/2. This means that the quantum fluctuations of the charge are large and that is reflected in the non-vanishing value of the variance $(\Delta Q_1)^2$. In the other case it is however the charge itself that is divided so that the charge in box 1 is $e/2$ with probability 1. This gives the vanishing variance of the charge operator Q_1. Obviously charge fractionalization, if this is a meaningful concept, should correspond closer to case B than to case A.

However, in a many-particle system the distinction between the cases is not so obvious. The reason is that the charge to be measured sits on the top of the many-particle ground state, and even if the mean value of the charge density of the ground state vanishes there are in general large fluctuations. If thus a local charge operator is introduced which measures the charge within a well-defined region of space this will for the many-particle states always have non-vanishing variance due to fluctuations in the charges along the boundary of the region. However, if the charge fluctuations are resolved in their frequency components, then the presence of an energy gap Δ for the charged excitations will effectively suppress the low frequency components of the fluctuations. This implies that in a low-energy approximation to the full theory, the charge fluctuations may be filtered out.

Let us restrict the discussion to the one-dimensional case, with the many-fermion system on the x-axis. A way to filter out the high frequency fluctuations is to define the local charge with a soft sampling profile for the charge. A specific way to do this is to introduce a Gaussian blur in the definition of the charge density.[1] We write the averaged density as

$$\tilde{\rho}(x) = \frac{1}{b\sqrt{\pi}} \int dx' e^{-(x-x')2/b^2} \rho(x') \tag{9}$$

with b as a softening parameter. The corresponding expression for the charge sampled in the interval $(-a/2, a/2)$ is

$$Q(a, b) = \int_{-a/2}^{a/2} \tilde{\rho}(x) = \int_{-\infty}^{\infty} dx f(x; a, b) \rho(x) \tag{10}$$

where the sampling function $f(x; a, b)$ is given by

$$f(x; a, b) = \frac{1}{2} \left[\operatorname{erf}\left(\frac{x + a/2}{b}\right) - \operatorname{erf}\left(\frac{x - a/2}{b}\right) \right] \tag{11}$$

with $\operatorname{erf}(x)$ as the error function. The sampling function is then essentially constant, with value 1 over the interval $(-a/2, a/2)$, and drops exponentially fast to zero, with b as a decay parameter, outside this interval, as shown in Fig. 4.

For the charge fluctuations $\Delta Q(a, b)$ the parameter b will effectively define a high frequency cutoff, while the energy gap Δ acts as a low frequency cutoff, so that if $1/b << \Delta/\hbar c$ there will be a suppression of the fluctuations in the whole frequency range. This can be demonstrated explicitly by considering charge fluctuations of the one-dimensional Dirac vacuum discussed earlier in the first example. In that case the energy gap of the system is defined by the fermion mass, $\Delta = mc^2$. With the charge operator $Q(a, b)$ defined as above, the variance $\Delta Q(a, b)$ is given by the following integral over Fourier modes,

[1] The space averaging actually suppresses the short wavelength rather than the high frequency contributions. In one dimension the difference is not important, but in higher dimensions an explicit frequency cutoff may be needed, as discussed in [28].

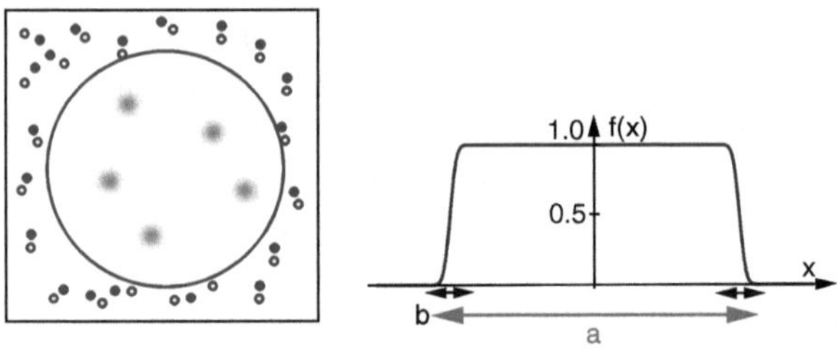

Fig. 4 Low frequency filter for the charge fluctuations. The charge fluctuations are to the *left* pictured as the presence of virtual particle-hole pairs in the ground state. The idea of a low frequency filter (inside the *circle*) is to suppress the fluctuations that are irrelevant for the low energy description, and thereby to identify the quasiparticles with sharply defined (fractional) charges. To the *right* the form of the particular sampling function with soft edges is shown, which is introduced in the text

$$\begin{aligned}\Delta Q(a,b)^2 &= \frac{1}{2\pi^2}\int_{-\infty}^{\infty} dk \int_{-\infty}^{\infty} dk' \left[1 - \frac{(mc/\hbar)^2 - kk'}{\sqrt{((mc/\hbar)^2 + k^2)((mc/\hbar)^2 + k'^2)}}\right] \\ &\quad\times \frac{\sin^2(a(k+k')/2)}{(k+k')^2} e^{-\frac{1}{2}b^2(k+k')^2} \\ &= \frac{1}{4\pi^2}\int_{-\infty}^{\infty} dK \int_{-\infty}^{\infty} d\kappa \left[1 - \frac{\kappa^2 - K^2 + 4(mc/\hbar)^2}{\sqrt{(\kappa^2 - K^2 + 4(mc/\hbar)^2)^2 + 8K^2(mc/\hbar)^2)}}\right] \\ &\quad\times \frac{\sin^2(aK/2)}{K^2} e^{-\frac{1}{2}b^2K^2}\end{aligned} \tag{12}$$

where in the last expression the substitutions $K = k + k'$ and $\kappa = k - k'$ have been done.

The exponential function effectively limits the K-integral so that $|K| \lesssim 1/b << mc/\hbar$, and keeping only the leading contribution in an expansion of the integrand in powers of $K\hbar/mc$, we get for the integral

$$\begin{aligned}\Delta Q(a,b)^2 &\approx \frac{(mc/\hbar)^2}{\pi^2}\int_{-\infty}^{\infty} \frac{d\kappa}{(\kappa^2 + 4(mc/\hbar)^2)^2} \int_{-\infty}^{\infty} dK \sin^2(aK/2) e^{-\frac{1}{2}b^2K^2} \\ &= \frac{\sqrt{\pi}\hbar}{32mc\,b}\left(1 - e^{-a^2/4b^2}\right)\end{aligned} \tag{13}$$

This expression shows that the variance of the local charge can be made negligibly small by choosing b sufficiently large, with $mc\,b/\hbar >> 1$.

Also the charge fluctuations of the soliton can be suppressed by the same smooth sampling function and in this sense the half-integral charge of the soliton is to be regarded as quantum mechanically sharp [19, 20].

It is clear from this result that the suppression of the charge fluctuation by the use of a smooth sampling function depends on the presence of a mass gap. Also this we may demonstrate explicitly, now by taking the limit $m \to 0$. In this limit the integrals in the definition of $\Delta Q(a,b)^2$ can be solved exactly

$$\begin{aligned}\Delta Q(a,b)^2 &= \frac{2}{\pi^2}\int_{-\infty}^{\infty} dK \frac{\sin^2(aK/2)}{K} e^{-\frac{1}{2}b^2K^2} \\ &= \frac{1}{2\pi^2}\left(\frac{a}{b}\right)^2 {}_2F_2\left(1,1;\frac{3}{2},2;-\frac{1}{2}\left(\frac{a}{b}\right)^2\right)\end{aligned} \tag{14}$$

with ${}_2F_2$ as a generalized hypergeometric function. For a large ratio a/b the variance shows a logarithmic behavior,

$$\Delta Q(a,b)^2 \approx \frac{1}{\pi^2}\ln\left(\frac{a}{b}\right) \tag{15}$$

The divergent behavior of the variance for $b \to 0$ is present also in the case $m \neq 0$, as the limit where the edges of the sampling function becomes sharp. However, with $m = 0$, it is the size a of the sampling region rather than the mass m that defines the low frequency cutoff . Thus, in the massless case there is an infrared divergence when $a \to \infty$ which is not cured by the presence of the smoothness parameter b.

This effect is relevant for the Luttinger liquid theory, due to the lack of an energy gap. Also there the ground state fluctuations cannot simply be suppressed by a smooth edge, and the discussion of whether local charges can be quantum mechanically sharp therefore depends on viewing the question of sharpness in a different way.

4 The One-Dimensional Luttinger Liquid

Let me now focus on a many-fermion system which is described by Luttinger liquid theory in the low energy approximation. The system is non-relativistic, and is assumed to be clean in the sense that no background potential affects the particles, except the one that confines them to one dimension. The fermions are considered as spinless, and if the interaction between the particles is neglected the ground state corresponds to the situation illustrated in Fig. 5 where all single particle states with energies below the Fermi energy ϵ_F are occupied and all the states above ϵ_F are unoccupied.

The interacting system will under quite general conditions have a ground state which is qualitatively like the that of the non-interacting system. Thus the Fermi sea, defined by the ground state, is only modified in the neighborhood of the Fermi surface, which in the one-dimensional case is restricted to two points. A low-energy approximation to the interacting theory is furthermore based on the assumption that only excitations limited to a an energy interval around the two Fermi points which is small in the sense $\Delta\epsilon << \epsilon_F$ are of importance. This restriction is indicated by

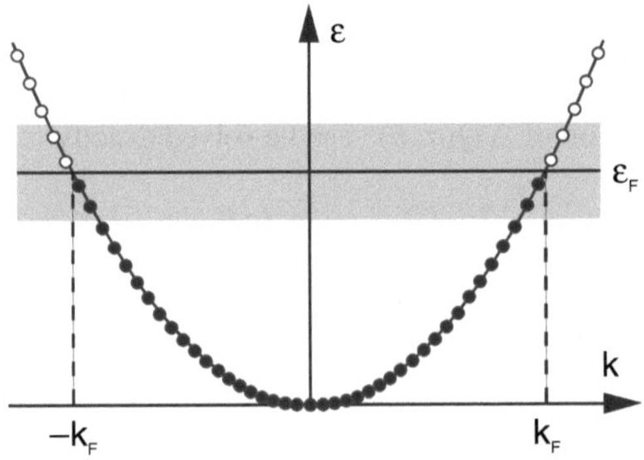

Fig. 5 The low-energy band of the one-dimensional fermion system. The figure shows the quadratic dispersion of the non-interacting system with all single-particle states occupied between the two Fermi points with momenta $\pm k_{\mathrm{F}}$ (*filled dots*) and all states outside these points unoccupied (*open dots*). In the low energy approximation excitations are restricted to energies close to the Fermi energy ϵ_{F} (*light band*). For the interacting system the picture is essentially the same, but then with interactions between the fermions which are either close to the same Fermi point or close to opposite points

the light blue band in the Fig. 5, and a linearized approximation to the dispersion formula $\epsilon = \hbar^2 k^2/2m$ is meaningful in this energy interval.

The linearized theory defines the Tomonaga-Luttinger model, which treats the fermions close to the two Fermi points as two different species [29, 30]. When the system is confined to a circle of length L the Hamiltonian is

$$\begin{aligned} H =& v_F \hbar \sum_{\chi,k} (\chi k - k_{\mathrm{F}}) : c^{\dagger}_{\chi,k} c_{\chi,k} : \\ &+ \frac{1}{2L} \sum_{\chi,q} \big(V_1(q) \rho_{\chi,q} \rho_{\chi,-q} + V_2(q) \rho_{\chi,q} \rho_{-\chi,-q} \big) \end{aligned} \tag{16}$$

with $\chi = \pm$ as a parameter which distinguishes the two species. The parameter v_{F} is the Fermi velocity and k_{F} the Fermi momentum. The interaction is here separated in two parts, with V_1 as the interaction between pairs of particles close to the same Fermi point, and V_2 as the interaction between particles at opposite Fermi points. A *local* interaction between the fermions introduces the restriction $V_1 = V_2$, however for more general (non-local) interactions the two interaction potentials may be different. Although the k quantum number is in the low energy approximation restricted to small deviations from $\pm k_{\mathrm{F}}$, this restriction is not essential since the low energy sector of the theory is not affected by an extension of the allowed values of k. This extended theory describes a system of massless Dirac particles.

One should note that the Hamiltonian (16) has no matrix element that changes the relative number of fermions at the two Fermi points. This implies that there are two conserved fermion numbers N_{χ}, that both take integer values.

The one-dimensional fermion theory is different from the theory in higher dimensions in one important aspect. In one dimension the theory can be fully described in terms of bosonic variables, which are essentially the charge and current densities. The *interacting* fermion theory can, quite remarkably, be written as a *free* boson theory of the form

$$H = \sum_{q \neq 0} \hbar \omega_q \, b_q^\dagger b_q + \frac{\pi \hbar}{2L} \left(v_N N^2 + v_J J^2 \right) \tag{17}$$

where the bosonic operators b_q and $b_q^\dagger$ are linear in the charge and current densities. The interactions are now hidden in the frequency ω_q, given by

$$\omega_q = \sqrt{\left(v_F + \frac{V_1(q)}{2\pi \hbar} \right)^2 - \left(\frac{V_2(q)}{2\pi \hbar} \right)^2} \, |q| \tag{18}$$

and in the two velocity parameters

$$v_N = v_F + \frac{1}{2\pi \hbar} \left(V_1(0) + V_2(0) \right) , \quad v_J = v_F + \frac{1}{2\pi \hbar} (V_1(0) - V_2(0)) \tag{19}$$

In the above expressions $N = N_+ + N_-$ is the total fermion number and $J = N_+ - N_-$ is the conserved current quantum number.

Under the assumption that the interactions are of sufficiently short range, the q dependence of the Fourier transformed potentials can be neglected, $V_1(q) \approx V_1(0)$ and $V_2(q) \approx V_2(0)$. The dispersion of the bosonic field is then linear in $|q|$ and the Hamiltonian can be given the following field theoretic form

$$H = \frac{u}{2} \pi \hbar \int_0^L dx \left[g^{-1} \left(\partial_x \Phi \right)^2 + g \left(\partial_x \Theta \right)^2 \right] \tag{20}$$

with $u = \sqrt{v_N v_J}$ and $g = \sqrt{v_J / v_N}$. In this formulation Θ and $\partial_x \Phi = \partial \Phi / \partial x$ (or alternatively Φ and $\partial_x \Theta$) are regarded as conjugate field variables, with the basic field commutator given as

$$\left[\Theta(x), \partial_x \Phi(x') \right] = \left[\Phi(x), \partial_x \Theta(x') \right] = \frac{i}{\pi} \delta(x - x') \tag{21}$$

The field $\partial_x \Phi$ can be identified with the fermion number density $\rho(x)$ of the original description while $\partial_x \Theta$ is proportional to the current density $j(x)$.

The fields $\Theta(x)$ and $\Phi(x)$ are related to the bosonic creation and annihilation operators in the following way

$$\Theta(x) = \Theta_0 + \frac{x}{L}J - i\sum_{q\neq 0}\frac{1}{\sqrt{2\pi Lg|q|}}\left(b_q e^{iqx} - b_q^\dagger e^{-iqx}\right)$$

$$\Phi(x) = \Phi_0 - \frac{x}{L}N + i\sum_{q\neq 0}\sqrt{\frac{g}{2\pi L|q|}}\,\mathrm{sgn}(q)\left(b_q e^{iqx} - b_q^\dagger e^{-iqx}\right) \tag{22}$$

with Θ_0 and Φ_0 are x-independent operators that generate changes in the fermion numbers N and J.

There is one important difference between the two representations of the Hamiltonian given by Eqs. (17) and (20). In Eq. (17) there is an explicit separation of the $q = 0$ contribution to the Hamiltonian, which depends on the conserved fermion numbers N and J, and the $q \neq 0$ contribution which involves the bosonic variables b_q and $b_q^\dagger$. In (20), on the other hand, the N and J dependent part of the Hamiltonian does not appear explicitly, but is hidden in the *zero mode*, which is the non-propagating $q = 0$ part of the theory. This mode is linked to topological properties of the fields, which are reflected in the quasi-periodic condition

$$\Theta(x+L) = \Theta(x) + J\,, \quad \Phi(x+L) = \Phi(x) - N \tag{23}$$

In the field theoretic description the fermions can therefore be interpreted as topological excitations of the (Φ, Θ) fields, and this gives an interesting, unified description of both the fermionic and bosonic variables. However, one of the points to be stressed is that for the discussion of sharpness of the fractional charges a distinction between the $q = 0$ and the $q \neq 0$ contributions needed [18].

One should note that in the linearized description (20) the two interactions V_1 and V_2 give rise to only one dimensionless interaction parameter g, while the remaining effect of the interactions is to rescale the velocity of propagation u. The case of non-interacting fermions, with $V_1 = V_2 = 0$, gives $g = 1$, and in the following I will refer to this value of g as describing the *non-interacting* system, even if in reality it requires only $V_2 = 0$. Also note that in the following I will refer to *charge* as equivalent to *fermion number* without including the charge unit e and I will measure fermion numbers relative to those of a fixed ground state, which then are assigned the values $N_\pm = 0$.

5 Chiral Separation and Fractional Charges

The fields $\Theta(x)$ and $\Phi(x)$ satisfy a one-dimensional wave equation, and can be separated in a natural way in two parts, its right- and left-moving components. These components, which can be defined as

$$\Theta_\pm(x) = \Theta(x) \mp \frac{1}{g}\Phi(x) \tag{24}$$

satisfy the linear differential equations

$$(\partial_t \pm u\partial_x)\Theta_\pm(x) = 0 \tag{25}$$

so that $\Theta_+(x)$ describes the right-moving and $\Theta_-(x)$ the left-moving modes. These two types of solutions of the field equation are also referred to as the positive and negative chirality modes.

The Fourier expansion of the chiral fields have the following form

$$\Theta_\pm(x) = \Theta_{0\pm} \pm \frac{2x}{gL} Q_\pm - i \sum_{\pm q>0} \sqrt{\frac{2}{\pi L g|q|}} \left(b_q e^{iqx} - b_q^\dagger e^{-iqx}\right) \tag{26}$$

with the zero mode operators

$$\Theta_{0\pm} = \Theta_0 \mp \frac{1}{g}\Phi_0\,, \quad Q_\pm = \frac{1}{2}(N \pm gJ) \tag{27}$$

These operators satisfy the g-independent commutation relations

$$\left[\Theta_{0i}, Q_j\right] = -\frac{i}{\pi}\delta_{ij} \tag{28}$$

with $i, j = \pm$.

The operators $Q_\pm$ have a natural interpretation as *chiral* charges, with Q_+ as the part of the total charge associated with the right-moving modes and Q_- as the part associated with the left-moving ones. In the non-interacting case, with $g = 1$, we have $Q_\pm = N_\pm$, which means that the two chiralities can be identified with the two types of fermions, with momenta close to either $+k_F$ or $-k_F$. The charges $Q_\pm$ then take, for both chiralities, integer values. In the interacting case, with $g \neq 1$, the situation is different. The distinction between the two types of chirality is no longer identical to the separation into two types of fermions, given by χ, and in general the chiral parts of the fermion number no longer take integer values. Thus a fermion injected into the system at one of the fermi points will split into two parts, each carrying a fraction of the fermion number to the right and to the left, respectively. The interesting question is whether we should view this chiral separation of the two parts of the charge as a charge fractionalization effect or rather as a splitting of the probabilities for an integer charge to move either to the right or to the left. This means that the question is whether the fractional charges (for $g \neq 1$) that move either to the right or to the left should be regarded as quantum mechanically sharp.

To discuss this question, it is useful to focus on a specific case, where a fermion is injected with momentum close to one of the Fermi point, e.g. $k \approx +k_F$. According to the above discussion it should dynamically be separated in its two parts, which move in opposite directions away from where it has been injected. The charges of the two chiral components in this case are $Q_\pm = \frac{1}{2}(1 \pm g)$. The initial state can be viewed as formed by acting with the fermion creation operator on the ground state, and the fermion creation operator can further be decomposed in two operators that create each of the chiral components of the state. The chiral creation operators, when expressed in terms of the chiral fields, have the form [14]

$$V_{\pm}(x) = \exp\left(i\pi\, Q_{\pm}\Theta_{\pm}(x)\right) \tag{29}$$

where V_+ creates the right-moving component and V_- the left-moving one. Furthermore, if we follow the reasoning of [14], the question of sharpness can be settled by examining the properties of these creation operators separately. Each of the operators creates, at the formal level, a chiral state with fractional charge, and by examining the algebra of the operators associated with each chirality the conclusion which is drawn in [14] is that the operators $V_{\pm}(x)$ create quasiparticles with sharply defined fractional charge which also satisfies fractional statistics.

However, there are reasons why we do not find this conclusion fully convincing, and that is a part of the motivation for our work [18], where we have examined the question of charge fractionalization in a different way. Let me briefly discuss why we do not find the arguments based on the discussion of the properties of the operators $V_{\pm}(x)$ convincing.

Clearly the state that we consider can be expressed in terms of the operators V_+ and V_-, but only when these are used in combination. When used separately, on the other hand, the operators are not well-defined. This is due to the periodic boundary conditions that the field operators have to satisfy, as a consequence of the fermionic character of the fundamental particles of the system. All well-defined operators on the Hilbert space of states have to satisfy these boundary conditions, and it is straight forward to check that the operators $V_{\pm}$ do not fulfill this obligation. In fact, since they are creation operators of *fractional* charge they necessarily do not preserve the spectrum of the charge operators N and J, which is another way to see that they violate the boundary conditions.

Let me also point out that the charge operators $Q_{\pm}$ measure *global* charges, in the sense that they depend only on the integrated charges (i.e., fermion numbers) N and J. As a linear combination of these two operators they in fact necessarily are sharply defined for any state which is an eigenstate of N and J, even if they take themselves fractional values. However, in the discussion of sharp fractional charges it is the *local* charges that are relevant rather than the global ones. Whereas the global charges, which correspond to the $q = 0$ components of the charge densities, are restricted by the boundary conditions of the fields, the local charges, which are determined by the $q \neq 0$ components, are not subject to these restrictions. Also note that the $q = 0$ components are not associated with any particular direction of motion and the division of the total charge into the chiral components therefore is somewhat arbitrary. This is different for the $q \neq 0$ components which are associated with a direction of motion.

I will then proceed to define local charge operators for each chirality. These can be extracted from the total charge density, which for $q \neq 0$ can be expressed in terms of the bosonic operators b_q and $b_q^\dagger$ in the following way,

$$\rho_q = \sqrt{\frac{L|q|g}{2\pi}} \left(b_q + b_{-q}^\dagger \right) \tag{30}$$

This expression contains also information about the chiral components of the charge density, since the time evolution is determined by the fact that b_q is a positive frequency operator and $b^\dagger$ is a negative frequency operator. With the frequency given by $\omega = u|q|$ the space and time dependent charge density is

$$\rho(x,t) = \sum_q \sqrt{\frac{L|q|g}{2\pi}} \left(b_q e^{i(qx-u|q|t)} + b_{-q}^\dagger e^{i(qx+u|q|t)} \right) \tag{31}$$

This gives for the positive and negative chirality components

$$\rho_q^+ = \begin{cases} \sqrt{L|q|g/2\pi}\, b_q & q > 0 \\ \sqrt{L|q|g/2\pi}\, b_{-q}^\dagger & q < 0 \end{cases} \qquad \rho_q^- = \begin{cases} \sqrt{L|q|g/2\pi}\, b_{-q}^\dagger & q > 0 \\ \sqrt{L|q|g/2\pi}\, b_q & q < 0 \end{cases} \tag{32}$$

At this point it may be useful to introduce a new variable Γ to characterize the two types of chirality, with $\Gamma = +$ for the right-moving and $\Gamma = -$ for left-moving modes. We note that in the non-interacting case ($g = 1$) Γ is identical to the quantum number χ which identifies the two types of fermions, while in the interacting case ($g \neq 1$) they are different. By use of this new variable, the above expressions for the chiral charge densities can be compactified

$$\rho_q^\Gamma = \sqrt{\frac{L|q|g}{2\pi}} \left[\Theta(\Gamma q)\, b_q + \Theta(-\Gamma q)\, b_{-q}^\dagger \right] \tag{33}$$

with $\Theta(q)$ as the Heaviside step function.

By use of the sampling function $f(x; a, b)$, introduced earlier in Eq. (11), the local chiral charge operators can be defined which measure charges in the interval $(-a/2, a/2)$,

$$Q_\Gamma(a,b) = \int dx f(x;a,b) \rho^\Gamma(x) = \frac{2}{L} \sum_{q>0} \sqrt{\frac{Lqg}{2\pi}} \frac{\sin(aq/2)}{q} e^{-\frac{1}{4}q^2 b^2} \left(b_{\Gamma q} + b_{\Gamma q}^\dagger \right) \tag{34}$$

We are interested in the expectation values and fluctuations of these for the state where an electron is injected with momentum close to one of the fermi points χk_{F}. To be more precise let us assume that the initial state after the injection has the form

$$|\psi\rangle = \int dx\, \phi(x)\, \psi_\chi^\dagger(x)\, |G\rangle \equiv \Psi_\chi^\dagger |G\rangle \tag{35}$$

$\psi^\dagger_\chi(x)$ as the fermion field for species χ, $\phi(x)$ is the wave function of the injected function, and $|G\rangle$ is the ground state of the interacting many-fermion system. $\Psi^\dagger_\chi$ is then the creation operator of the initial state which is characterized by fermion numbers $N_{\chi'} = \delta_{\chi\chi'}$. The expectation values of the chiral charge operators have the form

$$\langle Q_\Gamma(a,b)\rangle = \frac{2}{L}\sum_{q>0}\sqrt{\frac{Lqg}{2\pi}}\,\frac{\sin(aq/2)}{q}e^{-\frac{1}{4}q^2b^2}\left\langle G\left|\Psi_\chi\left(b_{\Gamma q}+b^\dagger_{\Gamma q}\right)\Psi^\dagger_\chi\right|G\right\rangle \quad (36)$$

The expectation value on the right-hand side of the equation is in fact straight forward to evaluate. This is because the ground state is annihilated by the operators b_q and because the commutators between these operators and the fermion creation and annihilation operators also have a simple form, as a consequence of the fact that b_q and $b^\dagger_q$ are linear in the charge densities. I will here skip the details and refer to [18] for these. Let me instead focus on the q dependence of the sum. We may be interested in choosing a sufficiently large so that the whole contribution from the fermion wave function is captured by the integral. That can be done in the following way. We first take the infinite length limit of the circle, $L\to\infty$, which makes the sum over q into an integral, and next take the limit $a\to\infty$. In this limit the function $\sin(aq/2)$ essentially suppresses the contribution of the integral to arbitrary small values of q. The corresponding expectation value for the chiral charges, denoted by $\bar{Q}_\Gamma$, can then be written as

$$\begin{aligned}\bar{Q}_\Gamma &= \lim_{a\to\infty}\left[\frac{1}{\pi}\int_0^\infty dq\frac{\sin(aq/2)}{q}e^{-\frac{1}{4}q^2b^2}\right]\lim_{q\to 0^\Gamma}\left[\sqrt{\frac{Lqg}{2\pi}}\left\langle G\left|\Psi_\chi(b_q+b^\dagger_q)\Psi^\dagger_\chi\right|G\right\rangle\right]\\ &= \lim_{q\to 0^\Gamma}\left[\frac{1}{2}\sqrt{\frac{Lqg}{2\pi}}\left\langle G\left|\Psi_\chi(b_q+b^\dagger_q)\Psi^\dagger_\chi\right|G\right\rangle\right]\end{aligned} \quad (37)$$

By comparing with the earlier expressions for the chiral charge densities, this expression can in fact be recognized as the following limit of the expectation value of the charge densities

$$\bar{Q}_\Gamma = \lim_{q\to 0}\left\langle\frac{1}{2}\left(\rho^\Gamma_q+\rho^{\Gamma\dagger}_q\right)\right\rangle \quad (38)$$

This expression should be compared to the mean value of the global charges, defined as $Q_\Gamma = \frac{1}{2}(N+\Gamma J)$, which we may identify as the $q=0$ components of the charge densities

$$\langle Q_\Gamma\rangle = \left\langle\rho^\Gamma_{q=0}\right\rangle \quad (39)$$

So the distinction between the total local charge defined by Eq. (38) and the global charge defined by Eq. (39) is the same as the distinction between the $q = 0$ component of the charge density and the $q \to 0$ limit [31].

In the case we consider here, with a fermion injected at one of the Fermi points, a further evaluation of the local charges shows that the two expectation values in fact are equal. For an initial state with $\chi = +$ we then have

$$\bar{Q}_\Gamma = \langle Q_\Gamma \rangle = \frac{1}{2}(1 + \Gamma g) \tag{40}$$

This result, which means that the transition $q \to 0$ is continuous, is easy to understand when we consider situations where the transition is *not* continuous. That happens when the initial state contains a constant background charge in addition to that of the ground state. Such a constant charge distribution will only affect the $q = 0$ component and not the $q \neq 0$ components of the charge density. In the case discussed here the fermion charge is injected locally and no constant background is created. I will discuss also another case, where a state with a purely right-going component is created. In such a case there has to be a compensating background charge and therefore a discontinuity in the transition $q \to 0$ for the charge density.

6 The Two-Dimensional Representation

In order to make clearer the distinction between the two variables χ and Γ and to illustrate the dynamical separation of the two chiral components of the initial state, it is useful to introduce a two-dimensional representation of the many-fermion system. The charge densities in the one- and two-dimensional representations are related by the mapping

$$\rho_2(x, y) = \mathcal{N} \int dx' \int d\xi e^{-\frac{1}{l^2}\left[(x-x')^2 + \frac{1}{4}\xi^2 - iy\xi\right]} \rho_1 \left(x' + \frac{\xi}{2}, x' - \frac{\xi}{2}\right) \tag{41}$$

with $\mathcal{N}$ as a normalization factor, and l as an unspecified constant with the dimension of length. The equation relates the *off-diagonal* matrix elements of the one-dimensional charge density ρ_1 to the *diagonal* matrix elements of the two-dimensional density ρ_2. The two-dimensional representation we may view as providing a phase-space description of the one-dimensional system, with the y coordinate corresponding to the rescaled momentum kl^2.

As discussed in [32] the two-dimensional description can also be viewed as corresponding to an alternative, physical realization of the theory. It describes a two-dimensional electron gas in a strong magnetic field, with l as the magnetic length. With the electrons restricted to the lowest Landau level, the electron gas is dynamically equivalent to one-dimensional system, and with a harmonic confinement potential in one direction of the plane, the Hamiltonian is equivalent to that of the one-dimensional fermion system with a quadratic dispersion. (The same correspondence has recently been applied in [17], where the possibility of detecting charge fractionalization on the edge of a quantum Hall system has been analyzed.)

The ground state of the non-interacting system corresponds to the situation where a band of electron states along the x-axis are fully occupied, with sharp transition to the unoccupied states along the edges of the band. Interaction between the electrons introduces a modification of the edge profile, but does not change the density of the interior away from full filling (of the lowest Landau level). The low energy approximation, in this picture, corresponds to a description of the dynamics restricted to edge excitations of the system. When the electron interaction is sufficiently long range to interconnect the two edges of the system, we have $g \neq 1$ for the interaction parameter, and this corresponds to the situation with a coupling between the edge excitations on the two sides of the band.

In the two-dimensional representation the parameter Γ can be associated with the horizontal (x)-direction, where $\Gamma = +$ corresponds to propagation to the right and $\Gamma = -$ corresponds to propagation to the left. The parameter χ, on the other hand, is associated with the vertical (y) direction, with $\chi = +$ corresponding to the upper edge and $\chi = -$ corresponding to excitations on the lower edge.

The situation is illustrated in Fig. 6, where an electron is injected close to $x = 0$ on one of the edges, pictured in the form of a modulation of the edge. In the case $g = 1$ (Fig. 6a) with no coupling between the edges, we have $\chi = \pm$ equivalent to $\Gamma = \pm$, and the modulation is therefore displaced with time in the positive x-direction for injection on the upper edge and in the negative direction for injection on the lower edge. In the interacting case, with $g \neq 1$, the situation is different due to the coupling between the two edges. The charge moving to the right on the upper edge will then produce a (smaller) image charge moving to the right on the lower edge (Fig. 6b). The ratio between these two charges is uniquely determined by the interaction parameter g. However, charge conservation on the lower edge implies that there must be a compensating charge of equal size but opposite sign that moves

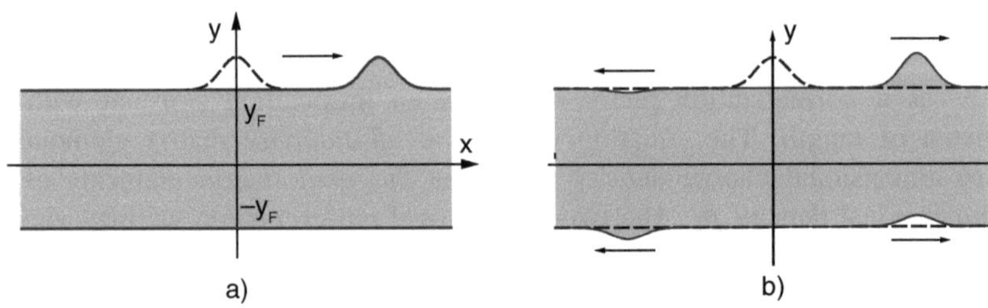

Fig. 6 Two-dimensional representation of the fermion system, with the y-coordinate proportional to the momentum k. The ground state corresponds to a state with constant density in a band between the Fermi momenta, here represented by $\pm y_F$. If a Fermion is injected in the non-interacting system ($g = 1$) at the *upper* edge (*dashed curve*), the excitation will propagate as a modulation of the edge to the *right*, as illustrated in the *left part* (**a**) of the figure. If there is an inter-edge interaction ($g \neq 1$), a fermion injected on the *upper* edge will instead separate in a *right*- and a *left*-moving mode, and each of these will have components on both edges, as illustrated in the *right* part (**b**) of the figure

to the left. This in turn has a mirror charge sitting on the upper edge that also moves to the left.

As indicated in the figure the initial charge is in this way separated in four parts. Let us denote these charges by $Q_{\chi\Gamma}$, so that $N_\chi = \sum_\Gamma Q_{\chi\Gamma}$ and $Q_\Gamma = \sum_\chi Q_{\chi\Gamma}$. It is interesting to note that these four components of the injected charge are all determined by charge conservation on each edge combined with the given ratio between the right(left)-going charge on the upper (lower) edge and its mirror component on the opposite edge. Thus, the ratios between the primary charge, defined by $\chi = \Gamma$ and its mirror charge, defined by $\chi = -\Gamma$ is given by

$$\frac{Q_{++}}{Q_{-+}} = \frac{Q_{--}}{Q_{+-}} = \frac{g+1}{g-1} \tag{42}$$

This fixed relation for the *local* mirror charges on the two edges can be ascribed to the fact that g acts as a mixing parameter between the two original chiralities $\Gamma = \pm$ when the interaction between the edges is turned on. With the given initial conditions $N_+ = 1$ and $N_- = 0$ we in addition have

$$Q_{++} + Q_{+-} = 1\,, \quad Q_{-+} + Q_{--} = 0 \tag{43}$$

These two sets of conditions give

$$\begin{aligned} Q_{++} &= \frac{1}{4g}(g+1)^2\,, & Q_{-+} &= \frac{1}{4g}(g^2-1) \\ Q_{--} &= -\frac{1}{4g}(g^2-1)\,, & Q_{+-} &= -\frac{1}{4g}(g-1)^2 \end{aligned} \tag{44}$$

which for the two chiral charges implies

$$\begin{aligned} Q_+ &= Q_{++} + Q_{-+} = \frac{1}{2}(1+g) \\ Q_- &= Q_{+-} + Q_{--} = \frac{1}{2}(1-g) \end{aligned} \tag{45}$$

This is consistent with the results of the more detailed calculation sketched in the previous section.

As already pointed out, in the case with a sudden injection of a fermion on one of the edges, there is no difference between the expectation values of the local and global charges. There is, however, a difference when we instead consider the quantum fluctuations of these variables. I will soon discuss this point further, but let me first discuss how to choose initial conditions so that only one chiral component with fractional charge will be present. I have argued that this cannot be done by introducing creation operators of the form (29), since these do not treat the background charge correctly. The point to note is that a single local excitation with fractional charge has always be accompanied by a compensating background charge.

A constructive way to introduce such an initial state is the following. We consider a fermion injected locally in the *non-interacting* system at one of the edges. Since $g = 1$ this excitation has well-defined chirality. As the next step we slowly (adiabatically) turn on the interaction so that $g \neq 1$. If this change is sufficiently slow it will not create a local excitation with opposite chirality (a back-scattered component) but will instead create a compensating background charge uniformly distributed over the system. The initial state can now be written

$$|\psi\rangle = U_g \Psi_\chi^\dagger |F\rangle \tag{46}$$

with $|F\rangle$ as the ground state of the non-interacting system and U_g as the operator that performs the adiabatic turning on of the inter-edge interaction. The evaluation of the two chiral components of the charge can be done in much the same way as the first case. This is so since U_g can be identified with the unitary operator that transforms the bosonic creation and annihilation operators of the non-interacting system into the corresponding operators of the interacting system [18]. I only quote the result for the expectation values of the local charges

$$\bar{Q}_+ = \sqrt{g}\,, \quad \bar{Q}_- = 0 \tag{47}$$

which shows that for a particle injected on the upper edge only a right-moving excitation is created. These expressions is to be compared with the corresponding ones for the *global* charges $Q_\pm$ defined in Eq. (27). They take the same values as before, since they are fixed by the fermion numbers $N_+ = 1, N_- = 0$, which are not affected by the interaction,

$$Q_+ = \frac{1}{2}(1+g)\,, \quad Q_- = \frac{1}{2}(1-g) \tag{48}$$

There is now a difference between the local charges (47) and the global charges (48) which can be ascribed to the presence of constant background charges on each edge. These have been created by the adiabatic process, under which the original integer charges on the two edges are separated in the fractional, local components and the compensating background charges. The situation is pictured in Fig. 7.

It is interesting to note that the non-integer charge carried by the positive chirality component in the adiabatic case is different from the charge of the same chirality in the case of sudden injection. This shows explicitly that even if non-integer charges may be created in the Luttinger liquid, the value of these charges are not uniquely determined by the theory, but rather depends on the initial conditions under which the fractionally charged excitations have been created.

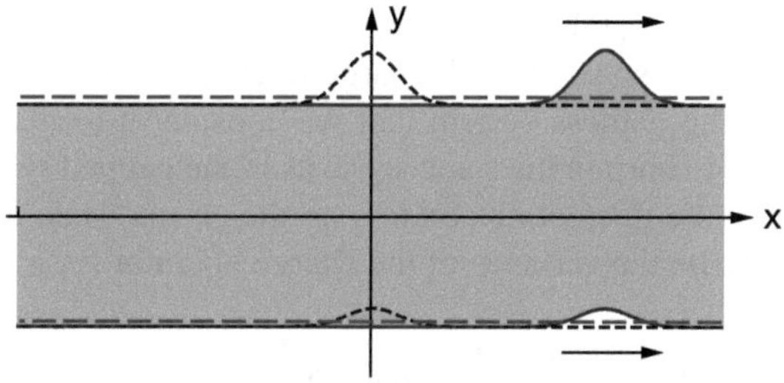

Fig. 7 Creation of a purely right-moving mode, shown in the two-dimensional representation. A fermion is injected into the non-interacting system on the upper edge and the interaction is adiabatically turned on. The created modulation of the edges (*dashed curves*), which carries a fractional charge, will propagate to the right. A compensating, constant background charge is created when the interaction is turned on, here illustrated by the difference between the original positions of the edges (*dashed straight lines*) and the positions of the edges with the excitation (*solid lines*)

7 Charge Fluctuations

Let me finally turn to the question of sharpness of the fractional charges that are created in the Luttinger liquid. I will then focus on the two examples discussed above, with sudden and with adiabatic injection of a fermion charge at one of the edges. Since the system is gapless we expect, in the same way as already discussed, that there are ground state fluctuations that cannot be suppressed by a smooth sampling function. This point can, in the bosonized representation of the theory, be checked for an arbitrary value of the interaction parameter g. The quantity to be evaluated is the variance

$$\Delta Q(a,b)^2 = \langle G|Q(a,b)^2|G\rangle \tag{49}$$

and this can be done in essentially the same way as sketched for the determination of the expectation values of the fractional charges. Again I refer to [18] for details, and simply quote the result

$$\Delta Q_\pm(a,b)^2 = \frac{g}{4\pi^2}\frac{a^2}{b^2}\, {}_2F_2(1,1;3/2,2;-a^2/2b^2) \equiv \frac{1}{2}\Delta Q_0(a,b)^2 \tag{50}$$

where $\Delta Q_0(a,b)^2$ is the variance of the full local ground state charge. We note that this variance is, except for the presence of g as a scaling parameter, precisely the same as previously found for the massless Dirac vacuum in one dimension. That is in fact not so surprising, since the linearized theory underlying the bosonized description can be viewed as describing a system of massless Dirac particles. For $a >> b$ the variance then has a logarithmic dependence on the ratio a/b, as earlier discussed.

The presence of undamped ground state fluctuations implies that *any* excited state of the system will also show such charge fluctuations for large values of a/b. This implies that in the gapless system that we consider a local charge cannot be quantum mechanically sharp in the same sense as in the gapped system. However, it i possible to measure the fluctuations *relative* to the ground state fluctuations in the following way. We write the variance of the charge operator for a general state as

$$\Delta Q_\Gamma(a,b)^2 = \frac{1}{2}\Delta Q_0(a,b)^2 + \Lambda_\Gamma(a,b) \tag{51}$$

where $\Lambda_\Gamma(a,b)$ measures the deviations from the ground state fluctuations. This deviation remains finite when $a/b \to \infty$ for any state which carries a finite excitation charge. Following [18] we may then *define* a chiral charge to be sharp when this deviation from the background variance vanishes for sufficiently large a.

The results we have found, for both cases, with either sudden injection of a fermion charge or with adiabatic switching on the interaction, is that the deviations $\Lambda_\Gamma(a,b)$ vanish in the limit $a \to \infty$. Thus, in both cases the fractional charges created dynamically by the chiral separation are sharp in the meaning given above. This is so even if the expectation values of the chiral charges in the two cases are different.

8 Concluding Remarks

The idea of charge fractionalization in Luttinger liquids is an interesting one, but as discussed in this talk the appearance of non-integer charges due to chiral separation is in some important respects different from charge fractionalization in the more "classic" sense. The main reason for this is that Luttinger liquids are gapless systems, where charge fluctuations of the many-particle background cannot be controlled by a smooth sampling function for the local charge operator, in the same way as in gapped systems. However, in the examples considered here, with a fermion injected close to one of the Fermi points, the fractional charges are sharp in the sense that the fluctuations are identical to the background fluctuations of the many-particle ground state. This is in contrast to what happens with more general initial states. In particular, if a particle is injected with a non-vanishing probability for exciting both edges, then there will be a finite addition to the variance of the chiral charges, which may be interpreted as statistical mixture of two different distributions of the charge.

As I have stressed, it is important in the discussion of sharpness of the charges to focus on *local* charge operators rather than the global ones. For a fractional charge the difference between these is a compensating charge, which in a compact space will be evenly distributed over the background, while in an open system may be collected at the remote edges of the system. For the charge density this means that the information about the fractional charge, and in particular the quantum fluctuations of the charge, sits in the Fourier components $q \neq 0$ rather than in the component $q = 0$. This distinction between the two types of contributions is also important since the $q = 0$ component is constrained by the conserved fermion numbers, and is insensitive to the inter-edge interactions, while the $q \neq 0$ components are not restricted in the same way but are sensitive to the interactions.

The point that fractional charges with different values can be created by choosing different initial conditions is also making the fractionalization effect in the Luttinger liquid different from the effect in the gapped systems. In those cases the topological properties of the systems impose restrictions on the allowed charge values, but in the Luttinger liquid there seems to be no similar restriction. The boundary conditions of the fields can be viewed as a topological constraint, but these are non-dynamical, fixed to the fundamental fermion charges, and do not impose constraints on the fractional, local charges. In fact, even in the non-interacting case $g = 1$ fractional charges of *arbitrary* values can be created as polarization charges induced by an external local potential [18]. If such a charge is attracted to a small region by the potential and then released by suddenly turning off the potential, then the charge will separate in the two chiral components, which will generally carry fractional charges with values determined by the strength of the potential. And these fractional charges will be quantum mechanically sharp in the same sense as in the two other examples [18]. In this case (with $g = 1$) the dynamical separation of the two different chiralities is *not* the cause of the fractionalization.

These results, that fractional charges of arbitrary value can be created that are sharp in the meaning given above, makes it difficult to identify these as charges of elementary quasiparticle excitations of the system. A possible conclusion to draw may then be that charge fractionalization in Luttinger liquids is a different, and simply a less interesting phenomenon than in gapped systems, where the fractionalization is associated with the appearance of quasiparticles with exotic properties. However, maybe one should be open for more interesting possibilities. Let us compare the situation for the Luttinger liquid with that of gapless systems described by Fermi liquid theory, in dimensions higher than one. In these systems the quasiparticle picture is of fundamental importance, even if the idea of charge fractionalization seems not to be a central issue. However, as discussed in [31], also in Fermi liquids the quasiparticles may be considered as carrying non-integer charges. Thus, if these are considered to evolve from the fundamental particles (electrons) through an adiabatic process, they seem to carry only a fraction of the electron charge. This can be shown in a perturbative analysis, where a compensating charge is stored in the background, in much the same way as discussed here [31]. For the Fermi liquids such a charge fractionalization seems, however, less important than the fact that the quasiparticles appear as weakly interacting particles that satisfy Fermi-Dirac statistics like the electrons.

As a final remark let me therefore suggest that there may be some additional conditions that are missing in the discussion of fractional charges in Luttinger liquids, which could be of importance for the application of the quasiparticle picture to this system. These conditions could imply that there are quasiparticles that carry specific, non-integer charges, but more importantly these quasiparticles could be different from the fundamental particles in the sense that they are weakly interacting and obey *fractional statistics* rather than the Fermi-Dirac statistics. This would bring the conclusion closer to ideas discussed in [33, 34] about relations between Luttinger liquid theory and a free gas of exclusion statistics particles.

Acknowledgement Thanks to Mats Horsdal and Hans Hansson for useful comments on the manuscript.

References

1. R. Jackiw, C. Rebbi, Solitons with fermion number 1/2, Phys. Rev. D **13**, 3398 (1976)
2. W.P. Su, J.R. Schrieffer, A.J. Heeger, Solitons in polyacetylen, Phys. Rev. Lett. **42**, 1698 (1979)
3. J. Goldstone, F. Wilczek, Fractional quantum numbers on solitons, Phys. Rev. Lett. **47**, 986 (1981)
4. A.J. Niemi, G.W. Semenoff, Fermion number fractionization in quantum field theory, Phys. Rep. **135**, 99 (1986)
5. R.B. Laughlin, Anomalous quantum Hall effect: An incompressible quantum fluid with fractionally charged excitations, Phys. Rev. Lett. **50**, 1395 (1983)
6. B.I. Halperin, Statistics of quasiparticles and the hierarchy of fractional quantized Hall states, Phys. Rev. Lett. **52**, 1583 (1984)
7. D. Arovas, J.R. Schrieffer, F. Wilzcek, Fractional statistics and the quantum Hall effect, Phys. Rev. Lett. **53**, 722 (1984)
8. C.-U. Hou, C. Chamon, C. Mudry, Electron fractionalization in two-dimensional graphenelike structures, Phys. Rev. Lett. **98**, 186809 (2007)
9. R. Jackiw, S.-Y. Pi, Chiral gauge theory for graphene, Phys. Rev. Lett. **98**, 266402 (2007)
10. X.-G. Wen, Int. J. Mod. Topological orders in rigid states, Phys. B **4**, 239 (1990)
11. M. Oshikawa, T. Senthil, Fractionalization, topological order and quasiparticle statistics, Phys. Rev. Lett. **96**, 060601 (2006)
12. I. Safi, Ann. Phys. Fr. **22**, 463 (1997)
13. M.P.A. Fisher, L.I. Glazman, in Mesoscopic Electron Transport, ed. by L. Kouwenhoven, G. Schoen, L. Sohn NATO Advanced Study Institute, Series E: Applied Sciences, vol 345 (Kluwer, Dordrecht, 1997)
14. K.-V. Pham, M. Gabay, P. Lederer, Fractional excitations in the Luttinger liquid, Phys. Rev. B **61**, 16 397 (2000)
15. K. Le Hur, B. I. Halperin, A. Yacoby, Charge Fractionalization in nonchiral Luttinger systems, Ann. Phys. (NY) **323**, 3037 (2008)
16. H. Steinberg, G. Barak, A. Yacoby, L.N. Pfeiffer, K.W. West, B.I. Halperin, K. Le Hur, Charge fractionalization in quantum wires, Nat. Phys. **4**, 116 (2008)
17. E. Berg, Y. Oreg, E.-A. Kim, F. von Oppen, Fractional charges on an integer quantum Hall edge, Phys. Rev. Lett. **102**, 236402 (2009)
18. J.M. Leinaas, M. Horsdal, T.H. Hansson, Sharp fractional charges in Luttinger liquids, Phys. Rev. B **80**, 115327 (2009)
19. S.A. Kivelson, J.R. Schrieffer, Fractional charge, a sharp quantum observable, Phys. Rev. B **25**, 6447 (1982)
20. R. Rajaraman, J.S. Bell, On solitons with half integral charge, Phys. Lett. B **116**, 151 (1982)
21. T. Jaroszewicz, Fermion-solenoid interactions: Vacuum charge and scattering theory, Phys. Rev. D **34**, 3128 (1986)
22. A.P. Polychronakos, Nucl. Phys. B **283**, 268 (1987)
23. E. Flekkøy, J.M. Leinaas, Int. J. Mod. Phys. **29**, 5327 (1991)
24. V.J. Goldman, B. Su, Resonance tunneling in the fractional quantum Hall regime: Measurement of fractional charge, Science **267**, 1010 (1995)
25. L. Saminadayar, D.C. Glattli, Y. Jin, B. Etienne, Observation of the e/3 fractionaly charged Laughlin quasiparticle, Phys. Rev. Lett. **79**, 2526 (1997)
26. R. de-Picciotto, M. Reznikov, M. Heiblum, V. Umansky, G. Bunin, D. Mahalu, Nature (Lond.) **389**, 162 (1997)

27. H. Kjønsberg, J.M. Leinaas, Charge and statistics of quantum Hall quasi-particles – A numericall study of mean values and fluctuations, Nucl. Phys. B **559** [FS], 705 (1999)
28. A.S. Goldhaber, S.A. Kivelson, Local charge versus Aharonov–Bohm charge, Phys. Lett. B **255**, 446 (1991)
29. S. Tomonaga, Prog. Theor. Remarks on Bloch's method of sound waves applied to many-fermion problems Phys. **5** 544 (1950)
30. J.M. Luttinger, An exactly soluble model of a many-fermion system, J. Math. Phys. **4**, 1154 (1963)
31. O. Heinonen, W. Kohn, Internal structure of a Landau quasiparticle wave packet, Phys. Rev. B **36**, 3565 (1987)
32. M. Horsdal, J.M. Leinaas, Explicit mapping between a 2D quantum Hall system and a 1D Luttinger liquid. I. Luttinger parameters, Phys. Rev. B **76**, 195321 (2007); Explicit mapping between a 2D quantum Hall system and a 1D Luttinger liquid. II. Correlation functions, Phys. Rev. B **76**, 195322 (2007)
33. S.B. Isakov, Quantum liquids of particles with generalized statistics, Phys. Lett. A **242**, 130 (1998)
34. Y.-S. Wu, Y. Yu, H.-X. Yang, Characterization of one-dimensional Luttinger liquids in terms of fractional exclusion statistics, Nucl. Phys. B **604**, 551 (2001)
35. F.D.M. Haldane, Luttinger liquid theory of one-dimensional quantum fluids: I. Properties of the Luttinger model and their extension to the general 1D interacting spinless Fermi gas, J. Phys. C: Solid State Phys. **14** 2585 (1981)

Part IV
Optical Properties of Nanostructures

Near-Field Optical Forces from Surface Plasmon Polaritons: Experiment and Theory

Kenneth B. Crozier

Abstract The propulsion of gold nanoparticles by surface plasmon polaritons (SPPs) is demonstrated experimentally. Numerical electromagnetic simulations, using three dimensional finite difference time domain (FDTD) modeling is carried out. These simulations predict that the optical forces are enhanced, due to the field enhancement provided by SPPs and the near-field coupling between the gold particles and the film. The physical origin of the coupling phenomenon is discussed, as it is an important part of the enhanced propulsion phenomenon. A comparison is made between the measured optical force, and the forces predicted by FDTD simulations. The results are shown to be in reasonable agreement.

Gold nanoparticles have broad applications in biotechnology [1]. This has recently resulted in increased attention to techniques for manipulating them by optical means. In biophysics, "handle" structures are often needed. Gold nanoparticles find themselves particularly suited for this function, as their chemical properties allow for custom functionalization. It is therefore of interest to develop improved methods for manipulating nanoparticles in a controlled fashion through optical forces, as this should translate into useful methods for biophysics. Trapping gold nanoparticles at lower laser powers than other types of nanoparticles should also be possible, as they have comparatively high polarizability.

Optical tweezers have been employed in many of the experiments at the forefront of biophysics over the past decade. They have been shown to be particularly useful for trapping dielectric particles. On the other hand, it has been difficult to trap metal nanoparticles in a stable fashion [2–4]. Metal nanoparticles can exhibit strong optical absorption. This means that stronger gradient forces are needed to overcome scattering and absorption forces for stable three dimensional (3D) trapping [3, 5]. Small gold nanoparticles are more suitable for optical trapping than large gold particles because scattering and absorption forces decrease faster than gradient force

K.B. Crozier (✉)
School of Engineering and Applied Science, Harvard University, Cambridge, MA 02138, USA
e-mail: kcrozier@seas.harvard.edu

A. Aldea, V. Bârsan (eds.), *Trends in Nanophysics*, Engineering Materials,
DOI 10.1007/978-3-642-12070-1_9, © Springer-Verlag Berlin Heidelberg 2010

with particle diameter. Svoboda and Block demonstrated in 1994 that 36 nm diameter gold beads could be trapped with a 7-fold higher trap stiffness than similarly sized polystyrene beads [5]. Hansen et al., have more recently expanded the range of sizes of gold nanoparticles that can be optically trapped. This was achieved by slightly overfilling the back aperture of the objective and by using a water immersion objective [6]. It was shown experimentally that gold nanoparticles with diameters ranging from 18 to 254 nm could be trapped with a single, tightly focused, laser beam. The optical trapping of silver nanoparticles has also been experimentally demonstrated [7].

Due to diffraction, the focal spots used by conventional optical tweezers to trap particles are no smaller than approximately half a wavelength. The gradient force achievable for a given input laser power is therefore limited. It is also the case that the nanoparticle being trapped may be considerably smaller than the trapping region. Compared to its size, therefore, the nanoparticle is not trapped in a precise fashion. Based on these considerations, various investigators have proposed the use of surface plasmon nanostructures for optical trapping, since these permit electromagnetic fields to be generated with subwavelength dimensions [8–12]. Garces-Chavez et al., excited surface plasmon polaritons (SPPs) on a gold film, and observed the self-organization of dielectric particles over a large area through the combined effect of optical and thermal convection forces [13]. Using a variant of an optical tweezer known as a photonic force microscope, Volpe et al., measured the enhancement of optical forces on dielectric particles by SPPs. The particles were 500 nm above the gold film. Therefore, they did not disturb the SPPs excessively [14]. As mentioned above, gold nanoparticles have high polarizability and are able to support surface plasmon resonance. We therefore expect that they should have stronger interaction with metallic surface plasmon resonant trapping structures than dielectric particles. In this work, as distinct from what has been done previously, we observe the enhanced manipulation of gold nanoparticles above a gold film. This is found by measuring the velocity of gold particles pushed by optical radiation forces arising from SPP excitation. The particle-film gap is small, meaning that the field intensity between them is enhanced. This provides a platform for studying the optical near-field force occurring for the case of two gold structures separated by a small gap. Previous studies on the propulsion of gold particles by the evanescent field above a dielectric surface [15–17] are different from what we report here. In those investigations, the coupling phenomenon was not involved. In addition, these reports did not employ surface plasmon polaritons on a flat metal film. Our results indicate that the system we study is highly effective at transferring momentum from photons to gold nanoparticles. Gaugiran et al., for example, employed where a high power density in a silicon nitride waveguide to propel gold nanoparticles [18]. The use of SPPs in our system, on the other hand, results in comparable propulsion velocities being achieved at much lower illumination intensities and over a larger area.

To excite surface plasmons on a gold film, we use the Kretschmann prism coupling geometry. The experimental configuration is shown in Fig. 1. The light source

Fig. 1 Experimental set-up

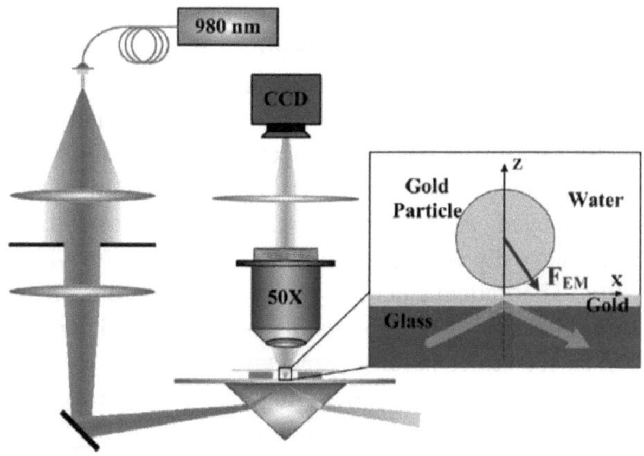

employed is a pigtailed semiconductor diode laser (980 nm) with a polarization-maintaining optical fiber output. Lenses are used to expand the beam. Spatial filtering by an iris enables control of the beam diameter, resulting in a beam with a diameter of ~6 mm in diameter. A lens ($f = 20$ cm) is then used to focus the beam loosely, which is then directed into a prism ($n = 1.51$) after being reflected by a mirror. The mirror sits on a rotation stage, permitting the incident angle to be controlled accurately over a wide range. From Gaussian beam theory, we estimate the resulting $1/e^2$ intensity contour of focused spot to have an elliptical shape of ~40 μm × 80 μm on the surface of the sample. A sample chamber containing gold colloids in water is placed on top of the prism with immersion oil in between. The sample chamber is composed of a glass microscope slide, an SU8 spacer, and a glass cover slide on top. The SU8 spacer, ~5° m thick, is fabricated on the glass cover slide. This spacer layer is intentionally made comparatively thin in order to minimize thermal convection caused by heating of the gold film by laser illumination [13]. Before the glass cover slide is stuck on to the glass microscope slide, gold and SiO_2 films are deposited on the glass microscope slide. Thermal evaporation is used to form the gold film, which is 55 nm thick. Atomic layer deposition is used to form the SiO_2 film (5 nm thick). The purpose of this film is to prevent adhesion between the gold nanoparticles and the gold film. The gold colloid solution is injected into the sample chamber without dilution after receiving it from the supplier (Ted Pella Corp.). The gold colloids are 250 nm diameter spheres with a reported standard deviation (20 nm). The gold colloids have a reported concentration of 3.6×10^8/ml. To track the positions of the gold nanoparticles, a dark-field scattering configuration is employed. The light scattered by the gold particles in the evanescent field is collected by a microscope objective (50×, Nikon CFI 60 LU Plan Epi ELWD infinity-corrected, NA = 0.55) and imaged onto a charge coupled device (CCD) camera (ImagingSource DMK 21F04).

Choosing the appropriate beam divergence involves balancing two considerations. Decreasing the divergence of the beam increases the coupling efficiency to SPPs. This has the undesirable effect, however, of decreasing the intensity of the focused spot. Considering these factors, we choose the divergence of the beam incident on the prism to be 0.8 degrees. We achieve this by focusing a 6 mm diameter beam with a 20 cm focal length lens. An SPP resonance with a sharp peaks and a reasonable intensity are therefore achieved simultaneously.

By measuring the reflectance of the glass/gold/SiO2/water stack, we are able to confirm the excitation of SPPs. A transverse magnetic (TM) polarized input beam is employed. As shown in Fig. 2, the experimental results exhibit a pronounced reflectance dip when the incident angle is around 64°. This is due to the excitation of SPPs. Theoretical calculations are shown on this plot. These take the dielectric constant of gold at $\lambda_0 = 980\,\text{nm}$ as $-44.4141 + 3.3065$ [19]. The experimental results can be seen to be in reasonable agreement with theory. Theory predicts a reflectance dip somewhat narrower than that measured. However, this can be expected because of the divergence of the incident beam. Using the CCD camera images, it is possible to track the positions of gold nanoparticles. This allows the particle velocity to be determined. We make the observations over an area that approximately corresponds to the $1/e^2$ intensity contour of the illumination beam. A nanoparticle in this evanescent field observation region follows a trajectory that is governed by a combination of propulsion from SPPs and Brownian motion. From the CCD camera image data, we find the average velocity of a nanoparticle while in the observation region using. The time at which the nanoparticle enters the evanescent field of the observation region is here denoted t_0. Similarly, t_{stay} is the period for which it stays in this region. Due to variation in intensity and hence scattering force across the beam, randomness in the positions at which particles enter the evanescent field, and Brownian motion, the average velocity varies between particles. By making observations of different particles, on the other hand, an average velocity in the observation region can be found. This averaged velocity, averaged over multiple particles, is plotted in Fig. 2 as a function of incident angle (black dots) for a fixed illumination field intensity of $6\times10^6\ \text{W/m}^2$ incident on the gold film. This corresponds to the average illumination intensity over the $1/e^2$ area of the beam. It is clear from Fig. 2 that particle velocity

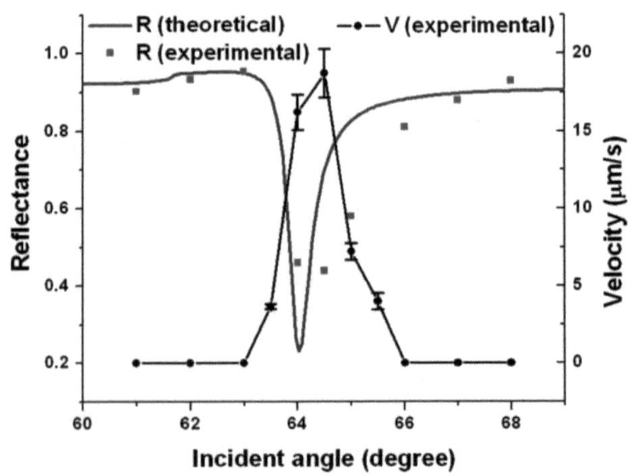

Fig. 2 Experimentally measured reflectance (*squares*) and particle velocity (*dots*) as a function of incident angle. For the velocity measurements, illumination intensity is $6 \times 10^6\text{W/m}^2$. Theoretically calculated reflectance is shown as the blue curve

peaks when SPPs are excited. The standard deviation of, obtained by measurements on different particles, are shown as error bars in this Figure. Brownian motion dominates when the incident angle is away from the SPP angle, or there is no gold film on the glass. For this case, we observe that the particles do not remain in the evanescent field for sufficient time for a velocity measurement to be taken. The velocity in that case should be very small, if not zero, and far smaller than that achieved when SPPs are excited. We perform the same experiment using a transverse electric (TE) polarized input beam, but do not observe any propulsion effect. In Fig. 3, we show a sequence of scattered light images of particles recorded by the CCD camera at a rate of 30 frames per second (fps). This sequence is of the data employed for determination of optical forces. The particle, indicated by an arrow, moves across the image field from right to left in successive frames. The particles circled with dashed lines are stuck on the surface, and do not move.

Particle tracking enables position tracking in two dimensions. Information on the third position dimension, out of the plane of the substrate, can be found from the CCD camera images by monitoring the brightness of the light scattered by a particle as a function of time. In this work, the coordinate axes are defined so that the sample plane is the x–y plane. The +z axis is therefore normal to the gold film and directed into the water. The laser illumination's plane of incidence is therefore the x-z plane. This means that the SPPs that are excited are therefore confined in the z direction, and propagate in the x direction. The forces experienced by the gold nanoparticles in our experiment include random thermal forces in all three dimensions, the optical scattering force in the x direction and the optical gradient force in the -z direction. In Fig. 4a, we show a typical trajectory of a particle in x-y plane. This is obtained by frame-by-frame image analysis of the CCD images. Due to the plane of incidence being the *x–z* plane, there is no optical force in the y direction. The motion of the gold nanoparticles in this direction is thus purely due

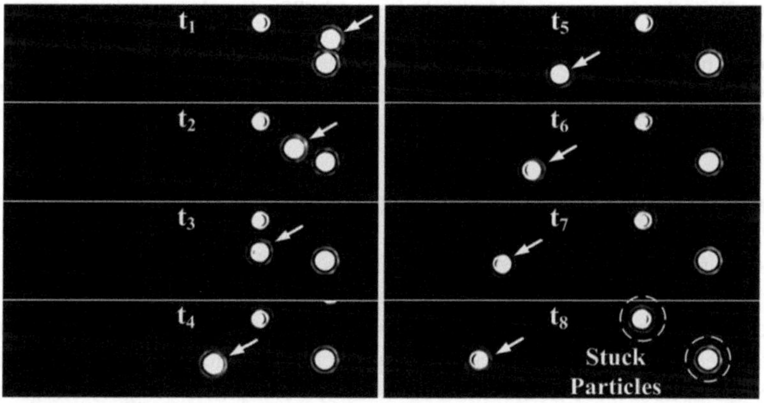

Fig. 3 Time sequence of scattered light images of gold nanoparticles, as recorded by CCD camera. The images have dimensions of $70\,\mu m \times 20\,\mu m$, and the time interval between images is 1/6 s. The gold nanoparticle denoted by the arrow moves right to left. Two nanoparticles, indicated by dashed circles, are stuck to the sample surface

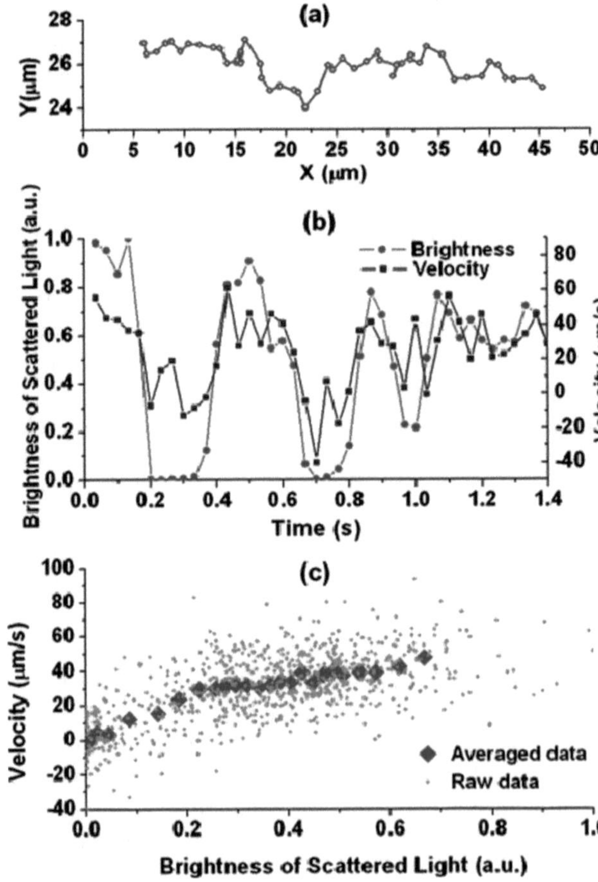

Fig. 4 (**a**) Typical trajectory of a gold particle driven by SPPs. Time between data points is 1/30 s. Particle moves from right to left. (**b**) Instantaneous brightness and velocity of a particle driven by SPPs. (**c**) Particle velocity and brightness scatter plot. Raw data (*dots*) are collected from a number of particles over a period of time. Averaged data (*diamonds*) are obtained by averaging brightness and velocity of 30 neighboring raw data points

to Brownian motion. Observation of the particle motion in this direction provides useful information on the fluid viscosity, as discussed later in further detail. The evanescent field has an exponential decay characteristic, meaning that the brightness of the light scattered by the gold particles allows monitoring of their positions in the z direction, which vary due to Brownian motion. Due again to the decay nature of the evanescent field, the variance of the particles' position in the z direction will modify the optical force in the x direction. When combined with Brownian motion in the x direction, this results in the particle velocity in the x direction being non constant. From the trajectory of the particle, averaged particle's velocity between two continuous frames can be simply calculated by. The brightness at the same instant in time is estimated by, where $P(t_0)$ is the brightness at time t_0, found from the CCD camera images. The result shows a strong correlation between velocity and brightness as given by Fig. 4b. By plotting velocity as a function of brightness using data collected from many particles, this can be seen more clearly. The relationship between brightness and velocity is not immediately obvious from examination of the raw experimental data (green dots), as shown in Fig. 4c. This is due to Brownian motion. This source of noise, however, can be reduced by performing a running average over neighboring points. Each averaged data point of Fig. 4c (red symbols) is obtained by averaging the brightnesses and velocities of 30 neighboring raw data points. It can be seen clearly that the averaged data shows that velocity increases with the brightness of the scattered light. This correlation between brightness and velocity is clear evidence that the observed propulsion is due to optical forces. The optical force is proportional to the total momentum change of the scattered photons. This means that it should be therefore expected that brighter particles, which scatter more light, would move at higher velocities. This is indeed the case. On the other hand, if the particles were being pushed by thermal convection flow accompanying SPP excitation, we would not expect an appreciable variation in particle velocity over the evanescent field decay length. We would therefore not expect to see the strong correlation between velocity and brightness that we observe in Fig. 4c. The same experiment is carried out with 150 nm diameter gold particles as an additional test. These smaller particles experience much weaker optical forces. This provides a way for determining whether particle movement might be due to water flow in the chamber arising from heating. No particle drift is observed, indicating a thermal convection free environment.

By varying the laser power, we perform experiments with different illumination intensities. This allows us to measure particle velocity as a function of illumination intensity. In Fig. 5, results are shown in which a fixed incident angle of 64.5° has been used. The results show that velocity is almost linearly proportional to input power. It is found that the velocity can reach as high as $\sim 50\,\mu\text{m/s}$ without the presence of thermal convection.

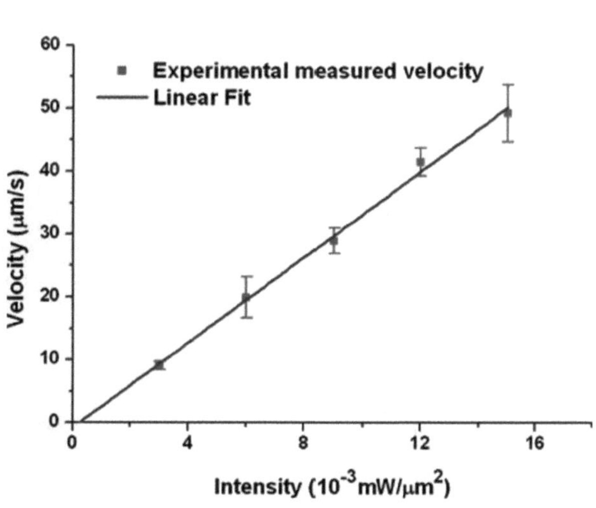

Fig. 5 Experimentally measured particle velocity as a function of illumination intensity. The incident angle is 64.5°

As a means for understand the observed optical propulsion phenomenon, we carry out numerical electromagnetics simulations to calculate the optical force F_{EM} that acts on the particles. Simulation parameters are chosen to match those of the experiments. The numerical simulations employ the 3D FDTD method to find the electric and magnetic fields on a surface enclosing the particle, with the Maxwell stress tensor then used to find the force. The software package FullWave from RSoft is employed. As shown in Fig. 6, the gold particle has a diameter of 250 nm, and the gold film is 55 nm thick. We choose the gap between the gold particle and film to be 10 nm. This is equal to the Debye length estimated for the experimental parameters used. It is believed that the main mechanism that prevents the gold particle from sticking to the surface is the electrostatic force between the negatively charged gold

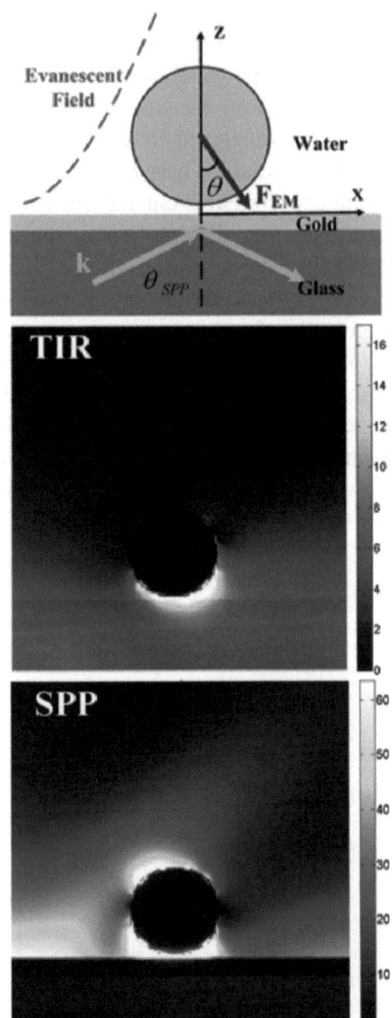

Fig. 6 Top: Schematic of optical manipulation of gold nanoparticle using SPPs. Middle and bottom: 3D FDTD calculated power flow distributions at y = 0 cross-section for TIR and SPP cases. There is no gold film on the glass substrate for the TIR case. For the SPP case, there is a gold film, 55 nm thick, on the glass substrate

particle and the negatively surface-charged SiO_2 film [20]. The gold particles are negatively surface charged to avoid aggregation, and the gold particle solution has a pH of ~9. The estimated concentration of ions in the solution is $\sim 10^{-4}$ mol/L. The Debye length is therefore ~10 nm. The precise separation between the gold nanoparticle and film is not known, but for the calculations we assume it to be the Debye length, as this is the characteristic length of the electrostatic interaction. We use a non uniform simulation grid, ranging from 2 nm near material interfaces to 10 nm further from these boundaries. This ensures accurate modeling of the particle, gap, and film. Quasi continuous wave (CW) illumination is used. The field components are sampled after the simulation reaches steady state. We also find the optical force by integrating the Maxwell stress tensor over different surfaces enclosing the particle and for different simulation times as an additional check. These simulations verify that the calculated optical force is stable. Both for the SPP case and the total internal reflection (TIR), case in which the sample consists of a glass substrate with no gold film, are numerically modeled. The typical power flow distributions, given by the magnitude of the time averaged Poynting vector, calculated for the TIR and SPP cases are shown in Fig. 6. It can be seen that the power flow magnitudes are much higher in the SPP case than the TIR case, which is due to the intense fields resulting from SPP excitation. The simulations also indicate that in the SPP case the gold particle disturbs the EM field much more strongly than it in the TIR case. In Fig. 7, we plot the calculated optical force as a function of incident angle. The angle of the force (defined in Fig. 6) is also shown in Fig. 7. The optical force is ~22 times larger for the SPP case than the TIR case at the SPP angle. If one considers just what would result from the enhancement in intensity of the SPP case by comparison to the TIR case, a force enhancement of ~10 times would be predicted. The simulation results also indicate a somewhat surprising result – for the SPP case the optical force is directed much more toward the sample than for the TIR case. This is remains true even if the incident illumination is far from the SPP excitation angle. This phenomenon is purely due to the interaction between gold film and gold particle. We plot the optical force on the gold particle as a function of gap size in Fig. 8.

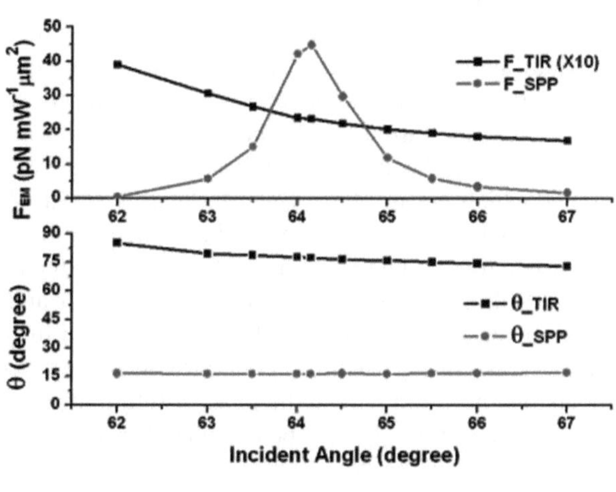

Fig. 7 Optical forces and their directions as a function of incident angles, as found by 3D FDTD simulations. Optical force is normalized to incident intensity

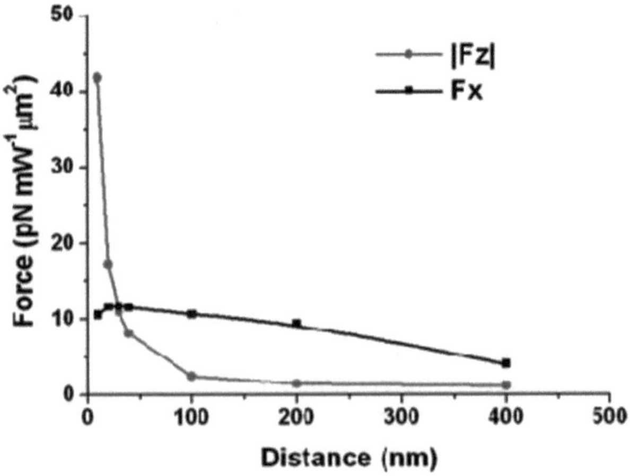

Fig. 8 Optical forces as a function of particle-film gap, found from 3D FDTD simulations

An attractive force F_z acts on the particle, pulling it down to the surface, and this force decreases sharply with increasing gap size. This behavior arises from the near field interaction between the gold film and the gold particle. One can gain insight into this by approximating the gold particle as a dipole near the gold film. This dipole induces an image dipole in the gold film. In the static approximation, the ratio between the induced dipole moment and the dipole moment of the gold nanoparticle is, where and are the relative permittivities of gold and water, respectively. Due to the high permittivity of gold at the operating wavelength of 980 nm, the ratio of the dipole moments is approximately unity. The interaction between the gold particle and the gold film can therefore be understood by analyzing the interaction between the two gold particles separated by a small gap. Theoretical investigations of the optical force between two coupled gold nanoparticles illuminated by a plane wave have been carried out [21–24]. These show the emergence of complex behaviors that depend on the wavelength, the polarization of the electromagnetic field, the size of the particles, and the size of the gap [21–24]. A strong attractive force is predicted by these studies when the electric field is across the gap and gold particles are far away from resonance. This is the case of our experiment, as 250 nm diameter gold particles resonate at about 1.3 μm. The strong attractive force decays rapidly with increasing gap size. The behavior of F_z shown in Fig. 8 is very similar to this. Intuitively, the attractive optical force comes from the interaction between two dipoles which are in phase when the gap is small. When the two particles come close to each other, a large field enhancement in the gap results, leading to a large increase in force. The FDTD simulations also predict that, as the particle approaches the gold film, the optical force F_x increases first and then decreases a little. The initial increase of F_x is due to the increased field intensity in the evanescent field. The decrease of F_x when the gold particle is very close to the surface is not well understood here and will be the topic of future investigations.

The optical force F_x can be calculated from the measured velocity. In order to do this, we know the drag coefficient μ of the particle in the solution, as the force and velocity are related by, where is the velocity. The drag coefficient is given by when the particle is far from the surface. Here, η is the dynamic viscosity and a is the radius of the particle, assumed to be spherical. For a 250 nm diameter sphere in water at room temperature, we have $6\pi\eta a = 2.4 \times 10^{-9}$ (N s/m). The accurate description of our experimental results, however, requires us to take temperature variation into account. The temperature could be higher than the room temperature because the gold film absorbs a large portion of the input laser power when the SPP is excited. This could have a significant effect on the determination of the optical force, as the dynamic viscosity of water is very sensitive to temperature. Analyzing the particles' Brownian motion, however, presents a means for estimating the temperature. Particles experience no optical force in the y direction. The diffusion coefficient measured in this direction can therefore be used to estimate the temperature of the water by the Einstein relationship, where is known as a function of temperature [25]. The incident angle is fixed at 64.5° in Fig. 9, and the measured diffusion coefficient (red symbols) can be seen to increase monotonically with input power because the temperature increases due to ohmic losses in the gold film. We can then find the optical force F_x from, as shown by the blue symbols of Fig. 9. From this Figure, it can be seen that the experimental results for F_x are in reasonable agreement with the predictions of the FDTD calculations.

To conclude, we carry out experiments in which we observed the enhanced propulsion of gold nanoparticles by SPPs. We carried out 3D FDTD simulations, in order to understand this phenomenon. From these calculations, we attribute the enhanced optical force to field enhancement due to SPP excitation and near-field coupling between the gold nanoparticles and the gold film. The field enhancement due to SPP excitation enables 250 nm diameter gold particles to be propelled at illumination field intensities as low as 3×10^6 W/m^2 at which the optical force is ~25 fN. A strong attractive force, keeping the gold nanoparticles in the evanescent field region of the SPP, results from near-field coupling between the gold nanoparticles and the gold film. By measuring the average velocity, and by characterizing the Brownian motion of the gold nanoparticles to find the drag coefficient, the optical

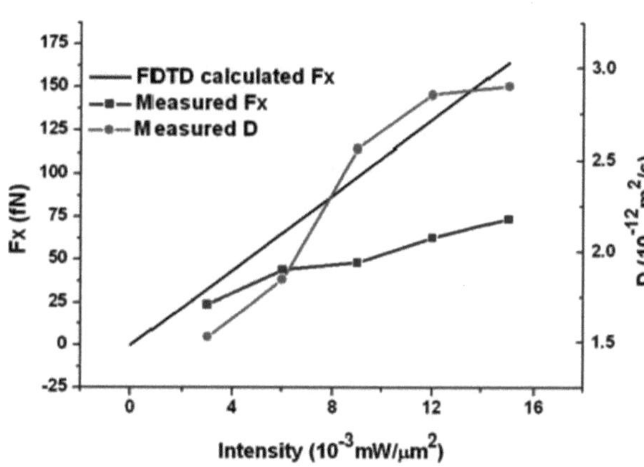

Fig. 9 Experimentally measured diffusion coefficient D, measured optical force Fx, and FDTD calculated optical force Fx as a function of incident intensity. Incident angle is fixed at 64.5°. Gold particles have diameters of 250 nm

force is experimentally estimated. Reasonable agreement between experimentally determined optical forces and the predictions of FDTD calculation is seen at low input power levels, with some deviation at higher input power levels.

Acknowledgements This work was carried out by Kai Wang and Ethan Schonbrun of the Crozier group at Harvard University. This work was supported by the Defense Advanced Research Project Agency (DARPA) and by the National Science Foundation (NSF). Fabrication work was carried out in the Harvard Center for Nanoscale Systems, which is supported by the NSF. For further details on this work, see Ref [26].

References

1. M.C. Daniel, D. Astruc, Chem. Rev. **104**, 293 (2004)
2. K.C. Neuman, Block, S,M, Rev. Sci. Instrum. **75**, 2787 (2004)
3. A. Ashkin, Science **210**, 1081 (1980)
4. K Sasaki, M. Koshioka, H. Misawa, N. Kitamura, H. Masuhara, Appl. Phys. Lett. **60**, 807 (1992)
5. K. Svoboda, S.M. Block, Opt. Lett. **19**, 930 (1994)
6. P.M Hansen, V.K Bhatia, N. Harrit, L. Oddershede, Nano Lett. **5**, 1937 (2005)
7. L. Bosanac, T. Aabo, P.M Bendix, L.B. Oddershede, Nano Lett. **8**, 1486 (2008)
8. L. Novotny, R.X Bian, X.S Xie, Phys. Rev. Lett. **79**, 645 (1997)
9. R. Quidant, D. Petrov, G. Badenes, Opt. Lett. **30**, 1009 (2005)
10. M. Righini, G. Volpe, C. Girard, D. Petrov, R. Quidant, Phys. Rev. Lett. **100**, 186804 (2008)
11. A.N Grigorenko, N.W. Roberts, M.R Dickinson, Y. Zhang, Nat. Photonics **2**, 365 (2008)
12. M. Righini, P. Ghenuche, S. Cherukulappurath, V. Myroshnychenko, F.J Garcia de Abajo, R. Quidant, Nano Lett. Online early access. doi:10.1021/nl803677x
13. V. Garces-Chavez, R. Quidant, P.J. Reece, G. Badenes, L. Torner, K. Dholakia, Phys. Rev. B **73**, 085417 (2006)
14. G. Volpe, R. Quidant, G. Badenes, D. Petrov, Phys. Rev. Lett. **96**, 238101 (2006)
15. L.N. Ng, M.N. Zervas, J.S. Wilkinson, Appl. Phys. Lett. **76**, 1993 (2000)
16. K. Sasaki, J. Hotta, Opt. Lett. **25**, 1385 (2000)
17. A.H.J. Yang, T. Lerdsuchatawanich, D. Erickson, Nano Lett. **9**, 1182 (2009)
18. S. Gaugiran, S. Getin, J.M. Fedeli, J. Derouard, Opt. Express **15**, 8146–8156 (2007)
19. E.D. Palik, Handbook of optical constants of solids II; Academic, Boston, MA (1991)
20. S.H. Behrens,, D.G. Grier, J. Chem. Phys. **115**, 6716 (2001)
21. A.S. Zelenina, R. Quidant, M. Nieto-Vesperinas, Opt. Lett. **32**, 1156 (2007)
22. L. Novotny, C. Henkel, Opt. Lett. **33**, 1029 (2008)
23. B. Sepulveda, J. Alegret, M. Kall, Opt. Express, **15**, 14914(2007)
24. S. Gaugiran, S. Getin, J.M. Fedeli, J. Derouard, Opt. Express, **15**, 8146 (2007)
25. J.R. Watson, R.S. Basu, J.V. Sengers, J. Phys. Chem. Ref. Data, **9**, 1255 (1980)
26. K. Wang, E. Schonbrun, K.B. Crozier, NanoLett. **9**, 2623 (2009)

Plasmon Spectra of Nano-Structures: A Hydrodynamic Model

I. Villo-Perez, Z.L. Mišković, and N.R. Arista

1 Introduction

In a seminal paper, published over half a century ago, R.H. Ritchie [1] used Bloch's hydrodynamic model (HDM) of an electron gas (EG) [2] to describe collective, or plasma excitations by fast electrons in thin metallic films. Ever since, this model played an important role as a phenomenological approach, which often gave qualitatively correct and physically transparent results for many interesting surface and bulk phenomena [3–11]. With the advent of nanotechnology, interest in plasmon excitations at the surfaces of nano-particles and at the metal-dielectric interfaces grew over the past few years at an unexpectedly fast rate, forming an entirely new research area named Plasmonics [12, 13]. Even in these modern developments where surface plasmon excitations are studied by means of the currently most sophisticated *ab initio* methods, such as Time-Dependent Density Functional Theory (TD-DFT) [14–16], the HDM continues to play an important role owing to its versatility in handling the often complex geometry and heterogeneous composition of nano-structures [15].

In essence, a HDM of the Bloch type [2] may be considered as a time-dependent Thomas-Fermi (TF) model, in which a quasi-free EG is described as a semi-classical charged fluid, superimposed on a continuous neutralizing positive-charge background. The dynamics of such an EG is governed by the continuity equation and the momentum-balance equation, which is amended by an internal pressure that stems from a quantum-mechanical expression for the kinetic energy of the EG in the TF picture. Whereas the long-range Coulomb interaction is treated in a self-consistent manner by coupling the momentum balance equation with the Poisson equation, the quantum, or Fermi pressure parameterizes the short-range effects due to statistical repulsion among electrons. Whereas in the classical limit, which is achieved by setting the Fermi pressure to zero, the linearized HDM amounts to a local description of the EG by means of a dispersionless dielectric function of Drude type, the inclusion of Fermi pressure gives rise to a dispersion of plasmon modes in the bulk of the EG.

I. Villo-Perez (✉)
Departamento de Electrónica, Universidad Politécnica de Cartagena, E-30202, Cartagena, Spain
e-mail: isidro.villo@upct.es

A. Aldea, V. Bârsan (eds.), *Trends in Nanophysics,* Engineering Materials,
DOI 10.1007/978-3-642-12070-1_10, © Springer-Verlag Berlin Heidelberg 2010

Furthermore, when dealing with a bounded EG, the surface problem is easily handled in the classical limit of the HDM by assuming a sharp surface, where the EG is abruptly truncated by an infinite barrier (IB) due to termination of the positive charge background, and by imposing the usual electrostatic boundary conditions: the continuity of the potential and the continuity of the normal component of the electric displacement field at the surface. We note that there exists abundant literature on using this, so-called local approach to studying surface plasmons in various structures of relevance in nano-technology, which will not be reviewed here [17–19].

However, the inclusion of Fermi pressure in the HDM causes conceptual problems when the EG is bounded, because the so-called additional boundary conditions (ABCs) of hydrodynamic type need to be invoked at the surface of the EG. In that respect, attempts to refine the issue of ABCs by considering smooth electron-density profiles at the surface caused a long debate in the context of using HDM to study the dispersion of surface plasmons [7, 9, 20, 21]. However, it seems that the debate about appropriate surface boundary conditions for a HDM with Fermi pressure has settled by accepting that the most reliable results may be obtained by adopting an IB picture where, in addition to the usual electrostatic boundary conditions, also the normal component of the fluid velocity must vanish [1, 8, 20].

In this review we shall first summarize several applications of the HDM to regions of EG, which are bounded by cylindrical and spherical surfaces, thus providing suitable descriptions for the spectra of both the volume and surface plasmons excited in metallic nanowires [22] and nano-particles [23, 24], respectively. We shall refer to such a model as a three-dimensional (3D) HDM because it describes the motion of an EG with three degrees of freedom, which is subject to the appropriate surface boundary conditions. Furthermore, a discussion will be provided on qualitative changes in the plasmon spectra as the size of such particles is reduced to the nano-scale [25]. This will be followed by a brief summary of the results for a planar slab within the 3D HDM [10, 11], which will aid us in introducing a two-dimensional (2D) HDM, suitable for thin layers of EG arising in carbon nanostructures. In the realm of a 2D HDM, issues related to the boundary conditions in the direction perpendicular to a thin layer are eliminated by restricting the motion of the EG to two degrees of freedom within such layers. On the other hand, we shall review recent developments in generalizing such a 2D HDM by introducing a two-fluid picture, suitable to describe different roles of carbon's σ and π valence electrons.

We shall neglect retardation effects by assuming that the phase speeds of all collective modes are much smaller that the speed of light. In addition, we shall use the gaussian electrostatic units throughout unless otherwise stated explicitly.

2 Three-Dimensional Hydrodynamical Model

Bloch's hydrodynamic model (HDM) gives the Navier-Stokes equation (the equation of motion), the Poisson equation, and the continuity equation [2]. Mathematically,

$$mn\left[\frac{\partial \mathbf{v}}{\partial t}+\nabla\left(\frac{1}{2}v^2\right)-\mathbf{v}\times(\nabla\times\mathbf{v})\right]=en\nabla\phi-\nabla P(n), \tag{1}$$

$$\nabla^2\phi=4\pi e(n-n_+), \tag{2}$$

$$\frac{\partial n}{\partial t}+\nabla\cdot(n\mathbf{v})=0, \tag{3}$$

where m is the (effective) electron mass, $-e$ the electron charge, $\phi(\mathbf{r},t)$ the electrostatic potential, $n(\mathbf{r},t)$ the volume density of electrons, $n_+(\mathbf{r})$ the ion density, $\mathbf{v}(\mathbf{r},t)$ the flow velocity of electron gas (EG), and $P(n)$ the Fermi pressure of EG, which is related with the electron density n by the equation of state $P(n)=\gamma n^{5/3}$ with $\gamma=(1/5)(3\pi^2)^{2/3}(\hbar^2/m)$.

Using the equation of state, we can write the pressure term in Eq. (1) as

$$\frac{1}{n}\nabla P(n)=\nabla\int_0^{n(\mathbf{r},t)}\frac{dP(n')}{n'}=\nabla I(\mathbf{r},t) \tag{4}$$

with $I(\mathbf{r},t)=(5/2)\gamma n^{2/3}(\mathbf{r},t)$.

On the other hand, as is usually assumed in the HDM, we consider the flow velocity $\mathbf{v}(\mathbf{r},t)$ to be irrotational [26], so that we can introduce a velocity-potential function $\Psi(\mathbf{r},t)$ by the relation

$$\mathbf{v}(\mathbf{r},t)=-\nabla\Psi(\mathbf{r},t). \tag{5}$$

Equations (4) and (5) permit us to write the equations of HDM in the form

$$\frac{\partial\Psi}{\partial t}=\frac{1}{2}|\nabla\Psi|^2-\frac{e}{m}\phi+\frac{1}{m}I, \tag{6}$$

$$\nabla^2\phi=4\pi e(n-n_+), \tag{7}$$

$$\frac{\partial n}{\partial t}=\nabla\cdot(n\nabla\Psi). \tag{8}$$

Following the usual linearization method we write: $n(\mathbf{r},t)=n_0(\mathbf{r})+n_1(\mathbf{r},t)+n_2(\mathbf{r},t)+\cdots$, $\phi(\mathbf{r},t)=\phi_0(\mathbf{r})+\phi_1(\mathbf{r},t)+\phi_2(\mathbf{r},t)+\cdots$, and $\Psi(\mathbf{r},t)=\Psi_1(\mathbf{r},t)+\Psi_2(\mathbf{r},t)+\cdots$, where $n_0(\mathbf{r})$, $\phi_0(\mathbf{r})$ are the equilibrium (static) electron density and corresponding static potential (the zeroth-order velocity-potential function Ψ_0 is null for the static case), the $n_1(\mathbf{r},t)$, $\phi_1(\mathbf{r},t)$ and $\Psi_1(\mathbf{r},t)$ terms are the first-order or linear contributions, and the following terms denote higher-order (non-linear) corrections. The linearization method is based on the assumption that $n_0>>|n_1|>>|n_2|$. In this case, we can write

$$I(\mathbf{r},t)=\frac{5}{2}\gamma(n_0+n_1+\cdots)^{\frac{2}{3}}=I_0(\mathbf{r})+I_1(\mathbf{r},t)+\cdots \tag{9}$$

with $I_0 = (5/2)\gamma n_0^{2/3}$ and $I_1 = (5/3)\gamma(n_1/n_0^{1/3})$. With this, Eqs. (6), (7) and (8) to zeroth-order become

$$0 = -e\phi_0 + \frac{5}{2}\gamma n_0^{\frac{2}{3}}, \tag{10}$$

$$\nabla^2\phi_0 = 4\pi e(n_0 - n_+), \tag{11}$$

and to first order

$$\frac{\partial\Psi_1}{\partial t} = -\frac{e}{m}\phi_1 + \frac{5}{3}\frac{\gamma}{m}\frac{n_1}{n_0^{1/3}}, \tag{12}$$

$$\nabla^2\phi_1 = 4\pi e n_1, \tag{13}$$

$$\frac{\partial n_1}{\partial t} = \nabla\cdot(n_0\nabla\Psi_1)\,. \tag{14}$$

Now, the general problem can be solved as follows. As a first step, given the ion density profile $n_+(\mathbf{r})$, the equilibrium electron density $n_0(\mathbf{r})$ and the electrostatic potential $\phi_0(\mathbf{r})$ can be determined using the system of equations to zeroth order. Subsequently, we can use the hydrodynamic equations to first order to calculate $n_1(\mathbf{r},t)$, $\phi_1(\mathbf{r},t)$, and $\Psi_1(\mathbf{r},t)$.

2.1 Equations for Homogeneous Medium

For simplicity, we consider n_0 to be constant (a homogeneous medium). In this case, the system of equations to solve (Eqs. (12), (13) and (14)) may be written as

$$\frac{\partial\Psi_1}{\partial t} = -\frac{\omega_\mathrm{p}^2}{4\pi e n_0}\phi_1 + \frac{\beta^2}{n_0}n_1, \tag{15}$$

$$\nabla^2\phi_1 = 4\pi e n_1, \tag{16}$$

$$\left(\beta^2\nabla^2 - \frac{\partial^2}{\partial t^2} - \omega_p^2\right)n_1(\mathbf{r},t) = 0. \tag{17}$$

Note that when n_0 is constant, using Eqs. (12), (13) and (14), we can obtain a separate equation for $n_1(\mathbf{r},t)$, it is the three-dimensional wave Eq. (17), where $\omega_\mathrm{p} = \sqrt{4\pi e^2 n_0/m}$ is the plasma frequency and $\beta = v_\mathrm{F}/\sqrt{3}$, with $v_\mathrm{F} = (\hbar/m)(3\pi^2 n_0)^{1/3}$ the Fermi velocity of the EG. The physical meaning of β is the speed of propagation of density disturbances in the EG. It should be noted that in the dielectric approach [27] the value of β is given by $\sqrt{3/5}v_\mathrm{F}$. The value of β obtained from the HDM ($\beta = v_\mathrm{F}/\sqrt{3}$) is considered adequate for low frequencies (system close to static equilibrium) while the value obtained from the dielectric response function is more adequate for high frequencies. For typical metals the values of β are of the order of one atomic unit (Bohr's velocity).

We note that Eq. (17) admits as solutions plane waves with wavevector k and frequency

$$\omega_{\text{bulk}} = \sqrt{\omega_{\text{p}}^2 + \beta^2 k^2}, \tag{18}$$

showing a dispersion effect on plasma waves due to the finite electron velocities represented by the value of β, which also produces a non-zero group velocity $v_{\text{g}} \equiv d\omega/dk$.

2.2 Boundary Conditions at Metal Surface

The system of Eqs. (15), (16) and (17) has to be completed by the boundary conditions. For metal embedded in vacuum, we can write:

(i) Continuity of the potential at the interface S

$$\phi_1^{(\text{metal})}(\mathbf{r}, t)\Big|_S = \phi_1^{(\text{vacuum})}(\mathbf{r}, t)\Big|_S. \tag{19}$$

(ii) Continuity of the normal derivative of the potential at the interface S

$$\left.\frac{\partial \phi_1^{(\text{metal})}(\mathbf{r}, t)}{\partial n}\right|_S = \left.\frac{\partial \phi_1^{(\text{vacuum})}(\mathbf{r}, t)}{\partial n}\right|_S. \tag{20}$$

(iii) Hydrodynamical condition of zero electron current normal to the interface S

$$\mathbf{J}_1^{(\text{metal})} \cdot \hat{n}\Big|_S = n_0 \mathbf{v}_1^{(\text{metal})} \cdot \hat{n}\Big|_S = 0 \iff \left.\frac{\partial \Psi_1^{(\text{metal})}}{\partial n}\right|_S = 0. \tag{21}$$

Using Eq. (15) the condition Eq. (21) can be written in a more convenient form,

$$\left.\frac{\omega_{\text{p}}^2}{4\pi e n_0} \frac{\partial \phi_1^{(\text{metal})}}{\partial n}\right|_S = \left.\frac{\beta^2}{n_0} \frac{\partial n_1^{(\text{metal})}}{\partial n}\right|_S. \tag{22}$$

In the following sections we solve the hydrodynamic equations for a cylindrical wire, a planar slab and a spherical ball. The plasmon modes are obtained in these three canonic geometries using the separation of variables method as follows: Firstly, we obtain $n_1(\mathbf{r}, t)$ solving the hydrodynamic wave Eq. (17), subsequently we obtain $\phi_1(\mathbf{r}, t)$ from Eq. (16) (Laplace'equation in the vacuum and Poisson in the medium). Finally, we obtain $\Psi_1(\mathbf{r}, t)$ from Eq. (15) and $v_1(\mathbf{r}, t)$ from Eq. (5).

3 Cylindrical Wire

We consider here the case of a cylindrical wire of radius a and infinite length, aligned along the z-axis, and surrounded by vacuum.

For the functions $n_1(\mathbf{r}, t)$ and $\phi_1(\mathbf{r}, t)$ we seek a solution in cylindrical coordinates (ρ, φ, z) of the type,

$$n_1(\mathbf{r}, t) = \begin{cases} N_\rho \, e^{im\varphi} \, e^{i(kz-\omega t)} & \rho < a \\ 0 & \rho > a \end{cases}, \tag{23}$$

$$\phi_1(\mathbf{r}, t) = \begin{cases} P_\rho \, e^{im\varphi} \, e^{i(kz-\omega t)} & \rho \leq a \\ L_\rho \, e^{im\varphi} \, e^{i(kz-\omega t)} & \rho \geq a \end{cases}. \tag{24}$$

Solving Eq. (17) we obtain for $n_1(\mathbf{r}, t)$,

$$N_\rho = n_0 A I_m(\alpha\rho), \tag{25}$$

where A is an arbitrary dimensionless constant, I_m is the modified Bessel function of the first kind and order m ($m = 0, 1, 2, \ldots$), and α is defined as

$$\alpha = \sqrt{k^2 - \frac{\omega^2 - \omega_\mathrm{p}^2}{\beta^2}}. \tag{26}$$

As we will see later, the solution Eq. (25) corresponds to surface plasmons for real values of α, and to volume plasmons for imaginary values. Hence, for volume plasmons α is of the form $\alpha = iq$ with q a real number. Using the identity $I_m(iq\rho) = i^m J_m(q\rho)$, we can write Eq. (25) in terms of the Bessel function $J_m(q\rho)$ which have real arguments for volume plasmon, and so, they are the natural functions for these modes.

Solving Eq. (16) we obtain for $\phi_1(\mathbf{r}, t)$,

$$P_\rho = \mathrm{B} I_m(k\rho) + \mathrm{C} I_m(\alpha\rho), \tag{27}$$

$$L_\rho = \mathrm{D} K_m(k\rho), \tag{28}$$

where B and D are constants, K_m is the modified Bessel function of the second kind and order m and C is given by

$$\mathrm{C} = -\frac{4\pi e n_0}{k^2 - \alpha^2}\mathrm{A}. \tag{29}$$

Applying the first and second boundary conditions (Eqs. (19) and (20)) at the surface ($\rho = a$) of the cylinder, we obtain

$$\mathrm{B} = \left[kaK_m'(ka)I_m(\alpha a) - \alpha a I_m'(\alpha a)K_m(ka)\right]\mathrm{C}, \tag{30}$$

$$\mathrm{D} = \left[kaI_m'(ka)I_m(\alpha a) - \alpha a I_m'(\alpha a)I_m(ka)\right]\mathrm{C}, \tag{31}$$

where the primes in $I_m'(x)$ and $K_m'(x)$ denote derivatives with respect to the argument.

The third boundary condition (Eq. (22)), yields a relation between ω and k, namely [22]

$$\frac{\omega^2}{\omega_\mathrm{p}^2} = kaI_m'(ka)K_m(ka)\left[1 - \frac{I_m(\alpha a)}{\alpha a I_m'(\alpha a)}\frac{kaK_m'(ka)}{K_m(ka)}\right], \tag{32}$$

where α is a function of ω and k according to Eq. (26).

The roots of the transcendental Eq. (32) provide the *dispersion relation* $\omega(k)$ of the modes, including both *surface plasmons* (for real values of α) and *volume plasmons* (for imaginary α). In fact, as will be shown, the solutions corresponding to imaginary values of α yield electron densities characterized by an oscillatory behavior inside the medium, while those corresponding to real values of α yield electron densities that are localized in the region close to the surface. In the non-dispersive limit they converge to the well know solutions for bulk and surface plasmons respectively [8, 22]. To show this, let us discuss the solutions of Eq. (32). We suppose now that the values of a, ω_p, β and m are given, and we take the value of k as the independent variable. Then, Eq. (32) admits as solution a numerable infinite set of values of ω, $\{\omega_{m0}, \omega_{m1}, \omega_{m2}, \ldots, \omega_{mn}, \ldots\}$, as illustrated in Fig. 1. The various curves in this figure show the values of $\omega_{mn}(k)$. So, we conclude that, for a given m, each value of n defines a particular dispersion relation, and each case is identified as a mode, explicitly the *mode mn*. Each mode defines a particular configuration for the

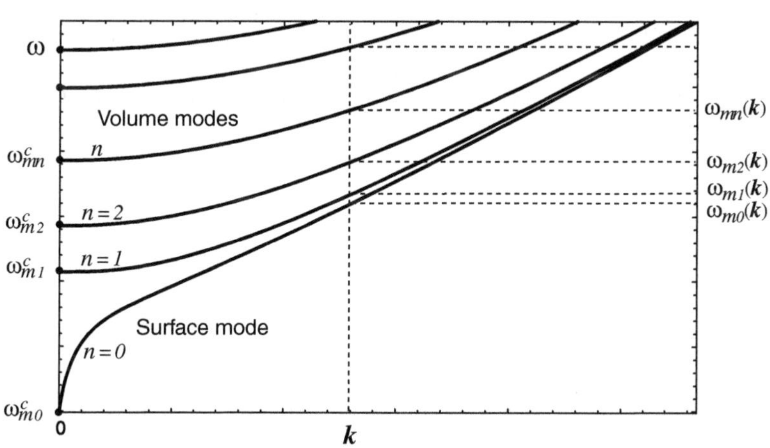

Fig. 1 Solutions of the dispersion relation for a cylindrical wire according to the hydrodynamic model. We use the notation $n = 0$ for the surface mode, and $n = 1, 2, \ldots$ for the volume modes

electric charge density, and as a consequence for the electrostatic potential. On the other hand, the root $(k, \omega_{mn}(k))$ fix the value of α according to Eq. (26) leading to two possible cases, namely: α a real number (surface plasmon) or an imaginary one (volume plasmon). In the following we use the notation $n = 0$ for surface modes, and $n = 1, 2, \ldots$ for volume modes.

As we show in Fig. 1, for each value of m there is only one surface mode ($n = 0$) and an infinite series of volume modes ($n = 1, 2, \ldots.$). Additionally, for each mode mn there is a characteristic frequency, this particular frequency is called *the cutoff frequency* and is denoted in Fig. 1 as ω^{c}_{mn}. The cutoff frequency is found by setting $k = 0$ in Eq. (32).

Here we already note an interesting difference with the dispersion relation obtained from the dielectric response model in the Drude's approximation [19]; in this case the frequency of the volume modes remains constant and equal to ω_p, while the surface plasmon frequency goes to a constant limit $\omega_{\mathrm{p}}/\sqrt{2}$ for large k. In the present case the asymptotic limit for $ka \gg 1$ is given by $\omega \sim \sqrt{\omega_{\mathrm{p}}^2 + \beta^2 k^2}$ in agreement with Eq. (32). This behavior will be discussed in detail in the following sections.

Although the present treatment assumes no limit in the values of k, it should be noted that some physical restrictions arise from the intrinsic properties of a free EG. In fact it may be shown [27] that collective oscillations become strongly damped and decay into single-electron excitations when the wave-vector approaches a critical value k_{c} which is of the order of $\omega_{\mathrm{p}}/v_{\mathrm{F}}$. Moreover, since retardation effects have not been considered, the range of validity of the present approach may de characterized by $\omega_{\mathrm{p}}/\mathrm{c} << k << \omega_{\mathrm{p}}/v_{\mathrm{F}}$. Since $\mathrm{c}/v_{\mathrm{F}} \sim 100$ the applicability range spans about two orders of magnitude of the variable k.

3.1 Scaling Property: Dimensionless Equations

For mathematical purposes it is convenient to transform the dispersion relations Eq. (32) in a more suitable form by introducing the following dimensionless variables: $\Omega = \omega/\omega_{\mathrm{p}}$, $x = ka$ and $\zeta = a\omega_{\mathrm{p}}/\beta$. Thus, in terms of these variables we can write Eq. (32) as:

$$\Omega^2 = x I'_m(x) K_m(x) \left[1 - \frac{I_m(y)}{y I'_m(y)} \frac{x K'_m(x)}{K_m(x)} \right], \tag{33}$$

where $y = \alpha a = \sqrt{x^2 - \zeta^2 \left(\Omega^2 - 1\right)}$.

This result shows a general scaling property of $\Omega = \omega/\omega_{\mathrm{p}}$ in terms of the variable $x = ka$. To explain the physical meaning of the parameter ζ we may write $\zeta = a/\lambda$ where $\lambda = \beta/\omega_{\mathrm{p}}$ represents a typical value of the distance the electrons can move in the characteristic time of the plasma oscillations; hence, ζ is the ratio between the transverse size of the system and the typical displacement of the electrons during an oscillation of the field. Therefore, it is the appropriate parameter to characterize the spatial dispersion effects.

3.2 The Non-Dispersive Limit

The parameter $\zeta = a\omega_\mathrm{p}/\beta$ plays an important role in the properties of the modes. In particular, for $\beta \to 0$ ($\zeta \to \infty$), corresponding to negligible dispersion, it may be expected to obtain the non-dispersive behavior of the Drude model, $\varepsilon(\omega) = 1 - \omega_\mathrm{p}^2/\omega^2$. Using some mathematical hypotheses, it is easy to obtain the limit for Eq. (33) in the non-dispersive case. The results is $\Omega = 1$ for the bulk modes while for the surface modes we have:

$$\Omega^2 = x I'_m(x) K_m(x). \tag{34}$$

In the next section we explore the fundamental mode $m = 0$. We will also illustrate the characteristics of surface and volume modes showing the radial dependence for some cases.

3.3 The Fundamental Mode, $m = 0$

For the mode $m = 0$ (axially symmetric) we have found all the roots of the dispersion relation Eq. (33) in the interval of relative frequencies $\Omega = \omega/\omega_\mathrm{p}$ in the range [0, 3]. The results are shown in Fig. 2. In this figure we have plotted the relative frequency $\Omega = \omega/\omega_\mathrm{p}$ as a function of the variable $x = ka$ for the value $\zeta = 10$. The

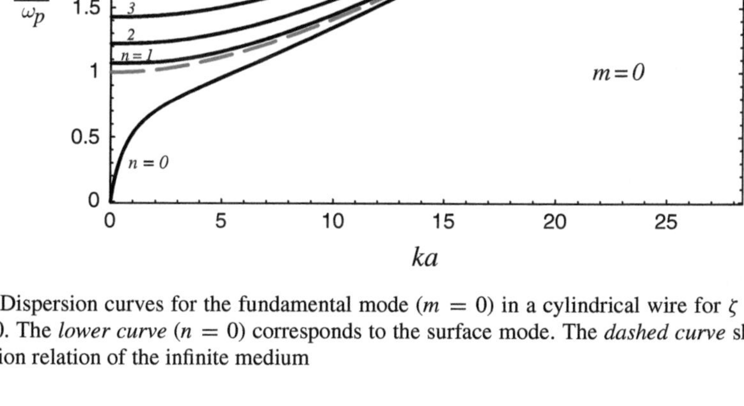

Fig. 2 Dispersion curves for the fundamental mode ($m = 0$) in a cylindrical wire for $\zeta = a\omega_\mathrm{p}/\beta = 10$. The *lower curve* ($n = 0$) corresponds to the surface mode. The *dashed curve* shows the dispersion relation of the infinite medium

dashed line in this figure is the frequency of the bulk modes according to Eq. (18). We discuss now the behavior of surface and volume modes separately.

3.3.1 The Surface Mode with $m = 0$

First we focus our attention on the surface mode ($n = 0$) which is represented by the lower curve in Fig. 2. Let us discuss the behavior of this mode for different values of ζ. In Fig. 3 we show the surface mode when the parameter ζ increases from 10 to 80. In this figure we can see the convergence to the dashed curve (the lowest curve of this set), which is the dispersion relation for the non-dispersive case, $\zeta = \infty$, and has been obtained from Eq. (34) taking $m = 0$. This limiting behavior is in agreement with calculations based on the dielectric approach [19]. The straight line in this figure, $\omega/\omega_p = 1/\sqrt{2}$, shows the plane-surface limit. Thus, we can observe that when $\zeta >> 1$ the non-dispersive behavior of the Drude model is obtained as expected.

Using the radial functions corresponding to Eqs. (25), (27) and (28) we show in Fig. 4 panel (a) the radial dependence of the electrostatic potential of the surface mode for $\zeta = 10, 20, 40, 80$. In this example we have taken the root corresponding to $x = 2$. The dashed line in this figure corresponds to the potential of the non-dispersive model $\zeta = \infty$. This figure also shows that for increasing values of ζ the potential produces a cusp behavior at $\rho = a$ as in previous models [19]. This is so because when ζ increases the induced electronic charge tends to accumulate at the surface as it is shown in panel (b) of this same figure (in the limit $\zeta = \infty$ we have a surface charge density described by a Dirac's delta function).

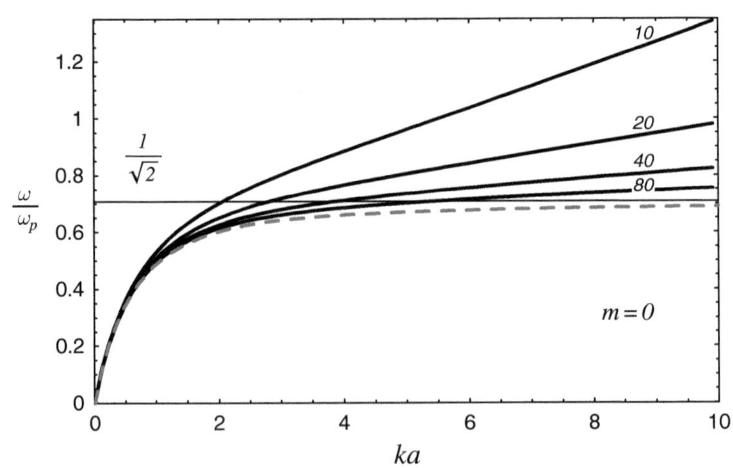

Fig. 3 Dispersion curves for the fundamental surface mode for several values of the parameter $\zeta = a\omega_p/\beta$ (10, 20, 40, 80). The *dashed curve* shows the dispersion relation of the non-dispersive model ($\zeta = \infty$)

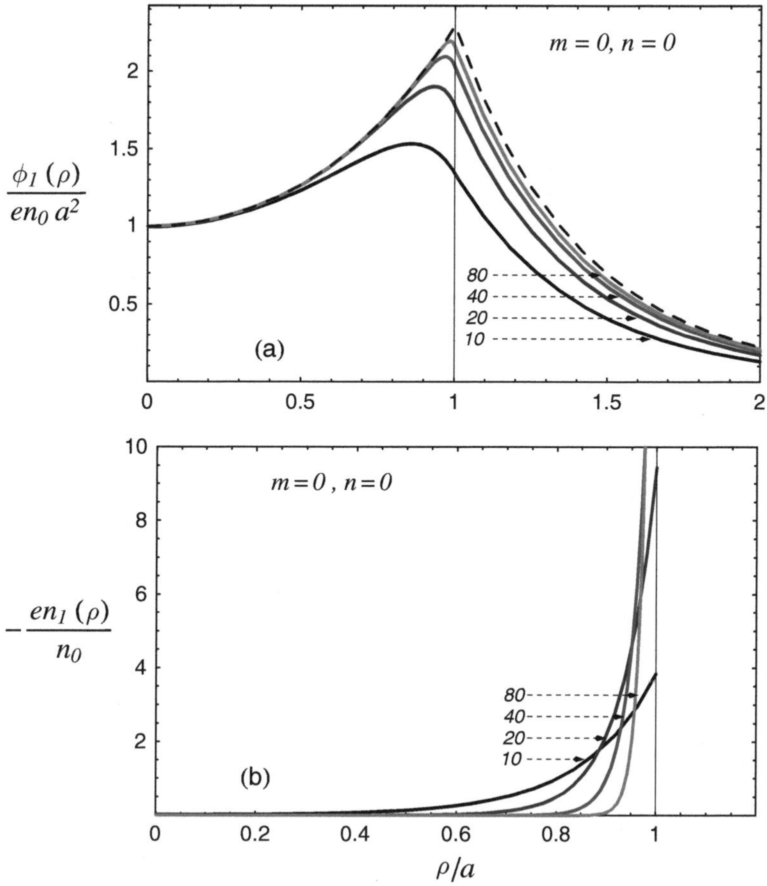

Fig. 4 Radial functions corresponding to the potential (panel (**a**)) and to the charge density (panel (**b**)) of the fundamental surface mode for several values of the parameter $\zeta = a\omega_p/\beta$ (10, 20, 40, 80). The *dashed curve* shows the potential of the non-dispersive model ($\zeta = \infty$). To plot these functions we have taken $x = ka = 2$

3.3.2 The Volume Modes with $m = 0$

To visualize how the volume modes change with increasing ζ, we show in Fig. 5 the radial dependence of the induced potential $\phi_1(\rho)$ for the modes $n = 1$ and $n = 2$ and for two different values of $x = ka$ (5 and 20). We can see that the difference with the non-dispersive case is bigger for $x = 20$, which corresponds to a larger value of k (for fixed values of $a = \zeta\beta/\omega_p$). Additionally, we can see that for $\zeta = \infty$ these modes are fully confined within the volume of the wire (as expected from previous dielectric calculations [19]) but this is not so for finite ζ values.

The global behavior of the modes of higher order is similar to the fundamental case ($m = 0$). These modes have been described in detail in [22].

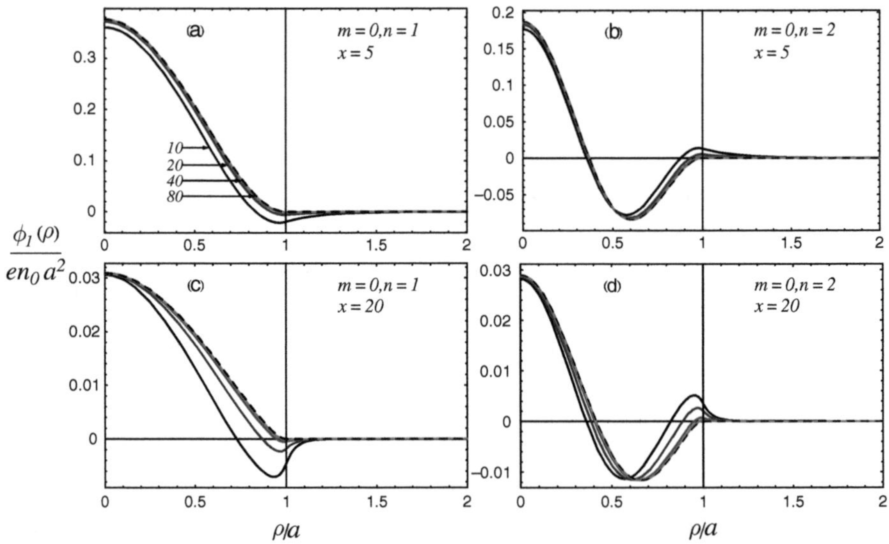

Fig. 5 Radial functions corresponding to the potential of the fundamental volume mode for several values of the parameter $\zeta = a\omega_p/\beta$ (10, 20, 40, 80). (**a**) sub-mode $n = 1$ with $x = ka = 5$, (**b**) sub-mode $n = 2$ for $x = 5$, (**c**) sub-mode $n = 1$ for $x = 20$ and (**d**) sub-mode $n = 2$ with $x = 20$. The dashed curve shows the potential of the non-dispersive model ($\zeta = \infty$)

4 Planar Slab

Consider a slab of thickness $2a$ along the z direction and infinitely wide in the x and y directions, and embedded in vacuum. For this geometry we look for a solution in cartesian coordinates $\mathbf{r} = \{x, y, z\}$ of the form

$$n_1(\mathbf{r}, t) = \begin{cases} 0 & z < -a \\ N_z\, e^{i(\mathbf{k}\cdot\mathbf{r}_\parallel - \omega t)} & -a < z < +a \\ 0 & z > +a \end{cases}, \tag{35}$$

$$\phi_1(\mathbf{r}, t) = \begin{cases} L_{z1}\, e^{i(\mathbf{k}\cdot\mathbf{r}_\parallel - \omega t)} & z \le -a \\ P_z\, e^{i(\mathbf{k}\cdot\mathbf{r}_\parallel - \omega t)} & -a \le z \le +a \\ L_{z2}\, e^{i(\mathbf{k}\cdot\mathbf{r}_\parallel - \omega t)} & z \ge +a \end{cases}, \tag{36}$$

where $\mathbf{k} = k_x\,\hat{\mathbf{e}}_x + k_y\,\hat{\mathbf{e}}_y$ and $\mathbf{r}_\parallel = x\,\hat{\mathbf{e}}_x + y\,\hat{\mathbf{e}}_y$.

Solving Eq. (17) we obtain for $n_1(\mathbf{r}, t)$,

$$N_z = n_0 \mathrm{A}_1 \cosh(\alpha z) + n_0 \mathrm{A}_2 \sinh(\alpha z) \tag{37}$$

with

$$\alpha = \sqrt{k^2 - \frac{\omega^2 - \omega_{\mathrm{p}}^2}{\beta^2}}. \tag{38}$$

As in the case of a cylindrical wire, the surface modes appear for real values of α, while volume modes correspond to imaginary values.

Solving Eq. (16) we obtain for $\phi_1(\mathbf{r}, t)$,

$$L_{z1} = \mathrm{D}_1 e^{kz}, \tag{39}$$

$$\begin{aligned} P_z &= \mathrm{B}_1 \cosh(kz) + \mathrm{C}_1 \cosh(\alpha z) \\ &\quad + \mathrm{B}_2 \sinh(kz) + \mathrm{C}_2 \sinh(\alpha z), \end{aligned} \tag{40}$$

$$L_{z2} = \mathrm{D}_2 e^{-kz}, \tag{41}$$

with $k = \sqrt{k_x^2 + k_y^2}$ and

$$\mathrm{C}_1 = -\frac{4\pi e n_0}{k^2 - \alpha^2} \mathrm{A}_1, \qquad \mathrm{C}_2 = -\frac{4\pi e n_0}{k^2 - \alpha^2} \mathrm{A}_2. \tag{42}$$

Applying the first and second boundary conditions (Eqs. (19) and (20)) at the surface ($z = \pm a$) of the slab, we find:

$$\begin{aligned} \mathrm{D}_1 &= \left[\sinh(ka)\cosh(\alpha a) - \frac{\alpha}{k}\cosh(ka)\sinh(\alpha a)\right] \mathrm{C}_1 \\ &\quad + \left[\frac{\alpha}{k}\sinh(ka)\cosh(\alpha a) - \cosh(ka)\sinh(\alpha a)\right] \mathrm{C}_2, \end{aligned} \tag{43}$$

$$\begin{aligned} \mathrm{D}_2 &= \left[\sinh(ka)\cosh(\alpha a) - \frac{\alpha}{k}\cosh(ka)\sinh(\alpha a)\right] \mathrm{C}_1 \\ &\quad - \left[\frac{\alpha}{k}\sinh(ka)\cosh(\alpha a) - \cosh(ka)\sinh(\alpha a)\right] \mathrm{C}_2, \end{aligned} \tag{44}$$

$$\mathrm{B}_1 = -\left[\cosh(\alpha a) + \frac{\alpha}{k}\sinh(\alpha a)\right] e^{-ka} \mathrm{C}_1, \tag{45}$$

$$\mathrm{B}_2 = -\left[\frac{\alpha}{k}\cosh(\alpha a) + \sinh(\alpha a)\right] e^{-ka} \mathrm{C}_2. \tag{46}$$

From the third boundary condition (Eq. (22)) we obtain the dispersion relation in a slab,

$$\alpha \sinh(\alpha a)\left[\frac{\omega^2}{\omega_{\mathrm{p}}^2}-\frac{1}{2}\left(1-e^{-2ka}\right)\left(1+\frac{k}{\alpha\tanh(\alpha a)}\right)\right]\mathrm{A}_1$$

$$\pm\alpha \cosh(\alpha a)\left[\frac{\omega^2}{\omega_{\mathrm{p}}^2}-\frac{1}{2}\left(1+e^{-2ka}\right)\left(1+\frac{k}{\alpha\coth(\alpha a)}\right)\right]\mathrm{A}_2=0. \tag{47}$$

To satisfy these equations it is necessary to consider separately two categories: even and odd modes. For an even mode $\mathrm{A}_2 = 0$ and

$$\frac{\omega^2}{\omega_{\mathrm{p}}^2}=\frac{1}{2}\left(1-e^{-2ka}\right)\left(1+\frac{k}{\alpha\tanh(\alpha a)}\right). \tag{48}$$

For an odd mode $\mathrm{A}_1 = 0$ and

$$\frac{\omega^2}{\omega_{\mathrm{p}}^2}=\frac{1}{2}\left(1+e^{-2ka}\right)\left(1+\frac{k}{\alpha\coth(\alpha a)}\right). \tag{49}$$

4.1 Scaling Property: Dimensionless Equations

Using the dimensionless variables $\Omega = \omega/\omega_{\mathrm{p}}$, $x = ka$ and $\zeta = a\omega_{\mathrm{p}}/\beta$ we can write Eqs. (48) and (49) in the form

$$\text{Even modes:}\quad \Omega^2=\frac{1}{2}\left(1-e^{-2x}\right)\left(1+\frac{x}{y\tanh(y)}\right), \tag{50}$$

$$\text{Odd modes:}\quad \Omega^2=\frac{1}{2}\left(1+e^{-2x}\right)\left(1+\frac{x}{y\coth(y)}\right), \tag{51}$$

where $y = \alpha a = \sqrt{x^2 - \zeta^2(\Omega^2 - 1)}$. As we explained before, y real corresponds to surface modes, while y imaginary corresponds to volume modes.

4.2 The Non-Dispersive Limit

For the non-dispersive limit $\beta \to 0$ ($\zeta \to \infty$) the Eqs. (50) and (51) yield the known results: $\Omega = 1$ for the volume modes while for the surface modes:

$$\text{Even modes:}\quad \Omega^2=\frac{1}{2}\left(1-e^{-2x}\right), \tag{52}$$

$$\text{Odd modes:}\quad \Omega^2=\frac{1}{2}\left(1+e^{-2x}\right). \tag{53}$$

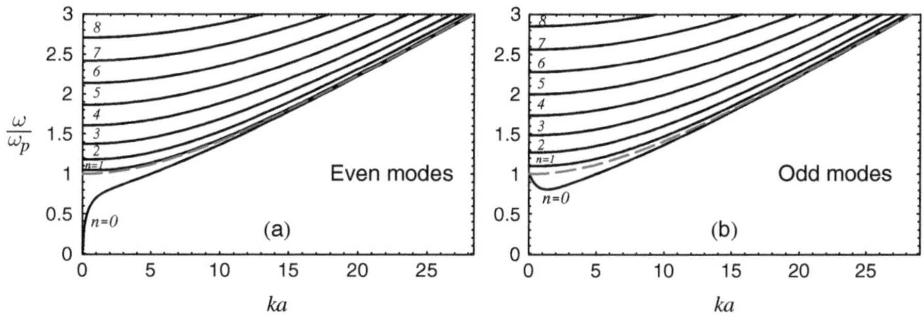

Fig. 6 Dispersion curves in a slab for $\zeta = a\omega_p/\beta = 10$. (**a**) Even modes. (**b**) Odd modes. The *lower curve* ($n = 0$) corresponds to the surface mode. The *dashed curve* shows the dispersion relation of the infinite medium

Let us consider now with some detail the properties of these modes. For the even and odd modes, we have found all the roots of the dispersion relation (Eqs. 50 and 51) falling in the frequency interval $0 \leq \Omega \leq 3$. The results are shown in Fig. 6, panel (a) for even modes and panel (b) for odd modes. One first difference between the even mode and the odd mode is that in this second case the cutoff frequency (frequency for $ka = 0$) is different from zero (in fact $\omega \to \omega_p$).

Let us now discuss the behavior of the surface and volume modes in a slab for different values of the parameter ζ. In Fig. 7 we observe the general behavior of the even mode when the parameter ζ increases from 10 to 80. For the even surface mode (panel (a)) we can observe the convergence to the non-dispersive case (Eq. (52)). For the volume mode (panel (b)) we can observe clearly that the dispersion curve goes to 1 when ζ increases. So, in the limit $\zeta = \infty$ the non-dispersive case (*dashed curve*, $\omega/\omega_p = 1$) is obtained. Figure 8 shows a similar behavior for odd modes.

Consider now the profile of the electrostatic potential for the even and odd modes in a slab. In Fig. 9 we show the induced potential $\phi_1(z)$ for the even mode (panel (a)) and for the odd mode (panel (b)) and for several values of ζ. Note that the odd modes have a node at the center of the slab; not so for the even mode.

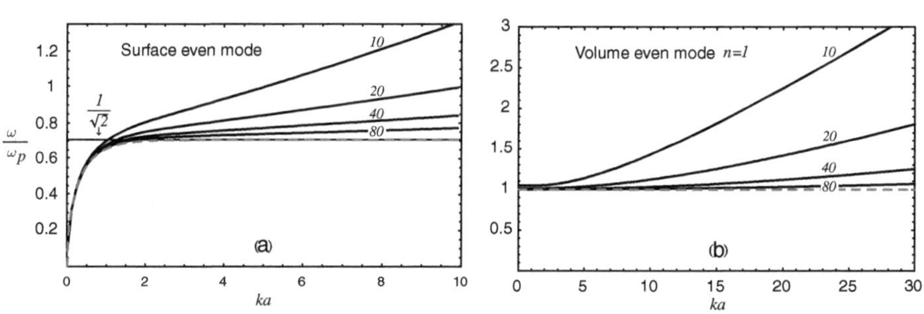

Fig. 7 Dispersion curves in a slab for several values of the parameter $\zeta = a\omega_p/\beta$ (10, 20, 40, 80), (**a**) Surface even mode. (**b**) Volume even mode $n = 1$. The *dashed curve* shows the dispersion relation of the non-dispersive model ($\zeta = \infty$)

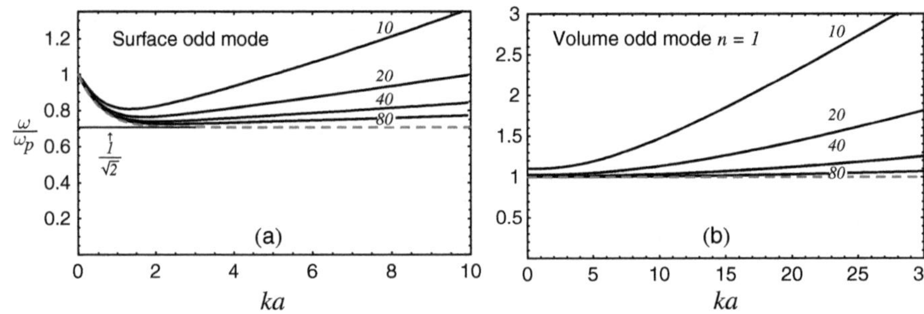

Fig. 8 Dispersion curves in a slab for several values of the parameter $\zeta = a\omega_p/\beta$ (10, 20, 40, 80), (**a**) Surface odd mode. (**b**) Volume odd mode $n = 1$. The dashed curve shows the dispersion relation of the non-dispersive model ($\zeta = \infty$)

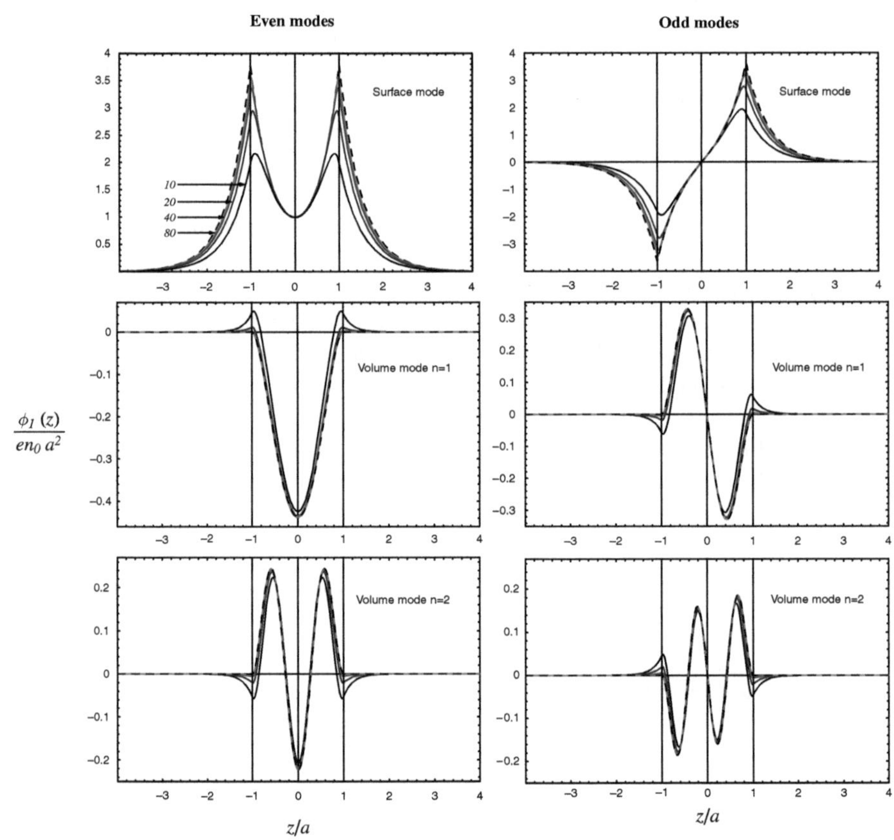

Fig. 9 Curves corresponding to the potential $\phi_1(z)$ in a slab for several values of the parameter $\zeta = a\omega_p/\beta$ (10, 20, 40, 80). The *dashed curve* shows the potential of the non-dispersive model ($\zeta = \infty$). To plot these functions we have taken $x = ka = 5$

These modes have been studied in detail in [10, 11, 28–30], where further information can be found.

5 Spherical Ball

Consider now a sphere of radius a embedded in vacuum. For this geometry we look for a solution in spherical coordinates (r, θ, φ) of the form,

$$n_1(\mathbf{r}, t) = \begin{cases} N_r Y_{lm}(\theta, \varphi)\, e^{-i\omega t} & r < a \\ 0 & r > a \end{cases}, \tag{54}$$

$$\phi_1(\mathbf{r}, t) = \begin{cases} P_r Y_{lm}(\theta, \varphi)\, e^{-i\omega t} & r \leq a \\ L_r Y_{lm}(\theta, \varphi)\, e^{-i\omega t} & r \geq a \end{cases}, \tag{55}$$

where Y_{lm} are the spherical harmonic functions.

Solving Eq. (17) we obtain for $n_1(\mathbf{r}, t)$,

$$N_r = n_0 \mathrm{A} \frac{I_{l+1/2}(\alpha r)}{\sqrt{\alpha r}}, \tag{56}$$

where A is a arbitrary dimensionless constant, I_l is the modified Bessel function of the first kind and order l ($l = 0, 1, 2, 3, \dots$), and α is defined in this geometry as

$$\alpha = \frac{1}{\beta} \sqrt{\omega_{\mathrm{p}}^2 - \omega^2}. \tag{57}$$

The surface modes appear for real values of α ($\omega_{\mathrm{p}} > \omega$), while volume modes correspond to imaginary values ($\omega_{\mathrm{p}} < \omega$). In this latter case, we can write $\alpha = iq$ with q a real number. For volume plasmon, making use of the identity $I_l(iq) = i^l \sqrt{2q/\pi}\, j_{l-1/2}$, the Eq. (56) can be written in terms of the spherical Bessel function of the first kind $j_l(qr)$.

Solving Eq. (16) we obtain for $\phi_1(\mathbf{r}, t)$,

$$P_r = \mathrm{B} r^l + \mathrm{C} \frac{I_{l+1/2}(\alpha r)}{\sqrt{\alpha r}}, \tag{58}$$

$$L_r = \mathrm{D} \frac{1}{r^{l+1}}, \tag{59}$$

where B and D are arbitrary constants and C is given by

$$\mathrm{C} = \frac{4\pi e n_0}{\alpha^2} \mathrm{A}. \tag{60}$$

Applying the first and second boundary conditions (Eqs. (19) and (20)) at the surface ($r = a$) of the sphere, we find:

$$\mathrm{B} = \frac{\alpha a}{2l+1}\frac{I_{l-1/2}(\alpha r)}{\sqrt{\alpha r}}\frac{1}{a^l}\mathrm{C}, \tag{61}$$

$$\mathrm{D} = -\frac{\alpha a}{2l+1}\frac{I_{l+3/2}(\alpha r)}{\sqrt{\alpha r}}a^{l+1}\mathrm{C}. \tag{62}$$

From the third boundary condition (Eq. (22)) we obtain the dispersion relation in a sphere,

$$\frac{\omega^2}{\omega_\mathrm{p}^2} = \left(1 + \frac{l+1}{l}\frac{I_{l+3/2}(\alpha a)}{I_{l-1/2}(\alpha a)}\right)^{-1}. \tag{63}$$

Here a distinction has to be made. In the case of a cylindrical wire of infinite length (same for a slab), the modes are propagating modes with wavevector k along the axis. These modes can propagate in the wire if they are appropriately excited. So, in this case, we have a relationship between angular frequency ω and wavevector k. In the case of a sphere, instead, the modes are simple oscillatory modes because a sphere acts like a cavity resonator [31]. So, in this case, we don't have a dispersion relation of the type $\omega(k)$, but a relationship between the angular frequencies ω of the various multipoles (corresponding to different l values) and the physical parameters that characterize the sphere: a, β and ω_p.

5.1 Scaling Property: Dimensionless Equations

Using the dimensionless variables $\Omega = \omega/\omega_\mathrm{p}$, and $\zeta = a\omega_\mathrm{p}/\beta$ we can write Eq. (63) as:

$$\Omega^2 = \left(1 + \frac{l+1}{l}\frac{I_{l+3/2}(y)}{I_{l-1/2}(y)}\right)^{-1}, \tag{64}$$

where $y = \alpha a = \zeta\sqrt{1-\Omega^2}$. As explained before, y real ($\omega < \omega_\mathrm{p}$) corresponds to surface modes, while y imaginary ($\omega > \omega_\mathrm{p}$) corresponds to bulk modes.

5.2 The Non-Dispersive Limit

For the non-dispersive limit $\beta \to 0$ ($\zeta \to \infty$) Eq. (64) yields the known result [3, 24]

$$\Omega^2 = \left(1 + \frac{l+1}{l}\right)^{-1}. \tag{65}$$

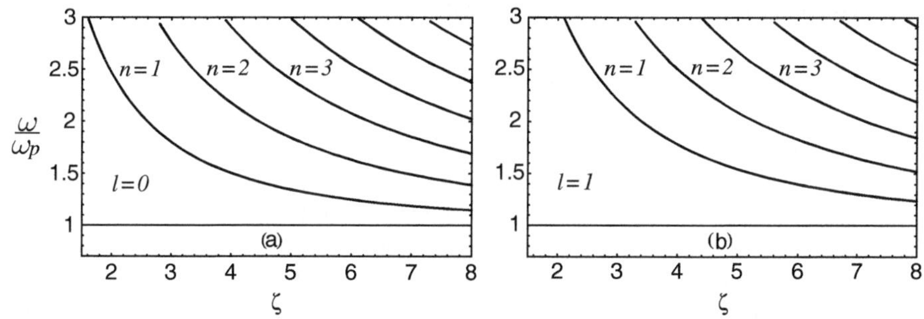

Fig. 10 Dispersion curves in a sphere for the volume modes (**a**) $l = 0$ and (**b**) $l = 1$ versus the parameter $\zeta = a\omega_{\mathrm{p}}/\beta$

A mathematical study shows that Eq. (64) has not solution for y real (surface modes) if $l = 0$. In other words, if $l = 0$ the surface mode is not possible. For volume modes the situation is different; for $l = 0$ there is an infinite number of volume modes ($n = 1, 2, 3, \ldots$) corresponding to imaginary values of y. Figure 10 shows the volume modes for $l = 0$ and $l = 1$ versus the parameter ζ. For $l \geq 1$, both, surface and volume modes, are present. In Fig. 11 we show the surface modes as a function of the parameter ζ for several values of l.

As it is clear from these figures, the frequency of the modes will depend very much on the radius of the particle, so that dispersion effects may become very large for particles of small sizes. Further information on these effects for spherical particles may be found in [23] and [24].

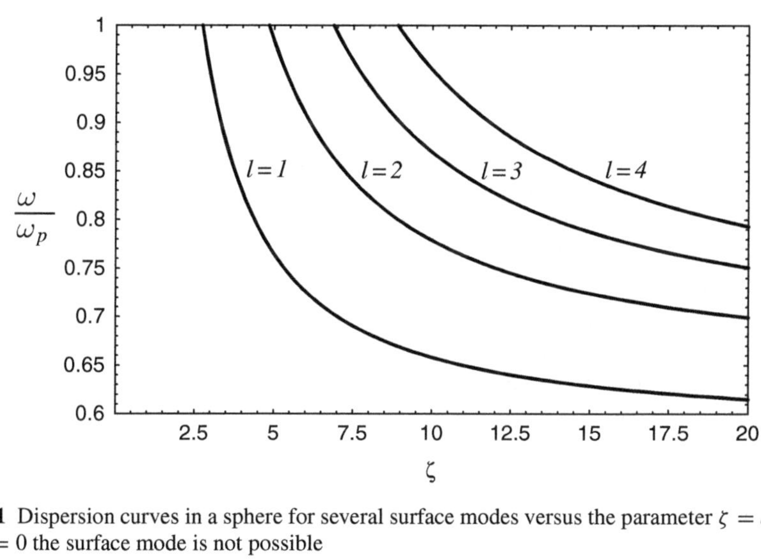

Fig. 11 Dispersion curves in a sphere for several surface modes versus the parameter $\zeta = a\omega_p/\beta$. For $l = 0$ the surface mode is not possible

6 Two-Dimensional Hydrodynamic Model

6.1 One-Fluid Model on a Plane

When the motion of an EG is bounded in one direction but remains quasi-free in the other two directions, then the description of its dynamics can be simplified by using a 2D version of the hydrodynamic model (HDM) that avoids the need for the hydrodynamical boundary conditions by assuming a zero electron current in the direction of bounded motion. Such a model was pioneered by Fetter [32, 33] in 1973, and was subsequently used to gain qualitative understanding of plasmon excitations in various quasi-2D electronic structures, including semiconductor inversion layers, quantum wells, and thin metallic films [34].

However, it is difficult to achieve a formal transition to the strict 2D limit for the EG, which continues to be one of the challenges in modern Density Functional Theory [35]. One may consider a thin planar slab of EG bounded by infinite barriers, with thickness b and an effective surface density (number of electrons per unit area) n_{2D}. Transition to a quasi-2D behavior of this EG is expected when $b < \sqrt{3\pi/(2n_{2D})} \equiv b_{\max}$, and the strict 2D limit may be achieved when $b \ll b_{\max}$ [35]. Then, one can readily deduce a dispersion relation for collective, or plasmon excitations in such 2D EG at wavelengths far exceeding the thickness b as follows. Referring to Sect. 4, classical limit of the 3D HDM for a slab of thickness $b \equiv 2a$ containing an EG with equilibrium volume density $n_{3D} = n_{2D}/b \equiv n$ yields two branches for even and odd collective modes at frequencies defined in Eqs. (52) and (53), respectively,

$$\Omega_{\mp}^2 = \frac{\omega_p^2}{2}\left(1 \mp e^{-bk}\right), \tag{66}$$

where $\omega_p = \sqrt{4\pi e^2 n_{3D}/m_*}$ is the bulk plasma frequency of the EG, m_* is effective electron mass, and k is the magnitude of the 2D wavevector $\mathbf{k} = k_x\,\hat{\mathbf{e}}_x + k_y\,\hat{\mathbf{e}}_y$. By assuming $bk \ll 1$, one recovers a high-frequency mode $\Omega_+ \approx \omega_p$ and a low-frequency mode $\Omega_- \approx \sqrt{2\pi e^2 k n_{2D}/m_*}$. These frequencies are well separated when $b \ll b_{\max}$, with the lower one exhibiting the square-root dispersion relation, which is a well-known signature of the reduced screening ability of a 2D EG [36].

The above plasmon dispersion relation can also be obtained directly from the classical limit of a 2D HDM for an EG with zero thickness and the equilibrium surface density n_{2D}. To develop such a model, we assume that the 2D EG occupies the $z = 0$ plane in a cartesian coordinate system with $\mathbf{r} = \{x, y, z\}$, and let $n(\mathbf{r}_\parallel, t)$ be the surface density of the EG around the point in plane with coordinates $\mathbf{r}_\parallel = \{x, y, 0\}$. Noting that the velocity field of a 2D EG, $\mathbf{u}(\mathbf{r}_\parallel, t)$, may only have in-plane components, one can write [32]

$$\frac{\partial n}{\partial t} + \nabla_\parallel \cdot (n\mathbf{u}) = 0 \tag{67}$$

$$m_* n \left[\frac{\partial \mathbf{u}}{\partial t} + \left(\mathbf{u} \cdot \nabla_\parallel \right) \mathbf{u} \right] = en \nabla_\parallel \phi|_{z=0} - \nabla_\parallel p(n) \tag{68}$$

$$\nabla^2 \phi = 4\pi e \left[n - n_+ \right] \delta(z), \tag{69}$$

where $\nabla_\parallel = \hat{\mathbf{e}}_x \partial/\partial x + \hat{\mathbf{e}}_y \partial/\partial y$, and $\phi|_{z=0} \equiv \phi(\mathbf{r}_\parallel, t)$ is the in-plane value of the self-consistent electrostatic potential $\phi(\mathbf{r}, t)$ that satisfies the full 3D Poisson equation, Eq. (69), with $n_+ = n_{2D}$ being the surface density of a uniform positive charge background. This latter equation can be solved by the Fourier transform with respect to coordinates, $\mathbf{r}_\parallel \to \mathbf{k}$, and time, $t \to \omega$, giving

$$\phi(\mathbf{r}, t) = -e \int \frac{d^2\mathbf{k}}{(2\pi)^2} \int \frac{d\omega}{2\pi} e^{i\mathbf{k}\cdot\mathbf{r}_\parallel - i\omega t} \frac{2\pi}{k} e^{-k|z|} \widetilde{n}_1(\mathbf{k}, \omega), \tag{70}$$

where $k = \|\mathbf{k}\|$ and $\widetilde{n}_1(\mathbf{k}, \omega)$ is the Fourier transform of the induced electron density $n_1 = n - n_+$. Taking the classical limit by setting the 2D pressure to zero, $p = 0$, and linearizing the continuity and the Navier-Stokes Eqs. (67) and (68), respectively, about the equilibrium state with $n = n_+$ and $\mathbf{u} = \mathbf{0}$, one can eliminate the electrostatic potential and the velocity to obtain $(\omega^2 - 2\pi e^2 k n_{2D}/m_*)\widetilde{n}_1 = 0$, which yields the above square-root plasmon dispersion relation for a 2D EG.

Note that Eqs. (67), (68), and (69) define a 2D HDM where all electrons are treated as belonging to a single fluid characterized by two parameters only: the equilibrium surface density n_{2D}, and the effective electron mass m_*. While this one-fluid (1-F) version of the 2D HDM is similar to the 3D HDM, Eqs. (68) and (69) reveal an intricate relation between the in-plane dynamics of the 2D EG and the electrostatic field that extends through the space. Another important difference in comparison to the 3D HDM stems from the expression for Fermi pressure in a 2D EG, $p(n) = \pi\hbar^2 n^2/(2m_*)$ [32], which gives the pressure force per electron in Eq. (68) as

$$\frac{1}{n} \nabla_\parallel p(n) = \pi \frac{\hbar^2}{m_*} \nabla_\parallel n(\mathbf{r}_\parallel, t), \tag{71}$$

which is *linear* in electron density n, unlike the situation in the 3D HDM. In view of that, it is straightforward to generalize the above analysis, and derive a plasmon dispersion relation for the planar 2D EG as

$$\omega_{2D}(k) = \sqrt{\frac{2\pi e^2}{m_*} n_{2D} k + \beta_{2D}^2 k^2}, \tag{72}$$

where $\beta_{2D} = v_F/\sqrt{2}$ comes from the Fermi pressure and represents the speed of propagation of density disturbances (or, sound speed) in the 2D EG, with $v_F = (\hbar/m_*)\sqrt{2\pi n_{2D}}$ being its Fermi speed. Note that this dispersion relation describes a crossover from the square-root, long wavelength mode of a 2D EG to a quasi-acoustic, short wavelength mode $\omega_{2D} \approx \beta_{2D} k$ for wavenumbers exceeding the critical value of $k_c \sim 2/a_B$ (with a_B being the Bohr's radius), which is notably independent of the electron density n_{2D}. One may assert that the applicability of a HDM at wavelengths shorter than a_B may be questionable, so that the term with β_{2D} in Eq. (72) would make a relatively unimportant contribution. While this may be so for the one-fluid (1-F) version of the 2D HDM, it will be shown below that the presence of such terms in a two-fluid (2-F) version of this model is essential for splitting of plasmon frequencies at long wavelengths. Further, we note that the expression $\beta_{2D} = v_F/\sqrt{2}$ holds for low frequencies, whereas the high-frequency result should be corrected so that $\beta_{2D} = (\sqrt{3}/2)v_F$ [37]. However, switching to this correction is not as critical in the 2D HDM as in a 3D HDM, because the dispersion relation in Eq. (72) implies rather low frequencies in the physically relevant long wavelength limit.

Finally, it is important to identify conditions under which one can justify using the above 2D HDM to describe a 2D EG that occupies a curved surface, such as cylinder or a spheroid. It can be shown that such a condition requires that the local radius of curvature R far exceeds the inverse of the Fermi wavenumber, $k_F = \sqrt{2\pi n_{2D}}$, of the planar 2D EG [38].

6.2 Two-Fluid Model for Carbon Nano-Structures

Discoveries of carbon nano-structures, involving the cage-like fullerene molecules, carbon nanotubes (CNTs) and the recently isolated graphene layers, brought realizations of one-atom thick layers of great practical interest. These sheets of carbon atoms can take various shapes (sphere, cylinder, plane), they interact with each other via van der Waals forces in multi-layered formations, and they couple electrostatically with the surrounding dielectric media. In the early spur of research activity concerned with carbon nano-structures, several versions of the 1-F 3D HDM were used in the configurations of spherical and cylindrical thin slabs for studying collective excitations in the C_{60} molecules [25, 39, 40] and in the single-walled CNTs (SWCNTs) [41], respectively. However, an effective thickness of the EG in these structures had to be invoked in the infinite-barrier version of 3D HDM as an empirical parameter, which was found to affect the obtained results for collective modes in a significant way [25, 39]. Moreover, considerations of smooth electron density profiles in the 3D HDM for both spherical [40] and cylindrical [41] slabs showed great sensitivity of the resulting plasmon frequencies to the form of these profiles.

On the other hand, the thickness of the EG in a single graphene layer can be estimated from the size of its π electron orbitals to be $b \approx 0.92$ Å, whereas the surface density of these electrons coincides with graphene's atomic density, $n_{at} = 38\ \text{nm}^{-2}$, giving $b_{max} \approx 3.5$ Å and $k_F \approx 1.55$ Å^{-1}. With the smallest radius of curvature in carbon nano-structures being that of the radius of C_{60}, $R_{min} \approx 3.5$ Å, one finds a fair margin of confidence that both inequalities $b < b_{max}$ and $R_{min} > 1/k_F$ hold well in these structures. Therefore, it is no surprise that the 2D HDM was quickly adopted in early discussions of collective excitations in carbon nano-structures represented by an infinitesimally thin layer of EG. Initially, this model was used in the classical limit on spheroidal configurations representing the C_{60} and C_{70} molecules [42], as well as in the concentric-layers configurations representing both the multi-walled CNTs (MWCNTs) [43–45], and the multi-shell (or, onion shaped) fullerene molecules [45]. This diversity of geometric configurations illustrates the versatility of the 1-F 2D HDM in various applications with no need for empirical parameters [43–45]. It should be then no surprise that various improvements of the 1-F 2D HDM continue to attract attention of many groups working in relatively diverse applications of CNTs and fullerene molecules [46–54].

However, the four valence electrons per atom in carbon nano-structures can hardly be treated as belonging to the same fluid. Three valence electrons are involved in the strong σ bonds characterizing the sp^2 hybridization within a layer of carbon atoms, whereas one electron occupies a weakly bound π orbital that is largely responsible for the conductivity properties of carbon nano-structures. While the π electron bands make graphene a zero-gap semiconductor, the so-called graphene nano-ribbons (GNRs), as well as CNTs may exhibit metallic or semiconducting character depending on the symmetry of their underlying atomic structure. On the other hand, the σ electron bands exhibit a large gap that gives rise to a characteristic high-frequency feature in the absorption spectra in various allotropic forms of carbon. Consequently, inspired by the empirical model of Cazaux [55], Barton and Eberlein [42] proposed a two-fluid (2F) version of the 2D HDM that treats the σ and π electrons as two classical fluids with equilibrium densities $n_\sigma^0 = 3\,n_{\rm at}$ and $n_\pi^0 = n_{\rm at}$, respectively, superimposed on the surface of a C_{60} molecule. By introducing an empirical restoring frequency for the σ electron fluid, the 2-F 2D HDM of Barton and Eberlein was shown to give rise to two groups of plasmons that explained well the two dominant absorption features in the frequency ranges of 5–10 and 15–30 eV, in both fullerene molecules [42, 56] and SWCNTs [57].

On the other hand, Mowbray et al. introduced the quantum, or Fermi pressure in the σ and π electron fluids and, without invoking any empirical parameters, they found that the electrostatic interaction between those fluids gives rise to two groups of high- and low-frequency plasmons for both SWCNTs [58] and MWCNTs [59] in similar ranges as those obtained by Barton and Eberlein [42]. Versatility of this version of the 2F-2D HDM was further exploited in studying the interactions of CNTs with dielectric media [38, 60], as well as in applications to the propagation of electromagnetic radiation through CNTs [61, 62].

6.2.1 Formulation of the Model for a MWCNT

We illustrate the 2F-2D HDM with Fermi pressure by considering a MWCNT that consists of N concentric cylindrical walls with radii $a_1 < a_2 < \cdots < a_N$ [59]. Using the cylindrical coordinates $\mathbf{r} = \{\rho, \varphi, z\}$ for an arbitrary point in space, we define $\mathbf{r}_j = \{a_j, \varphi, z\}$ to represent the coordinates of a point on the cylindrical surface $\rho = a_j$ (with $1 \le j \le N$). Similarly, $\nabla_j = \hat{\mathbf{e}}_z \frac{\partial}{\partial z} + a_j^{-1} \hat{\mathbf{e}}_\varphi \frac{\partial}{\partial \varphi}$ differentiates only *tangentially* to that surface. Let $n_j(\mathbf{r}_j, t)$ be the first-order perturbation of the electron number density (per unit area) on the jth wall, consisting of the density perturbations of the σ and π electron fluids, $n_j(\mathbf{r}_j,t) = n_{j,\sigma}(\mathbf{r}_j,t) + n_{j,\pi}(\mathbf{r}_j,t)$, and let the tangential velocity field of the electron fluids on that wall be similarly decomposed into $\mathbf{u}_{j,\sigma}(\mathbf{r}_j,t)$ and $\mathbf{u}_{j,\pi}(\mathbf{x}_j,t)$. Linearization of the continuity and the momentum-balance equations for each fluid on the jth wall about the unperturbed state with n_π^0, n_σ^0, and $\mathbf{u}_\pi^0 = \mathbf{u}_\sigma^0 = \mathbf{0}$, gives

$$\frac{\partial n_{j,\lambda}(\mathbf{r}_j,t)}{\partial t} + n_\lambda^0 \nabla_j \cdot \mathbf{u}_{j,\lambda}(\mathbf{r}_j,t) = 0, \tag{73}$$

$$\frac{\partial \mathbf{u}_{j,\lambda}(\mathbf{r}_j,t)}{\partial t} = \frac{e}{m_\lambda^*}\nabla_j \phi(\mathbf{r}_j,t) - \frac{\beta_\lambda^2}{n_\lambda^0}\nabla_j n_{j,\lambda}(\mathbf{r}_j,t), \tag{74}$$

where $\beta_\lambda = \left(\hbar/m_\lambda^*\right)\sqrt{\pi n_\lambda^0}$ with m_λ^* being the effective electron mass in the fluid λ, and where the index λ takes values σ and π. The total electrostatic potential in the system is a solution of a 3D Poisson equation in cylindrical coordinates, which can be expressed via Fourier-Bessel (FB) transform as [63]

$$\phi(\mathbf{r},t) = \sum_{m=-\infty}^{\infty} \int_{-\infty}^{\infty} \frac{dk}{(2\pi)^2} \int_{-\infty}^{\infty} \frac{d\omega}{2\pi}\, e^{ikz+im\varphi-i\omega t}\, \widetilde{\phi}(\rho,m,k,\omega), \tag{75}$$

where k is the wavenumber of collective modes propagating axially, and m is an integer describing the modes propagating around the circumference of a cylinder. The FB transform of the potential is given by the superposition of the FB transforms of the electron density perturbations $\widetilde{n}_j \equiv \widetilde{n}_{j,\sigma} + \widetilde{n}_{j,\pi}$,

$$\widetilde{\phi}(\rho,m,k,\omega) = -e\sum_{j=1}^{N} g(\rho,a_j;m,k)\, a_j\, \widetilde{n}_j(m,k,\omega), \tag{76}$$

where $g(\rho,\rho';m,k) \equiv 4\pi I_m(|k|\rho_<)K_m(|k|\rho_>)$ is the Coulomb Green function in cylindrical coordinates [63], with $\rho_< = \min(\rho,\rho')$ and $\rho_> = \max(\rho,\rho')$, while I_m and K_m are cylindrical Bessel functions of integer order m. After FB transforming the Eqs. (73) and (74) and eliminating $\widetilde{\mathbf{u}}_{j,\lambda}$, one can use Eq. (76) to obtain the following set of coupled equations for electron densities $\widetilde{n}_j$,

$$\widetilde{n}_j(m,k,\omega) + \chi_j^0(m,k,\omega)\sum_{\ell=1}^{N} g(a_j,a_\ell;m,k)\, a_\ell\, \widetilde{n}_\ell(m,k,\omega) = 0. \tag{77}$$

Here, the total density-density response function of the non-interacting σ and π electron fluids on the jth wall is given by $\chi_j^0 = \chi_{j,\sigma}^0 + \chi_{j,\pi}^0$, with

$$\chi^0_{j,\lambda}(m,k,\omega) = \frac{e^2 \frac{n^0_\lambda}{m^*_\lambda} q_j^2}{\beta_\lambda^2 q_j^2 - \omega^2}, \tag{78}$$

where $q_j \equiv q_j(m,k) = \sqrt{k^2 + m^2/a_j^2}$.

6.2.2 Plasmons in a SWCNT

For a SWCNT with radius $a_1 \equiv a$, the summation in Eq. (77) contains only one term with $\ell = j = 1$, so that one can readily derive two plasmon groups with frequencies $\omega_\pm(m,k)$ given by

$$\omega_\pm^2 = \frac{\omega_\sigma^2 + \omega_\pi^2}{2} \pm \sqrt{\left(\frac{\omega_\sigma^2 - \omega_\pi^2}{2}\right)^2 + \Delta^4}, \tag{79}$$

where ω_σ and ω_π are the plasma frequencies of the σ and π electron fluids that do not interact with each other, given by

$$\omega_\lambda^2 = \left[\beta_\lambda^2 + e^2 \frac{n^0_\lambda}{m^*_\lambda} a g(a,a;m,k)\right]\left(k^2 + \frac{m^2}{a^2}\right), \tag{80}$$

whereas the electrostatic coupling between these fluids is given by

$$\Delta^2 = e^2 \sqrt{\frac{n^0_\sigma}{m^*_\sigma}\frac{n^0_\pi}{m^*_\pi}}\, a g(a,a;m,k)\left(k^2 + \frac{m^2}{a^2}\right). \tag{81}$$

Choosing the example of a SWCNT with radius $a = 7$ Å, we set the effective masses to be equal to the bare electron mass, and show in Fig. 12 the two groups of plasmon dispersions from Eq. (79), $\omega_+(m,k)$ and $\omega_-(m,k)$, versus the longitudinal wavenumber k and for several angular modes $m = 0, 1, 2, 3$, and 4, along with the experimental points from the EELS measurements of plasmon energies in bundles of SWCNTs with the predominant values of radii being 7 Å [16, 64]. The experimental configuration is certainly more complex than the present model of an isolated SWCNT, so that the experimental data display some kind of an average over various (presumably lower valued) modes m. Nevertheless, those several dispersion curves, displayed in Fig. 12, show a good agreement with the experimental data, both in the magnitude and in the trend of the dispersion versus k. Given that no free parameters were used, this agreement lends confidence in the 2-F 2D HDM with Fermi pressure.

It is apparent in Fig. 12 that the splitting between the two groups of plasmon frequencies is rather substantial owing to a strong electrostatic interaction between the σ and π fluids in the 2-F 2D HDM. Formally speaking, this splitting is strictly a consequence of the presence of the pressure terms in this model. To illustrate this

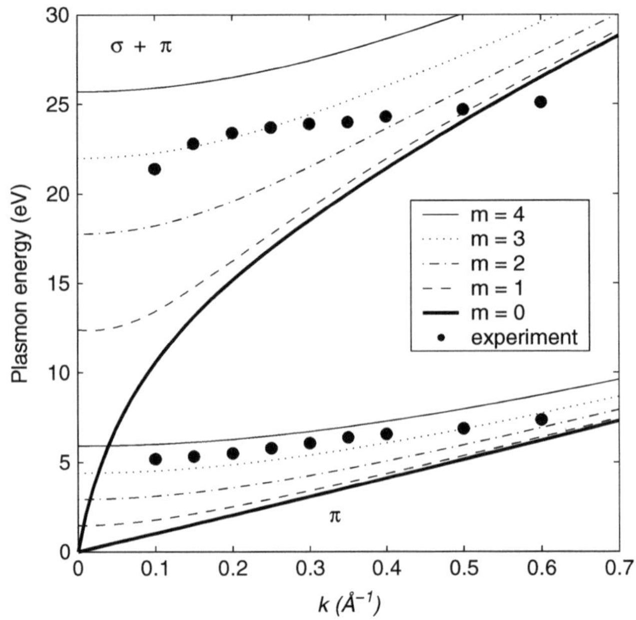

Fig. 12 Dispersion curves (in eV) $\omega_+(m,k)$ and $\omega_-(m,k)$ from Eq. (79), corresponding to the $\sigma+\pi$ and π plasmons in a single-wall carbon nanotube with radius $a = 7$ Å, shown versus longitudinal wavenumber k (in Å^{-1}) for several angular modes m [58]. Experimental points are from [16, 64]

point, we note that, if we set $\beta_\sigma = \beta_\pi = 0$ in Eq. (79), we find $\omega_- = 0$ and $\omega_+ = \sqrt{\omega_\sigma^2 + \omega_\pi^2}$, which becomes $\omega_+ = \sqrt{\left(e^2 n_0/m_*\right) a\, g(a,a;m,k)(k^2 + m^2/a^2)}$ if we further assume equal electron masses, $m_\sigma^* = m_\pi^* = m_*$, indicating that the σ and π fluids collapse into a single $\sigma+\pi$ fluid with the total density of $n_0 = n_\sigma^0 + n_\pi^0$ in the absence of pressure terms. Even when $\beta_\lambda \neq 0$ in Eq. (79), our calculations show that the dispersion curves ω_+ in the upper group in Fig. 12 are quite close to the corresponding 1-F frequencies, given by Eq. (80) with n_λ and m_λ^* replaced by n_0 and m_*, respectively. For that reason $\omega_+(m,k)$ are designated as the $\sigma+\pi$ plasmons. It is further interesting to note that the frequency $\omega_+(0,k)$ of the mode $m = 0$ exhibits a dimensional crossover from a 1D EG for $k \ll 1/a$, whence [44] $\omega_+^2(0,k) \approx \left(4\pi a e^2 n_0/m_*\right) k^2 \ln\left(1.123/(ka)\right)$, to the square-root dispersion of a 2D EG for $1/a \ll k \ll k_c \equiv 2/a_B$, where $\omega_+^2(0,k) \approx \left(2\pi e^2 n_0/m_*\right) k$, and finally approaches a quasi-acoustic dispersion for short wavelengths, $k > k_c$ (well outside the range of wavenumbers shown in Fig. 12), where $\omega_+(0,k) \approx \beta_+ k$ with $\beta_+ = \max(\beta_\sigma, \beta_\pi) = \beta_\sigma$, when the electron masses are assumed equal.

On the other hand, the dispersion curves $\omega_-(m,k)$ are grouped in Fig. 12 around much lower energies, corresponding to the oscillations of the π electron fluid alone, and are thus called the π plasmons [64]. This group generally exhibits weaker dispersions than the $\sigma+\pi$ plasmons. In particular, the $m = 0$ mode in the π-plasmon group exhibits a remarkably linear dispersion for long wavelengths, which can be derived analytically for sufficiently long wavelengths in the form $\omega_-(0, k) \approx v_a k$, where

$$v_a = \sqrt{\frac{\beta_\pi^2 \frac{n_\sigma^0}{m_\sigma^*} + \beta_\sigma^2 \frac{n_\pi^0}{m_\pi^*}}{\frac{n_\sigma^0}{m_\sigma^*} + \frac{n_\pi^0}{m_\pi^*}}}, \tag{82}$$

which becomes $v_a = (\hbar/m_*)\sqrt{3\pi n_0/8}$ for equal electron masses. While the acoustic speed, Eq. (82), is derived by using the standard series expansions for cylindrical Bessel functions of order $m = 0$ for $ka \ll 1$, it is remarkable how well the quasi-acoustic dispersion $\omega_-(0, k) \approx v_a k$ holds over the full range of wavenumbers shown in Fig. 12. Of course, one would expect that the π plasmon dispersion would approach $\omega_-(0, k) \approx \beta_- k$, where $\beta_- = \min(\beta_\sigma, \beta_\pi) = \beta_\pi$ for very short wavelengths, such that $k > k_c$. It should be further stressed that the result in Eq. (82) is independent of the nanotube radius a, and therefore it can be universally applied to MWCNTs (as shown below), or to graphene (in the limit of an "infinite" nanotube radius). We note that such universality in the quasi-acoustic behavior of the π plasmons at long wavelengths has been recently discussed in the context of comparison between the EELS data for SWCNTs and *ab initio* results for graphene [16].

Finally, it is of interest to outline how the above formulae for plasmon dispersion relations in the 2-F 2D HDM for a SWCNT can be adapted to describe planar and spherical 2D EG, corresponding to a single layer of graphene and a C_{60} molecule, respectively. The planar case is simply retrieved by assuming $a \to \infty$ in such a manner that the ratio m/a becomes a quasi-continuous variable, corresponding to the wavenumber for collective modes propagating around the circumference of a very wide SWCNT. Then, one should simply replace $ag(a, a; m, k) \to 2\pi/q$ with $q = \sqrt{k^2 + m^2/a^2}$ in the above formulae for a SWCNT. In the spherical case, note that solutions of the 3D Poisson equation are given in terms of spherical harmonics, $Y_{\ell m}(\varphi, \vartheta)$ [63], so that collective modes on C_{60} can be labeled by the usual indices ℓ and m. Consequently, the above expressions for a SWCNT need to be changed so that $g(a, a; m, k) \to 4\pi/(2\ell + 1)$ and $q^2 = k^2 + m^2/a^2 \to \ell(\ell + 1)/a^2$ where $a \approx 3.5$ Å is the radius of C_{60}.

6.2.3 Plasmons in a MWCNT

To illustrate the effects of multiple walls on collective modes in a MWCNT, we choose the example of a nanotube with the radius of the innermost cylinder $a_1 \approx 3.6$ Å and take the increments of the radii of the subsequent walls to be $a_{j+1} - a_j \approx 3.4$ Å for $1 \le j \le N-1$. By setting the determinant of the $N \times N$ matrix forming the left-hand sides of Eqs. (77) to zero, we obtain $2N$ positive roots for ω defining the plasmon dispersions which are clearly separated into a high-frequency, $\omega_{j,+}(m, k)$, and a low-frequency, $\omega_{j,-}(m, k)$, group (where $1 \le j \le N$) for each m, as shown in Fig. 13 for $N = 10$ and for $m = 0$ and 1. These groupings are rem-

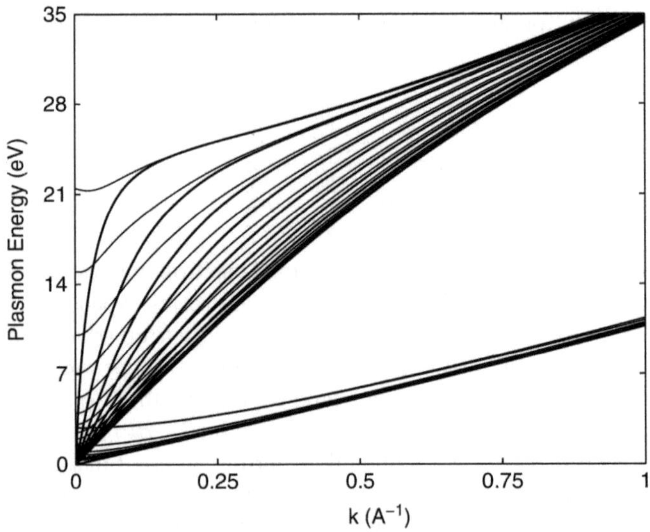

Fig. 13 Dispersion curves (in eV) for plasma oscillations in a $N = 10$ walled carbon nanotube, corresponding to the $\sigma+\pi$ (*upper group*) and π plasmons (*lower group*), shown versus longitudinal wavenumber k (in Å^{-1}) for angular modes $m = 0$ (*thick solid lines*) and $m = 1$ (*thin solid lines*) [59]

iniscent of the experimental plasmon branches describing longitudinal oscillations of $\sigma + \pi$ and π electrons on SWCNTs, discussed above [16, 64]. Interestingly, the low-frequency π plasmons in the case of a MWCNT are much more tightly packed than the high-frequency group, as seen in Fig. 13. In particular, the $m = 0$ modes of π plasmons are virtually degenerate for any (finite) number of nanotube walls, and they all exhibit the quasi-acoustic (linear) dispersion with the common phase speed, v_a, given by Eq. (82). This was anticipated from the independence of the expression Eq. (82) from nanotube radius.

We finally note that a form of the 2-F 2D HDM suitable for a multi-layer planar case (with not too large N) can be obtained from the above formulae for MWCNTs by taking the limit $a_j \to \infty$ so that each factor $q_j = \sqrt{k^2 + m^2/a_j^2}$ converges to a common continuous variable q describing the in-plane collective modes. Then, the formulae for MWCNTs need to be changed so that $a_\ell\, g(a_j, a_\ell; m, k) \to (2\pi/q) \exp\left(-q|a_j - a_\ell|\right)$. Note that the exponential factor in this expression describes a decrease in magnitude of the electrostatic coupling as the distance between the jth and ℓth layer increases.

6.3 Plasmon Hybridization with Dielectric Media

Carbon nano-structures are often placed in a closed proximity of various materials, e.g., CNTs can be grown inside the zeolite channels, or they can be coated by metallic layers for various purposes. Also, various materials can be efficiently intercalated into CNTs, including the notable examples of formation of metallic nanowires inside CNTs. Since electrical and optical properties of such composite structures are important in applications, it is of interest to study the effects of dielectric properties of the various neighboring media on plasmon excitations in CNTs, where the 2D HDM proved to be quite versatile [60]. Besides cylindrical structures, several interesting applications of the 2D HDM appeared studying effects of dielectric media in the spherical geometry, such as plasmons in a single-layer thin metallic shell around a dielectric sphere [65], or collective excitations in a pair of ionic double-layer structures separated by the phospholipid membrane in a spherical model of cell [66].

We shall illustrate the use of a 2-F 2D HDM in the special configurations of metallic coating of a CNT, or metallic nanowire embedded in a CNT, where one encounters a strong hybridization between the plasmon modes of the CNT and the plasmon modes occurring on the cylindrical surface of the neighboring metal due to electrostatic coupling between those two subsystems.

6.3.1 Qualitative Analysis

Although it would be quite appropriate to model the metal by a 3D HDM with dispersion, we shall limit here our discussion of such hybridization to the local approximation where any dispersion effects are neglected in the metal.

It was shown in Fig. 4(b) that the thickness of the induced electron density at the surface of a metallic wire of radius a becomes infinitesimally small in the classical, or non-dispersive limit of a 3D HDM [22]. This allows us to invoke the classical electrostatic boundary conditions [63], based on the assumption that the polarization of a metal wire is represented by a charge density of the form $\sigma_{\text{pol}}\delta\,(\rho - a)$ in cylindrical coordinates, which involves a delta function at its surface. While this assumption guarantees continuity of the electrostatic potential, as illustrated in Fig. 4(a) by the dashed line, the areal density of the polarization charge, σ_{pol}, can be determined by imposing the continuity of the radial component of the electrical displacement field across the metal surface, while describing the bulk of the metal, $\rho < a$, with a Drude-like, local dielectric function of the form $\varepsilon(\omega) = 1 - \omega_{\text{p}}^2/\omega^2$ [19]. A cylindrical cavity in a bulk metal can be treated in the same way [18].

Before embarking on developing the formalism, one can discuss hybridization that will occur between a SWCNT described by a 2-F 2D HDM and a metallic channel or wire, as depicted in Fig. 14. Note that a similar hybridization problem was studied for spherical metallic nano-shells by means of a dispersionless 3D HDM [67, 68]. Figure 14 is an energy-level diagram of plasmon hybridization in a composite SWCNT/dielectric system with the dielectric material represented by a metal wire or a metal channel, explained in three stages. First, plasmon hybridization occurs in the nanotube between the σ and π electron fluids, with frequencies ω_π and ω_σ respectively from Eq. (80), yielding the $\sigma + \pi$ and π plasmons, with frequencies ω_+ and ω_- respectively from Eq. (79), as shown in Figs. 12 and 15(a). Second, plasmon hybridization occurs between the $\sigma + \pi$ nanotube plasmon and the metal wire/channel plasmon branches with frequencies

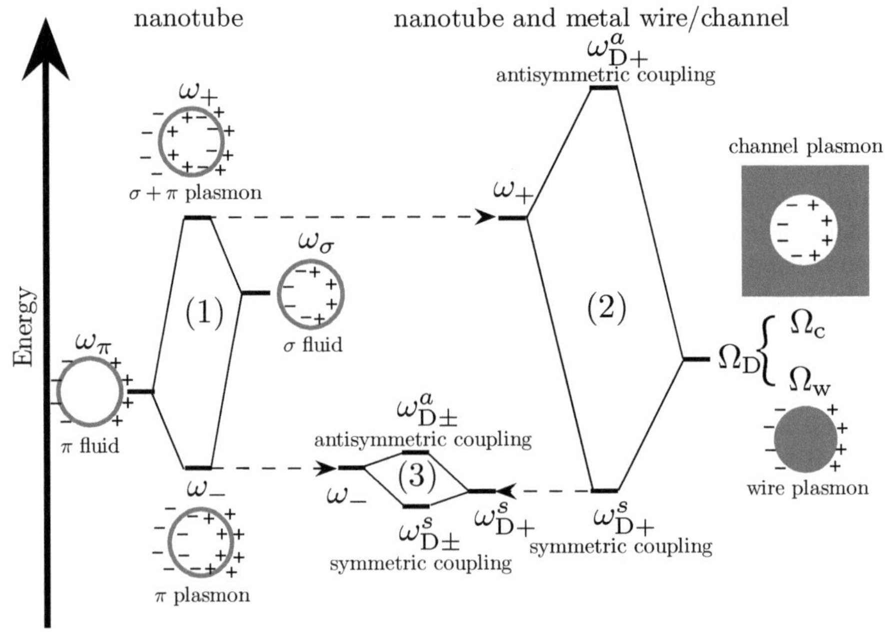

Fig. 14 Energy-level diagram describing plasmon hybridization in a SWCNT (*left*) between (1) the σ and π electron fluids yielding the $\sigma + \pi$ and π plasmons and in a nanotube and a dielectric material represented by a metal wire/channel (*right*), between (2) the $\sigma + \pi$ nanotube plasmon and the metal wire/channel plasmon branches, yielding antisymmetrically coupled (antibonding) and symmetrically coupled (bonding) metal-($\sigma + \pi$) plasmons, and between (3) the symmetrically coupled (bonding) metal-($\sigma + \pi$) plasmon and the π nanotube plasmon, yielding antisymmetrically coupled (antibonding) and symmetrically coupled (bonding) π-metal-($\sigma + \pi$) plasmons [60]

$\Omega_{w/c}$, as shown in Fig. 15(b), yielding antisymmetrically coupled (antibonding) and symmetrically coupled (bonding) metal-($\sigma + \pi$) plasmons with frequencies ω^a_{D+} and ω^s_{D+} respectively. Third, plasmon hybridization occurs between the symmetrically coupled (bonding) metal-($\sigma + \pi$) plasmon and the π nanotube plasmon, yielding antisymmetrically coupled (antibonding) and symmetrically coupled (bonding) π-metal-($\sigma + \pi$) plasmons, with frequencies $\omega^a_{D\pm}$ and $\omega^s_{D\pm}$, respectively.

The plasmon hybridization between the π and symmetrically coupled metal-($\sigma + \pi$) plasmons is a much weaker coupling than those between the nanotube fluids or the $\sigma + \pi$ and metal plasmons, due to the large energy separation between the π plasmon and the higher energy $\sigma + \pi$ and metal plasmons. The antisymmetrically and symmetrically coupled metal-($\sigma + \pi$) plasmon branches will thus have the same qualitative behavior as the plasmon branches obtained from a single-fluid model. Although in principle a coupling exists between the higher energy antisymmetrically coupled metal-($\sigma + \pi$) plasmon and the two lower energy antisymmetrically coupled and symmetrically coupled π-metal-($\sigma + \pi$) plasmons, this coupling is very weak due to the large energy separation between these modes [67].

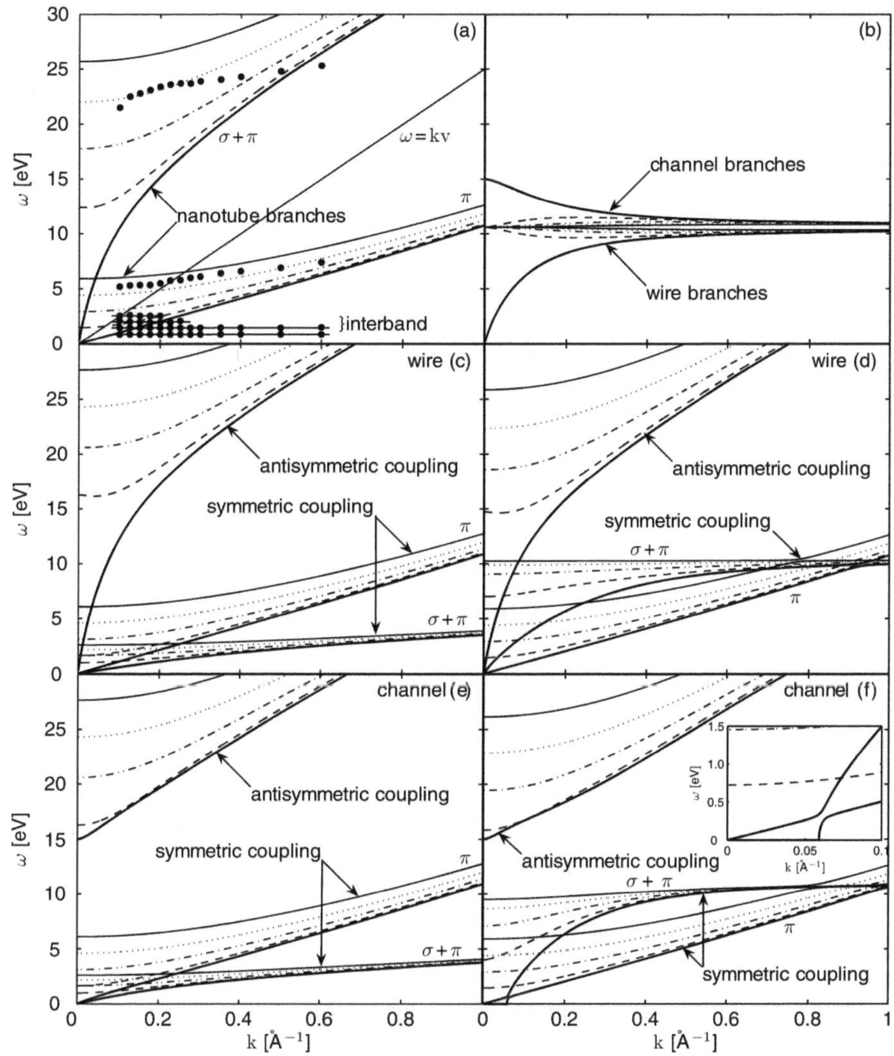

Fig. 15 The plasmon energy ω in eV versus longitudinal wave number k in Å^{-1} of: (**a**) a SWCNT of radius $a = 7$ Å from the 2-F 2D HDM [58] with experimental points (•) from [64], (**b**) an aluminum wire of radius $\rho_{\text{in}} = 7$ Å [19] and channel of radius $\rho_{\text{out}} = 7$ Å [18], a SWCNT of radius $a = 7$ Å encapsulating an aluminum wire of radius (**c**) $\rho_{\text{in}} = 7$ Å and (**d**) $\rho_{\text{in}} = 6$ Å, and a SWCNT of radius $a = 7$ Å encapsulated in an aluminum channel of radius (**e**) $\rho_{\text{out}} = 7$ Å and (f) $\rho_{\text{out}} = 8$ Å, plotted versus k in Å^{-1} for $m = 0$ (*thick solid line*), 1 (*dashed line*), 2 (*dash-dotted line*), 3 (*dotted line*), and 4 (*thin solid line*) [60]

6.3.2 SWCNT with Metal Wire/Channel

Using the cylindrical coordinates $\mathbf{r} = \{\rho, \varphi, z\}$, we assume that the inner region, $0 \le \rho < \rho_{\text{in}}$, is occupied by a material with dielectric function $\varepsilon_{\text{in}}(\omega)$, while the outer region, $\rho > \rho_{\text{out}}$, is occupied by a material with dielectric function $\varepsilon_{\text{out}}(\omega)$. A SWCNT of radius $\rho = a$ occupies the region between the two materials, $\rho_{\text{in}} < \rho < \rho_{\text{out}}$, which is assumed to be vacuum. A typical value for the gap between the SWCNT and the neighboring material can be estimated to be on the order of $\Delta\rho_g = a - \rho_{\text{in}} = \rho_{\text{out}} - a \sim 1$ Å. We further assume that surface polarization charge densities σ_{in} and σ_{out} are induced at the dielectric boundaries $\rho = \rho_{\text{in}}$ and $\rho = \rho_{\text{out}}$, respectively.

The total electrostatic potential in the system can be written as $\phi = \phi_{\text{nt}} + \phi_{\text{in}} + \phi_{\text{out}}$, where the three terms represent contributions coming from the induced electron density on the SWCNT, $-en_1$, and the charge densities σ_{in} and σ_{out}, respectively. So, the FB transform of the potential can be written as

$$\begin{aligned}\widetilde{\phi}(\rho, m, k, \omega) = & -eg(\rho, a; m, k)\, a\, \widetilde{n}_1(m, k, \omega) \\ & + g(\rho, \rho_{\text{in}}; m, k)\, \rho_{\text{in}}\, \widetilde{\sigma}_{\text{in}}(m, k, \omega) \\ & + g(\rho, \rho_{\text{out}}; m, k)\, \rho_{\text{out}}\, \widetilde{\sigma}_{\text{out}}(m, k, \omega). \end{aligned} \tag{83}$$

The polarization charge densities σ_{in} and σ_{out} can be eliminated by imposing the continuity of the radial component of electric displacement vector field at the dielectric boundaries,

$$\begin{aligned} \varepsilon_{\text{in}}(\omega) \left.\frac{\partial \widetilde{\phi}}{\partial \rho}\right|_{\rho=\rho_{\text{in}}^-} &= \left.\frac{\partial \widetilde{\phi}}{\partial \rho}\right|_{\rho=\rho_{\text{in}}^+}, \\ \left.\frac{\partial \widetilde{\phi}}{\partial \rho}\right|_{\rho=\rho_{\text{out}}^-} &= \varepsilon_{\text{out}}(\omega) \left.\frac{\partial \widetilde{\phi}}{\partial \rho}\right|_{\rho=\rho_{\text{out}}^+}. \end{aligned} \tag{84}$$

Following the derivation outlined above for a CNT in vacuum, one comes up with the equation

$$\frac{1}{\chi^0} + a\, g(a, a) + \frac{\rho_{\text{in}}}{\mathcal{E}_{\text{in}}} g(a, \rho_{\text{in}}) + \frac{\rho_{\text{out}}}{\mathcal{E}_{\text{out}}} g(a, \rho_{\text{out}}) = 0, \tag{85}$$

where $\chi^0 = \chi^0_\sigma + \chi^0_\pi$ is the density-density response function for the σ and π electron fluids on the SWCNT, defined in Eq. (78), and where we have dropped the (ω, m, k) variables. In Eq. (85), we have defined the auxiliary functions

$$\mathcal{E}_{\text{in}} \equiv \frac{\mathcal{R}_{\text{in}}^{-1}\mathcal{R}_{\text{out}}^{-1} - \rho_{\text{out}} g'(\rho_{\text{in}}, \rho_{\text{out}}) \rho_{\text{in}} g'(\rho_{\text{out}}, \rho_{\text{in}})}{\mathcal{R}_{\text{out}}^{-1} a g'(\rho_{\text{in}}, a) + \rho_{\text{out}} g'(\rho_{\text{in}}, \rho_{\text{out}}) a g'(\rho_{\text{out}}, a)}, \tag{86}$$

$$\mathcal{E}_{\text{out}} \equiv \frac{\mathcal{R}_{\text{in}}^{-1}\mathcal{R}_{\text{out}}^{-1} - \rho_{\text{out}}g'(\rho_{\text{in}}, \rho_{\text{out}})\rho_{\text{in}}g'(\rho_{\text{out}}, \rho_{\text{in}})}{\mathcal{R}_{\text{in}}^{-1}ag'(\rho_{\text{out}}, a) + \rho_{\text{in}}g'(\rho_{\text{out}}, \rho_{\text{in}})ag'(\rho_{\text{in}}, a)}, \tag{87}$$

where $g'(\rho_1, \rho_2)$ is the derivative of the function $g(\rho_1, \rho_2; m, k)$ with respect to the first argument ρ_1, and where

$$\mathcal{R}_{\text{in}} = \frac{1 - \varepsilon_{\text{in}}}{4\pi[1 - (1 - \varepsilon_{\text{in}})k\rho_{\text{in}}I_m'(k\rho_{\text{in}})K_m(k\rho_{\text{in}})]}, \tag{88}$$

$$\mathcal{R}_{\text{out}} = \frac{\varepsilon_{\text{out}} - 1}{4\pi[1 + (1 - \varepsilon_{\text{out}})k\rho_{\text{out}}I_m(k\rho_{\text{out}})K_m'(k\rho_{\text{out}})]}. \tag{89}$$

We solve Eq. (85) for ω and obtain eigen-frequencies of the collective modes in a combined system consisting of either a SWCNT encapsulating an aluminum wire (by setting $\varepsilon_{\text{out}} = 1$) or a SWCNT being encapsulated in an aluminum channel (by setting $\varepsilon_{\text{in}} = 1$), with the aluminum plasma frequency $\omega_p \approx 15.0$ eV. The results are shown in Fig. 15 for: (a) an isolated SWCNT [58], (b) an isolated wire/channel according to the results of [18, 19], a SWCNT with wire core for (c) $\Delta\rho_g = 0$ and (d) $\Delta\rho_g = 1$ Å, and a SWCNT within channel for (e) $\Delta\rho_g = 0$ and (f) $\Delta\rho_g = 1$ Å. Also shown in panel (a) is the experimental data of Pichler et al. [64] for an isolated SWCNT, as well as the so-called kinematic resonance condition $\omega = kv$, indicating frequencies below which charged particle moving paraxially with the SWCNT at speed v can excite collective modes in the system [60].

Note the elevation of the high energy antisymmetrically coupled metal-$(\sigma + \pi)$ plasmon branches ω_{D+}^a in the combined systems shown in Fig. 15(c–f), when compared with the $\sigma + \pi$ and metallic plasmon branches ω_+ and ω_D, shown in Fig. 15(a) and (b), respectively. Conversely, in Fig. 15(c) and 15(e) the antisymmetrically coupled π-metal-$(\sigma + \pi)$ plasmon branches $\omega_{D\pm}^a$ appear remarkably unchanged from the carbon nanotube quasi-acoustic π branches ω_- shown in Fig. 15(a). This is due to the large energy separation between the π branches ω_- and the metal wire and channel branches Ω_w and Ω_c. We also find in Fig. 15(c) and 15(e) that the presence of a metal wire or channel suppresses the symmetrically coupled π-metal-$(\sigma + \pi)$ plasma frequency branches $\omega_{D\pm}^s$ below the nanotube π branches ω_- shown in Fig. 15(a). Physically, this is because the electrostatic interaction between the nanotube electron fluids and the perfectly conducting metal boundary layers is quite strong in comparison with the electrostatic interaction between the σ and π electron fluids when the gap is zero, $\rho_{\text{out}} = \rho_{\text{in}} = a$.

It is remarkable that the nanotube π branches with frequencies ω_- and the antisymmetrically coupled metal-$(\sigma + \pi)$ channel/wire branches with frequencies ω_{D+}^a are independent of the size of the gap, $\Delta\rho_g = |a - \rho_{\text{in,out}}|$, as shown in Fig. 15(d) and (f) for $\Delta\rho_g = 1$ Å, whereas the symmetrically coupled metal-$(\sigma + \pi)$ branches with frequencies ω_{D+}^s are strongly affected by this gap. As the separation between the nanotube and the dielectric boundary $\Delta\rho_g$ increases, the coupling between the nanotube $\sigma + \pi$ plasmon branch and the metal wire/channel branches weakens. The symmetrically coupled metal-($\sigma + \pi$) plasmon branch ω^s_{D+} moves from an energy below that of the π plasmon branch for $\Delta\rho_g = 0$ to one crossing the π plasmon branch for $\Delta\rho_g \approx 1$ Å. The antisymmetrically and symmetrically coupled π-metal-($\sigma + \pi$) plasmons thus "bounce" between the plasma frequencies ω^s_{D+} and ω_-, due to an avoided crossing, illustrated in the inset of Fig. 15(f), maintaining a minimal separation of ≈ 0.2 eV. This avoided crossing induces a cutoff wavenumber k_{cut} for $m = 0$ in the symmetrically coupled metal-($\sigma + \pi$) channel plasmon branch with frequency $\omega^s_{D\pm}$, which occurs at $k_{\text{cut}} \approx 0.06$ Å^{-1} for $\Delta\rho_g = 1$ Å. Below this cutoff the symmetrically coupled π-metal-($\sigma + \pi$) plasmon branch is completely imaginary. Thus overdamping of this plasmon mode occurs at large wavelengths for nanotubes encapsulated in metal channels.

7 Summary

Surface and bulk plasmons have attracted great interest in recent years for their relevance in emerging areas such as subwavelength optics and nanophotonics [12, 69–73]. Thus for instance, surface modes confined in metallic nanowires are considered as alternatives for waveguides in highly miniaturized integrated optical devices [74, 75]. Also, the differences in dispersion relations for 1D, 2D and 3D conductors [38, 58–60, 76] are aspects of interest in current investigations of wire networks [77]. Studies related to the excitation, propagation and decay of bulk and surface modes in different geometries are currently considered among the processes of greatest interest for the development of new microelectronic networks, for the characterization of new materials and for related studies of geometrical or size effects in nanostructures [78–80]. Additionally, the growing interest on the properties of carbon nanotubes have stimulated a large number of studies on plasmon excitations in cylindrical tubules [81–86].

One of the aspects that acquires particular importance in systems with small dimensions is the question of dispersion effects, since it plays a critical role in the propagation of plasmon waves along wires, nanotubes, or across nanostructured devices. The hydrodynamical model provides a clear way to analyze the properties of surface and bulk plasmons and the related dispersion effects in systems with reduced dimensions. It provides the possibility of obtaining exact solutions for the induced fields and dispersion relations of the modes. For these reasons the method may become an important tool for further developments in the area.

The authors hope that the above overview of methods and recent developments will be found helpful by researchers who plan to start using the hydrodynamical model in their studies of plasmon excitations in nano-structures.

Acknowledgements The authors are grateful to all their collaborators who, over the years, have contributed to the results presented in this review.

References

1. R.H. Ritchie, Plasma losses by fast electrons in thin films. Phys. Rev. **106**, 874–881 (1957)
2. F. Bloch, Bremsvermögen von Atomen mit mehreren Elektronen. Z. Phys. A **81**, 363–376 (1933)
3. J. Crowell, R.H. Ritchie, Radiative decay of Coulomb-stimulated plasmons in spheres. Phys. Rev. **172**, 436–440 (1968)
4. F. Fujimoto, K. Komaki, Plasma oscillations excited by a fast electron in a metallic particle. J. Phys. Soc. Japan **25**, 1679–1687 (1968)
5. R.H. Ritchie, R.E. Wilems, Photon-plasmon interaction in a nonuniform electron gas. Phys. Rev. **178**, 372–381 (1969)
6. J.C. Ashley, O.H. Crawford, J.E. Turner, (eds): The interaction of swift particles and electromagnetic fields with matter (Proceedings of the symposium in honor of R.H. Ritchie. Oak Ridge, TN, 23–25 October 1994). Nucl. Instrum. Methods B **96**(Issues 3–4), 441–678 (1995)
7. A. Eguiluz, S.C. Ying, J.J. Quinn, Influence of the electron density profile on surface plasmons in a hydrodynamic model. Phys. Rev. B **11**, 2118–2121 (1975)
8. G. Barton, Some surface effects in the hydrodynamic model of metals. Rep. Prog. Phys. **42**, 963–1016 (1979)
9. C. Schwartz, W.L. Schaich, Hydrodynamic models of surface plasmons. Phys. Rev. B **26**, 7008–7011 (1982)
10. O.K. Harsh, B.K. Agarwal, Surface-plasmon dispersion relation for the inhomogeneous charge-density medium. Phys. Rev. B **39**, 8150–8152 (1989)
11. K. Dharamvir, B. Singla, K.N. Pathak, V.V. Paranjape, Plasmon excitations in a metallic slab. Phys. Rev. B **48**, 12330–12333 (1992)
12. W.L. Barnes, A. Dereux, T.W. Ebbesen, Surface plasmon subwavelength optics. Nature **424**, 824–830 (2003)
13. S.A. Maier, H.A. Atwater, Plasmonics: localization and guiding of electromagnetic energy in metal/dielectric structures. J. Appl. Phys. **98**, 011101 (2005)
14. A.G. Marinopoulos, L. Reining, A. Rubio, N. Vast, Optical and loss spectra of carbon nanotubes: depolarization effects and intertube interactions. Phys. Rev. Lett. **91**, 046402 (2003)
15. J.M. Pitarke, V.M. Silkin, E.V. Chulkov, P.M. Echenique, Theory of surface plasmons and surface-plasmon polaritons. Rep. Prog. Phys. **70**, 1–87 (2007)
16. C. Kramberger, R. Hambach, C. Giorgetti, M.H. Rummeli, M. Knupfer, J. Fink, B. Buchner, L. Reining, E. Einarsson, S. Maruyama, F. Sottile, K. Hannewald, V. Olevano, A.G. Marinopoulos, T. Pichler, Linear plasmon dispersion in single-wall carbon nanotubes and the collective excitation spectrum of graphene. Phys. Rev. Lett. **100**, 196803 (2008)
17. Z.L. Wang, Valence electron excitations and plasmon oscillations in thin films, surfaces, interfaces and small particles. Micron. **27**, 265–299 (1996)
18. N.R. Arista, M.A. Fuentes, Interaction of charged particles with surface plasmons in cylindrical channels in solids. Phys. Rev. B **63**, 165401 (2001)
19. J.L. Gervasoni, N.R. Arista, Plasmon excitations in cylindrical wires by external charged particles. Phys. Rev. B **68**, 235302 (2003)
20. G. Link, R. von Baltz, Hydrodynamic description of surface plasmons: Nonexistence of the unrestricted half-space solution. Phys. Rev. B **60**, 16157–16163 (1999)
21. J.F. Dobson, H.M. Le, High-frequency hydrodynamics and Thomas-Fermi theory. J. Mol. Struct. (Theocem) **501**, 327–338 (2000)
22. I. Villo-Perez, N.R. Arista, Hydrodynamical model for bulk and surface plasmons in cylindrical wires. Surf. Sci. **603**, 1–13 (2009)
23. N. Barberán, J. Bausells, Plasmon excitation in metallic spheres. Phys. Rev. B **31**, 6354–6359 (1985)
24. M.T. Michalewicz, Identification of plasmons on small metallic particles. Phys. Rev. B **45**, 13664–13670 (1992)

25. M.T. Michalewicz, M.P. Das, Collective electronic excitations on C60 molecule. Solid State Commun. **84**, 1121–1125 (1992)
26. J.J. Monaghan, Collective oscillations in many electron atoms. Aust. J. Phys. **26**, 597–606 (1973)
27. D. Pines, *Elementary Excitations in Solids*, (Benjamin New York, NY 1963)
28. M. Sunjic, A.A. Lucas, Multiple plasmon effects in the energy-loss spectra of electrons in thin films. Phys. Rev. B **3**, 719–729 (1971)
29. P. Ahlqvist, Surface plasmons in a thin film. Phys. Scr. **26**, 217–224 (1982)
30. C.Z. Li, Y.H. Song, Y.N. Wang, Wake effects and energy loss for charged particle moving above a thin metal film. Phys. Rev. A **79**, 062903 (2009)
31. J.A. Stratton, *Electromagnetic Theory*, (McGraw-Hill, New York, NY 1941)
32. A.L. Fetter, Electrodynamics of a layered electron gas. 1. Single layer. Ann. Phys. **81**, 367–393 (1973)
33. A.L. Fetter, Electrodynamics of a layered electron gas. 2. Periodic array. Ann. Phys. **88**, 1–25 (1974)
34. E. Zaremba, H.C. Tso, Thomas-Fermi-Dirac-von Weizsäcker hydrodynamics in parabolic wells. Phys. Rev. B **49**, 8147–8162 (1994)
35. L.A. Constantin, J.P. Perdew, J.M. Pitarke, Collapse of the electron gas to two dimensions in density functional theory. Phys. Rev. Lett. **101**, 016406 (2008)
36. J.-N. Chazalviel, *Coulomb Screening by Mobile Charges*, (Birkhauser, Boston, MA 1999)
37. G.S. Atwal, N.W. Ashcroft, Relaxation of an electron system: conserving approximation. Phys. Rev. B **65**, 115109 (2002)
38. D.J. Mowbray, *Hydrodynamic Modelling of the Electronic Response of Carbon Nanotubes*, (unpublished) (University of Waterloo, Waterloo, ON 2007)
39. D. Ostling, S.P. Apell, A. Rosen, Theory for collective resonances of the C60 molecule. Europhys. Lett. **21**, 539–544 (1993)
40. D. Ostling, S.P. Apell, G. Mukhopadhyay, A. Rosen, Collective resonances of the C60 molecule: effects of electron-density profile. J. Phys. B: At. Mol. Opt. Phys. **29**, 5115–5125 (1996)
41. B. Vasvari, Collective resonances in carbon nanotubes. Phys. Rev. B **55**, 7993–8003 (1997)
42. G. Barton, C. Eberlein, Plasma spectroscopy proposed for C-60 and C-70. J. Chem. Phys. **95**, 1512–1517 (1991)
43. P. Longe, S.M. Bose, Collective excitations in mettalic graphene tubules. Phys. Rev. B **48**, 18239–18243 (1993)
44. C. Yannouleas, E.N. Bogachek, U. Landman, Dimensionality crossovers of the sigma plasmon in coaxial carbon nanotubes. Phys. Rev. B **50**, 7977–7980 (1994)
45. C. Yannouleas, E.N. Bogachek, U. Landman, Collective excitations of multishell carbon microstructures: Multishell fullerenes and coaxial nanotubes. Phys. Rev. B **53**, 10225–10236 (1996)
46. T. Stockli, J.M. Bonard, A. Chatelain, Z.L. Wang, P. Stadelmann, Collective oscillations in a single-wall carbon nanotube excited by fast electrons. Phys. Rev. B **64**, 115424 (2001)
47. Y.N. Wang, Z.L. Miskovic, Interactions of fast ions with carbon nanotubes: self-energy and stopping power. Phys. Rev. A **69**, 022901 (2004)
48. L. Wei, Y.N. Wang, Electromagnetic wave propagation in single-wall carbon nanotubes. Phys. Lett. A **333**, 303–309 (2004)
49. D.J. Mowbray, S. Chung, Z.L. Miskovic, F.O. Goodman, Y.N. Wang, Dynamic interactions of fast ions with carbon nanotubes. Nucl. Instrum. Methods Phys. Res. B **230**, 142–147 (2005)
50. R.F. Tuktarov, R.F. Akhmet'yanov, E.S. Shikhovtseva, Y.A. Lebedev, V.A. Mazunov, Plasma oscillations in fullerene molecules during electron capture. JETP Lett. **81**, 171–174 (2005)
51. D.P. Zhou, Y.H. Song, Y.N. Wang, Z.L. Miskovic, Coulomb explosions and stopping of molecular ions channeled through carbon nanotubes. Phys. Rev. A **73**, 033202 (2006)
52. D. Borka, S. Petrovic, N. Neskovic, D.J. Mowbray, Z.L. Miskovic, Influence of the dynamical image potential on the rainbows in ion channeling through short carbon nanotubes. Phys. Rev. A **73**, 062902 (2006)

53. G. Miano, F. Villone, An integral formulation for the electrodynamics of metallic carbon nanotubes based on a fluid model. IEEE Trans. Antennas Propagation **54**, 2713–2724 (2006)
54. C. Javaherian, B. Shokri, Guided dispersion characteristics of metallic single-wall carbon nanotubes. J. Phys. D: Appl. Phys. **42**, 055307 (2009)
55. J. Cazaux, Graphite dielectric constant in ultraviolet. Solid State Commun. **8**, 545–547 (1970)
56. D.A. Gorokhov, R.A. Suris, V.V. Cheianov, Electron-energy-loss spectroscopy of the C-60 molecule. Phys. Lett. A **223**, 116–122 (1996)
57. X.D. Jiang, Collective plasmon excitations in graphene tubules. Phys. Rev. B **54**, 13487–13490 (1996)
58. D.J. Mowbray, Z.L. Miskovic, F.O. Goodman, Y.N. Wang, Interactions of fast ions with carbon nanotubes: two-fluid model. Phys. Rev. B **70**, 195418 (2004)
59. S. Chung, D.J. Mowbray, Z.L. Miskovic, F.O. Goodman, Y.N. Wang, Dynamic interactions of fast ions with multiwalled carbon nanotubes. Radiat. Phys. Chem. **76**, 524–528 (2007)
60. D.J. Mowbray, Z.L. Miskovic, F.O. Goodman, Ion interactions with carbon nanotubes in dielectric media. Phys. Rev. B **74**, 195435 (2006)
61. A. Moradi, H. Khosravi, Collective excitations in single-walled carbon nanotubes, Phys. Rev. B **76**, 113411 (2007)
62. M. Nejati, C. Javaherian, B. Shokri, B. Jazi, The single-wall carbon nanotube waveguides and excitation of their sigma plus pi plasmons by an electron beam. Phys. Plasmas **16**, 022108 (2009)
63. J.D. Jackson, *Classical Electrodynamics*, 2nd edn. (Wiley, New York, NY 1975)
64. T. Pichler, M. Knupfer, M.S. Golden, J. Fink, A. Rinzler, R.E. Smalley, Localized and delocalized electronic states in single-wall carbon nanotubes. Phys. Rev. Lett. **100**, 4729 (1998)
65. C.Z. Li, Y.H. Song, Y.N. Wang, Energy loss of a charged particle moving outside a nanodielectric sphere covered with infinitesimally thin metal film. Nucl. Instrum. Methods B **267**, 3129–3132 (2009)
66. O.I. Fisun, 2D-plasmon excitation and nonthermal effects on microwaves on biological membranes. Bioelectromagnetics **14**, 57–66 (1993)
67. E. Prodan, C. Radloff, N.J. Halas, P. Nordlander, A hybridization model for the plasmon response of complex nanostructures. Science 302, 419–422 (2003)
68. E. Prodan, P. Nordlander, Plasmon hybridization in spherical nanoparticles. J. Chem. Phys. **120**, 5444–5454 (2004)
69. T.W. Ebbesen, H.J. Lezec, H.F. Ghaemi, T. Thio, P.A. Wolff, Extraordinary optical transmission through sub-wavelength hole arrays. Nature (London) **391**, 667–669 (1998)
70. H. Ditlbacher, A. Hohenau, D. Wagner, U. Kreibig, M. Rogers, F. Hofer, F.R. Aussenegg, J.R. Krenn, Silver nanowires as surface plasmon resonators. Phys. Rev. Lett. **95**, 257403 (2005)
71. N. Kroo, Z. Szentirmay, H. Walter, Sub-wavelength microscopy of surface plasmon oscillations and their statistical properties. Surf. Sci. **582**, 110–116 (2005)
72. L.I. Chelaru, F.J.M. Heringdorf, In situ monitoring of surface plasmons in single-crystalline Ag-nanowires. Surf. Sci. **601**, 4541–4545 (2007)
73. K. Kusová, F. Charra, G. Schull, I. Pelant, Plasmon modes in light emission from silver nanoparticles induced by a scanning tunneling microscope. Surf. Sci. **602**, 345–348 (2008)
74. R.M. Dickson, L.A. Lyon, Unidirectional plasmon propagation in metallic nanowires. J. Phys. Chem. B **104**, 6095–6098 (2000)
75. M.I. Stockman, Nanofocusing of optical energy in tapered plasmonic waveguides. Phys. Rev. Lett. **93**, 137404 (2004)
76. M.F. Lin, K.W.K. Shung, Elementary excitations in cylindrical tubules. Phys. Rev. B **47**, 6617–6624 (1993)
77. F. Parage, M.M. Doria, O. Buisson, Plasma modes in periodic two-dimensional superconducting-wire networks. Phys. Rev. B **58**, R8921–R8924 (1998)
78. W.L. Barnes, Surface plasmon-polariton length scales: a route to sub-wavelength optics. J. Opt. A: Pure Appl. Opt. **8**, S87–S93 (2006)
79. L.S. Mukina, M.M. Nazarov, A.P. Shkurinov, Propagation of THz plasmon pulse on corrugated and flat metal surface. Surf. Sci. **600**, 4771–4776 (2006)

80. V. Chegel, Y. Demidenko, V. Lozovski, A. Tsykhonya, Influence of the shape of the particles covering the metal surface on the dispersion relations of surface plasmons. Surf. Sci. **602**, 1540–1546 (2008)
81. T. Stöckli, Z.L. Wang, J.M. Bonard, P. Stadelmann, A. Châtelain, Plasmon excitations in carbon nanotubes. Phil. Mag. B **79**, 1531–1548 (1999)
82. D. Taverna, M. Kociak, V. Charbois, L. Henrard, Electron energy-loss spectrum of an electron passing near a locally anisotropic nanotube. Phys. Rev. B **66**, 235419 (2002)
83. D. Taverna, M. Kociak, V. Charbois, L. Henrard, O. Stephan, C. Colliex, Simulations of electron energy-loss spectra of an electron passing near a locally anisotropic nanotube. J. Electron Spectrosc. Relat. Phenomena **129**, 293298 (2003)
84. G. Gumbs, A. Balassis, Comparison of the stopping power of plasmons and single-particle excitations for nanotubes. Phys. Rev. B **71**, 235410 (2005)
85. A. Balassis, G. Gumbs, Plasmon instability in energy transfer from a current of charged particles to multiwall and cylindrical nanotube arrays based on self-consistent field theory. Phys. Rev. B **74**, 045420 (2006)
86. S. Segui, J.L. Gervasoni, N.R. Arista, Plasmon interference in nanotubes and nanowires. Surf. Sci. **601**, 4169–4174 (2007)

Part V
Magnetic Nanophases; Magnetic and Non-magnetic Nanocomposites

Magnetic Nanocomposites at Microwave Frequencies

Jaakko V.I. Timonen, Robin H.A. Ras, Olli Ikkala, Markku Oksanen, Eira Seppälä, Khattiya Chalapat, Jian Li, and Gheorghe Sorin Poraoanu

Abstract Most conventional magnetic materials used in the electronic devices are ferrites, which are composed of micrometer-size grains. But ferrites have small saturation magnetization, therefore the performance at GHz frequencies is rather poor. That is why functionalized nanocomposites comprising magnetic nanoparticles (e.g. composed of Fe, Co) with dimensions ranging from a few nm to 100 nm, and embedded in dielectric matrices (e.g. silicon oxide, aluminium oxide) have a significant potential for the electronics industry. When the size of the nanoparticles is smaller than the critical size for multidomain formation, these nanocomposites can be regarded as an ensemble of particles in single-domain states and the losses (due for example to eddy currents) are expected to be relatively small.

Here we review the theory of magnetism in such materials, and we present a novel measurement method used for the characterization of the electromagnetic properties of composites with nanomagnetic insertions. We also present a few experimental results obtained on composites consisting of iron nanoparticles in a dielectric matrix.

1 Introduction

For a long time have ferrites been the best choice of material for various applications requiring magnetic response at radio frequencies (RF). In recent times, there has been a strong demand both from the developers and the end-users side for decreasing the size of the modern-day portable communication devices and to add new functionalities that require access to broader communication bands or to other bands than those commonly used in communication between such devices. All this should be achieved without increasing power consumption; rather, a decrease would be desired. The antenna for example is a relatively large component of modern-day communication devices. If the size of the antenna is decreased by a certain

---

J.V.I. Timonen (✉)
Molecular Materials, Department of Applied Physics, School of Science and Technology, Aalto University, PO Box 15100, FI-00076 AALTO, Finland
e-mail: jaakko.timonen@tkk.fi

A. Aldea, V. Bârsan (eds.), *Trends in Nanophysics*, Engineering Materials, 257
DOI 10.1007/978-3-642-12070-1_11, © Springer-Verlag Berlin Heidelberg 2010

factor, then the resonance frequency of the antenna is increased by the same factor [1]. As a result, in order to compensate this increase in the resonance frequency, the antenna cavity may be filled with a material in which the wavelength of the external radiation field is reduced by the same factor. The wavelength λ inside a material of relative dielectric permittivity ϵ and the relative magnetic permeability μ is given by $\lambda = \lambda_0/\sqrt{\epsilon\mu}$ where λ_0 is the wavelength in vacuum. Hence, it is possible to decrease the wavelength inside the antenna – and therefore also the size of the antenna – by increasing the permittivity or the permeability or the both. Once the size reduction is fixed – that is, $\epsilon\mu$ is fixed – the relative strength between the permittivity and the permeability needs to be decided. It is known that the balance between these two affects the bandwidth of the antenna. Generally speaking, high-ϵ and low-μ materials decrease the bandwidth of the microstrip antenna while low-ϵ and high-μ materials keep the bandwidth unchanged or even increase it [2].

Typical high-μ materials are magnetically soft metals, alloys, and oxides. Of these, metals and alloys are unsuitable for high-frequency applications since they are conducting. On the other hand, non-conducting oxides – such as the ferrites mentioned above – have been used and are still being used in many applications. Their usefulness originates from poor conductivity and the ferrimagnetic ordering. But ferrites are limited by low saturation magnetization which results in a low ferromagnetic resonance frequency and a cut-off in permeability below the communication frequencies [3]. The ferromagnetic resonance frequency has to be well above the designed operation frequency to avoid losses and to have significant magnetic response. However, modern standards such as the Global System for Mobile communications (GSM), the Wireless Local Area Network (WLAN), and the Wireless Universal Serial Bus (Wireless USB) operate in the Super High Frequency (SHF) band or in its immediate vicinity [4]. The frequency range covered by the SHF band is 3–30 GHz and it cannot be accessed by the ferrites whose resonance frequency is typically of the order of hundreds of MHz [3]. Hence, other kinds of materials need to be developed for the applications mentioned.

The important issue related to the miniaturization by increasing the permittivity and/or the permeability is the introduced energy dissipation. In some contexts, losses are good in a sense that they reduce the resonance quality factor and hence increase the bandwidth. The cost is increased energy consumption which goes to heating of the antenna cavity. In general, several processes contribute to losses in magnetic materials. At low frequencies, the dominant loss process is due to hysteresis: it becomes less important as the frequency increases, due to the fact that the motion of the domain walls becomes dampened. The eddy current loss plays a dominant role in the higher-frequency range: the power dissipated in this process scales quadratically with frequency. In this paper will have a closer look at this source of dissipation, which can be reduced in principle by using nanoparticles instead of bulk materials. Another important process which we will discuss is ferromagnetic resonance (due to rotation of the magnetization).

All these phenomena limit the applicability of standard materials for high-frequency electronics. However, the SHF band may be accessed by the so called magnetic granularmaterials. A granular material is composed of a non-conducting matrix with small (metallic) magnetically soft inclusions. Such composites have both desired properties; they are non-conducting and magnetically soft. Granular materials are of special interest at the moment since the synthesis of extremely small magnetic nanoparticles has taken major leaps during the past decades. Especially the synthesis of monodisperse FePt nanoparticles [5] and the synthesis of shape and size controlled cobalt nanoparticles [6] have generated interest because these particles can be produced with a narrow size distribution. In addition, small nanoparticles exhibit an intriguing magnetic phenomenon called superparamagnetism. Superparamagnetic nanoparticles are characterized by zero coercivity and zero remanence which can lead to a decrease in loss in the magnetization process [7].

There have been numerous studies investigating dielectric and magnetic responses of different granular materials. For example, an epoxy-based composite containing 20% (all percentages in this article are defined as volume per volume) rod-shaped CrO_2 nanoparticles has been demonstrated to have a ferromagnetic resonance around 8 GHz and relative permeability of 1.2 [8]. Similarly, a multimillimetre-large self-assembled superlattice of 15 nm FeCo nanoparticles has been shown to have a ferromagnetic resonance above 4 GHz [9].

This raises the interesting question of whether it would be possible in general to design novel nanocomposite materials with specified RF and microwave electromagnetic properties, aiming for example at very large magnetic permeabilities and low loss at microwave frequencies. Such properties should arise from the interparticle exchange coupling effects which, for small enough interparticle separation, extends over near-neighbour particles, and from the reduction of the eddy currents associated with the lower dimensionality of the particles. In this paper, we aim at evaluating the feasibility of using magnetic polymer nanocomposites as magnetically active materials in the SHF band.

The structure of the paper is the following: in Sect. 2 we review briefly the physics of ferromagnetism in nanoparticles, namely the existence of single-domain states (Sect. 2.1), ferromagnetic resonance and the Snoek limit (Sect. 2.3), and eddy currents (Sect. 2.3). In Sect. 3 we discuss theoretically issues such as the requirements stated by thermodynamics on the possibility of dispersing nanoparticles in polymers (Sect. 3.1). A set of rules governing the effective high-frequency magnetic response in magnetic nanocomposites is developed in Sect. 3.2. Then we describe the experimental details and procedures used to prepare and characterize the nanocomposites (Sect. 4). We continue to Sect. 5 where we first discuss a measurement protocol which allow us to measure the electromagnetic properties of the iron nanocomposites (Sect. 5.1). Finally, as the main experimental result of this paper, magnetic permeability and dielectric permittivity spectra between 1–14 GHz are reported in Sect. 5.2 for iron-based nanocomposites (containing Fe/FeO nanoparticles in a polystyrene matrix) as a function of the nanoparticle volume fraction. This paper ends with a discussion (Sect. 6) on how to improve the magnetic performance in the SHF band.

2 Magnetism in Nanoparticles

Magnetic behavior in ferromagnetic nanoparticles is briefly reviewed in this section c.f. [10–12]. The focus is especially in the so called single-domain magnetic nanoparticles which lack the typical multi-domain structure observed in bulk ferromagnetic materials. The topics to be discussed are: (A) when does the single-domain state appear, (B) what is its ferromagnetic resonance frequency, and (C) what are the sources of energy dissipation in single-domain nanoparticles.

2.1 Existence Criteria for the Single-Domain State

A magnetic domain is a uniformly magnetized region within a piece of ferromagnetic or ferrimagnetic material. Magnetic domains are separated by boundary regions called the domain walls (DW) in which the magnetization gradually rotates from the direction defined by one of the domains to the direction defined by the other. The domain wall thickness (d_{DW}), which depends on the material's exchange stiffness coefficient (A) and the anisotropy energy density (K), extends from 10 nm in high-anisotropy materials to 200 nm in low-anisotropy materials. The domain thickness, on the other hand, depends more on geometrical considerations. For example, in one square centimeter iron ribbon, 10 μm thick, the domain wall spacing is of the order of 100 μm. The spacing increases if the thickness is reduced. Reducing the thickness over a critical value leads to the complete disappearance of the domain walls. That state is called the single-domain (SD) state. Between multidomain and single-domain states there may be a vortex state: this is not discussed however here. Similarly, the domains in spherical nanoparticles vanish below a certain diameter which is of the order of few nanometers or few tens of nanometers. In hard materials this diameter ($d_{\mathrm{SD,HARD}}$) can be estimated to be roughly ([11], p. 303):

$$d_{\mathrm{SD,HARD}} \approx 18\frac{\sqrt{AK}}{\mu_0 M_{\mathrm{S}}^2}, \tag{1}$$

where M_{S} is the saturation magnetization and μ_0 is the vacuum permeability. The equation is based on the assumption that the magnetization follows the energetically favorable directions (easy axes or easy planes) defined by the anisotropy. The single-domain diameter given by Eq. (1) should be always compared to the domain wall thickness given by ([11], p. 283)

$$d_{\mathrm{DW}} = \pi\sqrt{\frac{A}{K}}. \tag{2}$$

If the diameter of the particle is less than the wall thickness, it is obvious that it cannot support the wall. The condition $d_{\mathrm{SD,HARD}} > d_{\mathrm{DW}}$, leads to the criterion

$$Q \stackrel{\text{def}}{=} \frac{18}{\pi} \frac{K}{\mu_0 M_S^2} > 1. \tag{3}$$

On the other hand, in magnetically soft nanoparticles the magnetization does not necessary follow the easy directions. In the perfectly isotropic case, that is $K = 0$, the surface spins are oriented along the spherical surface and a vortex core is formed in the center of the particle if the particle is above the single-domain limit. The single-domain diameter ($d_{SD,SOFT}$) of a perfectly isotropic nanoparticle is given by ([11], p. 305),

$$d_{SD,SOFT} \approx 6\sqrt{\frac{A}{\mu_0 M_S^2}\left[\ln \frac{d_{SD,SOFT}}{a} - 1\right]}, \tag{4}$$

where a is the lattice constant. This equation can be solved by the iteration method. The single-domain diameter is more difficult to estimate if the anisotropy is non-zero but does not meet the requirement of Eq. (3). In that case, the single-domain diameter is likely to rest between the values predicted by Eqs. (1) and (4).

Single-domain diameters, domain wall thicknesses and other relevant physical quantities for selected ferromagnetic metals are shown in Table 1. The additional surface-induced anisotropy has been neglected. The uniaxial hexagonal close packed (HCP) cobalt is the only strongly anisotropic material with $Q \approx 1$. The body centered cubic (BCC) iron and the FCC nickel fall in between hard and soft behavior.

Table 1 The saturation magnetization (M_S) [13], the anisotropy energy density K [13, 14], the Q-factor Eq. (3), the single-domain diameter in the hard material approximation $d_{SD,HARD}$ Eq. (1), single-domain diameter in the isotropic material limit $d_{SD,SOFT}$ Eq. (4), and the domain wall width d_{DW} Eq. (2), for iron, cobalt, and nickel. The exchange stiffnesses used in the calculations are from [15]

	M_S (emu/cm^3)	K (erg/cm^3)	Q	$d_{SD,HARD}$ (nm)	$d_{SD,SOFT}$ (nm)	d_{DW} (nm)
Iron (BCC)	1707	4.8×10^5	0.075	5	89	63
Cobalt (HCP)	1440	4.5×10^6	0.996	26	169	26
Nickel (FCC)	485	-5.7×10^4	0.110	13	173	113

2.2 *Ferromagnetic Resonance and the Snoek Limit*

The two major processes contributing to the magnetization change are the domain wall motion and the domain rotation. The resonance frequency of the domain wall motion is typically less than the resonance frequency of the domain rotation. Hence, the only process active in the highest frequencies is the domain rotation which is associated with the ferromagnetic resonance (FMR).

The natural[1] ferromagnetic resonance was first explained by Snoek to be the resonance of the magnetization vector ($\vec{M}$) pivoting under the action of some energy anisotropy field ($\vec{H}_{\mathrm{A}}$) [16]. The origin of the anisotropy is not restricted. It can be induced, for example, by an external magnetic field, magnetocrystalline anisotropy or shape anisotropy. It is common to treat any energy anisotropy as if it was due to an external magnetic field.

The motion of the magnetization around in the anisotropy field is described by the Landau-Lifshitz equation [17],

$$\frac{d\vec{M}}{dt} = -\nu(\vec{M} \times \vec{H}_{\mathrm{A}}) - \frac{4\pi\mu_0\hat{\lambda}}{M_{\mathrm{S}}^2}(\vec{M} \times (\vec{M} \times \vec{H}_{\mathrm{A}})), \tag{5}$$

where $\hat{\lambda}$ is the relaxation frequency (not the resonance frequency) and ν is the gyromagnetic constant given by ([10], p. 559)

$$\nu = g\frac{e\mu_0}{2m} \approx 1.105 \times 10^5 g(\mathrm{mA}^{-1}\mathrm{s}^{-1}) \approx 2.2 \times 10^5 \mathrm{mA}^{-1}\mathrm{s}^{-1}, \tag{6}$$

where g is the gyromagnetic factor (taken to be 2), e is the magnitude of the electron charge and m is the electron mass.

If the Landau-Lifshitz equation is solved, one obtains the resonance condition ([10] p. 559)

$$f_{\mathrm{FMR}} = (2\pi)^{-1}\nu H_{\mathrm{A}}, \tag{7}$$

where f_{FMR} is the resonance frequency and H_{A} is the magnitude of the anisotropy field.

For example, for HCP cobalt the magnetocrystalline anisotropy energy density (U_{A}) is given by ([10], p. 264)

$$U_{\mathrm{A}} = K\sin^2\theta \approx K\left(\theta^2 - \frac{1}{3}\theta^4 + \ldots\right), \tag{8}$$

---

[1] "Natural" is used so that this resonance is distinguished from the dimensional resonance associated with standing waves within a sample of finite size. The dimensional resonance takes place when the end-to-end length (L) of the sample times two is equal to an integer multiple of the wavelength of the radiation (λ) in the material. The same is mathematically expressed as $L = n\lambda/2 = n/2\lambda_0/\sqrt{\epsilon\mu}$ where λ_0 is the wavelength in vacuum and n is a positive integer. The resonance frequency (f_r) is given by $f_r = c/\lambda_0 = nc/2L\sqrt{\epsilon\mu}$ where c is the speed of light. For example, the first dimensional resonance in a 10 mm long sample with $\epsilon\mu = 9$ is approximately 5 GHz. The dimensional resonance can be avoided in experiments by carefully estimating the product $\epsilon\mu$ and designing the sample length accordingly.

where θ is the angle between the easy axis and the magnetization. The energy density due to an imaginary magnetic field is given by ([10], p. 264)

$$U_{\mathrm{A}} = -\mu_0 H_{\mathrm{A}} M_{\mathrm{S}} \cos\theta \approx -\mu_0 H_{\mathrm{A}} M_{\mathrm{S}} \left(1 - \frac{1}{2}\theta^2 + \dots\right). \tag{9}$$

By comparing the exponents one obtains

$$H_{\mathrm{A}} = \frac{2K}{\mu_0 M_{\mathrm{S}}} \approx \frac{0.62\ \mathrm{T}}{\mu_0}, \tag{10}$$

and from Eq. (7)

$$f_{\mathrm{FMR}} = (2\pi)^{-1} \nu H_{\mathrm{A}} \approx 17\ \mathrm{GHz}. \tag{11}$$

It is tempting to use nanoparticles with as high anisotropy as possible in order to maximize the FMR frequency. Unfortunately, the permeability decreases with the increasing anisotropy; for uniaxial materials the relative permeability μ is given by ([10], p. 493),

$$\mu = 1 + \frac{\mu_0 M_{\mathrm{S}}^2 \sin^2\theta}{2K}. \tag{12}$$

It is easy to show that that Eqs. (7),(10), and (12) lead to

$$\langle\mu\rangle \cdot f_{\mathrm{FMR}} = \frac{\nu M_{\mathrm{S}}}{3\pi}, \tag{13}$$

where $\langle\mu\rangle$ is the angular average of the relative permeability (which we assume much larger than the unit). This equation is known as the Snoek limit. It is an extremely important result since it predicts the maximum permeability achievable with a given FMR frequency as a function of the saturation magnetization. It can be shown to be valid for both the uniaxial and cubic materials (taken that $K > 0$). Some values for the maximum relative permeability as a function of the FMR frequency and the saturation magnetization are shown in Table 2.

It has been found out that the Snoek limit can be exceeded in materials of negative uniaxial anisotropy [18]. In that case, the magnetization can rotate in the easy plane perpendicular to the c-axis. Such materials obey the modified Snoek limit ([10], p. 561)

$$\mu \cdot f_{\mathrm{FMR}} = \frac{\nu M_{\mathrm{S}}}{3\pi} \sqrt{\frac{H_{\mathrm{A1}}}{H_{\mathrm{A2}}}}, \tag{14}$$

where H_{A1} is the anisotropy field along the c-plane (small) and H_{A2} is the anisotropy field out of the c-plane (large). One such material is the Ferroxplana [12].

Table 2 Maximum relative permeability (μ) Eq. (13) achievable in cubic and uniaxial materials with positive anisotropy as a function of the saturation magnetization (M_S) and the FMR frequency (f_{FMR})

	Saturation magnetization $\mu_0 M_S$				
f_{FMR} (GHz)	0.1 T	0.3 T	0.5 T	1.0 T	2.0 T
0.1	19.7	57.7	94.3	187.60	374.2
0.5	4.7	12.2	19.7	38.3	75.6
1.0	2.9	6.6	10.3	19.7	38.3
2.0	1.9	3.8	5.7	10.3	19.7
5.0	1.4	2.1	2.9	4.7	8.5

2.3 Eddy Currents and Other Sources of Loss

Magnetic materials can dissipate energy through various processes when magnetized. When the oscillation period of the external driving field is long, the main sources of loss are the processes that contribute to the hysteresis. The hysteresis loss is linearly proportional to the frequency of the driving field since the loss during one complete hysteresis cycle (*B-H* loop) is proportional to the area within the cycle (assuming that the hysteresis loop does not change with the frequency). The main contribution to the hysteresis comes from the domain wall motion and pinning and a smaller contribution is due to the magnetization rotation and domain nucleation. The domain wall motion is damped as the frequency is increased over the domain wall resonance so that only the magnetization rotation persists to the highest frequencies. In addition to the domain rotation hysteresis, the loss in the SHF band stems also from the electrical currents induced by the changing magnetic field inside the particles.

A change in the magnetic field (B) inside a piece of material with finite resistivity (ρ) induces an electric field which generates an electric current as stated by the Faraday's law. This current is called eddy current. It dissipates energy into the sample through the electrical resistance. For example, the averaged loss power $\langle P \rangle$ in a spherical nanoparticle of radius r can be calculated to be[2]

[2] Assume that a spherical nanoparticle (radius r, resistivity ρ), is placed in an alternating magnetic field so that the magnetic field inside the particle is B. According to the Faraday's equation $2\pi x E(x) = -\pi x^2 dB/dt$ where x is the distance from the particle center and $E(x)$ is the electric field. The differential current dI circulating around the cylindrical shell at the distance x is given by $dI = 2\pi x E(x)/dR$ where the differential resistance dR is given by $dR = \rho 2\pi x/h(x)\,dx$ where $h(x)$ is the height of the cylindrical shell and dx is the thickness of the shell. Now the dissipated power can be calculated from $P = \int 2\pi x E(x)\,dI = (\pi/\rho)(dB/dt)^2 \int_0^r x^3\sqrt{r^2 - x^2}dx$. Changing the integration variable from x to $r\sin\varphi$ simplifies the integral to $P = (\pi/\rho)(dB/dt)^2 r^5 \int_0^{\pi/2}\left(\sin^3\varphi - \sin^5\varphi\right)d\varphi$ where the integral part is equal to 2/15. Assuming that the magnetic field is given by $B = \hat{B}\sin 2\pi f$ leads to the result given in Eq. (15).

$$\langle P \rangle = \frac{2\pi}{15}\frac{1}{\rho}r^5\left\langle\left(\frac{dB}{dt}\right)^2\right\rangle = \frac{4\pi^3}{15\rho}r^5\left(f\hat{B}\right)^2, \tag{15}$$

where f is the frequency of the driving field and $\hat{B}$ is the amplitude of the oscillating component of the total magnetization. From Eq. (15) it is obvious that the loss power per unit volume increases as r^2, indicating that the loss can be decreased by using finer nanoparticles. Notice that the loss power will vanish above the FMR resonance since there cannot be magnetic response above that frequency, that is $\hat{B} \to 0$.

For example, the loss power per unit volume (p) in cobalt nanoparticles can be calculated to be

$$p \approx 32\left(\frac{r}{\text{nm}}\right)^2\left(\frac{f}{\text{GHz}}\right)^2\left(\frac{\hat{B}}{\text{T}}\right)^2\frac{\text{W}}{\text{cm}^3}. \tag{16}$$

If the volume of the magnetic element is 0.1 cm^3, the radius of the nanoparticles 5 nm, and $\hat{B} = 1.8$ T one obtains 0.26 mW for loss power.

It has been shown that this simple approach is inadequate to describe the eddy current loss in materials containing domain walls [10]. The eddy currents in multidomain materials are localized at the domain walls, which leads to a roughly four-times increase in the loss. However, since there are no walls present in single-domain nanoparticles and the magnetization reversal can take place by uniform rotation, this model is considered here to be adequate in describing the eddy current loss in single-domain nanoparticles.

One more matter to be addressed is the penetration depth of the magnetic field into the nanoparticles. Because the eddy currents create a magnetic field counteracting the magnetic field that induced the eddy currents, the total magnetic field is reduced when moving from the nanoparticle surface towards its core. The depth (s) at which the magnetic field is reduced by the factor 1/e is called the skin-depth and it is given by ([10], p. 552),

$$s = \sqrt{\frac{2\rho}{\omega\mu\mu_0}}. \tag{17}$$

For example, from Eq. (17) the skin-depth for cobalt ($\rho = 62\ n\Omega$m and $\mu = 10$) at 1 GHz is 1.3 μm and at 10 GHz 400 nm. Hence, cobalt nanoparticles that are less than 100 nm in diameter would already be on the safe side. The situation is rather different in typical ferrites for which $\rho \approx 10^4\ \Omega$m and $\mu = 10^3$, giving 5 cm for the skin depth. Therefore ferrites can be used in the bulk form in near-microwave applications.

3 Magnetic Polymer Nanocomposites

In the simplest form a polymer nanocomposite is a blend of small particles (the diameter is less than 100 nm) incorporated in a polymeric matrix. Polymer nanocomposites are characterized by the convergence of three different length scales: the average radius of gyration of the polymer molecules (R_G), the average diameter of the nanoparticles ($2r$), and the average nearest-neighbor distance between the particles (d), as shown in Fig. 1. In such composites, the polymer chains may not adopt bulk-like conformations [19]. Associated with this, there can be a change in the polymer dynamics which can lead to either an increase or a decrease in the glass transition temperature. Furthermore, the nanoparticles bring their own flavor to the nanocomposite – magnetism, in our particular case.

The most severe problem faced in polymer nanocomposites is the aggregation of nanoparticles. The thermodynamic stability of the nanoparticle dispersion has been addressed in the recent literature experimentally, theoretically and through computer simulations. The experiments have showed that nanoparticles aggregate even at small particle volume fractions – less than 1% in many compositions [20]. Theoretical considerations and computer simulations have revealed that the quality of the nanoparticle dispersion depends delicately on the balance between the entropic and the enthalpic contributions – quite similarly as in polymer blends [21]. The solution for the dispersion dilemma has been pursued by modifying the nanoparticle surface, changing the architecture and size of the polymer and by applying alternative processing conditions.

The simulation results and the theoretical arguments presented in the literature are often difficult to interpret. Furthermore, they do not take into account the magnetic interactions in magnetic nanocomposites. The aim of Sect. 3.1 is to analyze the factors affecting the dispersion quality of magnetic nanoparticles in nonmagnetic polymers. Section 3.2 discusses the effective magnetic response of such nanocomposites.

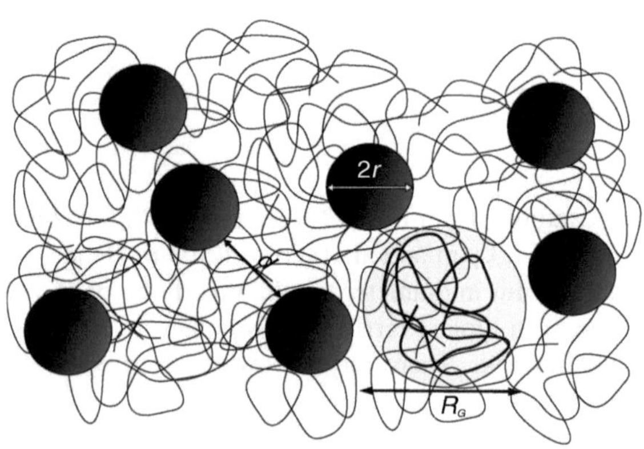

Fig. 1 A schematic illustration of a polymer nanocomposite. The average radius of the nanoparticles ($2r$) (*filled dark circles*), the average radius of gyration of the polymer molecules (R_G) (the *thick black line* inside the *filled light-gray circle*) and the average nearest-neighbor distance (d) between the nanoparticles are of the same magnitude

3.1 Factors Affecting the Nanoparticle Dispersion Quality

3.1.1 Attractive Interparticle Interactions

There has been considerable interest in modifying chemically the nanoparticle surface towards being more compatible with the polymer [22, 23]. Especially important surface modification techniques are the grafting-techniques. They involve either a synthesis of polymer molecules onto nanoparticle surface (grafting-from) or attachment of functionalized polymers onto the the nanoparticle surface (grafting-to). The advantage of the grafting-techniques is that they can make the nanoparticle surface not only enthalpically compatible with a polymer, but the grafted chains also exhibit similar entropic behavior as the surrounding polymer molecules. One disadvantage is that these techniques require precise knowledge of the chemistry involved.

It is well-established that a monolayer of small molecules attached to the nanoparticle surface is not enough to significantly enhance the quality of the dispersion even if the surface molecules were perfectly compatible with the polymer – that is, they were identical to the constitutional units of the polymer. This is due to the fact that the London dispersion force[3] acting between the nanoparticles is effective over a length which increases linearly with the nanoparticle diameter. This is proven in the following.

The London dispersion energy (U_{LONDON}) between two identical spheres, diameters $2r$, separated by a distance d was first shown by Hamaker to be [24],

$$U_{\mathrm{LONDON}} = -\frac{A_{121}}{6}\left[\frac{(2r)^2}{2(2r+d)^2} + \frac{(2r)^2}{2\left((2r+d)^2-(2r)^2\right)} + \ln\left(1-\frac{(2r)^2}{(2r+d)^2}\right)\right], \tag{18}$$

where A_{121} is the effective Hamaker for the nanoparticles (phase 1) immersed in the polymeric matrix (phase 2). The Hamaker constants are typically listed for two objects of the same material in vacuum from which the effective value can be calculated by using the approximation [24]

$$A_{121} \approx \left(\sqrt{A_{11}} - \sqrt{A_{22}}\right)^2, \tag{19}$$

where A_{11} is the Hamaker constant for the nanoparticles and A_{22} is the Hamaker constant for the medium. The typical effective Hamaker constant for metal particles immersed in organic solvent or a polymer is approximately $25\cdot 10^{-20}$ J. By using this value, the London potential Eq. (18) is plotted for 5 nm metal particles in Fig. 2a and for 15 nm particles in Fig. 2b. The distance $\left(d_{k_{\mathrm{B}}T,\mathrm{LONDON}}\right)$ over which the London

[3]Despite of its name it is an attractive force.

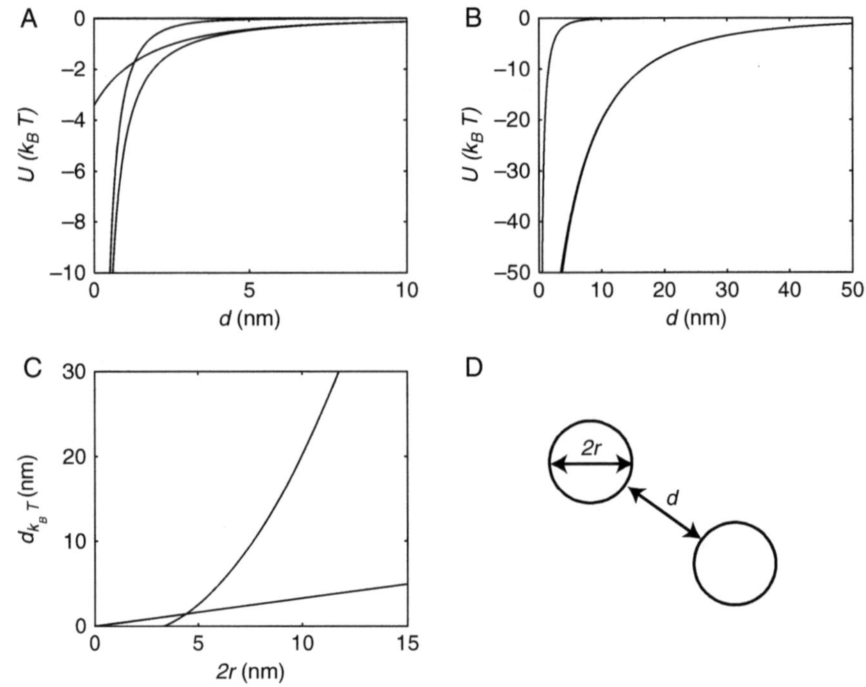

Fig. 2 Comparison between the London dispersion force and the magnetic dipolar interaction between two identical metal nanoparticles. (a) The reduced London potential Eq. (18) (*grey thin curve*), the magnetic dipolar energy Eq. (24) (*black thin curve*) and the total interaction energy (*black thick curve*) between two 5 nm metal nanoparticles. (b) The same for two metal particles 15 nm in diameter. The magnetic dipolar energy curve is overlapping with the total interaction curve. (c) The distance between the particle surfaces as a function of the particle diameter when the interaction energy is comparable to the thermal energy. The *black line* corresponds to the magnetic interaction Eq. (25) and the grey to the London dispersion Eq. (20). (d) Schematic illustration and definition of the used variables

dispersion force is effective can be estimated by setting the interaction energy equal to the thermal energy and by solving for the distance. The result is[4]

$$d_{k_{\mathrm{B}}T,\mathrm{LONDON}} = (\alpha - 1) \cdot 2r \approx \frac{2r}{3}, \tag{20}$$

where α is a constant in excess of unity and typically around 1.33 for metals immersed in organic medium. This linear dependence is shown in Fig. 2c. Typically,

[4]By setting $U_{\mathrm{LONDON}} = -k_{\mathrm{B}}T$ and defining a reduced variable $\alpha = 1 + d_{k_{\mathrm{B}}T,\mathrm{LONDON}}/2r$ one can rewrite Eq. (18) as $6k_{\mathrm{B}}T/A_{121} = 1/(2\alpha^2) + 1/[2(\alpha^2 - 1)] + \ln(1 - 1/\alpha^2) \equiv f(\alpha)$. This equation can be solved for α by plotting $y = f(\alpha)$ and $y = 6k_{\mathrm{B}}T/A_{121}$ and by locating the point of intersection. The effective distance can be calculated by inserting the obtained intersection point into the equation defining the reduced variable.

nanoparticles are covered with a monolayer of alkyl chains ranging up to 20 carbon-carbon bonds in length. Even if the chains were totally extended and rigid, their length would be only roughly 2 nm. Such a shielding layer can protect only nanoparticles less than 12 nm in diameter from aggregation.

Fortunately, the thermodynamic equilibrium is not solely dependent on the enthalpy which always drives the system towards the phase separation. The additional component is entropy which opposes the separation. The Gibbs free energy (G) which determines the thermodynamic stability in the constant temperature and the constant pressure is given by $G = H - TS$, where H is enthalpy and S is entropy. The entropic term per unit volume in a mixture of nanoparticles and small molecular weight solvent molecules can be estimated to be[5]

$$-\frac{TS}{V} = -\frac{k_B T}{V_S}\left[\ln\left(\frac{x}{x-\phi}\right) + \frac{\phi}{x}\ln\left(\frac{x-\phi}{\phi}\right)\right] \approx -\frac{k_B T}{V_S}\frac{\phi}{x}\ln\frac{x}{\phi}, \tag{21}$$

where V_S is the volume of the solvent molecule, ϕ is the volume fraction of the nanoparticles and x is the volume ratio between a nanoparticle and a solvent molecule. In the case of $x = 1$ the equation properly reduces to

$$-\frac{TS}{V} = -\frac{k_B T}{V_S}\left[-\phi\ln\phi - (1-\phi)\ln(1-\phi)\right], \tag{22}$$

which corresponds to the entropy of mixing between two molecules of the same size.

For example, the volume of a toluene molecule is approximately 0.177 nm^3 and the volume of a 10 nm nanoparticle is 524 nm^3. In that case $x \approx 3000$. Equation (21) states that the entropy of mixing is reduced by a factor 1/1300 in a 1% nanocomposite when compared to a situation in which both the nanoparticles and the solvent molecules were of the same size. Without a proof, it is suggested that the magnitude of the entropy is even less when the nanoparticles are mixed with polymer molecules. The suggestion is justifiable due to the entropic restrictions introduced by covalent bonding between the monomer units.

If the nanoparticles are magnetic, they interact with each other more strongly than non-magnetic nanoparticles. The magnetic dipolar interaction energy (U_M) between two particles, $2r$ in diameter, is given by [10]

$$U_M = \frac{\mu_0}{4\pi(d+2r)^3}\left(3(\mathbf{m}_1\cdot\hat{\mathbf{r}})(\mathbf{m}_2\cdot\hat{\mathbf{r}}) - \mathbf{m}_1\cdot\mathbf{m}_2\right), \tag{23}$$

---

[5] Assume that an arbitrary lattice of N sites is filled with N_1 nanoparticles and N_2 solvent molecules. Each nanoparticle incorporates x lattice sites and each solvent molecule one lattice site. Then the number of microstates (Ω) is approximately $\Omega \approx N!/((N-N_1)!N_1!)$. Notice that $N - N_1 \neq N_2$ in contrast to the mixing theory of small molecules of same size. The Eq. (22) is obtained from the definition of entropy $S = k_B \ln\Omega$ by simple algebraic manipulation and by assuming that the density of the nanoparticles is low ($N_1 \ll N$).

where $\hat{\mathbf{r}}$ is the unit vector between the particles, d is the distance between the particle surfaces and $\mathbf{m}_1$ and $\mathbf{m}_2$ are the magnetic moments of the particles. Assuming that the particles are magnetically single-domain, their saturation magnetization is M_S and that the magnetization vectors are parallel to each other and to the unit vector, the interaction energy is reduced to

$$U_M = -\frac{8\pi}{9}\frac{r^6}{(d+2r)^3}\mu_0 M_S^2. \tag{24}$$

Similarly to the effective distance of the London dispersion force, one can derive the distance at which the magnetic energy is comparable to the thermal energy. It is given by

$$d_{k_B T,\mathrm{MAGNETIC}} = \left(\frac{8\pi}{9}\frac{\mu_0 M_S^2}{k_B T}\right)^{\frac{1}{3}} r^2 - 2r. \tag{25}$$

To give an example, the magnetic interaction energy Eq. (24) is drawn for two pairs of cobalt particles, 5 nm and 15 nm in diameter, in Fig. 2a and b, respectively. The interaction between the 5 nm particles is dominated by the London dispersion potential and only weakly modified by the magnetic interaction. In the case of the 15 nm particles, the magnetic interaction is effective over a distance of 50 nm, rendering the London attraction negligible. In order to shield magnetic nanoparticles from such a long-ranging interaction with a protective shell is unpractical. First of all, the maximum achievable nanoparticle volume fraction $\left(\hat{\phi}_{\mathrm{MAGNETIC}}\right)$ is limited by the shielding. If the shielding layer volume is not taken to be a part of the nanoparticle volume, the maximum achievable volume fraction (neglecting entropic considerations) is proportional to

$$\hat{\phi}_{\mathrm{MAGNETIC}} \propto \frac{r^3}{\left(d_{k_B T,\mathrm{MAGNETIC}} + 2r\right)^3} \propto r^{-3}. \tag{26}$$

On the other hand, the maximum volume fraction $\left(\hat{\phi}_{\mathrm{LONDON}}\right)$ limited by shielding against the London attraction does not depend on the nanoparticle size:

$$\hat{\phi}_{\mathrm{LONDON}} \propto \frac{r^3}{\left(d_{k_B T,\mathrm{LONDON}} + 2r\right)^3} = \mathrm{const.} \tag{27}$$

Second, the shielding against the magnetic dipolar attraction by using the conventional grafting techniques is difficult due to the enormous length required from the grafted chains.

Based on the considerations presented in this section, it is unlikely that a uniform dispersion of magnetic nanoparticles of decent size can be achieved by using the conventional shielding strategy. The magnetic interaction starts to dominate the free energy when the magnetic nanoparticles are 10 nm in diameter or larger. Furthermore, the entropic contribution decreases approximately as x^{-1} where x is the volume of the nanoparticle relative to the volume of the solvent molecule. Hence, the dispersion dilemma needs to be approached from some other point of view than the conventional shielding strategy.

3.1.2 Effect of the Polymer Size, Architecture, and Functionalization

A general dispersion strategy proposed by Mackay et al. suggests that the quality of a nanoparticle dispersion is strongly enhanced if the radius of gyration of the polymer is larger than the average diameter of the nanoparticle [20]. The radius of gyration (R_G) for a polymer molecule which is interacting neutrally with its surroundings is given by $R_G \approx \sqrt{C/6}\sqrt{\mathrm{N}}\mathrm{a}$ where N is the number of monomers, a is the length of a single monomer and C is the Flory ratio. For the polystyrene that for example we use the equation yields 16 nm for the radius of gyration ($C \approx 9.9$, $N \approx 2400$ and $a \approx 0.25$ nm). It is based on the assumption that small particles can be incorporated within polymer chains easily but large particles prevent chains from achieving their true bulk conformations. In other words, large particles stretch the polymer molecules and hence introduce an entropic penalty. Pomposo et al. have verified the Mackay's proposition in a material consisting of polystyrene and crosslinked polystyrene nanoparticles [25]. Such a system is ideal in a sense that the interaction between the polymer matrix and the nanoparticles is approximately neutral. That emphasizes the entropic contribution to the free energy. However, if the main contribution to the free energy is enthalpic, as it is in magnetic nanocomposites, one should use the Mackay's proposition with a considerable care. The entropic enhancement is most likely much smaller than the enthalpic term, rendering the improvement in the dispersion quality negligible.

One other remedy for the dispersion dilemma is to replace the linear polymer by a star-shaped one. It has been shown both theoretically [21] and experimentally [26] that it can lead to a spontaneous exfoliation of a polymer-nanoclay composite. It has been also demonstrated that replacing polystyrene in a polystyrene-nanoclay composite by a telechelic hydroxyl-terminated polystyrene results in exfoliation. Since the polymer-nanoclay composites are geometrically different from the polymer-nanoparticle composites, one cannot directly state that these techniques would also work with polymer-nanoparticles composites.

3.2 *Effective Magnetic Response*

The effective relative permeability of a nanocomposite containing spherical magnetic inclusions can be determined from several different effective medium theories (EMT) [27]. The two most popular are the Maxwell-Garnett formula

$$\mu = 1 + 3\phi \frac{\mu_{\mathrm{NP}} - 1}{\mu_{\mathrm{NP}} + 2 - \phi\left(\mu_{\mathrm{NP}} - 1\right)}, \tag{28}$$

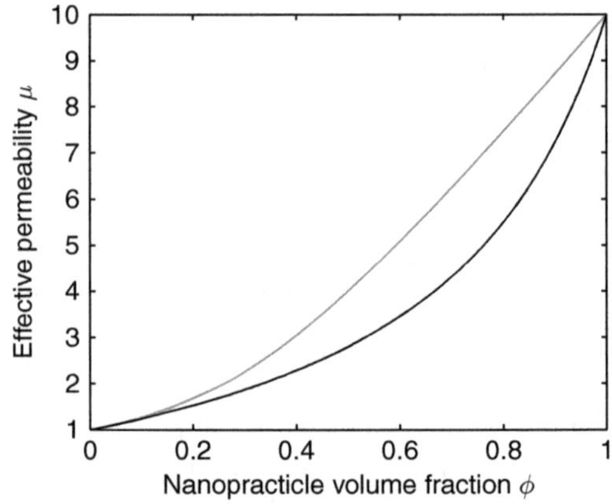

Fig. 3 The effective relative permeability of a nanocomposite containing spherical magnetic inclusions ($\mu_{\mathrm{NP}} = 10$) as a function of the nanoparticle volume fraction. The Maxwell-Garnett theory prediction (*black line*) was obtained from Eq. (28) and the Bruggeman theory prediction (*grey line*) from Eq. (29)

and the symmetric Bruggeman formula

$$\frac{\mu_{\mathrm{NP}} - \mu}{\mu_{\mathrm{NP}} + 2\mu}\phi + \frac{1-\mu}{1+2\mu}(1-\phi) = 0, \tag{29}$$

where μ is the effective relative permeability, μ_{NP} is the relative permeability of the nanoparticles and ϕ is the nanoparticle volume fraction. The effective relative permeability of a nanocomposite containing spherical particles ($\mu_{\mathrm{NP}} = 10$) is plotted in Fig. 3 according to both Eqs. (28) and (29). Below 20% filling, the dependence of the permeability on the volume fraction is approximately linear. However, the rate of the linear increase is not as high as would be expected for homogeneous mixing. The Bruggeman theory has been shown to agree with the experiments with similar materials as studied in this article [28]. Before using the Bruggeman theory one needs to know what is the permeability of the nanoparticles. For uniaxial single-grain particles it is ([10], p. 439)

$$\mu_{\mathrm{NP,UNIAXIAL}} = 1 + \frac{\mu_0 M_{\mathrm{S}}^2 \sin^2\theta}{2K} \tag{30}$$

and for cubic particles

$$\mu_{\mathrm{NP,CUBIC}} = \begin{cases} 1 + \frac{\mu_0 M_{\mathrm{S}}^2 \sin^2\theta}{2K}, & K > 0 \\ 1 - \frac{3\mu_0 M_{\mathrm{S}}^2 \sin^2\theta}{4K}, & K < 0 \end{cases}. \tag{31}$$

where θ is the angle between the easy axis and the external field. Permeabilities of some ferromagnetic metals are calculated in Table 3. It should be pointed

Table 3 The saturation magnetization M_S [13], the anisotropy energy density (K) [13, 14], and the calculated relative permeabilities μ_{NP} (from Eqs. (30) and (31)) for selected ferromagnetic metals. $\langle\mu_{NP}\rangle$ refers to the calculation where the permeability has been averaged over the isotropic distribution of the easy axes

	M_S (emu/cm^3)	K (erg/cm^3)	μ_{NP} ($\theta = \pi/2$)	$\langle\mu_{NP}\rangle$
Iron (BCC)	1707	4.8×10^5	39	26
Cobalt (HCP)	1440	4.5×10^6	4	3
Nickel (FCC)	485	-5.7×10^4	40	27

out that once again the surface anisotropy has been neglected, and that it is most likely that the experimentally determined permeabilities are smaller than those in Table 3.

The effective magnetic response of a polymer nanocomposite containing single-domain nanoparticles can be determined from the following rules:

- The FMR frequency determines the high-frequency limit of the magnetic response. The FMR frequency is determined from the effective anisotropy field by using Eq. (7).
- The permeability below the FMR is dispersion-free since the only magnetization process taking place in single-domain particles is the domain rotation (which is associated with the FMR).
- The magnitude of the permeability is determined from the Bruggeman theory, Eq. (29).
- The permeability of the nanoparticles – which is used in the Bruggeman theory – is determined from the effective anisotropy energy density and the saturation magnetization according to the Eqs. (30) and (31).

The anisotropy used in the calculations should be the true total anisotropy: the sum of the (bulk) magnetocrystalline anisotropy, the surface anisotropy, and the anisotropy due to magnetic field. Especially if the bulk anisotropy is small, the surface anisotropy can be the dominant term. Since the experimental data on the surface anisotropy is scarce, it has been neglected in the analysis so far.

4 Preparation and Characterization

In this section we describe the experimental details and procedures used to prepare and characterize the nanocomposites.

4.1 High Volume Fraction Nanocomposites for High-Frequencies

Nanocomposites containing iron nanoparticles for the SHF band characterization were made according to the following procedure. First, a desired amount of

Table 4 Compositions of the nanocomposites prepared for electromagnetic characterization

Designation	Nanoparticle type	φ (%)
PS/QS-Fe 5%	Quantum sphere iron	5
PS/QS-Fe 10%	Quantum sphere iron	10
PS/QS-Fe 15%	Quantum sphere iron	14.7

nanoparticles (provided by Quantum Sphere, from now on abbreviated QS) were weighted and mixed with 15 ml of toluene (Fluka, purity better than 99.7%). The desired amount of polystyrene was added and allowed to dissolve before vigurously sonicating the solution to break nanoparticle aggregates. The toluene was allowed to evaporate, resulting in a dark polystyrene-like film which was then collected.

Using this method, we have prepared nanocomposites of iron nanoparticles, with three different concentrations, 5, 10 and 15% (see Table 4).

4.2 Transmission Electron Microscopy: Structural Analysis

The nanoparticles were imaged with a TEM (FEI Company model Tecnai G2 BioTwin) in bright field at the acceleration voltage of 120 kV. Before imaging the alignment of the microscope was checked and corrected. The image was recorded with a digital camera (Gatan model UltrascanTM 1000) and its contrast and brightness was adjusted after acquisition. An image of iron nanoparticles is shown in Fig. 4.

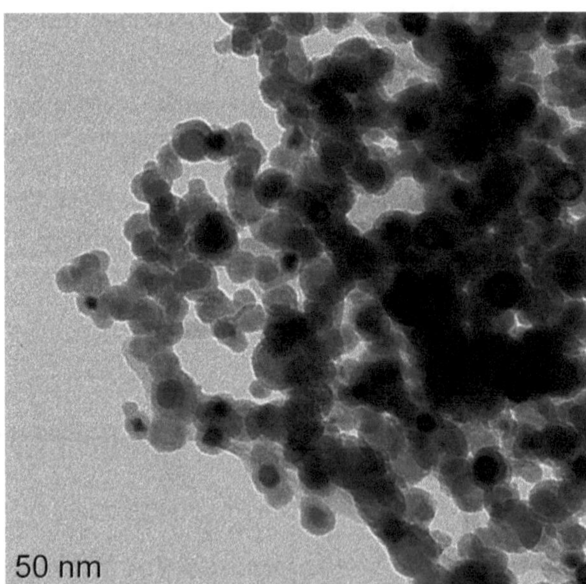

Fig. 4 Bright field TEM images of the quantum sphere iron (scale bar is 50 nm)

4.3 Magnetometry: Low-Frequency Permeability

Static hysteresis loops (the magnetization versus the applied field) of the nanoparticles were measured with a Superconducting Quantum Interference Device (SQUID) magnetometer (Quantum Design model MPMS XL) at 300 K. Roughly 1 mg of the nanoparticles was encapsulated in a piece of aluminum foil (approximately 100 mg) and attached to the plastic straw sample holder with Kapton tape. The permeability was extracted from the measured magnetization curve by fitting a straight line to the low-field part of the curve. The demagnetizing factor was approximated to be zero because the nanoparticles were compressed into flat layers inside the aluminum wrap and the layer surface was aligned along the external field.

4.4 X-Ray Diffraction: Structure of the Nanoparticles

The nanoparticle structure was analyzed with XRD. The diffraction intensities of the nanoparticles were measured as a function of the diffraction angle 2θ with a diffractometer (PANalytical model X'Pert PRO MRD) using Cu K_α radiation (wavelength of 0.154056 nm) at room temperature. The XRD patterns of QS-Fe nanoparticles is shown in Fig. 5. The samples were prepared by filling a circular cavity (35 mm in diameter and 0.7 mm high) bored into an acrylic glass plate with the particles. The powder was compressed and smoothed with a piece of a silicon wafer. The adhesion between the powder and the plate was sufficient to hold the powder within the cavity even though the plate was turned vertically for the measurement. The sample was scanned from 30° to 90° for one hour. The resulting data was processed by first stripping off the peaks due to the Cu K_{α_2} radiation and by filtering the background noise. The data was smoothed if the signal-to-noise ratio was poor. Second, the Lorentzian function was fitted to all peaks using the (self-implemented) Gauss-Newton algorithm. The performance of the algorithm was excellent in the case of well-defined peaks, but vague peaks had to be fitted manually. From the fitted peaks the angle, the FWHM and the intensity (integrated over the peak area) were extracted. Based on these values, the composition of nanoparticles was determined. Furthermore, the coherently scattering domain size was estimated from broadening of the FWHM. The natural width of a peak due to diffractometer was determined by measuring an annealed silicon powder sample and assuming that the coherently

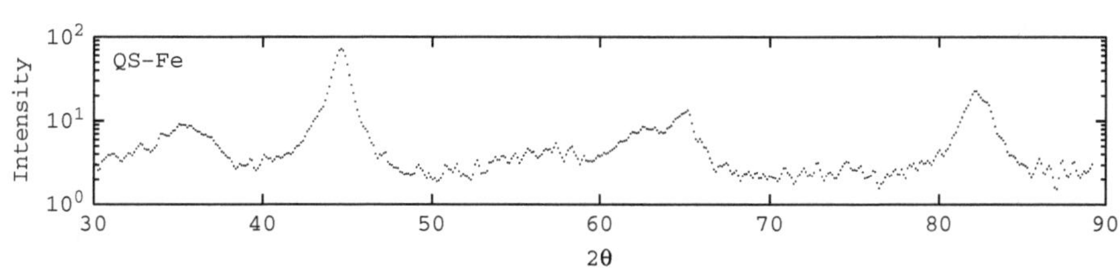

Fig. 5 XRD spectrum of iron nanoparticles

scattering domains were so large that their contribution to the broadening of the FWHM was negligible. The broadening due to the lattice strain was assumed to be minimal.

4.5 Summary

To summarize our results (see Table 5), we find that The Quantum Sphere iron (QS-Fe) nanoparticles are roughly 20–30 nm in diameter and most of the particles exhibit a core-shell structure. Based on the XRD analysis, the core is suggested to comprise 9 nm BCC iron crystallites and the shell 3 nm FeO crystallites. The composition was determined to be a 50–50% balance between the oxide and the metal phases. The measured saturation magnetization (125 emu/g) is in rough agreement with the metal volume fraction estimated from XRD and the saturation magnetization given in literature for pure iron (218 emu/g) [13].

Table 5 Summary of the nanoparticles and their properties. Particle diameters d were estimated from the TEM images and the crystalline composition and the average crystallite diameters d_{CRYST} from the XRD measurement

	d nm	M_S (emu/g)	μ	Crystal (nm)	φ (%)	d_{cryst} (nm)
Quantum sphere iron	20–30	125	12.3	Fe (BCC)	50 ± 5	9 ± 1
				FeO	50 ± 5	3 ± 1

5 High-Frequency Properties

5.1 Coaxial Airline Technique: Permittivity and Permeability in the SHF Band

A broadband coaxial airline method developed in [29] was used to measure the complex permittivity and the complex permeability of magnetic composites in the superhigh-frequency band (SHF). The technique involves measurement of the reflection parameters S_{11} and S_{22}, the transmission parameters S_{12} and S_{21}, and the group delay through a sample inserted inside a 7 mm precision coaxial airline. The measurement was done by connecting the coaxial airline to a vector network analyzer (Rohde and Schwarz ZVA40) using a pair of high-performance cables (Anritsu 3671K50-1). Prior to the measurement, the errors due to the loss and reflection in the cables, connectors and the network analyzer were removed by performing a SOLT calibration up to both ends of the RF cables.

The sample required in the coaxial waveguide measurement is a cylinder, 7.00 mm in diameter, with a 3.04 mm hole in the middle. Its thickness can be adjusted between 4 and 10 mm in order to avoid the dimensional resonance. The samples were made by hot-pressing each nanocomposite inside a polished 7 mm hole drilled through a steel plate. Prior to the pressing, the plate and the nanocomposites inside the holes were sandwiched between two sheets of poly(ethylene terephthalate) and further between two solid steel plates. The assembly was inserted into a hot-press (Fontijne model TP 400) at 160°C and kept there for two minutes. After the nanocomposite had softened, a 400 kN force was applied over the plates. After waiting for another two minutes, the pressure was released and the plate system was disassembled. The holes containing the softened and compressed nanocomposites were refilled and the pressing was done again. The filling was repeated one more time. After the third pressing, the assembly was placed between two metal plates cooled with circulating water under a 400 kN force. After the plates had cooled down to room temperature, the pressure was released and the plates were disassembled. Before detaching the solidified cylindrical samples, the plate with the samples was sandwiched between two 5 mm thick steel plates with 3.1 mm holes exactly above and under of each of the 7 mm holes in the central plate. All the three plates were aligned with respect to each other and clamped together. The assembly was fixed under a vertical boring machine and holes were drilled through the nanocomposites through the guiding 3.10 mm holes in the upper plate. The used drill bit was 2.9 mm in diameter since it was found out that the drilling produced holes 0.1–0.2 mm wider than the drill bit. After drilling, the construction was disassembled and the samples were detached by gently pushing them out of the holes. If necessary, the pellets were finished by carefully removing any imperfections with sandpaper. The sample dimensions were measured with a caliper.

Below we briefly present the measurement method; the mathematical relations between the S-parameters and the material parameters are given. More discussions, including detailed error analysis of the method we use, are presented in [29]. Basically, the method is developed based on the multiple reflection model, as shown schematically in Fig. 6.

When the wave arrives at the first interface at $z = 0$, the reflection and transmission occurs. This means part of the wave is reflected with a coefficient Γ, and part of it is transmitted with a coefficient T_{21}. The transmitted wave then travels through the second medium and gets reflected again at the second interface with a coefficient

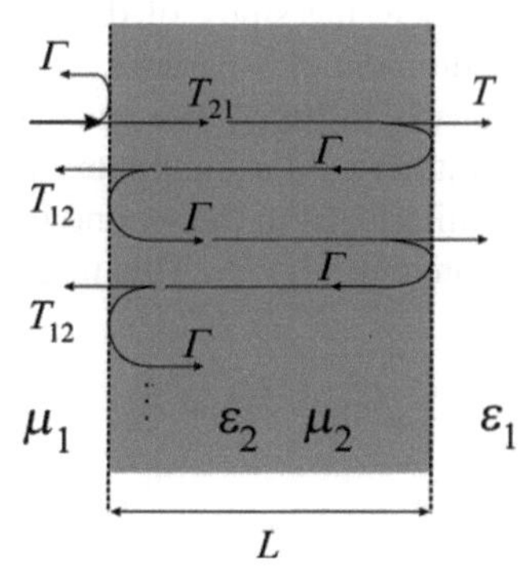

Fig. 6 The model of multiple reflection between two interfaces. Figure republished with permission (©IEEE 2009) from [29]

Γ while part of it is transmitted through the second interface with a coefficient T_{12}. It can be seen from Fig. 6 that this transmission and reflection continuously occurs (ideally) an infinite number of times or until the wave has lost all of its energy.

To find the total reflection coefficient in this model, we need to sum up all the reflected waves. The superposition of waves can be calculated in the same way as the summation of vectors in which both amplitude and phase must be considered. We know that a wave traveling a distance L through the second medium has a propagation factor given by

$$P = e^{-\gamma_2 L}, \tag{32}$$

where $\gamma_2 = i\omega/v_2 = i\omega n_2/c$.

The total reflection coefficient can then be expressed as follows

$$\Gamma_{\text{tot}} = \Gamma + T_{21}T_{12}\Gamma P^2 + T_{21}T_{12}\Gamma^3 P^4 + \ldots = \frac{\Gamma(1-P^2)}{1-\Gamma^2 P^2}, \tag{33}$$

where

$$\Gamma = \frac{1-\sqrt{\frac{\epsilon_2\mu_1}{\epsilon_1\mu_2}}}{1+\sqrt{\frac{\epsilon_2\mu_1}{\epsilon_1\mu_2}}}, \tag{34}$$

and

$$T_{12} = 1+\Gamma = \frac{2}{1+\sqrt{\frac{\epsilon_1\mu_2}{\epsilon_2\mu_1}}} = \sqrt{\frac{\epsilon_2\mu_1}{\epsilon_1\mu_2}}\, T_{21}. \tag{35}$$

Similarly, the total transmission coefficient in terms of Γ and P is

$$T_{\text{tot}} = \frac{P\left(1-\Gamma^2\right)}{1-\Gamma^2 P^2}. \tag{36}$$

In practice, the study of discontinuities within a transmission line is done via the measurement of S-parameters. Considering the measurement setup as illustrated schematically in Fig. 7, we can see that when a wave travels from the first port to the first interface, it accumulates a phase change of $-\gamma_1 L_1$, where $\gamma_1 = i\omega n_1/c \approx i\omega/c$. Similarly, from the second interface to the second port, it will pick up another phase change of $-\gamma_1 L_2$. This means

$$S_{21} = S_{12} = e^{-\gamma_1(L_1+L_2)}T_{tot} = e^{-\gamma_1(L_1+L_2)}\frac{P(1-\Gamma^2)}{1-\Gamma^2 P^2}, \tag{37}$$

$$S_{11} = e^{-2\gamma_1 L_1}\Gamma_{tot} = e^{-2\gamma_1 L_1}\frac{\Gamma\left(1-P^2\right)}{1-\Gamma^2 P^2}, \tag{38}$$

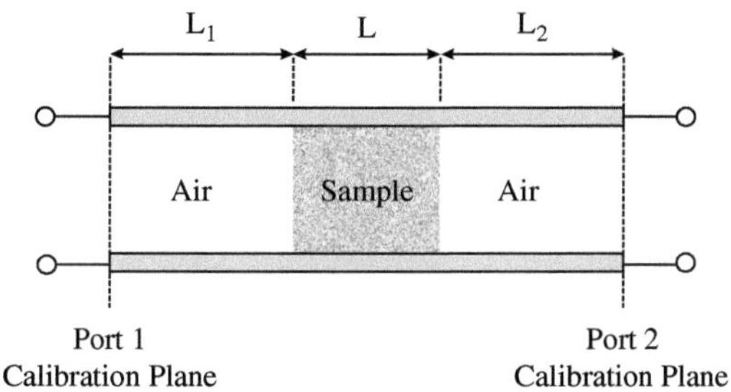

Fig. 7 The diagram of a transmission line containing two interfaces and the planes at which scattering parameters are measured. Figure republished with permission (©2009 IEEE) from [29]

and

$$S_{22} = e^{-2\gamma_1 L_2}\Gamma_{\text{tot}} = e^{-2\gamma_1 L_2}\frac{\Gamma\left(1-P^2\right)}{1-\Gamma^2 P^2}. \tag{39}$$

In principle, there are many ways to obtain the material parameters based on the above equations. The method presented here is chosen because it does not require the measurement of L_1 and L_2; as a result, material parameters can be accurately determined.

The algorithm proceeds further by first defining two reference-plane invariant quantities, namely

$$A = \frac{S_{11}S_{22}}{S_{21}S_{12}} = \frac{\Gamma^2}{(1-\Gamma^2)^2}\frac{(1-P^2)^2}{P^2}, \tag{40}$$

and

$$B = e^{2\gamma_1(L_{\text{air}}-L)}(S_{21}S_{12} - S_{11}S_{22}) = \frac{P^2-\Gamma^2}{1-\Gamma^2 P^2}. \tag{41}$$

Next, Eq. (41) is solved for P^2,

$$P^2 = \frac{B+\Gamma^2}{1+B\Gamma^2}. \tag{42}$$

Then, simply by substituting P^2 into (40), a new expression for A is obtained,

$$A = \frac{\Gamma^2(1-B)^2}{(B+\Gamma^2)(1+B\Gamma^2)}, \tag{43}$$

which gives us

$$\Gamma^2 = \frac{-A(1+B^2)+(1-B)^2}{2AB} \pm \frac{\sqrt{-4A^2B^2+\left[A(1+B^2)-(1-B)^2\right]^2}}{2AB}, \tag{44}$$

where the sign in this equation is chosen so that $|\Gamma| \leq 1$. As we can see, these expressions for P^2 and Γ^2 are reference-plane invariant.

In the next step, another quantity is defined, namely

$$R = \frac{S_{21}}{S_{21}^o} = \frac{e^{\gamma_1 L} P(1-\Gamma^2)}{1-P^2\Gamma^2}. \tag{45}$$

Substituting P^2 from Eq. (42) into Eq. (45), we get a new expression for P,

$$P = R\frac{1+\Gamma^2}{1+B\Gamma^2} e^{-\gamma_1 L}. \tag{46}$$

By using Eqs. (44) and (46), we can determine the other constitutive parameters of materials, for example the complex index of refraction,

$$n = n' + in'' = \sqrt{\mu_r \epsilon_r} = \frac{1}{\gamma_1 L} \ln\left(\frac{1}{P}\right). \tag{47}$$

Similar to the Nicolson-Ross Weir algorithm, [30, 31], the method requires the evaluation of group delay for choosing the correct result. But, it should be noted that, only the real part of n, in Eq. (47), is multi-valued, the imaginary part is not, *i.e.* every root provides the same n''. So measuring the imaginary part of the index of refraction does not require the evaluation of the group delay. This concept could be practically useful for examples when energy loss resonances are studied.

In case of non-magnetic materials, determining the complex permittivity from $\epsilon_r = n^2$ provides a better alternative relative to the NRW method. This is because, this way, one does not need to calculate the relative wave impedance $z = (1+\Gamma)/(1-\Gamma)$, which exhibits high errors at and around the Bragg resonance frequencies [32].

As discussed in [29], extra steps must be done if this method is applied to measure materials with unknown magnetic properties. One way to do so, is to simply use one of the roots $\pm\Gamma$ of Eq. (44), and simultaneously plot the spectra of both ϵ_r and μ_r. Then, based on chemical analysis, the permeability spectra can be extracted. This algorithm is based on the fact that the sign of Γ only swaps the values of permittivity and permeability.

5.2 Nanocomposites: Permittivity and Permeability in the SHF Band

We now present our experimental results corresponding to a nanocomposite comprising iron nanoparticles (20–30 nm, BCC) in polystyrene (PS/QS-Fe) (Figs. 8, 9, and 10).

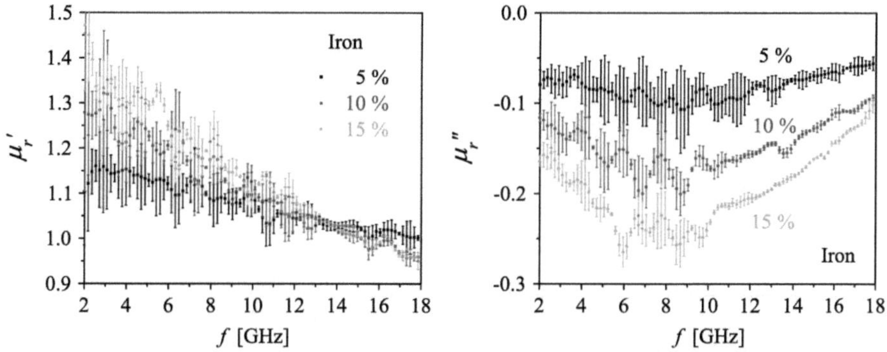

Fig. 8 The complex relative permeability (real part in the *left figure* and imaginary part in the *right figure*) of PS/QS-Fe nanocomposites between 2 and 12 GHz. The three sets of points correspond to samples with 5% (squares), 10% (dots), and 15% (triangles) magnetic inclusions

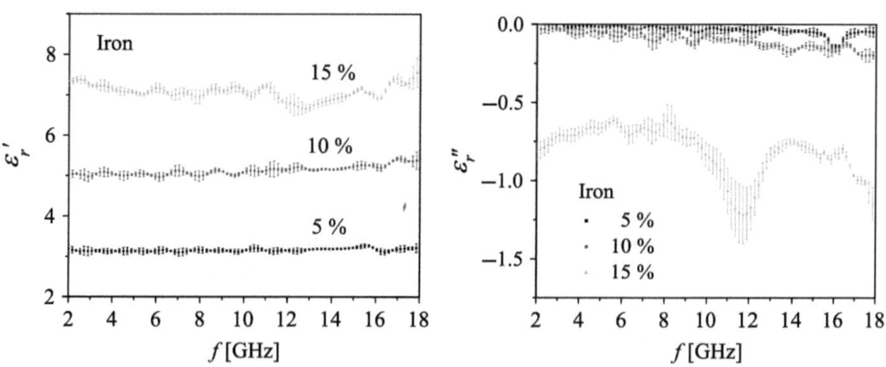

Fig. 9 The complex relative permittivity (real part in the *left* figure and imaginary part in the *right* figure) of PS/QS-Fe nanocomposites between 2 and 12 GHz. The three sets of points correspond to samples with 5% (squares), 10% (dots), and 15% (triangles) magnetic inclusions

All the composites 5, 10 and 15% exhibited mild ferromagnetic resonances between 6 and 8 GHz. These resonances correspond to the anisotropy fields between 0.20 and 0.27 T. The expected anisotropy field calculated from bulk BCC iron magnetocrystalline anisotropy is 56 mT. This large difference may be explained by additional anisotropy components due to surface effects or particle-particle interactions. In the case of surface anisotropy the broadening of the resonance peak would be due to the finite size distribution of the nanoparticles, and in the case of particle-particle interactions due to variations of the polarizing field due to irregular spatial arrangement and orientation of the particles. The Snoek limit (Eq. 13) predicts that no higher relative magnetic permeability than 8.5 can be achieved at 5 GHz in (positive) uniaxial and cubic materials which are either bulk or composites containing spherical inclusions. According to the Bruggeman theory (Eq. 29) the effective relative permeability is at maximum one sixth of 8.5 in a nanocomposite

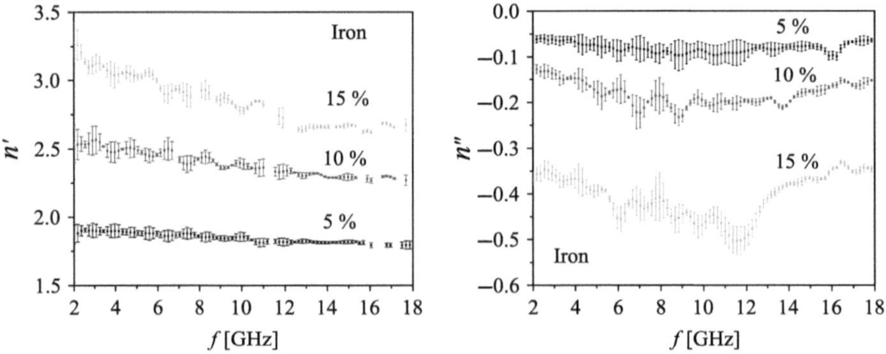

Fig. 10 The complex index of refraction (real part in the *left* figure and imaginary part in the *right* figure) of PS/QS-Fe nanocomposites between 2 and 12 GHz. The three sets of points correspond to samples with 5% (squares), 10% (dots), and 15% (triangles) magnetic inclusions

containing less than 15% magnetic inclusions. We find a relative permeability of the order $\mu = 1.3$ in the 15% nanocomposite, which is already pushing the Snoek limit. The exact determination of whether the Snoek limit has been exceeded depends on the nature of the anisotropy which would require precise knowledge of the surface contribution.

The permeabilities (both real and imaginary) in all composites were roughly constants over the measurement range. While the real part increased roughly linearly with the volume fraction, there was much larger jump in the imaginary part from 10 to 15% than from 5 to 10%. This behavior could be attributed to the increase in conductivity due to exceeding the percolation threshold.

The relative permittivities were observed to increase with the nanoparticle volume fraction from approximately 2.5, which is a typical value for polystyrene. The imaginary parts were observed to be significantly larger than 10^{-4} (a typical value for pure polystyrene). The increase in both the real part and the imaginary part is understood to be due to the electrical polarizability of the nanoparticles in the electric field. The dispersion for the 5% and the 10% composites was also comparatively lower than for the 15% sample (Fig. 10).

6 Conclusions: How to Improve the Performance in the SHF Band

The high-frequency magnetic performance is always a compromise between the permeability and the FMR frequency. Increasing the magnetic anisotropy, no matter wherefrom it originates, decreases the permeability but increases the FMR frequency. Only increasing the saturation magnetization increases both the permeability and the FMR frequency (Eq. 13). Hence, the saturation magnetization should be maximized while a compromise needs to be done with the anisotropy. Further degree of freedom stems from the shape of the magnetic inclusions. It is well known that the resonance frequency of an arbitrary magnetic body with the demagnetization factors N_x, N_y and N_z is given by the Kittel formula [12]

$$f_{\mathrm{FMR}} = (2\pi)^{-1}\nu\sqrt{\left[H_{\mathrm{A}} + (N_x - N_z)\,M_{\mathrm{S}}\right]\left[H_{\mathrm{A}} + \left(N_y - N_z\right)M_{\mathrm{S}}\right]}, \tag{48}$$

from which the well known resonance formulas for bulk, film, rod and sphere can be derived (ν is the gyromagnetic constant, defined as in Eq. (6)). In the cases of spheres ($N_x = N_y = N_z = 1/3$) and bulk material ($N_x = N_y = N_z = 0$), the resonance frequency is directly proportional to the anisotropy field (as was assumed in Subsection 2.2). In the infinite rod limit ($N_x = N_y = 1/2$, $N_z = 0$) the FMR frequency is linearly proportional to the saturation magnetization (assuming that $M_{\mathrm{S}} \gg H_{\mathrm{A}}$) and in the thin film limit ($N_x = N_y = 0$, $N_z = 1$) to the square root of the saturation magnetization and the anisotropy field (assuming that $M_{\mathrm{S}} \gg H_{\mathrm{A}}$). Even in the case of HCP cobalt, which has a high anisotropy field of $\mu_0 H_{\mathrm{A}} \approx$ 0.63 T, the highest resonance frequency is obtained in the non-isotropic geometries, namely the infinite rod and the thin film. The same conclusion is valid for the BCC iron ($\mu_0 H_{\mathrm{A}} \approx 56$ mT), and the FCC nickel ($\mu_0 H_{\mathrm{A}} \approx 16$ mT). However, due to the surface anisotropy, in the end the anisotropy field in nanoscale rods, spheres and films can be much larger than in bulk. It should be understood that the FMR frequency depends on the shape of the magnetic inclusions, but obtaining qualitative results from calculations without knowing the surface anisotropy is not possible (as already pointed out in Sect. 2.2). In addition, the FMR frequency depends on particle-particle interactions.

The volume fraction of the inclusions has obviously an effect on both the resonance frequency and the permeability. The permeability of a nanocomposite with spherical inclusions can be calculated directly from the Bruggeman theory (Sect. 3.2) but in all the other cases the Bruggeman equation must be solved iteratively and self-consistently with the Landau-Lifshitz equation [33]. The results from such calculations indicate that (1) the FMR frequency of a nanocomposite containing spherical inclusion does not depend on the volume fraction and (2) in all other cases the FMR frequency smoothly varies from the single-inclusion limit to the homogeneous bulk limit. Hence, the set of rules for predicting the high-frequency magnetic performance stated in Sect. 3.2 are valid only for spherical nanoparticles. As argued above, the FMR resonance is the lowest in the bulk and spherical particle limits. The results presented in Sect. 5.2 agree with the literature in a sense that the FMR frequency was found out to vary only a little with the nanoparticle volume fraction. The small variation might be due to slight deviations from ideal spherical form or due to the aggregation of the nanoparticles.

The magnetic performance in nanocomposites could be improved by taking all the above considerations into account when designing the material. In addition, it is preferential to use monodisperse, single-crystal nanoparticles in order to observe well-defined resonance peaks. Without such information, quantitative evaluation of the magnetic performance is difficult. Also, the surface effects such as the surface anisotropy have to be taken into account since nanoparticles have a huge surface-to-volume ratio compared to bulk materials. Because the magnitude and the symmetry of the surface anisotropy is difficult to calculate, it cannot be taken in practice into consideration before the measurement. Instead, it is the deviation of the observed FMR frequency from the value expected from bulk magnetocrystalline anisotropy that indicates the magnitude of the surface anisotropy. After having decided the target permeability and resonance frequency and having approximated the type and the volume fraction of the magnetic inclusions, one still needs to find the appropriate processing route which can lead to such a nanocomposite. As argued in this article, and also accepted in literature, homogeneous blends of plain nanoparticles in polymers are almost impossible to achieve even at the lowest filling ratios. In the high-frequency applications the role of the nanoparticles is not just an additive since the practical volume fractions (from the application perspective) begin from 10%. Hence, the nanoparticles should be a supporting part of the nanocomposite – not an additive.

Suppressing the large permittivity and dielectric loss will be a difficult task. The imaginary losses can be reduced by using single-crystal nanoparticles in which the conduction electron scattering is suppressed. Tackling the real part is much more difficult, since all metallic nanoparticles are highly conductive and their polarizability should of the same order.

Acknowledgements This work was supported by the Finnish Funding Agency for Technology and Innovation (TEKES). G.S.P. would like to acknowledge also partial support from the Academy of Finland (Acad. Res. Fellowship 00857 and projects 129896, 118122, and 135135). K.C. wishes to thank the Thailand Commission on Higher Education for financial support.

References

1. Y. Shirakata et al., High permeability and low loss Ni-Fe composite material for high-frequency applications. IEEE Trans. Magn. **44**, 2100 (2008)
2. R.C. Hansen, M. Burke, Antennas with magneto-dielectrics. Microw. Opt. Technol. Lett. **26**, 75 (2000)
3. R.A. Waldron, *Ferrites: An Introduction for Microwave Engineers*, (Van Nostrand, London, 1961)
4. Super high frequency (2009), http://en.wikipedia.org/wiki/Super_high_frequency
5. S. Sun et al., Monodisperse FePt nanoparticles and ferromagnetic FePt nanocrystal superlattices. Science **287**, 1989 (2000)
6. V.F. Puntes et al., Colloidal nanocrystal shape and size control: the case of cobalt. Science **291**, 2115 (2001)
7. S. Moerup, M.F. Hansen, Fundamentals and Theory, in *Handbook of Magnetism and Advanced Magnetic Materials*, ed. by H. Kronmueller, S. Parkin (Wiley, New York, NY, 2007)
8. L.Z. Wu et al., Studies of high-frequency magnetic permeability of rod-shapes CrO2 nanoparticles. Phys. Stat. Sol. **204**, 755 (2006)
9. C. Desvaux et al., Multimillimetre-large superlattices of air-stable iron-cobalt nanoparticles. Nat. Mater. **4**, 750 (2005)
10. S. Chikazumi, *Physics of Ferromagnetism* (Oxford University Press, New York, 1997)
11. R.C. O'Handley, *Modern Magnetic Materials* (Wiley, New York, NY, 2000)

12. C. Kittel, *Introduction to Solid State Physics* (Wiley, New York, NY, 1971)
13. C.M. Sorensen, *Magnetism in Nanoscale Materials in Chemistry*, ed. by K.J. Klabunce (Wiley-IEEE, New York, NY, 2001)
14. S.J. Steinmuller et al., Effect of substrate roughness on the magnetic properties of thin fcc Co films. Phys. Rev. B **76**, (2007)
15. P.K. George, E.D. Thompson, Exchange stiffness in hexagonal cobalt. Phys. Rev. Lett. **24**, 1431 (1970)
16. J.L. Snoek, Dispersion and absorption in magnetic ferrites at frequencies above one Mc/s. Physica **15**, 207 (1948)
17. L. Landau, E. Lifshitz, On the theory of the dispersion of magnetic permeability in ferromagnetic bodies. Phys. Z. Sowjetunion 8, (1935)
18. G.H. Jonker et al., Philips Tech. Rev. **18**, 1956 (1956)
19. R. Krishnamoorti, R.A. Vaia, Polymer nanocomposites. J. Polymer Sci. B: Polymer Phys. **45**, 3252 (2007)
20. M.E. Mackay et al., General strategies for nanoparticle dispersion. Science **311**, 1740 (2006)
21. A.C. Balazs, S. Chandralekha, Effect of polymer architecture on the miscibility of polymer/clay mixtures. Polymer Int. **49**, 469 (2000)
22. E. Glogowski et al., Functionalization of nanoparticles for dispersion in polymers and assembly in fluids. J. Polymer Sci. A: Polymer Chem. **44**, 5076 (2006)
23. A.H. Latham, M.E. Williams, Controlling transport and chemical functionality of magnetic nanoparticles. Acc. Chem. Res. **41**, 411 (2008)
24. J. Israelachvili, *Intermolecular and Surface Forces* (Academic, London, 1985)
25. J.A. Pomposo et al., Key role of entropy in nanoparticle dispersion: polystyrene-nanoparticle/linear-polystyrene nanocomposites as a model system. Phys. Chem. Chem. Phys. **10**, 650 (2007)
26. C. Barnes et al., Spontaneous formation of an exfoliated polystyrene-clay nanocomposite using a star-shaped polymer. J. Am. Chem. Soc. **126**, 8118 (2004)
27. Sihvola, *Electromagnetic Mixing Formulas and Applications* (The Institution of Electrical Engineers, 1999)
28. J.H. Paterson et al., Complex permeability of soft magnetic ferrite/polyester resin composites at frequencies above 1 MHz. J. Magn. Magn. Mater. **196–197**, 394 (1999)
29. K. Chalapat, K. Sarvala, J. Li, G.S. Paraoanu, Wideband reference-plane invariant method for measuring electromagnetic parameters of materials. IEEE Trans. Microw. Theory Tech. **57**, 2257 (2009)
30. A.M. Nicolson, G.F. Ross, Measurement of the intrinsic properties of materials by time-domain techniques. IEEE Trans. Instrum. Meas. **IM-19**, 377–382 (1970)
31. William B. Weir, Automatic measurement of complex dielectric constant and permeability at microwave frequencies. Proc. IEEE, **62**, 33–36 (1974)
32. A.-H. Boughriet, C. Legrand, A. Chapoton, Noniterative stable transmission/reflection method for low-loss material complex permittivity determination. IEEE Tran. Microw. Theory Tech. **45**, 52–57 (1997)
33. R. Ramprasad et al., Fundamental limits of soft magnetic particle composites for high frequency applications. Phys. Stat. Sol. **233**, 31 (2002)
34. H.P. Klug, L.E. Alexander, *X-Ray Diffraction Procedures* (Wiley, New York, NY, 1954)
35. Å. Björck, Numerical methods for least squares problems, (SIAM, Philadelphia, 1996)

Magnetic Nanocomposites for Permanent Magnets

F. Tolea, M. Sofronie, A. Birsan, G. Schinteie, V. Kuncser, and M. Valeanu

Abstract The influence of different crystallization conditions on the microstructure and magnetic hardening of RE-Fe-B amorphous ribbons with different Fe concentrations and Pr and Nd as rare earth elements was analyzed. The microalloying effects of Zr and Ti substitution on the evolution of crystallization process and the magnetic hardening was also discussed. The final aim of this experimental study was to obtain performing exchange spring magnets with as much as lower content of the expensive RE material.

1 Introduction

Exchange-spring magnets (or spring magnets) consist of interfaced hard and soft magnetic nano-phases that are coupled by exchange interactions. Essentially, spring magnets combine the high anisotropy typical to a hard magnetic phase (e.g. $Nd_2Fe_{14}B$, with high uniaxial magneto-crystalline-anisotropy, K_u) with the large magnetization typical to a soft magnetic phase (e.g. α-Fe or Fe_3B, with high magnetization at saturation, M_s). The magnetic properties of RE-Fe-B/α-Fe (or Fe_3B) nanocomposite magnets (where RE is a Rare Earth), are deeply related to their microstructure. The optimal microstructure of magnetic nanocomposites consists of uniformly distributed hard and soft magnetic nanophases, the nano-size dimension of the two interconnected phases being a critical parameter for an enhanced magnetic coupling [1–3]. Evidently, the cause of the magnetic strengthening (hardening) is the exchange interaction at the interface between the soft and hard phases. Since the exchange interaction is a short range interaction, the dimension of the soft phase nano-clusters has to be less than twice the width of the domain wall in the hard phase (the hardening's condition) [2, 4]. In principle, it is difficult to control the phase dimensions, but this becomes possible by starting from the amorphous

F. Tolea (✉)
National Institute of Materials Physics (INFM), 105 Atomistilor Street, 077125 Magurele-Bucharest, Romania
e-mail: felicia@infim.ro

A. Aldea, V. Bârsan (eds.), *Trends in Nanophysics,* Engineering Materials,
DOI 10.1007/978-3-642-12070-1_12,

state and applying suitable annealing treatments and/or different substitutions with alloying elements like Cu, Si, V, Nb, W, Al, Dy, etc. [5, 6].

Even if they are the most usual exchange spring magnets, the Nd-Fe-B + α-Fe nanocomposites are not suitable to be used at low temperature. This is because the hard phase undergoes a spin reorientation at 135 K and develops a planar magnetocrystalline anisotropy which destroys the exchange coupling. Pr-Fe-B + α-Fe nanocomposites are available for low temperature purposes [4, 7].

This work reports on the effect of thermal treatments on the exchange coupling and microstructure of nanocomposites based on $RE_2Fe_{14}B + x\%Fe$ (RE = Nd and Pr and x less than 35). The influence of Zr and Ti substitution in $Nd_7Fe_{81}B_{12}$ [8], on the formation of the nanocomposite material for permanent magnets, is also considered. The final aim of this experimental study was to obtain performing exchange spring magnets with as much as lower content of the expensive RE material.

2 Experimental Set-Up

Ingots with nominal compositions $RE_2Fe_{14}B + x\%$ Fe, where RE is Nd or Pr and $x = 5, 10, 15, 20, 25, 30, 35$ were prepared by arc melting. The amorphous ribbons were obtained by the melt spinning technique in argon atmosphere. Different annealing treatments were performed in vacuum, for 2 until 15 min, at temperatures between 650°C and 811°C. We introduce the following notation of the samples: A1 for Nd based sample with 5% additional Fe, A2 for Nd based sample with 10% additional Fe,. . .,A7 for Nd based sample with 35% additional Fe. Pr based sample with similar additional Fe content as in the previous description were denoted by C1, C2,. . .,C7, respectively.

For structural information, room temperature X-ray diffractions (XRD) were performed by using a Seifert Diffractometer with Cu Kα radiation. The microstructure information has been completed by Transmission Electron Microscopy (TEM).

The structural influence of Ti and Zr substitution on $Nd_7Fe_{81}B_{12}$, $Nd_7Fe_{81}Ti_2B_{10}$, $Nd_7Fe_{79}Zr_2B_{12}$ compositions was studied in situ with synchrotron radiation, by energy-dispersive X-ray diffraction at high temperature (HASYLAB, F2.1 beamline).

The Curie temperatures were acquired via the Faraday method, using a horizontal Weiss balance. The hysteresis loops were obtained by the extraction method at room temperature, in fields up to 1.8 T, applied along the length of the ribbons.

3 Results and Discussions

3.1 $Nd_2Fe_{14}B$ and $Pr_2Fe_{14}B + x\%$ Fe

The crystallization behavior of the amorphous material imposes the subsequent microstructure of the nanocomposite. Figure 1 shows the diffraction patterns of sample A3, as cast (or as quenched (AQ)) and after two different thermal treatments.

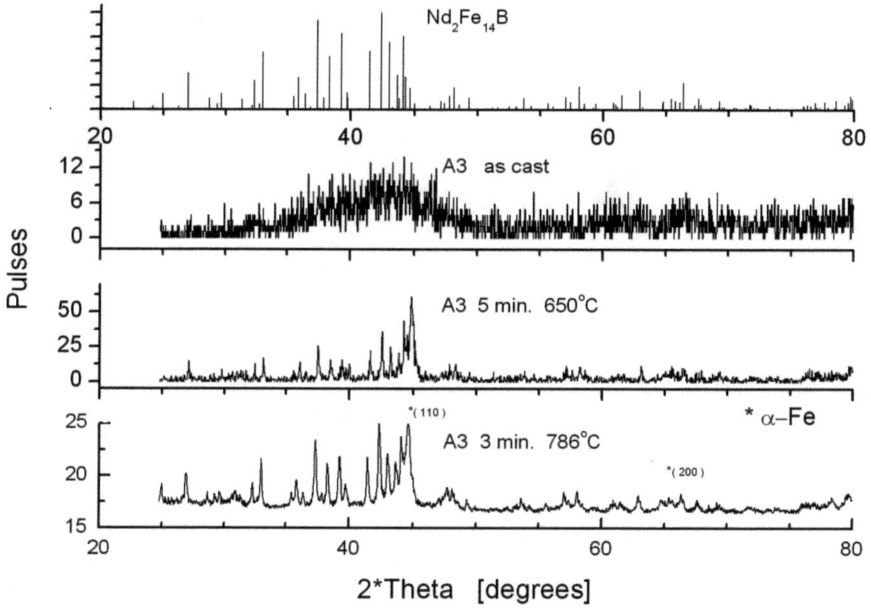

Fig. 1 The diffraction patterns of sample A3, in different annealing conditions

For the as cast ribbons, the XRD pattern confirms their amorphous state. Annealing treatments at temperature higher than 600°C promote the crystallization of both hard – and soft – phases. A higher temperature treatment, even shorter, induces a better refinement of the crystalline structure, maintaining the phase's dimensions in range of nanometers. After a short treatment at 786°C, the mean grain sizes of a-Fe and $Nd_2Fe_{14}B$ phases, evaluated from X-ray broadening analysis using the Scherrer formula, are $\sim$ 5.5 and 11.0 nm, respectively.

The TEM analysis of sample A3 reveals an amorphous matrix for the as quenched sample. One can notice that the thermal treatment (TT) promotes the crystallization, evidenced by the grain formation. At higher temperature the grains' boundaries become more evident (see Fig. 2).

Thermo-magnetic measurements performed on A3 sample in the as quenched state revealed a magnetic order-disorder transition of about 150°C, temperature assumed to be the Curie temperature of the amorphous phase. The magnetization of the sample is negligible above the mentioned temperature. The thermomagnetic measurement is, in fact, equivalent to a thermal annealing during which crystallization is induced. Therefore, by heating the sample, as soon as the crystallization begins with the formation of a crystalline phase which presents magnetic order above the crystallization temperature, the magnetization starts to increase. That is happened at $\sim$550°C on the analyzed samples, where the formation of the hard ($Nd_2Fe_{14}B$) and the soft (α-Fe) magnetic phases is expected. It is worth to notice that $Nd_2Fe_{14}B$ enter the paramagnetic state at temperatures above 312°C whereas the soft phase, α-Fe in this case, enter the paramagnetic state above 740°C. Therefore, the magnetization observed between 550°C and 740°C along the heating curve

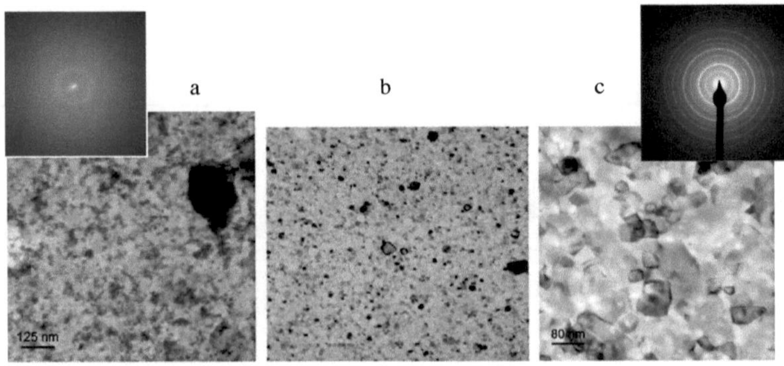

Fig. 2 Bright field transmission electron micrographs of the A3 ($Nd_{10}Fe_{85}B_5$) alloy, in the following states: as quenched (**a**), annealed at 650°C for 5 min (**b**) and respectively at 750°C for 3 min (**c**) and the corresponding selected area electron diffractions shown in inset for (**a**) and (**b**)

has to be due to only the α-Fe phase. Further on, by cooling the already crystallized sample, the magnetization reveals two augmentations which are attributed to the magnetic ordering of the two phases: the soft magnetic phase α-Fe, at about 750°C, and the hard magnetic phase $Nd_2Fe_{14}B$, at about 312°C (see Fig. 3).

For A3 TT 10 min/757°C the heating curve indicates directly the two order-disorder transitions corresponding to the two magnetic phases. One should notice that even this sample in the starting state is not yet totally crystallized. The magnetization is continuously increasing before the Curie temperature of the hard phase, suggesting that the process of refining the crystalline structure and grain growth is going on. The thermo-magnetic cooling curve is almost identical with that for the AQ sample (see also Fig. 3), suggesting the similar crystalline state of the two samples after the thermo-magnetic heating up to 800°C.

Magnetic measurements were performed by extraction method at room temperature, in fields up to 1.8 T, applied along the length of the ribbons.

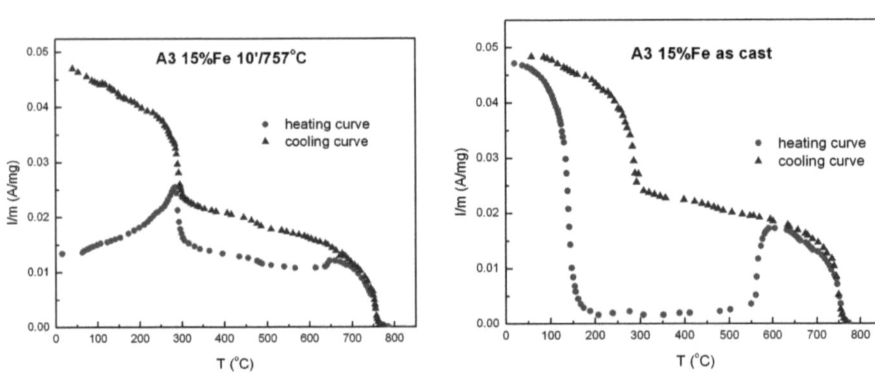

Fig. 3 Thermo-magnetic measurements performed in vacuum on A3 sample in as-quenched state (*right side*) and on A3 TT 10 min/757°C (*left side*)

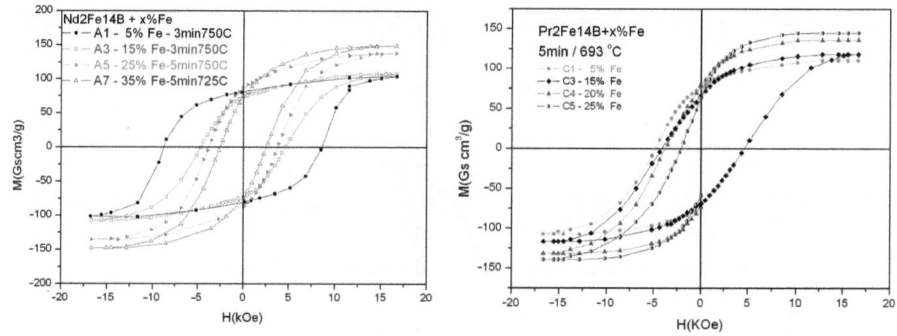

Fig. 4 Histerezis loops for A_1 A_3 A_5 and A_7 samples with Nd as rare earth (*left side*) and for C_1, C_3, C_4 and C_5 samples with Pr as rare earth (*right side*)

The hysteresis loops for A1, A3, A5 and A7 samples annealed at 750°C and 725°C, for either 3 or 5 min are presented in Fig. 4. The experimental data evidence that by increasing the iron content, the saturation magnetization increases and the coercive field decreases, suggesting the simultaneous augmentation of the relative amount of the soft magnetic phase and the reduction of the magnetic hardness. The same tendency is observed for samples with Pr as rare earth (see also Fig. 4). This general trend is also observed from the numerical parameters of all the analyzed Nd based samples presented in Table 1.

As can be observed from data in Table 1, the annealing conditions are very critical in regard to optimal size and crystalline structure of the involved phases, with direct influence on the magnetic properties of the nanocomposites. The effect of the thermal treatment on the magnetic behavior of sample A_3 (amorphous $Nd_{10}Fe_{85}B_5$

Table 1 Characteristics of the A_1–A_7 samples, where: treatment = annealing conditions, remanent magnetization = M_R, coercive field = H_c, saturation magnetization = M_S

Sample	Treatment (time/temperature)	M_S(Gs*cm^3/g)	M_R(Gs*cm^3/g)	H_c(kOe)	M_R/M_S
A1	5 min/650°C	113	77	8.7	0.68
	3 min/750°C	103	80	8.8	0.78
	2 min/790°C	116	86	7.6	0.74
A2	5 min/650°C	120	69	4.4	0.57
	3 min/750°C	117	81	5.3	0.69
	2 min/790°C	120	88	5.3	0.74
A3	5 min/650°C	123	54	2.9	0.44
	3 min/750°C	115	75	4.5	0.65
	2 min/790°C	118	73	3.3	0.61
A4	5 min/650°C	131	68	3.2	0.55
	3 min/750°C	128	70	3.8	0.55
	2 min/790°C	122	69	3.4	0.56
A5	5 min/650°C	139	63	3.1	0.45
	5 min/750°C	137	70	3.7	0.51
A6	5 min/650°C	141	53	2.1	0.37
	5 min/750°C	138	71	2.8	0.52
A7	5 min/650°C	144	52	1.1	0.36
	5 min/750°C	148	75	2.5	0.51

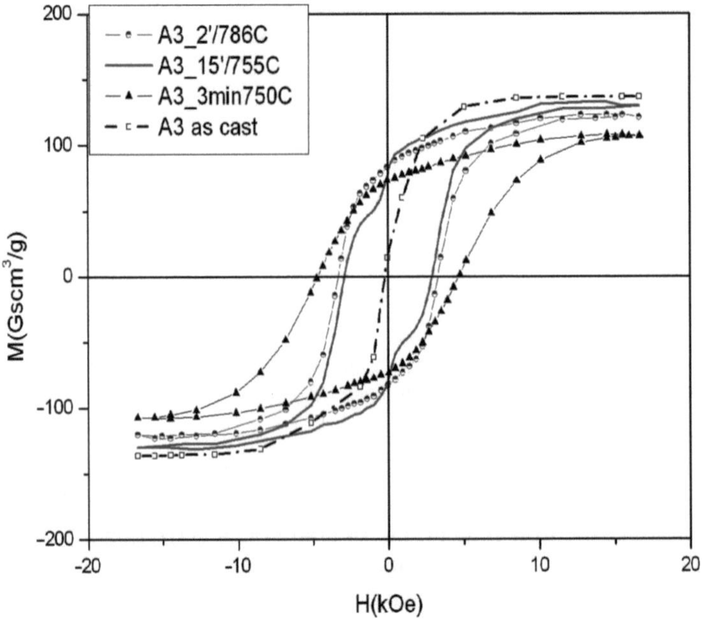

Fig. 5 Histerezis loops for A_3 sample in AQ state and annealed for 3 min at 750°C, 2 min at 786°C and 15 min at 760°C. The AQ state shows a negligible coercive field whereas the sample annealed for 15 min at 760°C presents a mixture of two independent ferromagnetic phases with different hardness (constricted loop). The best coupling in sample A_3 involves a thermal treatment of 3 min at 750°C

in the AQ state) is evidenced also via Fig. 5. On a hand, the amorphous phase shows the expected negligible coercive field. Oppositely, when the thermal treatment is too long or the temperature is too high, the dimension of the soft magnetic phase exceeds the critical dimensions for an optimal exchange coupling between the soft and the hard phase grains. Consequently, the two magnetic phases are only weakly coupled and a small kink appears in the demagnetization curve. Only optimal treatments lead to suitable coupling with increased coercive field, remanent magnetization, and M_R/M_S ratio.

The best-achieved exchange couplings among the Nd based samples (see Table 1) have been obtained for samples A_5 annealed at 750°C for 5 min and A_4 annealed 5 min at 725°C, for which the energy product, $(BH)_{max}$, defined as the maximum product of the flux density, B, and the corresponding opposing field, H, is higher than 90 KJ/m^3.

3.2 $Nd_7Fe_{81}B_{12}$ with Zr and Ti Substitutions

An interesting crystallization process was evidenced in some Nd based compositions more rich in boron, for which the soft magnetic phase is expected to be

Fe_3B [6, 9]. The influence of Zr and Ti substitution on the evolution of the crystallization process and magnetic hardening was studied, by starting from the parent compound $Nd_7Fe_{81}B_{12}$. The temperature-induced crystallization processes on amorphous ribbons with nominal compositions $Nd_7Fe_{81}B_{12}$, $Nd_7Fe_{81}Ti_2B_{10}$ and $Nd_7Fe_{79}Zr_2B_{12}$ were studied in situ using synchrotron radiation in high-temperature energy-dispersive X-ray diffraction experiments performed at HASYLAB (F2.1 beam line).

Figure 6 shows the diffraction patterns of the substitution-free sample in the temperature range from 300 – 850°C. It is worth noticing that the Nd fluorescence lines at 36.84, 37.36, 42.27 and 43.32 keV are superposed to the diffraction spectra. The crystallization process starts at 600°C with the precipitation of the $Nd_2Fe_{23}B_3$ metastable phase, which was detected as the predominant one even at the highest temperature. The first diffraction peaks of the $Nd_2Fe_{14}B$ hard phase and of the soft Fe_3B/α-Fe phase become visible only above 700°C. The reason for such a high decomposition temperature is related to the sample composition, very closed to the stoichiometry of the $Nd_2Fe_{23}B_3$ phase, as well as to a lack of composition gradient around the metastable phase grains.

On the other hand, the high value of the decomposition temperature induces a non-uniform microstructure, unsuitable for an exchange coupled magnet, as

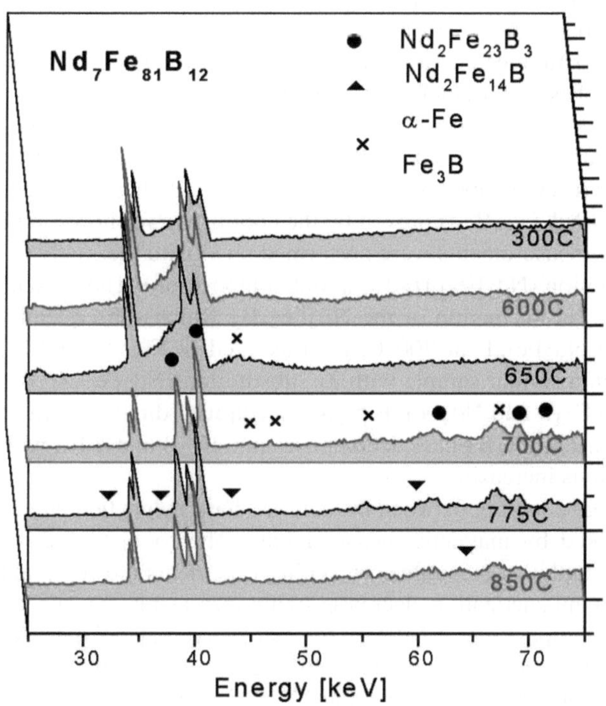

Fig. 6 In situ high temperature energy dispersive XRD measurements performed on the rapidly solidified $Nd_7Fe_{81}B_{12}$ sample

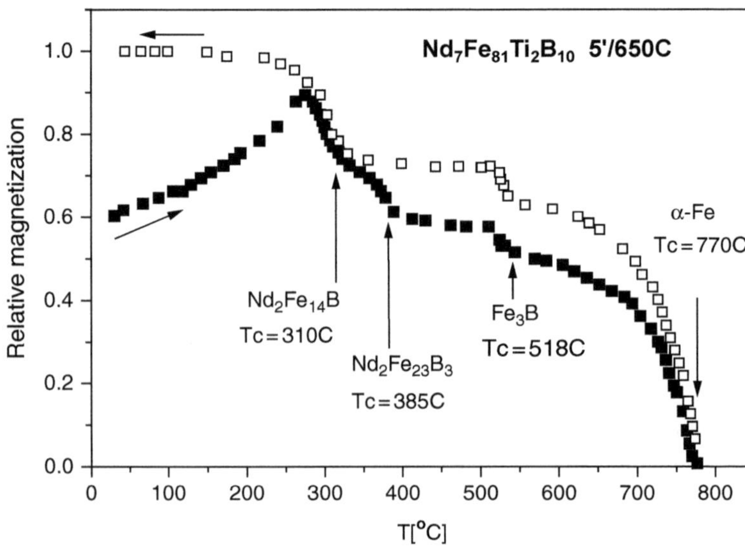

Fig. 7 Thermo-magnetic measurements for the free substitution sample $Nd_7Fe_{81}B_{12}$

supported by additional thermo-magnetic measurements. The thermo-magnetic measurements performed with a Weiss-type susceptometer on a sample with Ti substitution, ex situ annealed at 650°C for 5 min (Fig. 7) confirm the presence of the following phases: (i) $Nd_2Fe_{14}B$ with a Curie temperature of 310°C, (ii) the metastable $Nd_2Fe_{23}B_3$ with a Curie temperature of 385°C, (iii) the Fe_3B with a Curie temperature of 518°C and finally (iv) the α-Fe with 770°C Curie temperature. However, after reaching approx. 800°C, the $Nd_2Fe_{23}B_3$ has fully decomposed into Fe_3B/α-Fe and $Nd_2Fe_{14}B$, as proven by the cooling curve shown in Fig. 7.

Usual XRD measurements were performed on ex-situ thermally treated samples with Ti substitution ($Nd_7Fe_{81}Ti_2B_{10}$ initial composition). The crystallization starts at 600°C, with the formation of the $Nd_2Fe_{23}B_3$ which subsequently decomposes into Fe_3B and $Nd_2Fe_{14}B$ at 700°C, as illustrated in Fig. 8a and b, respectively. Surprisingly, in a similar sample with Zr substitution ($Nd_7Fe_{79}Zr_2B_{12}$ initial composition), both Fe_3B and $Nd_2Fe_{14}B$ phases precipitate directly from the amorphous phase (Fig. 8c). The Fe_3B phase decomposes into Fe_2B and α-Fe, when the annealing temperature is increasing.

The exchange coupling between the soft and hard magnetic phases in these samples was checked by magnetic measurements. The room temperature hysterezis loops of the samples after optimal annealing treatments are represented in Fig. 9. The saturation magnetization decreases with increasing the annealing temperature, due to decomposition of $Nd_2Fe_{23}B_3$, as observed also from XRD pattern and thermo-magnetic measurements. The highest coercive field of $Nd_7Fe_{81}B_{12}$ sample was obtained after a thermal treatment at 775°C for 5 min. However, the loop shows the typical behavior of a hard/soft mixture without exchange coupling. The intergranular exchange interaction is evidenced in $Nd_7Fe_{81}Ti_2B_{10}$ sample, annealed for 2 min at 700°C. Finally, the best exchange coupling is obtained for $Nd_7Fe_{79}Zr_2B_{12}$

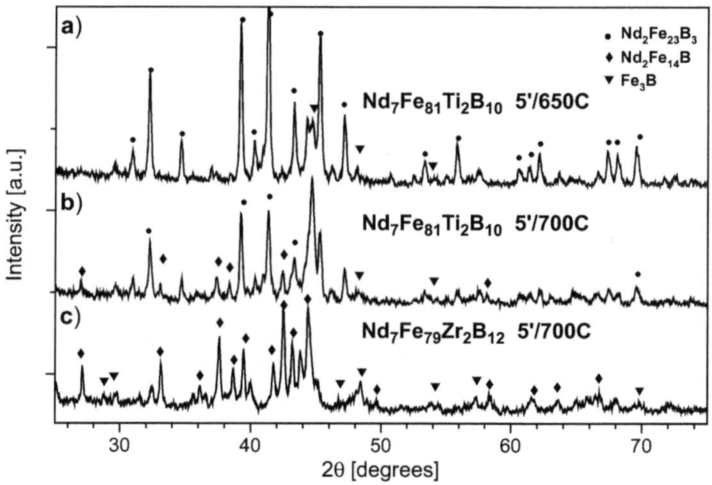

Fig. 8 Conventional XRD patterns for $Nd_7Fe_{81}Ti_2B_{10}$ (**a** and **b**) and $Nd_7Fe_{79}Zr_2B_{12}$ (**c**)

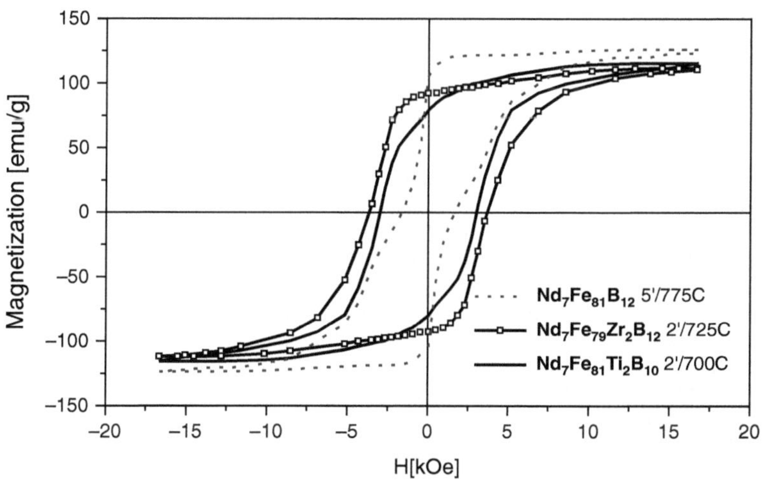

Fig. 9 Hysteresis loops performed at room temperatures after optimized annealing treatments

sample annealed at 725°C for 2 min. This sample presents an energy product of 87.2 kJ/m^3.

4 Conclusions

It is known that the exchange coupling and the microstructure of magnetic nanocomposites are deeply related to the thermal treatment. Our magnetic measurements indicate that optimal magnetic hardening occurs for thermal treatments at higher temperatures and shorter annealing times applied to initially amorphous ribbons.

Suitable thermal treatments applied to amorphous compounds of compositions corresponding to the combination $RE_2Fe_{14}B+x\%Fe$ (RE = Nd or Pr) with or without Ti or Zr substitutions can lead to performing exchange spring magnets of lower costs – with reduced RE concentration. The best exchange coupling is obtained in the Nd based series with $x = 25\%$ Fe addition, annealed for 5 min at 750°C (on sample A_5). This sample presents an energy product of 92 KJ/m^3.

Corroborating detailed crystallographic, Mossbauer spectroscopy and magnetometry studies, it can be proven that optimal magnetic properties are obtained for a microstructure in which hard and soft magnetic phases co-exist together with a small amorphous phase [10, 11].

We also analyzed the 2% Ti substitution for B in $Nd_7Fe_{81}B_{12}$ which leads to the undesirable formation of the metastable $Nd_2Fe_{23}B_3$ phase with a high decomposition temperature, requiring an inconveniently high thermal treatment temperature. Oppositely, 2% of Zr substituting Fe atoms changes the sequence of crystallization process, promoting the direct formation of $Nd_2Fe_{14}B/Fe_3B$ nanocomposities with improved magnetic properties [8].

Acknowledgements The authors wish to thank Ministry of Education and Research for financial support by national program PN2 72-186.

References

1. E.F. Kneller, R. Hawing, IEEE Trans. Magn. **27**, 3588 (1991)
2. R. Skomski, J.M.D. Coey, Phys. Rev. B **48**, 15812 (1993)
3. R. Fischer, T. Leineweber, H. Kronmuller, Phy. Rev. B **57**, 10723 (1998)
4. Z. Chen, H. Okumurra, G.C. Hadjipanayis, Q. Chen, J. Appl. Phys. **89**, 2299 (2001)
5. T. Schrefl, J. Fidler, IEEE Trans. Magn. **35**, 3223 (1999)
6. D.H. Ping, K. Hono, H. Kanekiyo, S. Hirosawa, Acta Mater, **18**, 4641 (1999)
7. D. Goll, H. Kronmuller, *Proceedings of the 15th International Workshop on Rare-Earth Magnets and Their Applications 30 August–3 September*, Dresden, Germany, 1998, p. 189
8. A. Jianu, M. Valeanu, D.P. Lazar, F. Lifei, C. Bunescu, V. Pop, J. Mag. Mag. Mat. **272–276**, 1493–1494 (2004)
9. D.H. Ping, Y.Q. Wu, K. Hono, J. Mag. Mag. Mat. **239**, 437 (2002)
10. V. Kuncser, M. Valeanu, F. Lifei, D. Predoi, P. Palade, G. Schinteie, O. Crisan, G. Filoti, J. Mag. Mag. Mat. **272–276**, 797–799 (2004)
11. V. Kuncser, M. Valeanu, F. Lifei, A. Jianu, W. Kappel, G. Filoti, Phys. Stat. Sol. C **1**(12), 3503–3506 (2004)

Magnetic Configuration and Relaxation in Iron Based Nano-Particles: A Mössbauer Approach

V. Kuncser, G. Schinteie, R. Alexandrescu, I. Morjan, L. Vekas, and G. Filoti

Abstract The difficulty to stabilize a definite crystalline structure or phase, due to a large gradient of defects along the particle diameter, is unfortunately specific to particles of nanometer size. Therefore, atypical spin configurations and magnetic anisotropies as well as enhanced magnetic relaxations via thermal excitations are expected in systems of fine nano-particles. The actual work reports on various possibilities for a comprehensive characterization of the magnetic configuration and magnetic relaxation mechanisms of nano-particles, by corroborating the powerful method of Mössbauer spectroscopy with complementary magnetic and structural techniques. The capabilities of temperature and field dependent Mössbauer spectroscopy to provide valuable information about spin blocking temperatures, effective anisotropy constants, amount of uncompensated spins inside of particle and magnetic phase composition are critically discussed for both powder like systems and nano-particles dispersed in different solvents.

1 Introduction

Iron based magnetic nano-particles are subject of an intense research effort, due to both their fundamental scientific interest and capability for novel applications. Nano-scale metallic particles or nano-oxides are widely studied because of their potential applications in microelectronics (data storage devices) and bio-medicine (magnetic resonance imaging, targeted drug delivery, a.c. magnetic field-assisted cancer therapy). Maghemite ($\gamma - Fe_2O_3$) is a well-known iron oxide which basically crystallizes in the cubic spinel structure and can be considered as Fe^{2+} deficient magnetite [1–3]. The spinel ferrite structure (general formula AB_2O_4) consists of a face centered cubic oxygen (O) array with 8 formula units and with divalent or trivalent transition metal cations on the 8 tetrahedral (A) and 16 octahedral (B) positions [1]. There are two structural limits: (i) the normal spinel ferrite, where Fe^{3+}

V. Kuncser (✉)
National Institute for Materials Physics (INFM), 105 Atomistilor Street,
077125 Magurele-Bucharest, Romania
e-mail: kuncser@infim.ro

A. Aldea, V. Bârsan (eds.), *Trends in Nanophysics*, Engineering Materials,
DOI 10.1007/978-3-642-12070-1_13,

ions occupy all the 16 octahedral (B) positions whereas a divalent transition metal cation, M^{2+}, occupies the 8 tetrahedral (B) positions, and (ii) the inverse spinel ferrite, where Fe^{3+} ions occupy all the 8 tetrahedral (A) positions whereas the 16 octahedral (B) positions are equally shared by Fe^{3+} and M^{2+} cations. Magnetite is typically an inverse spinel, with Fe^{2+} ions on the B positions. Maghemite contains only Fe^{3+} cations which however are insufficient to fill all the A and B positions, the cation vacancies (absence of Fe^{2+}) inducing a trigonal local distortion, although the crystal symmetry is cubic. Another very known Fe oxide is hematite, which has a crystal structure similar to that of orthorhombic corundum, with a close-packed oxygen lattice and Fe^{3+} cations on octahedral positions. However in the incipient stages, hematite is less stable compared with maghemite and magnetite, and very thin oxide nanoparticles (average size below 10 nm) show mainly the spinel structure. There is a well defined interest in dispersing Fe oxide nanoparticles in different matrices with the purpose to form nano-composites with particular desired properties. Nanoparticles of maghemite have been dispersed in DNA [4], polyaniline [5], dimethyl sulfoxide [6], etc. Other Fe oxides were dispersed in silica matrix [7] or polyvinyl-alcohol [8]. Such magnetic nanocomposites have been intensively investigated for their potential applications in magnetic recording and magneto-optics [9], color imaging [10], bioprocesses [11] or catalysis [12]. Magnetic oxide nanoparticles can be also dispersed in a carrier liquid such as water, liquid organic compounds or oils, giving rise to a so called ferrofluid or magnetic fluid. The magnetic properties of all the above systems derive mainly from the magnetic properties of the embedded nanoparticles. Further on, the low dimensionality of the constituent magnetic nanoparticles provides a peculiar magnetic behavior of the system, connected with both magnetic relaxation phenomena and anomalous spin structure on the particle surface.

Briefly, there are two key properties which define the magnetic response of a nanoparticle: (i) the maximum value of the magnetic moment of the nanoparticle and (ii) the time averaged projection of the magnetic moment along the applied field. Related to property (i), the present case corresponds to very fine particles, with a size lower than the limit where the magnetic energy can be minimized by breaking up the particle volume into domains. Hence, they are magnetic mono-domains. Estimated single domain size for quasi-spherical particles are larger than 14 nm for Fe metallic nanoparticles and larger than 100 nm for Fe oxide nanoparticles [13]. The maximum value of the magnetic moment of the nanoparticle (particles of similar size are compared) is firstly imposed by the crystalline structure of the particle, the involved atomic magnetic moments and the type of magnetic coupling. In this regard, α-Fe with its bcc structure and ferromagnetic coupling (magnetization at saturation, M_S, of about $1.7 \cdot 10^6$ A/m, crystalline anisotropy constant K, of about $12 \cdot 10^4$ J/m^3) should present the most convenient magnetic response and hence, should be the most suitable candidate for applications. Unfortunately, it is very difficult to protect such fine particles against oxidation and therefore, most of the applications involve either stable Fe oxide nanoparticles or metallic Fe cores covered by protective shells which can be either oxides, carbides, corrosion resistant compounds or noble metals. The magnetic structure of the spinel ferrite is determined by the antiferromagnetic coupling between the magnetic ions on each of the A and B sites via oxygen mediated super-exchange interactions. In case of ideal magnetite, the net magnetic moment per formula unit is controlled by the spin contribution of the Fe^{2+} ions on the B site after compensation of the two spins of Fe^{3+} on the A and B sites which are anti-parallel coupled. In case of maghemite, the net magnetic moment per formula unit is due to the unequal occupation of the A and B sites by Fe^{3+}. Therefore both compounds are ferri-magnetically ordered ($M_S = 0.5 \cdot 10^6$ A/m and $K = -1.2 \cdot 10^4$ J/m^3 for magnetite and $M_S = 0.43 \cdot 10^6$ A/m and $K = -2 \cdot 10^4$ J/m^3 for maghemite). The magnetic moment per particle can be easily obtained by multiplying the magnetization at saturation with the volume of the particle. Hematite nanoparticles shows anti-ferromagnetic order at room temperature and is beyond real interest for systems with magnetic response.

However, all the above assumptions related to the magnetic structure start to fail for nanometer sized particles. Firstly, the crystalline structure of spinel-ferrite nanoparticles becomes much defected and it is difficult to make distinction between the effectively defect magnetite-like structure and the intrinsic defect maghemite structure. Hence, both the crystalline and the magnetic structure of the particle become difficult to be determined, mainly for particles of size lower than 10 nm. Conceptually, in this case, the most expected physical situation is to have a phase gradient along the particle diameter, with a magnetite-like phase at the centre, which evolves to a maghemite-like at the particle surface. From the magnetic point of view, it is not so important to know the exact structure of the particle, but is rather more important to estimate the amount of uncompensated spins (or magnetic moments) per particle, what is directly related to its magnetic response. Secondly, for small enough sizes, the particles enter the regime of super-paramagnetic relaxation which is timescale-dependent due to the stochastic nature of the thermal energy. The magnetic anisotropy energy of a spherical magnetic particle is KV, with K the magnetic anisotropy energy constant and V the volume of the particle. For nanoparticles of mean dimension lower than 20 nm, the thermal energy, $k_B T$, can approach, already at usual temperatures, the anisotropy energy. Hence, the magnetic moment of the particle can be thermally activated, passing over the barrier energy. Therefore, the time averaged magnetic moment (the projection along the field direction is considered) becomes temperature dependent. Quantitatively, for a particle with uniaxial anisotropy, the energy $E(\theta)$ required for changing the direction of the total magnetic moment (or the total spin) by an angle θ from the magnetic easy axis, is given by:

$$E(\theta) = KV \sin^2 \theta \qquad (1)$$

The theoretical treatment of the thermally induced fluctuations is usually divided into two regimes, depending on the ratio $r = k_B T/KV$: (i) the regime of the collective excitations, for $r < 0.1$ and (ii) the regime of the super-paramagnetic relaxation, for $r > 0.1$ [14]. In the regime of collective excitations, the local spins fluctuate by small angles θ around the equilibrium direction (e.g. the magnetic easy axis), leading to an almost linear decrease of the magnetization versus temperature.

In the super-paramagnetic relaxation regime, the spins jump by $\theta = 180°$ along the easy-axis direction with a fluctuation time, τ, given in the simplest form by the Neel model [15]:

$$\tau = \tau_0 \exp(KV/k_B T) \tag{2}$$

The observed spin relaxation process depends actually on the value of τ as compared to the characteristic measurement time of the method, τ_m. If $\tau \ll \tau_m$, the magnetic relaxation appears so fast that the particle system behaves as a paramagnet, with a giant spin per particle (the super-paramagnetic regime), whereas for $\tau \gg \tau_m$ the relaxation appears so slow that quasi-static properties are observed (the spin blocked regime). The temperature where $\tau = \tau_m$ is defined as the blocking temperature, T_B, and, therefore, the value of the measured blocking temperature depends on the time window of the experimental method. It is worth mentioning that the system behaves as a paramagnet above T_B (with respect to the measurement technique which is used), although the transition to the real paramagnetic state (associated to the bulk material of the nano-particle) appears at a higher temperature. Physically, the difference between the supar-paramagnetic state and the real paramagnetic state resides from the fact that in the super-paramagnetic state all the spins inside the particles jump coherently (all together) along the easy axis of magnetization whereas in the paramagnetic state they jump independently and in random directions.

Because in both the above mentioned fluctuating regimes, the parameter $r = k_B T/KV$ plays a key role, a suitable analysis of the corresponding experimental data on magnetic relaxation may provide valuable information on the product KV. At a first glance, by using either X ray diffraction (XRD) or Electron Microscopy, an average volume of the particle can be obtained. Based on this information, the anisotropy constant can be extracted from the above product. However, in practice the issue is not so straightforward, because of the different volumes which can be attributed to nanoparticles. Let us assume for the clearness of the discussion a nanoparticle with spherical shape. One may define firstly a physical volume (or a morphological one), which is obtained directly from electron microscopy. Secondly, due to the strongly defected surface of the particle, a suitable crystalline structure might be defined along a smaller diameter than the above mentioned physical one. Such smaller diameter will be approached by the so called structural coherence length of the nanoparticle, which may be estimated from the line-width of the X ray diffraction peaks. Finally, due to the mentioned surface defects, also the magnetic order can be perturbed and a magnetic dead layer may be formed on the particle surface. A magnetic volume, excluding the magnetic dead shell might be also defined, and this is the volume which is taken into account at the estimation of the particle anisotropy energy barrier, KV. Hence, a careful attention has to be paid when using the volume provided by a structural or microscopic tool, for the product KV estimated via a magnetic relaxation measurement. More adequately would be to obtain information of the anisotropy constant by an additional magnetometry measurement and then to subtract valuable information concerning the magnetic volume of the particle and to compare it with the earlier mentioned physical or structural volume. Dealing often with quite complex magnetic systems, presenting either core-shell type nanoparticles or a simple co-existence of different magnetic phases (e.g. part of the nanoparticles belong to one phase and another part to a different phase), a microscopic method giving information on the magnetic phase composition is strongly required. The large capabilities of Mössbauer spectroscopy to provide valuable information related to most of the above mentioned aspects, especially when it is used in conjunction with other magnetic, structural or morphologic evaluations, will be briefly presented in the following.

2 About ^{57}Fe Mössbauer Spectroscopy

The ^{57}Fe Mössbauer spectroscopy is a powerful technique giving information on both the Fe phase composition as well as on the local electronic configurations and magnetic interactions involving Fe atoms [1]. It is based on the emission and absorption of a gamma-ray photon without loss of energy due to recoil of the nucleus (in solid state there is always a certain fraction of such events of recoil-free photon absorptions/emissions). That is, a photon which is emitted via a transition between two nuclear levels, can probabilistically promote a resonant absorption between the same nuclear levels. A Mössbauer source containing ^{57}Co radioactive nuclei embedded in a Rhodium matrix provides a ready supply of excited ^{57}Fe nuclei in their nuclear state with $I = 3/2$. By transition to the ground state with $I = 1/2$, the source emits gamma ray photons of 14.4 keV which are incident on a Fe containing sample (absorber). Due to the natural abundance, 2% of the Fe nuclei in the absorber are ^{57}Fe isotopes and may produce resonant absorptions in the condition that they have the same electron surroundings as in the source. It is worth mentioning that the nuclear levels can be very slightly perturbed via the interactions of the surrounding electrons with the nucleus. Such interactions, called hyperfine interactions, are different in the source and in the absorber and therefore, they can cancel the resonant absorption. The resonance can be obtained only by compensating the energy of the photon incident on the absorber by the difference in the perturbation energy of the source and absorber, respectively. Taken as reference the electron configuration around ^{57}Fe in the source, the condition of resonance will depend on the electron configuration around ^{57}Fe in the sample (absorber). Hence, the Mössbauer spectrum related to the nuclear gamma resonant absorptions will contain a significant amount of information concerning the contribution of different types of interactions influencing the electron configurations in the sample. Experimentally, in classical Mössbauer spectroscopy, the energy of the incident photon emitted by the source is tuned over the resonance conditions of the absorber via the Doppler effect. Practically, a periodic variation of the source-absorber relative velocity is imposed in this regard. In usual experiments, relative velocities up to 10 mm/s are enough in order to cover all nuclear resonances of ^{57}Fe due to hyperfine interactions appearing in the absorber. In these terms, the Mössbauer spectrum accounts for the dependence of the intensity of the gamma-ray beam transmitted through the absorber versus the relative velocity between source and absorber. Firstly, the resonant absorption may not occur at zero relative velocity, as expected for similar ^{57}Fe nuclei in the source and absorber, respectively, but may be slightly shifted. This shift, known as isomer shift, gives valuable information on the Fe valence, coordination, spin state, etc., in the absorber. Secondly, one may not necessarily observe just a resonant absorption but a number of 2 (called doublets) or 6 (called sextets) absorption lines, in the Fe case, when working with non-polarized radiation. The hyperfine parameters corresponding to such spectral components are called quadrupole splitting or hyperfine magnetic field, respectively. The quadrupole splitting is due to the interaction of the electric quadrupole moment of the excited ^{57}Fe nucleus with the electric field gradient generated by a non-spherical distribution of the electric charge (either of the own electrons or of the surrounding anions) in the absorber. This hyperfine parameter gives information on the asymmetries in the electron configurations as well as on the local distortions. Finally, the six-line splitting, known also as a magnetic splitting, is caused by the interaction between the nuclear spin and the local magnetic field generated at the nucleus by the surrounding electrons. This magnetic hyperfine field, which can be obtained from the magnetic splitting, is proportional with the Fe (metallic, ionic, covalent, coordinative state) net magnetic moment and has a sign which is opposite to it. Therefore, by following the evolution of the magnetic splitting versus temperature, the most reliable information on magnetic relaxation phenomena can be obtained. It is worth to mention that the presence of the magnetic ordering as well as the magnitude of the magnetic moment may be obtained by Mössbauer spectroscopy without applying any external field. On the other hand, if an external magnetic field is applied on the sample (absorber), it can be vectorially composed with the existing internal field at nucleus (the magnetic hyperfine field), producing an effective field via vector summation. The effective field is this time obtained from the magnetic splitting of the Mössbauer spectrum. Magnetic hyperfine fields of magnetic iron states are of order of tenths of Tesla. Applied fields of order of Tesla are required in order to influence sensibly the magnetic hyperfine splitting. Excepting some peculiar cases, for such values of the applied field, the magnetic moments of various species of Fe become aligned along the field. Hence, by increasing the applied field, the effective field can increase or decrease, depending on the orientation of the Fe magnetic moments with respect to the applied field (e.g. along or opposite). The direction of the magnetic hyperfine field or magnetic moment with respect to the direction of the gamma radiation is accounted by the ratio between the intensity of the second and the third absorption peak, $R = I_2/I_3$ ($I_2 = I_5$ and $I_3 = I_4$), in the sextet component. Finally, it is worth to underline that all this information can be obtained for each Fe configuration in the absorber, by optimally fitting the Mössbauer spectrum using its spectral components (singlets, doublets and sextets) via suitable programs. Mössbauer hyperfine parameters for each Fe phase are derived and analyzed with respect to their variation versus temperature, applied magnetic field, pressure or other external excitations. The Fe phase composition and local distribution in the sample is straightforward obtained from the relative area of the spectral components.

3 Magnetic Fluids

A magnetic fluid, also known as a ferrofluid, consists of dispersed ferro/ferri- magnetic nanoparticles in a carrier liquid, mainly water or oil. Since the discovery of nanofluids, there was a growing interest in the theoretical and experimental characterization as well as in the specific applications of such complex magnetic systems. Applications in the sealing technology, viscoelasticity, heat transfer or magnetically controlled thermal flow [16–21] and also in biomedicine [22–24], are to be mentioned. Further applications include printer inks, magneto-rheological fluids and shock absorbers [25, 26]. An excellent revue on fundamental aspects, processing procedures and applications of ferrofluids can be found in [27].

In order to obtain a stable dispersion of the nanoparticles in non-polar or polar solvents, the inter-particle magnetic interactions are suppressed by specific ways. The dispersion of ferrite-like magnetic particles in water (a polar solvent) can be ensured by coating the particles with a double layer of surfactants whereas a single layer of surfactant is enough for assuring the dispersion of the nanoparticles in a non-polar solvent. However, the clustering process can not be completely avoided in water-based ferrofluids [28]. In spite of non-complete canceling of the attractive inter-particle interactions (considered as dipole-dipole type) they should be considered quite weak, due to the steric repulsion induced by the double coating layers. In the following, the attractive interactions are considered to induce only perturbations to the main anisotropy energy of the particles. Ferrite-like nano-particles in a ferrofluid can magnetically relax by two different mechanisms: (i) the Néel relaxation, discussed above, which is characterized by the internal reorientation of the magnetization with respect to an internal energy barrier, and (ii) the Brownian relaxation due to the rotational diffusion of the particle in the liquid. They depend on many other internal parameters as well as on the cluster dimension and are often showing competing processes. A way to separate the two contributions, for a correct interpretation of the experimental results, is to cut off one of the two components by choosing suitable experimental conditions. For example, the Brownian relaxation can be removed by freezing the magnetic liquid at low temperatures. Temperature dependent Mössbauer spectroscopy can be applied in this situation, providing information on particle phase composition, local structure and symmetry, local magnetic interactions inside the particle and Néel-type relaxation phenomena. The magnetic characterization becomes comprehensive in this situation but the whole information concerning the magneto-hydro-dynamics of the ferrofluid is then lost. The mechanisms which are relevant for the migration and assamblage of magnetic nanoparticles in the liquid matrix, can be studied by optical methods (e.g. [29]).

Different types of ferrofluids consisting of either magnetite or cobalt ferrite nanoparticles coated with double layers of surfactants (miristic acid, lauric acid and dodecyl-benzene-sulphonic-DBS-acid) and dispersed in different liquids (water, methyl-ethyl-cetone (MEC), n-butanole (C_4OH) and n-pentanole (C_5OH)) have been prepared and characterized by Mössbauer spectroscopy. We will present a brief comparison between relaxation phenomena in some of the analyzed samples. The actually considered samples, which are presented in Table 1, have been chosen in

Table 1 The characteristics of the analysed ferrofluids

Sample label	Carrier liquid	Particle type	Coating double layer
CF1	H_2O	$CoFe_2O_4$	Oleic acid + oleic acid
CF2	H_2O	$CoFe_2O_4$	Myristic acid + myristic acid
MS1	H_2O	Fe_3O_4	Lauric acid + lauric acid
MS2	H_2O	Fe_3O_4	Myristic acid + DBS acid
ML1	MEC	Fe_3O_4	Oleic acid + DBS acid
ML2	C_4OH	Fe_3O_4	Oleic acid + DBS acid

order to compare: (i) the influence of the surfactants on cobalt ferrite based ferrofluids (CF1 and CF2), (ii) the influence of surfactants on magnetite based ferrofluids (MS1 and MS2) and (iii) the influence of dispersing liquids on magnetite based ferrofluids (ML1 and ML2).

The Mössbauer spectra acquired at different temperatures on sample CF1 are shown in Fig. 1. Even at the lowest temperature of 4.2 K, where magnetic relaxation is expected to be completely suppressed, the spectrum consists of a broad sextet pattern presenting a line-width which is much larger than for well crystallized phases.

This suggests on a hand, the presence of the magnetic order and on the other hand, a situation with distributed Fe sites, as usually observed in low size nanoparticles. The mean magnetic hyperfine field at low temperature is about 51.5 T (these value corresponds to all the analyzed samples), close to typical hyperfine fields in magnetite or cobalt ferrites. Except for the fitting procedure with distributed Fe positions, the 4.2 K Mössbauer spectra can be suitably fitted by two strongly overlapping crystalline sextets with hyperfine fields of about 50 T and 52 T, respectively. The relative areas of the two sextets are close to 50%-50%. Comparatively, the 4.2 K Mössbauer spectra of magnetite based samples (MS and ML type) give for the relative area of the two components, values close to 35–65%. These results can

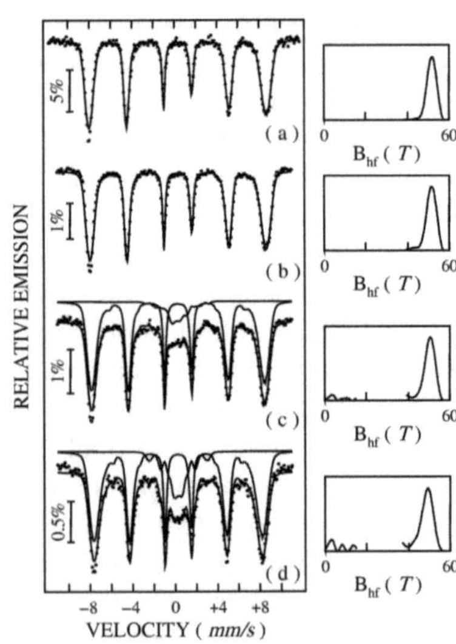

Fig. 1 Mössbauer spectra of sample CF1, collected at different temperatures: 4.2 K (**a**), 80 K (**b**), 126 K (**c**) and 180 K (**d**). The corresponding distributions of probability for the magnetic hyperfine field are presented on the *right side*

be interpreted in the terms of the above presented spinel structure, confirming in fact the presence of the two compounds. It is well known that magnetite is normally an inverse spinel, with all of Fe^{3+} ions in tetrahedral sites, and Fe^{2+} and Fe^{3+} ions in octahedral positions. At low temperature, below the Verwey transition, Fe^{2+} sites give rise to the sextet with a lower hyperfine field, whereas the Fe^{3+} sites give rise to the sextet with the higher hyperfine field (5 uncompensated electrons in Fe^{3+} and only 4 in Fe^{2+}). The double number of Fe^{3+} ions with respect to the Fe^{2+} ions in magnetite, is in agreement with the ratio of 35/65 between the relative area of the spectral components belonging to Fe^{2+} and Fe^{3+} respectively. However, the ratio between the two hyperfine fields belonging to the two iron states (Fe^{2+} and Fe^{3+} respectively) is much higher than the expected value of 4/5, proving peculiar electron delocalization processes in such nano-particles. On the other side, Co ferrite is an almost inverse spinel with an equal amount of Fe^{3+} on each octahedral and tetrahedral positions. This time, the different values of the hyperfine field belong to the different octahedral and tetrahedral positions leading to slightly different electron configurations around Fe (also the isomer shifts confirm this assignment). Hence, the almost equal number of Fe^{3+} ions on the two positions is in agreement with the 50–50% relative area of the two sextets in the 4.2 K Mössbauer patterns of cobalt ferrite based ferrofluids. Excepting the case of alternative fitting of the spectra at very low temperature, where the phase composition information is the most transparent due to the lack of magnetic relaxation, the fitting procedure with Fe distributed sites is more convenient for getting out the suitable information in the case where relaxation mechanisms are occurring. It could be affirmed in the following that a suitable interpretation of the temperature dependent relaxation phenomena, observed via the temperature evolution of the Mössbauer spectra, may provide information on the particle size distribution and effective anisotropy constant of the magnetic nano-particles.

According to the above mentioned aspects related to the magnetic relaxation of a fine nano-particle, the thermal fluctuations can be divided into two regimes. In the regime of the collective excitations ($r < 0.1$, where $r = kT/KV$), the coupled spins of the particle oscillate coherently along the magnetic easy axis. In a first approximation, the time averaged magnitude of the Fe spin, and hence, of the Fe magnetic hyperfine field, is given by [14, 30]

$$B_{\text{hf}} = B_0 (1 - \frac{k_B T}{2KV}) \tag{3}$$

with B the hyperfine field at temperature T and B_0 the hyperfine field in the static regime (e.g. at very low temperature). That is, the magnetic hyperfine field decreases almost linearly with the temperature, the slope of the dependence being inversely proportional with the energy barrier KV. In case of size distributed nano-particles (any real system presents a size distribution of relative size dispersion $\delta D/D$, with D the average diameter and δD the widts of the size distribution), each nanoparticle of a certain size presents its own linear decrease of the hyperfine field. Therefore, at a given temperature the system exhibits a distribution of hyperfine fields with the width dependent on the particle size distribution. If ΔB is the distribution width and $<B_0>$ is the average hyperfine field, the following relation between the reduced distribution width $\Delta B/<B_0>$ and $\delta D/D$ can be written according to [30]:

$$\Delta B_{\rm hf}/<B_0> = 3\frac{k_B}{2KV}\frac{\delta D}{D}T \tag{4}$$

By exploiting the linear dependencies of $<B_{\rm hf}>/<B_0>$ and $\Delta B_{\rm hf}/<B_0>$ versus temperature, specific to the regime of collective excitations, both the anisotropy energy and relative size dispersions can be evaluated from Mössbauer spectroscopy.

In the regime of super-paramagnetic relaxation ($r > 0.1$, where $r = kT/KV$), the thermal energy is large enough to switch the coupled spins of the particle between opposite directions along the easy axis. It results in a fast decrease of the time averaged Fe magnetic moments, and hence of the Fe average magnetic hyperfine field. For fast relaxation (short enough relaxation time), the average hyperfine field becomes zero and the initial low temperature sextet may collapse to either a singlet or a doublet (both considered as paramagnetic patterns). The blocking temperature can be derived from the Mössbauer spectra in this relaxation process, with the condition to impose a criterion establishing the spectral mark where the super-paramagnetic regime starts. In case of temperature dependent hyperfine field distributions, the blocking temperature can be defined as the temperature where the averaged hyperfine field decreases at half of its maximum variation range [31]. Once the blocking temperature is known and assuming specific values for the Mössbauer time window, τ_m, as well as for τ_0, the anisotropy energy may be also estimated from relation (2) with $T = T_{\rm B}$.

The general aspect of the Mössbauer spectra presented in Fig. 1 suggests that mainly the regime of collective excitations is manifesting for the involved temperatures. It can be observed that the hyperfine field distributions are centered at lower values and the width of the main peak becomes larger, when increasing the temperature. Above 120 K, a second but less intense central component is observed in the Mössbauer spectra as developing from the lower temperature sextet. It was interpreted as related to a reduced fraction of super-paramagnetic particles among the main fraction behaving in the regime of collective excitations. This result could indicate a bimodal type of size distribution for the magnetic nanoparticles, with a much less intense peak of low size (between 8 and 15% relative contribution, depending on the sample). Assuming a blocking temperature of about 100 K and an anisotropy constant of order of 10^4 J/m^3, the dimension of particles belonging to this second peak is of the order of 3–4 nm. For the rest of particles assigned to the main lobe, the mean particle size was estimated via X ray diffraction at 11(1) nm. It is worth mentioning that this average size is observed for all the analyzed systems of nanoparticles listed in Table 1.

Temperature dependent Mössbauer spectra of samples ML1 and ML2 are shown in Figs. 2 and 3, in order to put in evidence and to emphasize the different magnetic relaxations appearing in various samples.

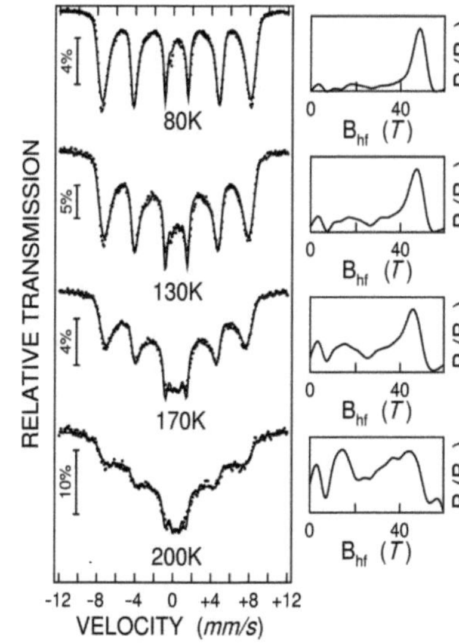

Fig. 2 Mössbauer spectra of sample ML1, collected at different temperatures. The corresponding distributions of probability for the magnetic hyperfine field are presented on the *right side*

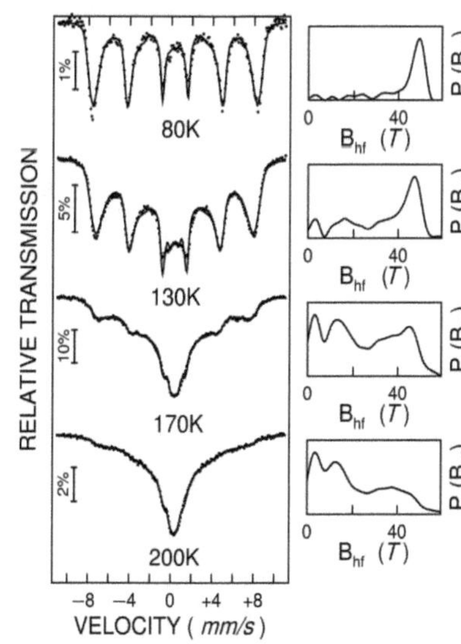

Fig. 3 Mössbauer spectra of sample ML2, collected at different temperatures. The corresponding distributions of probability for thc magnetic hyperfine field are presented on the *right side*

The dependencies of the ratio $<B_{hf}>/<B_0>$ versus temperature derived from the Mössbauer spectra are shown in Fig. 4, for all samples presented in Table 1.

Samples CF1 and CF2 show clearly a linear decrease of the above ratio versus temperature up to about 200 K (Fig. 4a) and therefore, below this temperature they can be considered in the regime of collective excitation. The anisotropy energy KV can be estimated from the slopes of these dependences, according to relation (3) and further, the anisotropy constant can be obtained, by assuming the structural coherence length (as approximation for the average particle size) of 11(1) nm, as deduced by XRD.

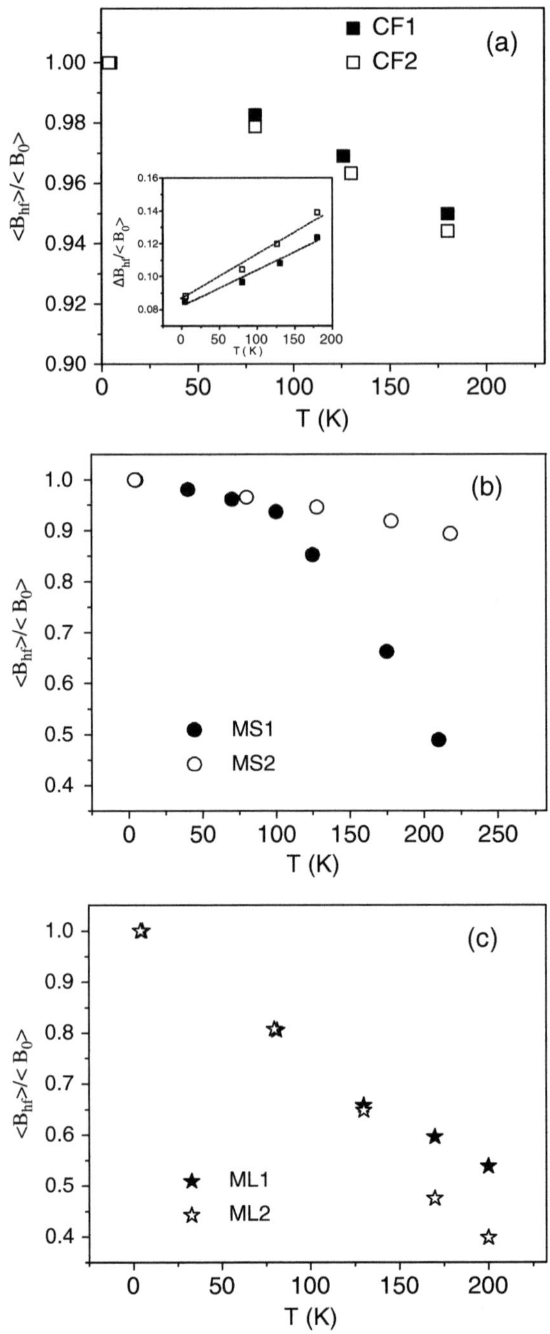

Fig. 4 The temperature dependence of the reduced average hyperfine fields versus temperature for all samples from Table 1 (CF1 and CF2 in (**a**), MS1 and MS2 in (**b**) and ML1 and ML2 in (**c**)). Fig. 4 (**a**) shows in the inset the linear dependences of the reduced distribution width versus temperature

The effective anisotropy constant of sample CF1 is $3.8(3) \cdot 10^4\ \mathrm{J/m^3}$ and of sample CF2 is $3.0(2) \cdot 10^4\ \mathrm{J/m^3}$. Both values are a few times lower than the anisotropy constant of bulk cobalt ferrite.

The variation of the relative width of the hyperfine field distribution versus temperature, for the same CF1 and CF2 samples is presented in the inset of Fig. 4a. The linear dependencies suggest that the relaxation regime is in agreement with the limits of validity for relation (4), what can provide the relative particle size dispersion, $\delta D/D$ (spherical particles are assumed). For both systems the relative particle size dispersion is 0.25(2) within the error limits, confirming in fact the electron microscopy data providing almost similar size dispersions for the analyzed particles.

Concerning the other systems, only a brief qualitative analysis will be performed, by comparing the dependencies in Fig. 4. All the magnetite based systems present a faster magnetic relaxation, according to the sharper decrease of the reduced hyperfine field versus temperature (that is principally accepted starting from the much lower value of the bulk anisotropy constant of magnetite compared with the one of cobalt ferrite). The faster relaxation appears in nanoparticles dispersed in MEC and C_4OH. For magnetite nanoparticles dispersed in water, the relaxation phenomena are very sensitive to the type of the surfactant, which is not the case for the cobalt ferrite nanoparticles dispersed in water.

4 Iron Based Nanopowders Prepared by Laser Pyrolysis

Among the successful methods to produce Fe based nano-particles is the laser pyrolysis [32] of a gas mixture containing pentacarbonyl (as precursor), air (as oxidizer, when Fe oxide particles are envisaged) and ethylene (as sensitizer). Comparing with the soft chemistry synthesis methods, the laser pyrolysis is a single step method offering high purity products formed in a very narrow reaction zone, well defined by the focused laser spot. Due to the millisecond time scale in which particles are formed, the nucleated particles are rapidly quenched to narrow size distributions with average diameters of less than 10 nm. The properties of the nanoparticle systems can be controlled in different ways: the composition via the gas mixture and the laser spot area, the size distribution via the pressure in the reactor chamber and via the laser power (a CO_2 laser is used) [33], etc. Particles with controlled core-shell structures may be formed by using special geometries of the reaction chamber entrance apertures of the gases. Usually, concentric apertures with the Fe precursor flowing axially, ethylene flowing concentrically and a third confining agent flowing in the next concentric nozzle are used when targeting core-shell structures. Taking into account the variety of nanoparticulate systems which can be obtained, a large number of investigation techniques are required in order to perform their comprehensive characterization. Mössbauer spectroscopy is one of the most powerful with respect to information concerning the phase composition of the system and its magnetic behavior. Two examples will be chosen in this regard.

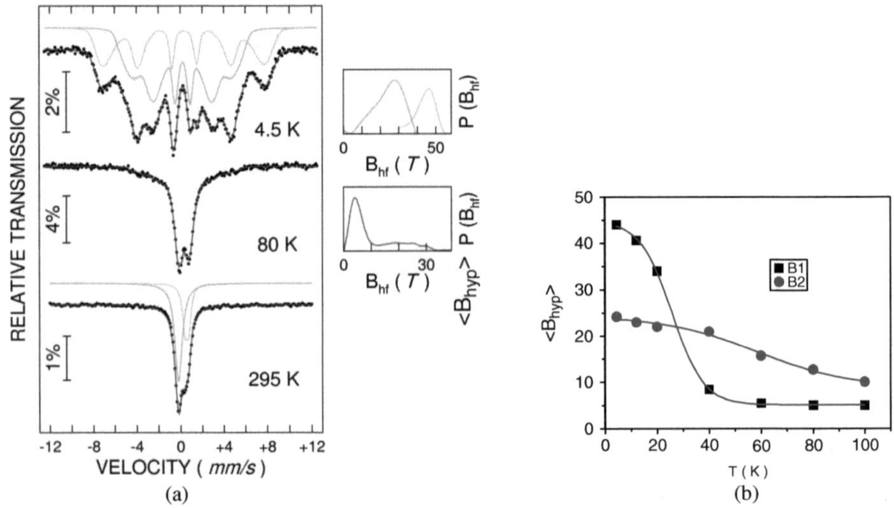

Fig. 5 (**a**) Temperature dependent Mössbauer spectra of sample NP1. Hyperfine field distributions are shown on the *right side*. (**b**) Average hyperfine fields of the two phases (Fe oxide- B1- and Fe carbide-B2-) versus temperature

Sample NP1 was obtained at a diameter of the central aperture of 0.55 mm whereas sample NP2 at a diameter of the central aperture of 1.5 mm. The Mössbauer spectra of each sample, collected at three temperatures are shown in Figs. 5 and 6.

The 4.5 K Mössbauer spectrum of sample NP1 (Fig. 5a, *left side*) was best fitted via two spectral components with distributed magnetic hyperfine fields.

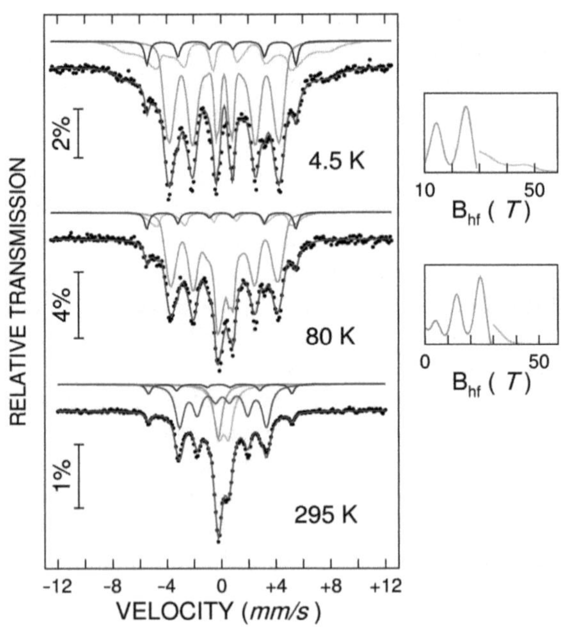

Fig. 6 Mössbauer spectra of sample NP2 acquired at different temperatures. Hyperfine field distributions are shown on the *right side*

The first distribution, belonging to the most external sextet, is centered on about 48 T whereas the second one, belonging to the internal sextet, on about 27 T (see Fig. 5a, graphs on the *right side*). By the specific hyperfine parameters and the arguments exposed in the introductory part, the first sextet is assigned to a defected Fe spinel oxide, consisting of a mixture (solid solution) of maghemite and magnetite. The sextet with the mean average hyperfine field of 27 T at 4.5 K is assigned to a carbide phase (almost cementite, Fe_3C). The relative content of the two phases, obtained from the relative spectral contributions of the two distributed components is 35% Fe oxide and 65% Fe carbide. At 80 K both phases are magnetically almost relaxed, proving blocking temperatures lower than 80 K and hence, nano-particles of average size lower than 5 nm. Finally, at 295 K, in the pure super-paramagnetic states, the hyperfine parameters of the two central patterns confirm again the Fe oxide and Fe carbide phases. The dependencies of the average hyperfine fields associated to Fe in the two phases, versus temperature, are shown in Fig. 5b. It is worth to remember that the blocking temperature may be estimated from the half of the maximum decrement of the average hyperfine field. A blocking temperature of about 30 K is derived for the Fe oxide nano-particles and of 60 K for Fe carbide nano-particles. The presence of two distinct blocking temperatures proves clearly that the sample consists of two different types of nano-particles, every one with its own size distribution, and does not involve nanoparticle with a core-shell structure. In this regard, Mössbauer spectroscopy is more sensitive compared to electron microscopy which allows to derivate only an uni-modal size distribution with an average particle size of about 3.5 nm.

The 4.5 K Mössbauer spectrum of sample NP2 (Fig. 6) are best fitted with three Mössbauer components. Two of them involve distributed hyperfine fields (the probability distribution is shown on the right side of the figure) whereas the third is a crystalline sextet with hyperfine parameters specific to metallic Fe (hyperfine field of 34 T and negligible isomer shift and quadrupolle splitting). Concerning the distributions, similar to the previous case, the most external sextet was assigned to an Fe oxide phase whereas the internal one to the Fe carbide phase. The hyperfine field distribution of the carbide phase shows two distinct peaks, one at about 15 T and another one at about 27 T, inferring that this phase consists mainly of Fe_5C_2 and Fe_3C. According to the relative spectral area, the phase composition is 27% Fe oxide, 67% Fe carbide and 6% metallic Fe. Contrary to the previous sample, the magnetic relaxation develops much slowly, the sextet associated to the oxide phase collapsing firstly, namely just above 80 K. At 295 K, the oxide nano-particles are already super-paramagnetic, as well as a fraction of 15% from the carbide (see the two central doublets in the corresponding Mössbauer spectrum). The rest of 6% of metallic Fe and 50% carbide are ferromagnetic at room temperature and form nano-particles with core-shell structures. These east ones are the most sensitive to an applied field and therefore, are most suitable for magnetic applications at room temperature. Related to the response to an applied magnetic field, another aspect which has to be considered is the value of the magnetic moment of the nanoparticle.

Considering nano-particles of similar size, the highest magnetic moment per particle involves a ferromagnetic order of the constituent material. However, in case of spinel oxide nano-particles, a ferrimagnetic coupling is present. Having in mind also the high degree of structural disorder in nanometer size particles, leading to both a mixture of spinel phases as well as to an expected enhanced spin disorder on the particle surface, the most useful parameter is related to the amount of uncompensated spins per particle. An interesting Mössbauer experiment providing such information is briefly presented in Fig. 7. The structural and morphologic characterization of sample NP3 was presented in detail in [34], where it was discussed with respect to other two samples obtained by laser pyrolysis at different laser powers, ranging between 35 and 55 W. Sample NP3, prepared by using a laser power of 45 W, consists of Fe oxide nano-particles with spinel-like structure, showing a narrow size distribution and an average size of about 6 nm. In the absence of the applied magnetic field, the Mössbauer spectrum at 3 K was fitted via a distribution of magnetic hyperfine fields, presented on the right side of the figure. The unimodal distribution is quite large and has a maximum at about 52 T. These features stand for maghemite-magnetite type nano-particles with defect spinel structure. If a field of 5 T is applied, the hyperfine field distribution transforms into a bimodal one, with an intense local maximum at about 47 T and a less intense one at about 57 T. According to a previously presented discussion, the two effective fields have to correspond to Fe spins opposite to the external field ($B_1 = 47T$) and along the external field ($B_2 = 57T$), respectively. In both cases the effective field is obtained by subtracting/adding the extern field (5 T) to the hyperfine field (52 T). This procedure allows evidencing separately the spins pointing along or opposite to the applied field and hence, to evaluate the relative fraction of uncompensated spins as the difference between the distribution area of the left side peak and the right side peak in the hyperfine (effective) field distribution, which area is normalized to unit.

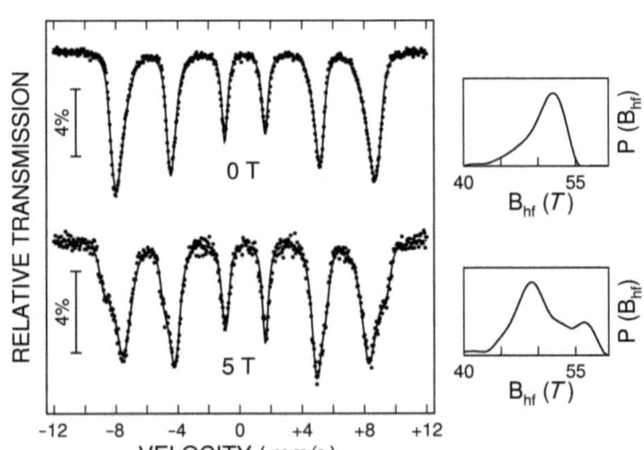

Fig. 7 Mössbauer spectra of sample NP3, collected at 3 K without and in the presence of an applied field of 5 T. The corresponding hyperfine field (effective field) distributions are shown on the *right side*

5 Conclusions

Important aspects related to the local magnetism and magnetic relaxation of Fe based nano-particles, were briefly emphasized. The ability and specificity of Mössbauer spectroscopy to solve delicate questions related to the magnetic behavior of such complex systems were evidenced. Methodological problems and available information provided by temperature dependent Mössbauer spectra in regime of magnetic relaxation were carefully considered in case of magnetic fluids. Further examples with the aim to relate the magnetic response of the system to its phase composition and spin configuration were presented in case of nano-powders obtained by laser pyrolysis.

Acknowledgements The authors wish to thank Ministry of Education and Research for financial support by national program PN2 71-083 and by core program PN09450103.

References

1. N.N. Greenwood, T.C. Gibb, *Mössbauer Spectroscopy* (Chapman and Hall, London 1971)
2. R.M. Cornell, U. Schwertman, *The Iron Oxides* (VCH, Weinheim 1996), p. 198
3. S.M. Yusuf, J.M. De Teresa, M.D. Mukadam, J. Kohlbrecher, M.R. Ibarra, J. Arbiol, P. Sharma, S.K. Kulshreshtha, Phys. Rev. B **74**, 224428 (2006)
4. P. Dutta, A. Manivannan, M.S. Seehra, N. Shah, G.P. Huffman, J. Appl. Phys. **99**, 08H105 (2006)
5. P. Dallas, N. Moutis, E. Devlin, D. Niarchos, D. Petridis, Nanotechnology **17**, 5019 (2006)
6. D. Guin, S.V. Manorama, S Radha, A.K. Nigam, Bull. Mater. Sci. **29**, 617 (2006)
7. D. Predoi, V. Kuncser, M. Zaharescu, A. Jitianu, M. Crisan, W. Keune, B. Sahoo, G. Filoti, M. Raileanu, J. Opt. Adv. Mater. **8**(2), 518 (2006)
8. P. Obreja, D. Cristea, E. Budianu, R. Rebigan, V. Kuncser, M. Bulinski, G. Filoti, Prog Solid State Chem **34**, 103 (2006)
9. S. Onodera, H. Kondo, T. Kawana, MRS Bull. **21**, 35 (1996)
10. R.F. Ziolo, E.P. Gionnelis, B.A. Weinstein, M.P. O'Horo, B.N. Ganguly, V. Metrota, M.W. Russel, D.R. Huffman, Science **257**, 219 (1992)
11. L. Nixon, C.A. Koval, R.D. Noble, G.S. Staff, Chem. Mater. **4**, 117 (1992)
12. C.R.F. Lund, J.A. Dumisec, J. Phys. Chem. **86**, 130 (1982)
13. C.M. Sorensen, in *Nanoscale Materials in Chemistry* ed. by Kenneth J. Klabunde (Wiley, New York, NY, 2001), P. 291
14. M.F. Thomas, C.E. Johnson, in *Mössbauer Spectroscopy* ed. by D.P.E Dickson, F.J Berry (Cambridge University Press, Cambridge, 1986), P. 192
15. L. Neel, Ann. Geophys. **5**, 9 (1949)
16. J. Popplewell, Phys. Technol. **15**, 150 (1984)
17. R.L. Bailey, J. Magn. Magn. Mater. **39**, 178 (1983)
18. S. Odenbach, J. Phys. Cond. Matter **16**(32), 1135 (2004)
19. L.P. Orlov, V.E. Fertman, Magnetohydrodynamics **164**, 400 (1980)
20. S. Odenbach, Int. J. Mod. Phys B **14**(16), 1615 (2000)
21. J.V. Vadasz, S. Govender, P. Vadasz, Int. J. Heat Mass Transfer **48**(13), 2673 (2005)
22. J. Roger, J.N. Pons, R. Massart, A. Halbreich, J.C. Bacri, Eur. Phys. J. AP **5**, 321 (1999)
23. A. Halbreich, E.V. Groman, D. Raison, C. Bouchaud, S. Paturance, J. Magn. Magn. Mater. **248**(2), 276 (2002)

24. M.K. Prasad, D. Panda, S. Singh, M.D. Mukadam, S.M. Yusuf, D. Bahadur, J. Appl. Phys. **97**(10), 100903 (2005)
25. S. Sem, M. Manciu, F.S. Manciu, Appl. Phys. Lett. **75**(10), 1479 (1999)
26. A.Yu Zubarev, L.Yu Iskakova, Physica A **343**, 65 (2004)
27. L. Vekas, Adv. Sci. Technol. **54**, 127 (2008)
28. L. Vekas, D. Bica, O. Marinica, M. Rosa, V. Socoliuc, F. Stoian, J. Magn. Magn. Mater. **289**, 50 (2005)
29. A.C. Galca, Ellipsometric studies of anisotropic nanoscaled media: Ferrofluids in magnetic fields and anodized aluminium, PhD thesis, University of Twente, Enschede, The Netherlands 2006
30. S Morup, J. Magn. Magn. Mater. **37**, 39 (1983)
31. V. Kuncser, G. Schinteie, B. Sahoo, W. Keune, D. Bica, L. Vekas, G. Filoti, J. Phys. Cond. Matter **19**, 016205 (2007)
32. R. Alexandrescu, I. Morjan, I. Voicu, D. Dumitrache, L. Albu, I. Soare, G. Prodan, Appl. Surf. Sci. **248**, 138 (2005)
33. G. Filoti, V. Kuncser, G. Schinteie, P. Palade, I. Morjan, R. Alexandrescu, D. Bica, L. Vekas, Hyp. Int. **199**, 55 (2009)
34. I. Morjan, D. Dumitrache, R. Alexandrecu, R. Birjega, C. Fleaca, I. Voicu, L. Gavrila, I. Soare, G. Filoti, V. Kuncser, G. Prodan, V. Ciupina, L. Vekas, NSTI-Nanotech 2007, www.nsti.org, ISBN 1420063766 **4**, (2007)

Properties of Ag/PVP Hydrogel Nanocomposite Synthesized In Situ by Gamma Irradiation

Ž. Jovanović, A. Krklješ, S. Tomić, V. Mišković-Stanković, S. Popović, M. Dragašević, and Z. Kačarević-Popović

Abstract Metal nanoparticles embedded in crosslinking polymer matrices i.e. hydrogels are novel class of materials which have attracted great attention due to applications in catalysis, photonics, optic, pharmaceutics and biomedicine. This work describes novel, simple and facile radiolytic synthesis of silver nanoparticles (AgNPs) within the poly(N-vinyl-2-pyrrolidone) (PVP) hydrogel. The hydrogel matrix was previously obtained by gamma irradiation induced crosslinking, while the in situ reduction of Ag^+ ions was performed using strongly reducing species formed under the radiolysis of water. Absorption spectrum of Ag/PVP hydrogel shows the presence of surface plasmon band with maxima at 400 nm, which confirms formation of AgNPs (diameter less than 10 nm). Swelling properties of synthesized hydrogels, neat PVP and Ag/PVP nanocomposite, were investigated in the SBF (simulated body fluid) solution at 37°C. Obtained results show that Ag/PVP hydrogel nanocomposite has higher equilibrium swelling compared with neat PVP hydrogel. Moreover, kinetics parameters were calculated from the swelling curves. The fluid transport mechanism is non-Fickian for both PVP and Ag/PVP hydrogels, meaning that both diffusion and polymer relaxation control the fluid transport.

1 Scientific Background

1.1 Hydrogels as Biomaterials

Hydrogels are polymer networks that are able to swell in aqueous solutions with a multiple uptake of solvent relative to their own volume or mass. Hydrogels are being considered as most promising biomaterials in form of contact lens, burn dressing, artificial corneas, soft tissue substances, drug delivery systems, constituents of hybrid-type organs etc. The polymer gels have a very low modulus of elasticity,

Z. Kačarević-Popović (✉)
Vinča Institute of Nuclear Sciences, University of Belgrade, PO Box 522, 11001 Belgrade, Serbia
e-mail: zkacar@vinca.rs

A. Aldea, V. Bârsan (eds.), *Trends in Nanophysics,* Engineering Materials,
DOI 10.1007/978-3-642-12070-1_14,

and therefore cause minimal mechanical irritation. They are chemically stable in a living environment and permeable to low molecular weight metabolites, which are formed at the boundary between living and non-living tissue. Thus they moderate the physical abnormality of the boundary [1]. Hydrogels are applied from tissue expanders to coatings for penile and testicular implants, urology devices and materials as well as for wound dressing. Even in diabetic foot wound therapy, in some cases, a tissue expander can be used for resolving the open wound. Hydrogels can be used as balk materials, coatings or can be grown and dispersed in water as colloidal particles with a crosslinked polymer framework. On the other hand nearly synchronous with the advent of the prosthetic and other medical devices, strategies were developed to attempt to minimize the risk of infections because of clinical and economic significance. Nosocomial or hospital related bacterial infections are estimated to be the fifth-leading cause of death in the United States after heart disease, cancer, stroke, and pneumonia or flu. The Centers for Disease Control estimate that nosocomial infections cost hospitals more than 2300 dollars per patient for diagnosis and treatments. For example, twenty-five percent of hospitalized patients in the United States undergo indwelling urinary catheterization, which is associated with a risk of bacteriuria of approximately 5% per day of catheterization. Urinary tract infections (UTI), largely because of urinary catheters, account for up to 40% of all nosocomical infections are complicated by bacteremia in up to 4% of cases, result in 1–3.8 extra hospital days, and increase hospital costs by more than 500 million dollars annually [2].

1.2 Nano Ag in Biomedical Application

In recent years nanoscale antibacterial materials as novel antimicrobial species have been seen as promising candidates for application owing to their high surface to volume ratio and their novel physical and chemical properties on the nanoscale level. Many kinds of nanometer–sized antibacterial materials such as TiO_2, ZnO, MgO, chitosan, calamine, copper and silver have been reported on in this area. Among them nanocrystalline silver has been proved to be most effective antimicrobial agent, since silver and its compounds have powerful antimicrobial capability and broad inhibitory biocidal spectra for microbes, including bacteria, viruses and eukaryotic microorganism. Enhanced antibacterial properties of nanocrystalline silver have been demonstrated both in vitro and in vivo. Silver (Ag) nanoparticles show excellent antimicrobial activity by binding both to microbial DNA causing loss of its replication ability, preventing bacterial replication and to the sulfhydryl groups of the metabolic enzymes of the bacterial electron transport chain causing their inactivation [3]. Higher concentrations of Ag^+ ions have been shown to interact with cytoplasmic components and nucleic acids. In addition micromolar levels of Ag^+ ions are known to uncouple respiratory electron transport from oxidative phosphorylation [4]. Moreover, metal nanoparticles such as silver provide interesting research area because they have a close lying conduction and valence bands in which electrons move freely. Due to their ability to absorb and scatter the light and biocompatible nature they are therefore well suited for nanoparticles optical and biological application. Moreover, in 2005 investigators first reported that Ag nanoparticles prevent HIV-1 infection. Findings suggest that the particles interacted with the viral particles in a size-dependent manner, and behaved as lectins-proteins that interact predominantly with glycosylated loci-probably due to the preferential binding of the nanoparticles to the cysteine-rich moieties of the gp120 glycoprotein present in the viral envelope. The interaction of Ag nanoparticles and gp120 prevented viral fusion with cells, therefore preventing infection [5].

Silver does not interfere with therapeutic properties of medical products. Silver can be safely used even for patients who have diseases like *Diabetes Mellitus* that interfere with wound healing. Namely, in modern management of certain types of open wounds (e.g. traumatic, thermal or chronic wounds) dressing play a major role, since the moist, warm and nutritious environment of wound beds provides an ideal condition for microbial growth. Bacterial colonization and subsequent infection can interfere with the wound healing process by producing various substances (e.g. toxins, proteases and proinflammatory molecules) which may cause an excessive and prolonged inflammatory response of the host tissues. Ideal antimicrobial dressings should have a number of key attributes including provision of a moist environment to enhance healing and broad spectrum antimicrobial activity, including activity against antibiotic-resistant bacteria (e.g. methicilin-resistant *Staphylococus aureus* (MRSA)). A recent resurgence of the use of silver-based dressing has been ascribed to their broad spectrum antibacterial activity as well as to a lesser possibility for inducing bacterial resistance than antibiotics [6].

In Fig. 1 treatment of diabetic foot wound with the silver containing topical dressing at the Institute of Endocrinology, Diabetes and Metabolic Diseases, University

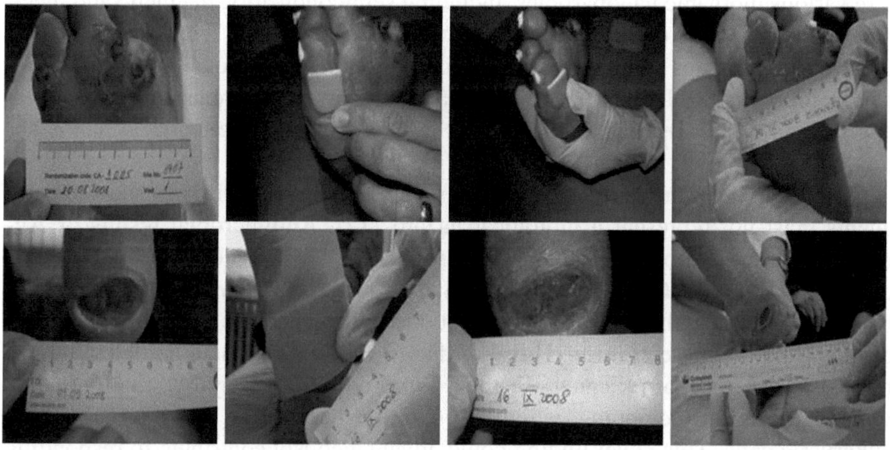

Fig. 1 Treatment of the diabetic foot wound with the silver containing dressing at the Institute of Endocrinology, Diabetes and Metabolic Diseases, University Clinical Centre of Serbia, Belgrade

Clinical Centre of Serbia, Belgrade is shown. Impaired wound healing is common complication of *Diabetes Mellitus*. Healing in patients with *Diabetes Mellitus* is characterized by reduced tensile strength of wounds when compared with controls, suggesting either defective matrix production or deposition. In the human mammal, diminished perfusion resulting from the presence peripheral arterial disease as well as decreased sensory nerve function caused by peripheral and autonomic neuropathy may contribute to impaired healing, as reported by [7]. It is presumed that diabetic complications result from periods of poor glycemic control. However, aberrant growth factor expression or factors secondary to *Diabetes Mellitus*, such as advanced glycation and crosslinking of matrix protein, may also be involved. The recent emergence of nanotechnology has provided a new therapeutic modality in silver nanoparticles for healing wounds.

1.3 Hydrogel Based Nanocomposites with Silver Nanoparticles

Recent research efforts are heading for exploiting the in situ synthesis of noble metal nanoparticles within the polymeric network architectures and the synthesized products using these approaches are leading to new hybrid-composite systems in chemistry and engineering sciences. Three-dimensional hydrogels (either in balk or in the form of colloidal nanoparticles) are appropriate templates for producing noble metal nanoparticles over conventional non-aqueous and polymeric systems for biomedical applications in view of their exceptional compatibility with biological molecules, cells and tissues. In this way the carrier-hydrogel system acts as nanoreactors that, in addition, immobilize nanoparticles and provide an easy handling.

The technical progress in polymer micro- or nanoreactors facilitates the development of the fabrication technology of metal nanoparticles. Nanoreactor chemistry is a versatile method by which a single template can be used to synthesize various inorganic nanoparticles, including semiconductors, magnetic and metal nanoparticles, and to control the size, concentration and spatial location at the same time, which provides an important novel route of fabricating metal/organic polymer nanocomposites [8]. In the literature there are only few reports regarding these novel approach to synthesizing of Ag nanoparticles using hydrogel networks as nanoreactors [9–11].

To find applications in the biomaterials field, both the stabilizing and reducing agents mast not represent a biological hazard [12]. Antimicrobial results of Ag-polysaccharice hydrogel nanocomposite show that these nanocomposite systems display a very effective bactericidal activity toward both $Gram+$ and $Gram-$ bacteria. However, the hydrogel does not show any citotoxic effect toward three different eukaryotic cell lines. This is a due to the fact that nanoparticles, immobilized in the gel matrix, can exert their antimicrobial activity by simple contact with the bacterial membrane, while they can not be uptaken and internalized by eukaryotic cells. This novel finding could advantageously contribute to responding to the growing concerns on the toxicity of nanoparticles and facilitate the use of biocompatible Ag-hydrogel composites in the preparation of biomaterials. In general hydrogel technologies use a reservoir system of ionic silver in nanoreactive hydrogel polymer system and the silver ions release is controlled to provide antimicrobial activity without disrupting the body's production of cells required during the healing process [13].

2 Characterization and Determination of Swelling and Diffusion Characteristics of In Situ Radiolytic Synthesized Ag/PVP Hydrogel Nanocomposite in SBF Solution

2.1 Introduction

Irradiation has been recognized as highly suitable tool to aid in the formation of hydrogels. The radiation process has various advantages, such as easy process control, the possibility of joining hydrogel formation and sterilization in one technological step, the lack of necessity for initiators and crosslinkers, which are possibly harmful and difficult to remove. The radiation technique is clean, because it does not require any extra substances, does not leave some unwanted residues, and does not need any further purification. These qualities make irradiation the method of choice in the synthesis of hydrogels. On the other hand, although gamma irradiation has proven to be a powerful tool for synthesis and modification of materials, not so many studies have been reported concerning the radiolytic formation of metal nanoparticles in hydrogel matrix. Since performances of Ag nanoparticles have a considerable association with the structure, surface modality, size and size distribution, a strict control is needed. It has been shown that the morphology, particle size distribution, stability and properties of Ag nanoparticles are strongly dependent on the method of preparation and specific experimental conditions. Most of the technologies used to incorporate Ag into polymeric matrixes involve either chemical workups such as reduction, or synthesis of complex Ag compounds, mixing performed Ag particles with polymers or complicated physical techniques, such as sputtering, and plasma deposition.

The radiolytic method is very is suitable for generating metal particles in solution. The radiolytically generated species, solvated electrons and secondary radicals exhibit strong reduction potentials, and consequently metal ions are reduced at each encounter. The control of particle size is achieved by the use of capping agents such as polymers, which are present during the formation of metal clusters. Polymer molecules interact with the growing metal particles, inhibiting the aggregation process [14].

The aim of our work is systematically developing synthetic strategies for incorporation of nano-Ag in hydrogel networks by gamma irradiation for possible biomedical application, using liquid filled cavities in hydrogels as nanoreactors (template synthesis). The incorporation of template technologies with the synthesis techniques such as gamma ray irradiation methods is expected to produce more elaborated Ag nanoparticles and more diversified nanoelevated Ag.

Nano-Ag is successfully incorporated in hydrogel matrix such as PVA, and poly(BIS-co-HEMA-co-IA) copolymer hydrogel [16, 17] The electrostatic interaction (i.e. ion-dipole) between the electron reach (either) oxygen atoms of polar hydroxyl groups, or/and carbonyl groups and electropositive transition metal cations was factor affecting the Ag grain morphologies and helped in modulating and controlling the size of the nanoparticles and supplying better stabilization with the Ag nanoparticles. Moreover, it was shown for the first time that PVA$^{\bullet}$ radicals in PVA hydrogel, obtained by gamma irradiation, have sufficient reducing ability to produce Ag nanoparticles in swollen polymer matrix, as shown in Fig. 2. The further work on the biological potential of these materials is in progress.

In this study we use poly(*N*-vinyl-2-pyrrolidone) (PVP) as matrix component to synthesize Ag nanoparticles by gamma irradiation. PVP is a water-soluble polymer, which exhibits a series of interactions toward dissolved small molecules and ions in aqueous solutions. Aqueous PVP solutions are permanently gelled when crosslinked by irradiation with gamma rays [18]. PVP hydrogel has excellent transparency and biocompatibility and has been used as main component in temporary skin covers or wound dressings [19].

The mechanism of stabilizing of Ag nanoparticles by PVP polymer originates from the property of PVP homopolymer whose individual structural units contain a hydrophobic vinyl group and a hydrophilic cyclic amide group. PVP is positively charged in aqueous solution due to protonation of amide nitrogen atoms. The anions associate with amphiphilic PVP remaining in the silver sol. The resulting Ag-cation containing nano-Ag particles are stabilized by capping effect of the PVP. Ag-cation containing nano-Ag is a Ag/PVP nanocomposite that releases Ag-ions very slowly. The hydrophilic amide groups of PVP are bound to the surface of the

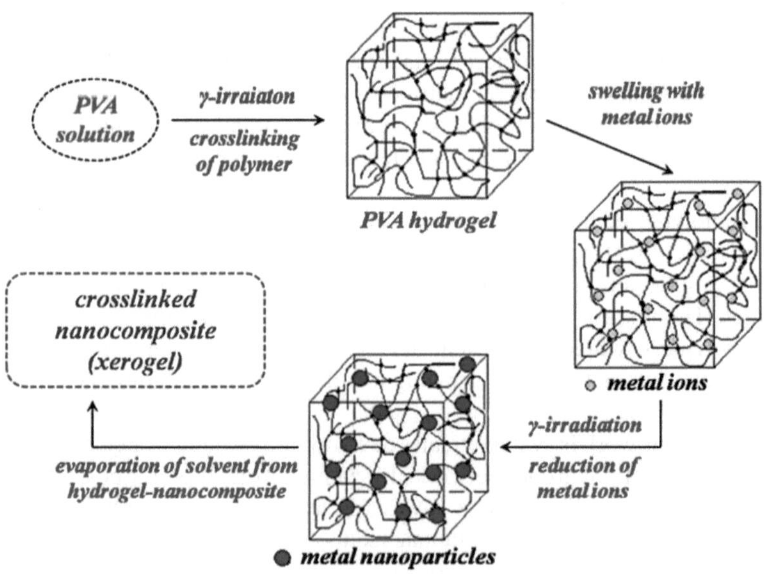

Fig. 2 Synthesis of Ag nanoparticles in PVA hydrogel by gamma irradiation [15]

Ag nanoparticles through strong affinity of N and O atoms for transition metallic clusters, whereas the polyvinyl backbone of PVP forms a hydrophobic domain, which surrounds the silver nanoparticles to prevent agglomeration. In general PVP serves as both a coordinating agent and a stabilizing agent; it also plays an important role in controlling the size and shape of the silver nanoparticles [4]. The aim of the present study is to investigate the swelling behavior and the diffusion of SBF solution in Ag/PVP hydrogel nanocomposite prepared in situ by gamma irradiation compared to pure PVP hydrogel.

2.2 Experimental

2.2.1 Materials

All chemicals, $AgNO_3$ (*Lachner*), 2-propanol (*Zorka Pharma*) and PVP (*Sigma-Aldrich*, $M_w = 360000$), were of analytical grade and they were used as received. Water from Millipore Milli-Q system was used in all experiments. Ar gas was of high purity (99.5%).

2.2.2 The Synthesis of Ag/PVP Hydrogel Nanocomposite

The Synthesis of Hydroges

The polymer was completely dissolved at 80°C under the stirring for 6 h. In order to remove O_2, the PVP solution were bubbled with Ar in sealed air-tight containers before being exposed to gamma rays. Gamma irradiation was performed in ^{60}Co radiation facility, at room temperature and a dose rate of 0.5 kGy h^{-1} up to the total radiation dose of 25 kGy for crosslinking of 10 wt. % PVP solution

Assemby of Ag Nanoparticles in PVP Hydrogel Network

The nanocomposite was prepared by swelling the crosslinked polymer sample with $AgNO_3$ and with 0.2 M 2-propan solutions in water for 24 h. Swelling of Ar-saturated gels was carried out in tightly closed containers for 24 h at room temperature in the dark. Only gel free of voids was employed. Reduction reaction is performed by gamma irradiation, in ^{60}Co radiation facility, at room temperature to absorbed dose of 8 kGy for reduction of 3.9×10^{-3} M Ag^+ ions, at a dose rate of 9 kGy h^{-1}.

2.2.3 Methods of Characterization

Optical and Microscopy Characterization

The UV-Vis absorption spectra of PVP hydrogel and Ag/PVP hydrogel nanocomposite were recorded using Thermo Scientific Evolution 600 spectrophotometer. Transmission electron microscopy (TEM) measurements were performed using JEOL 100 CX Electron Microscope operated at 100 kV.

Crosslinking Degree of as Prepared PVP Hydrogel

After the process of drying to the constant mass, the obtained hydrogel was extracted in water at the 45°C during 6 h, in two cycles, in order to remove the non-crosslinked polymer. The weight fraction of gel, W_g in %, was calculated as:

$$W_g(\%) = W_g / W_0 \times 100 \tag{1}$$

where W_g is the weight of dry gel after extraction and W_0 is the initial weight of the dry gel. The calculated weight fraction of gel was 94 %.

2.2.4 Swelling Studies

The xerogel discs were immersed in an excess of Kokubo's Simulated Body Fluid (SBF) solution [20], with the pH value of 7.1, to obtain equilibrium swelling at 37°C. The progress of the swelling process was monitored gravimetrically. The equilibrium degree of swelling (q_e) was calculated as:

$$q_e = W_e - W_0 / W_0 \tag{2}$$

where W_e is the equilibrium mass of the swollen gel and W_0 is the initial weight of the dry gel, as referred before [21, 22].

2.3 Results and Discussion

The formation of the silver nanoparticles in the hydrogel network was indicated by the appearance of a yellow colour, as shown in Fig. 3.

2.3.1 TEM Microscopy

To probe into the morphologies and size of Ag nanoparticles, TEM micrographs of Ag nanoparticles are shown in Fig. 4. It is obvious that these Ag nanoparticles assume spherical–like morphologies in appearance at nanoscale levels. As it was already mentioned, Ag^+ can combine with oxygen and nitrogen atom existing in the hydrogel networks via a weak coordinating bond forming O-Ag^+ and/or N-$Ag^+\sigma$ coordinating bonds, and thus astricting Ag^+ particles. Therefore Ag ions may uniformly be fixed and distributed in the hydrogel networks.

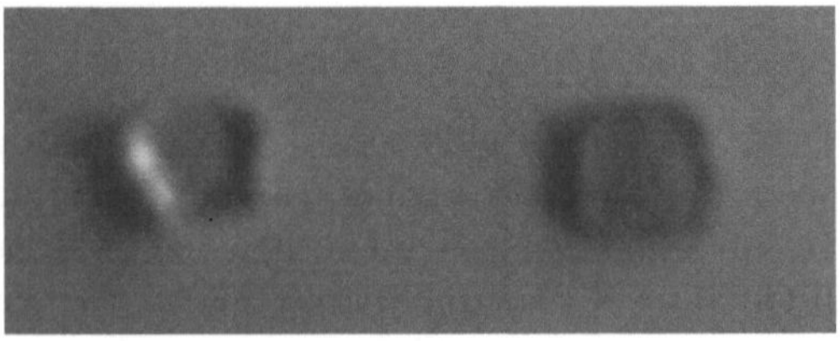

Fig. 3 Photograph of PVP hydrogel (*left*) and Ag/PVP hydrogel nanocomposite (*right*)

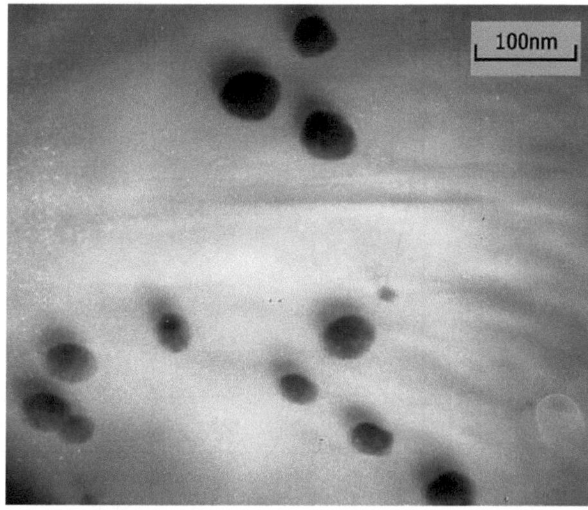

Fig. 4 TEM micrograph of Ag/PVP nanocomposite

A three step mechanism for the growth of silver nanoparticles synthesized by gamma irradiation and stabilized by PVP polymer was proposed: (i) silver ions interact with PVP, (ii) nearby silver atoms that have been reduced by gamma irradiation aggregate at close range (the primary nanoparticles), and (iii) nearby primary nanoparticles coalescence with other primary nanoparticles or interact with PVP molecules to form larger aggregates (secondary nanoparticles) [23]. The similar mechanism of particles assembly is probably active in this system and the observed nanoparticles by TEM analysis are the aggregates of primary particles synthesized in situ by gamma irradiation.

2.3.2 UV-Vis Spectroscopy

The presence of embedded silver nanoparticles within the gel macromolecular network is confirmed in addition by using UV-Vis spectral studies.

Figure 5 represents the absorption spectra of pure PVP hydrogel and Ag/PVP hydrogel nanocomposite. The absorption spectrum of Ag/PVP nanocomposite has surface plasmon absorption band peaking at 405 nm, which confirms formation of AgNPs (diameter less than 10 nm).

2.3.3 Swelling Studies

Swelling of the crosslinked polymers in chosen solvent is the most important parameter for swelling studies. Moreover, preliminary studies in buffered solution of pH similar to the pH of biological fluids are very important for the application of hydrogels as biomaterials. In general, a fundamental relationship exists between the swelling of a crosslinked polymer and solvent. The intake of xerogels, PVP and

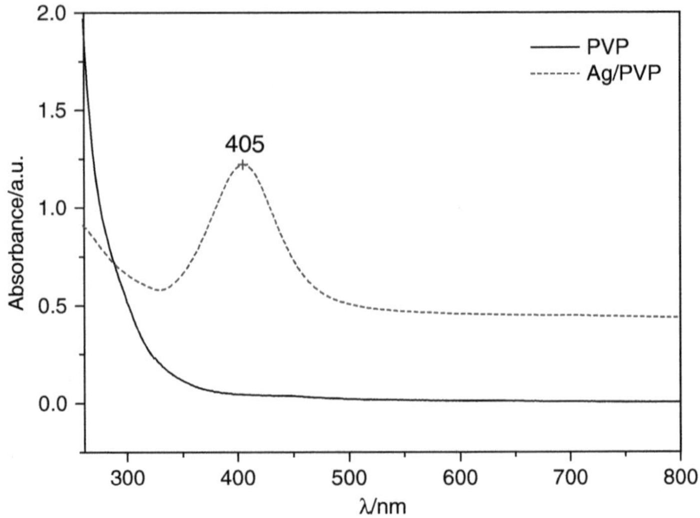

Fig. 5 The absorption spectra of hydrogels, PVP and Ag/PVP nanocomposite

Ag/PVP nanocomposite, were followed for a long period of time in SBF solution at 37°C. Swelling curves are plotted and presented in Figs. 6 and 7.

As can be seen from Fig. 6, swelling capabilities of PVP hydrogels (xerogel) and Ag/PVP hydrogel (xerogel) nanocomposite are increased by time, reaching constant swelling (equilibrium swelling) after a certain period of time. Another observation is that the swelling of the gels is slightly greater for the gel containing Ag nanoparticles.

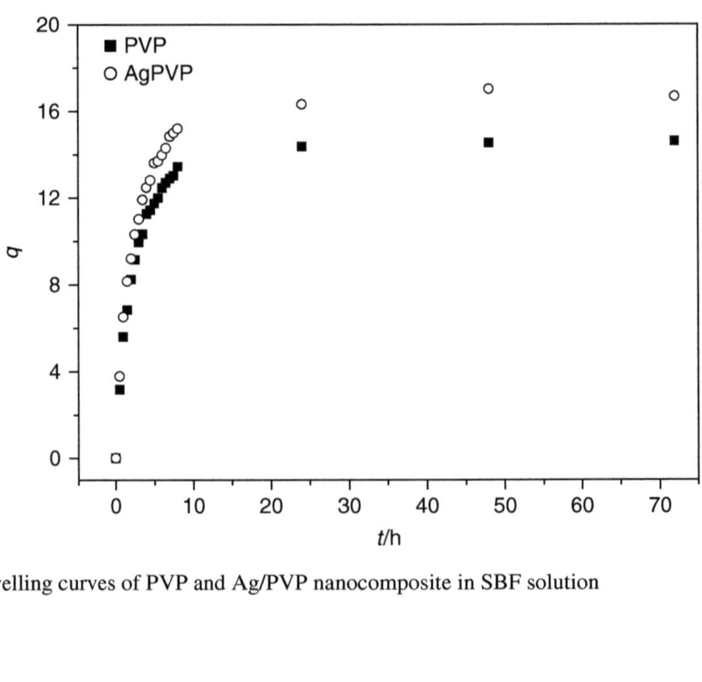

Fig. 6 Swelling curves of PVP and Ag/PVP nanocomposite in SBF solution

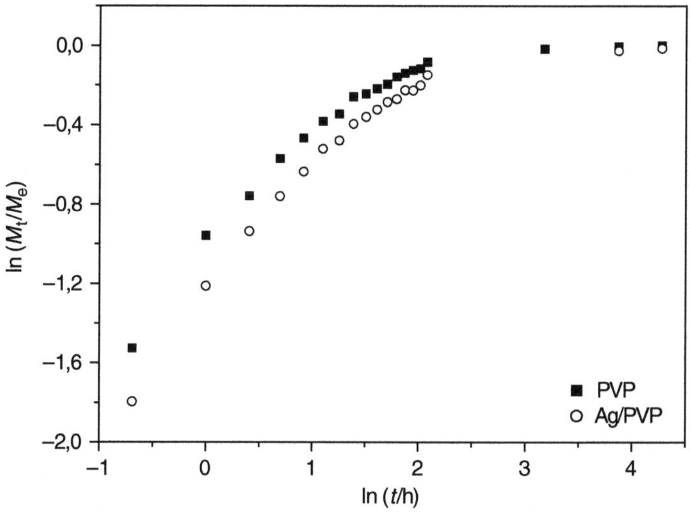

Fig. 7 Swelling curves of PVP and Ag/PVP hydrogels in SBF at 37°C

Many factors can influence swelling behaviour of hydrogels. Some of them are related to the effect of additives on the structure of water (or solvent) and interaction between polymer and solvent molecules. Namely, the Mark-Houwink exponent a, depends on the nature of solvent, temperature and the molecular weight of the polymer. At the theta temperature, a is equal to 0.50. This exponent has been found as 0.55 in Mark-Houwink equation for the system PVP/H_2O, which is close to 0.50, showing the characteristic of a theta solvent. The additives can change the hydrogen bonded structure of water and the molecular association of the water soluble polymer in aqueous media, as well as the swelling behaviour of crosslinked PVP chain [24]. The observed slightly increased swelling capacity of Ag/PVP nanocomposite compared to pure PVP probably can be ascribed to existence of remnant Ag^+ ions in the in situ synthesized nanoparticles in hydrogel, which can be well dissolved in water changing the structure of SBF medium.

On the other hand, it was observed that the optical properties of metal sol prepared in water or other matrices, particularly of Ag (when preparation utilized strong reductant or when strong reducing conditions are achieved with gamma irradiation) can be changed during the aging in absence of oxygen. It is well known that colloidal Ag particles can store several hundreds of electrons and can be negatively charged or cathodic polarized. Discharge of the sol, by reduction of the water to hydrogen limits the attainable negative charge of the particles. The upper limit of the cathodic polarisation of the silver particles lies around −0.4 V NHE where hydrogen evolution begins [25]. So, in addition to other factors causing different swelling behaviour of Ag/PVP hydrogel nanocomposite compared to pure PVP hydrogel, possibility of the charging of Ag nanoparticles in polielectrolyte SBF (Kukubo) solution may change swelling affinity of nanocomposite.

Diffusion

When a dry hydrogel is brought into contact with water, water diffuses into the hydrogel and the hydrogel swells. Diffusion involves the migration of water in pre-existing of dynamically formed spaces among hydrogel chains, while swelling of the hydrogel involves a larger scale segmental motion, resulting ultimately in an increase in the separation distance among the hydrogel chains. To investigate the diffusion model of the gel, kinetic parameters were calculated from the first part of the swelling curves [26], using Eq. (3):

$$M_{\mathrm{t}}/M_{\mathrm{e}} = kt^n \tag{3}$$

where M_{t} and M_{e}, calculated from Eq. (2), are the quantities of fluid absorbed at time t and at the equilibrium, respectively, k is a kinetic constant and n is the diffusion exponent. The logarithmic form of Eq. (3) was used to calculate n and k, from the slope and intercept, respectively. This equation is applicable to the initial stages of swelling, where a linear fit of the data was observed. The diffusion coefficient, D, can be calculated [27] from Eq. (4):

$$D = k^{1/n}\pi\delta^2/16,\ \mathrm{u\ cm^2s^{-1}} \tag{4}$$

where δ stands for the thickness of the dried hydrogel (xerogel). The values of q_{e} (SBF solution at 37°C), n, k and D, for investigated samples are listed in Table 1. It is evident that q_{e} value of Ag/PVP nanocomposite hydrogel shows higher swelling. As it can be seen, the introducing Ag nanoparticles into the PVP hydrogel lead to higher swelling degree. This may be, in addition to properties of nanoparticles itself, as previously discussed, the consequence of change in dimensions of voids in xerogels. Namely, the silver nanoparticles are formed inside the network and they occupy the voids of the matrix. Making these voids (dynamically formed spaces among hydrogel chains) different, the presence of nanoparticles enables the swelling in the larger extent, comparing to the pure PVP hydrogel.

The information about the physical mechanism controlling fluid adsorption can be gained from the value of the diffusion exponent. In the case of cylindrical geometry the transport mechanism can be Fickian ($n < 0.5$), non-Fickian or anomalous ($0.5 < n < 1$), Case II transport ($n = 1$) and Super case II transport ($n > 1$). The fluid transport mechanism is non-Fickian for both PVP and Ag/PVP hydrogels, meaning that both diffusion and polymer relaxation control the fluid transport [28].

Table 1 The values of the swelling degree, q_{e}, kinetic constant, k, the diffusion exponent, n and diffusion coefficient, D, for crosslinked PVP and crosslinked Ag/PVP nanocomposite

Hydrogel sample	q_{e}	n	k	$D/\mathrm{cm^2s^{-1}}$
PVP	14.76	0.6478	0.3584	1.329 E-5
Ag/PVP	17.35	0.6315	0.3513	1.262 E-5

3 Conclusion

Obtained results indicated that gamma irradiation is suitable for *in situ* generation of Ag nanoparticles in PVP hydrogel matrix. Swelling properties of synthesized hydrogels, neat PVP and Ag/PVP nanocomposites, investigated in the SBF (simulated body fluid) solution at 37°C show that Ag/PVP hydrogel nanocomposite has higher equilibrium swelling compared with neat PVP hydrogel. Kinetics parameters calculated from the swelling curves reveal that the fluid transport mechanism is non-Fickian for both PVP and Ag/PVP hydrogels, meaning that both diffusion and polymer relaxation control the fluid transport

Acknowledgements This work is financed by the International Atomic Energy Agency, Vienna, project CRP: F23028 contract No. 15384.; Ministry of the Science and Technological development, Republic of Serbia (Contract Nos. 19027 and 142061). Željka Jovanović, Ph. D. student, is grateful to the Ministry of the Science and Technological development, Republic of Serbia for the financial support.

References

1. K.R. Park, Y.Ch. Nho, J. Appl. Polym. Sci. **91**, 1612–1618 (2003)
2. M.E. Rupp, T. Fitzgerald, N. Marion, V. Helger, S. Puumala, J.R. Anderson, P.D. Fey, 12th Annual meeting of the society for health care epidemiology of america, Solt Lake City, Utah, April 7–9 2002, Abstract no 272.
3. V. Thomas, M.M. Yllapu, B. Sreedhar, S.K. Bajpai, J. Coll. Interf. Sci. **315**, 389–395 (2007)
4. J.-S. Kim, J. Ind. Eng. Chem. **13**(4), 718–722 (2007)
5. J.L. Elechiguerra, J.L. Burt, J.R. Morones, A. Camacho-Brabado, X. Gao, H.H. Lara, M.J. Yacaman, J. Nanobiotechnol. **3**, 1 (2005)
6. P. Rujitanaroj, N. Pimpha, P. Supaphol, Polymer **49**, 4723–4732 (2008)
7. M. Mishra, H. Kumar, K. Tripathi, Dig. J. Nanomater. Biostruct. **3**(2), 49–54 (2008)
8. Y.-L. Luo, Q.-B. Wei, F. Xu, Y.-S. Chen, L.-H. Fan, C.-H. Zhang, Mater. Chem. Phys. **118**(15), 329–336 (2009)
9. N.S. Saturkar, J.Z. Hjlt, Acta Biomater. **4**, 11–16 (2008)
10. Y.M. Mohan, K. Lee, T. Premkumar, K.E. Geckeler, Polymer **48**, 158 (2007)
11. F. Furno, K.S. Morley, B. Wong, B.L. Sharp, J. Antimicrob. Chemother. **54**, 1019–1024 (2004)
12. A. Travan, Ch. Pelillo, I. Donati, E. Marsich, M. Benincasa, T. Scarpa, S. Semeraro, G. Turco, R. Gennaro, S. Paoletti, Biomacromolecules **10**(6), 1429–1435 (2009)
13. D. Tobler, L. Warner, Med. Devices Diagn. Ind. **27**(5), 164–169 (2005)
14. A. Kekljes, M. Marinović-Cincovoć, Z. Kačarević-Popović. J. Nedeljković, Eur. Polym. J. **43**, 2171–2176 (2007)
15. Report of the 1st RCM on nanoscale radiation engineering of advanced materials for potential biomedical applications. (Vienna, Austria, 30 March–3 April 2009), 127–134
16. Z. Kačarević-Popović, S. Tomić, A. Krklješ, M. Micic, E. Suljovrujić, Radyolytic synthesis of Ag-poly(BIS-co-HEMA-ci-IA) nanocomposites. Radiat. Phys. Chem. **76**, 2661–2665 (2007)
17. A. Krklješ, J.M. Nedeljković, Z.M. Kačarević-Popović, Polym. Bull. **58**, 271 (2007)
18. H. Kaplan, A. Guner, J. Appl. Polym. Sci. **78**, 994–1000 (2000)
19. S. Benamer, M. Mahlous, Nucl. Instr. Meth. Phys. Res. B **248**, 284–290 (2006)
20. T. Kokubo, H. Kushitani, S. Sakka, T. Kitsugi, T. Yamamuro, J. Biomed. Mater. Res. **24**, 721 (1990)
21. C.L. Bell, N.A. Peppas, J. Control Release **37**, 277–280 (1995)

22. N.A. Peppas, Pharm. Acta. Helv. **60**, 110–111 (1985)
23. H.S. Shin, H.J. Yang, S.B. Kim, M.S. Lee, J. Coll. Interf. Sci. **274**, 890–894 (2004)
24. H.C. Can, B.K. Denizli, S. Kavlak, A. Guner, Radiat. Phys. Chem. **72**, 483–488 (2005)
25. V. Vodnik, J. Nedeljković, J. Serb. Chem. Soc. **65**(3), 195–200 (2000)
26. J.A. Stammen, S. Williams, D.N. Ku, R.E. Guldberg, Biomaterials **22**, 799 (2001)
27. R.S. Harland, N.A. Peppas, Pol. Bull. **18**, 553–556 (1987)
28. P.L. Ritger, N.A. Peppas, J. Control Release **5**, 23 (1987)

Part VI
Nanofluids and Flows at Nanoscale

Revealing Magnetite Nanoparticles Aggregation Dynamics – A SLS and DLS Study

Dan Chicea

Abstract Fe_3O_4 nanoparticles in aqueous suspensions are not stable but aggregate, tremendously changing the rheological properties of the nanofluid. Modified version of both the Static Light Scattering (SLS) setup and of Dynamic Light Scattering (DLS) setup experiment were used to monitor Fe_3O_4 nanoparticle aggregation in aqueous diluted suspension. The experiments are described in detail and the variation of the average aggregate diameter in time is presented in this work.

1 Introduction

The heat transfer properties of a fluid were found to be considerably enhanced by adding a small amount of nanoparticles [1]. Such a suspension is currently named a nanofluid; it is a relatively new notion and was first mentioned by Choi in 1995 [2].

The nanoparticles have a continuous, irregular motion in nanofluids, which is the effect of several factors such as gravity, Brownian force, buoyant force and friction force between fluid and the particles. The irregular nanoparticle motion in the fluid is the cause the remarkable enhancement of heat transfer properties of the nanofluids [3–6]. The irregular motion directly depends of the particle dimension, therefore the particle size distribution dictates the rheological properties of the nanofluid.

In aqueous suspensions nanoparticles aggregate forming clusters of colloids with an isotropic shape, most probably caused by van der Waals attractions and weak magnetic attractions, which are both of the order of kT [7]. Cobalt nanoparticle are reported to aggregate in "bracelets" [8] which are distinct from the micrometer-sized rings created by rapidly evaporating films of dispersed nanoparticles, with regard to ring size (typically 5–12 particles and 50–100 nm in diameter). Reference [9] describes the preparation of robust micrometer size ring structures on mica surfaces. Ring shaped clusters of Co-PFS were patterned onto a thin gold film sputtered onto

D. Chicea (✉)
Department of Physics, University Lucian Blaga of Sibiu, Dr. Ion Ratiu Str. 7-9,
Sibiu 550012, Romania
e-mail: dan.chicea@ulbsibiu.ro

A. Aldea, V. Bârsan (eds.), *Trends in Nanophysics,* Engineering Materials,
DOI 10.1007/978-3-642-12070-1_15, © Springer-Verlag Berlin Heidelberg 2010

a silicon wafer that had been primed with a 5 nm layer of titanium as is reported in [10]. The clusters mentioned in [10] have diameters between 0.6 and 12 μm.

Nanoparticles are extensively considered for biomedical applications because the living cells have dimensions of the order of microns and parts of the order of tens to hundreds of nanometers. Proteins are even smaller, having dimensions around 5 nm. For this reason it was natural to imagine that nanoparticle structured materials can be used in many ways to investigate, to modify living cells or to deliver certain substances or drugs to them without perturbing much the cells. Thus many practical applications were developed in the last years and are nicely presented in [11, 12] and in many other review papers, not cited here.

The shape of the nanoparticle is more often spherical but it can be cylindrical or plate-like; other shapes are possible. The size and size distribution are crucial in some cases, for example if penetration through a pore structure of a cellular membrane is required. Fe_3O_4 nanoparticles are of special interest for biomedical applications because they are not toxic and they can be metabolized by living organisms. They present a major inconvenient though, as they aggregate very fast in diluted aqueous suspension. As the human body fluids are aqueous solutions, transporting Fe_3O_4 nanoparticles through human blood to a tumor or to a specific organ using magnetic field might create serious problems as the nanoparticles aggregate during the transportation process through the blood vessels. Once big aggregates are present, the rheological properties of the suspension are different and the suspension ceases to be a nanofluid any more.

Moreover, the size and size distribution are becoming extremely critical when quantum-sized effects are used to control material properties, therefore establishing a fast technique for measuring the size of the nanoparticles in suspension and for monitoring the aggregates formation rate is of interest.

The most accurate and also the most expensive technique used to obtain a very high resolution image of an emulsion containing nanoparticles is the Transmission Electron Microscopy (TEM). The image is processed and the size distribution of the nanoparticles is the outcome. An example of using TEM is [13] but other hundreds of papers could be cited to have successfully used the technique to characterize the nanoparticle size distribution.

Other currently used methods for nanoparticle sizing are Atomic Force Microscopy (AFM), and ferromagnetic resonance (FMR), for magnetic nanoparticles. A comparison of the TEM with the AFM results is presented in [14]. The results in [13] reveal that the AFM measured nanoparticle diameter appears to be reduced with 20% and the standard deviation appears to be increased with 15%. The differences in the diameter and in the standard deviation findings were associated with the AFM tip and the nanoparticle concentration on the substrate. The AFM technique, as the TEM, requires expensive equipment and a delicate, time consuming procedure to prepare the sample.

Another technique currently used in measuring the nanoparticle diameter is the X-ray powder diffraction. In the seminal paper [15] it is stated that when subject to X ray beam "every crystalline substance gives a pattern; the same substance always gives the same pattern; and in a mixture of substances each produces its pattern independently of the others." The powder diffraction method is thus ideally suited for characterization and identification of polycrystalline phases. Today about 50,000 inorganic and 25,000 organic single component, crystalline phases, diffraction patterns have been collected and stored as standards. Moreover, the areas under the peak are related to the amount of each phase present in the sample. Most important for particle sizing, the Scherrer equation, obtained in [16], is used frequently in X-Ray analysis, particularly powder diffraction, of materials. It relates the peak full width at half maximum of a specific phase of a material to the mean crystallite size of that material. This technique assumes that the nanoparticle size is the same as the size of the crystallite, which is not the case for magnetic nanofluids, which form clusters in certain conditions.

While the TEM above mentioned offers the best resolution, the sample requires specific preparation and can not be used for monitoring nanoparticle aggregation dynamics, simply because the very thin sample must be placed in vacuum. The X ray diffraction and the AFM techniques require a solid sample even if they are carried on in air at atmospheric pressure, but are slow, in respect of the time scale of nanoparticle aggregation, therefore they are not suited for investigating the aggregation process.

A lower cost alternative to these techniques, and to others not mentioned in this introductory part, is based on the fact that the nanoparticles have a continuous, irregular motion in nanofluids, which is the effect of several factors such as Brownian force, buoyant force, friction force between fluid and the particles and gravity, which becomes significant for micron sized particles but can be neglected for nanometer sized particles [9, 10, 17]. The method is called Dynamic Light Scattering (DLS) or Photon Correlation Spectroscopy (PCS) and its physical principles are explained in [18–20] and other papers published subsequently. Details on a modified version of DLS are presented in the next section, together with the results on monitoring the aggregates dimension in time during a dilution process.

A different approach consists of recording the far field and performing a statistical analysis of the speckled image. The speckled image appears as a result of the interference of the wavelets scattered by the scattering centers (SC hereafter), each wavelet having a different phase and amplitude in each location of the interference field. The image changes in time as a consequence of the scattering centers complex movement of sedimentation and Brownian motion giving the aspect of "boiling speckles" [18, 19]. In papers like [20], an optical set-up is used to measure the correlation function in the near-field, and reveals the near-field speckle dependence on the particles size. The work reported in [21, 22] uses a transmission optical set-up to measure the far field parameters like contrast and speckle size revealing that they are related to the average particle diameter. Reference [23] revealed a strong variation of the average speckle size and contrast with the concentration of the scattering centers. In a diluted aqueous suspension as aggregates are formed, both the concentration and the size of the scattering centers change in time, therefore these far field parameters are not suited for monitoring the nanoparticle aggregation process.

Another system that contains scattering centers that aggregate and presents a particular interest in medicine is the human blood. For several decades the basic method of studying platelets aggregation has been the Born technique based on the increase in light transmission [24]. In the last years light scattering techniques were extensively used as alternative methods. The first idea was to use the information offered by the backward scattered light [25], technique that works well in the single scattering regime, recording the light scattered backward by separate particles or their aggregates. Other development reported in the literature was applied to aggregation studies [26].

Another interesting procedure for monitoring the particle aggregation in human rich platelet plasma was presented in [27] and consists of monitoring the scattered light intensity variation at a certain small angle in 1.5–4° range. The intensity variation is quantitatively modeled in terms of the two parameters first order Hill function to describe the platelets aggregation kinetics in [27].

An alternative approach used the time dependent forward scattered light recorded using a photodiode array in the angular range 1–15° to describe the kinetics of platelet aggregation [28, 29]. Two angular domains with qualitatively different behaviors were clearly evidenced by the reported measurements. Below 6°, the signal given by the photodiodes increases as the platelets turn into aggregates while the signal for higher angles 6–15° decreases. The effect is explained by de dependence of the light scattering anisotropy by the size of the scattering object.

A different approach is the Static Light Scattering (SLS) which is a technique that measures the intensity of the scattered light to obtain the average molecular weight of a macromolecule like a polymer or a protein. Measurement of the scattering intensity at many angles enables calculation of the root mean square radius, also called the radius of gyration [30]. For static light scattering experiments, a high intensity monochromatic light, usually a laser, is launched in a solution containing the macromolecules. Usually many detectors are used to measure the scattering intensity at many angles simultaneously. Data analyses is carried on using Guinier plot, Kratky plot, Debye plot or Zimm plot, complete or partial.

In this work a modified version of Static light scattering, using the Henyey-Greenstein phase function is used and presented in detail. Unlike the methods briefly described here, the variation of the scattering anisotropy parameter with the diameter of the scattering center (SC hereafter), as results from Mie calculation, is used to describe the aggregation process in a diluted nanofluid. Details on the theoretical phase function and on the experimental setup and results are presented in the fourth section.

The next sections present the recipe used in nanofluid synthesis, the modified version of the DLS setup and data processing procedure with results, the modified version of the SLS, results and a conclusion.

2 Nanofluid Preparation

The nanofluid preparation does not require any special equipment and was carried on at room temperature, which was 22°C. The reagents used were: $FeCl_3 \cdot 6H_2O$, $FeCl_2 \cdot 4H_2O$, ammonium hydroxide ($NH_3[aq]$) 25%, citric acid ($C_6H_8O_7$), all produced by Merck, Darmstadt. All chemicals used were p.a. grade.

The solutions were prepared just before the determinations, in order to prevent their contamination with the atmospheric oxygen. The concentrations were 1M for $FeCl_3$, 2M for $FeCl_2$, 0.5 M for ammonium hydroxide and 25% for citric acid. 4mL of the 1M $FeCl_3$ solution and 1 mL of the 2M $FeCl_2$ solution were poured into a 150 mL beaker. While continuously stirring the iron chloride solution, 50 mL of the 0.5 M ammonium hydroxide were slowly added, dropwise, in such a way that the process lasted for 5 min. A black precipitate formed during the slow addition, the precipitate being magnetite ($Fe_2 + Fe_3 + 2O_4$). A strong magnet was placed under the beaker. It accelerated the precipitation of magnetite out of the solution, and the water become clear. Keeping the magnet on the bottom of the beaker, the excess water was decanted.

The magnetite was rinsed three times by adding deionized water, using the magnet to settle the magnetite, and discarding the clear water, to remove the excess ammonium hydroxide from the particles. A viscous fluid was the result.

At this stage of the synthesis procedure the nanofluid, which in this case is a ferrofluid, was stabilized. This was accomplished by adding 1 mL of the 25% citric acid and mixing the ferrofluid for 2 min. Overall the chemical reaction was [31]:

$$2FeCl_3 - FeCl_2 - 8NH_3 - 4H_2O - Fe_3O_4 - 8NH_4Cl \quad (1)$$

The output was a viscous, dark brown fluid having the volume ratio of 33%. Details on nanofluid synthesis and rheological properties are presented in [31].

3 The Dynamic Light Scattering Technique (DLS)

Dynamic light scattering (DLS) is a well established technique for measuring particle size over a size range from nanometers to microns. As previously stated, the light scattered by a suspension presents fluctuations [18, 19]. By placing a detector at a certain angle and recording the scattered light intensity a time series is recorded. As proved in [32, 33] the width of the autocorrelation function of the time series is proportional to the diffusion coefficient, which, on its turn, depends of the particle diameter. This leads to a fast procedure for measuring the particle diameter. An improved version is described further on in this subchapter.

The early experimental works [34, 35] and the later theoretical treatises [36–38] proved the assumption that the power spectrum of the intensity of the light scattered by particles in suspension can be linked to the probability density function (hereafter PDF). This link between the PDF and the power spectrum is a consequence of the translation of the relative motion of the scattering particles into phase differences of the scattered light. Thus spatial correlations are translated into phase correlations, which is manifested in the usage of the Wiener-Khintchine-Theorem, relating the power spectrum to the autocorrelation of a process. The phase correlations lead to fluctuations in the intensity of the scattered light recorded using a detector and a data acquisition system, in a typical experimental setup as presented in Fig. 1.

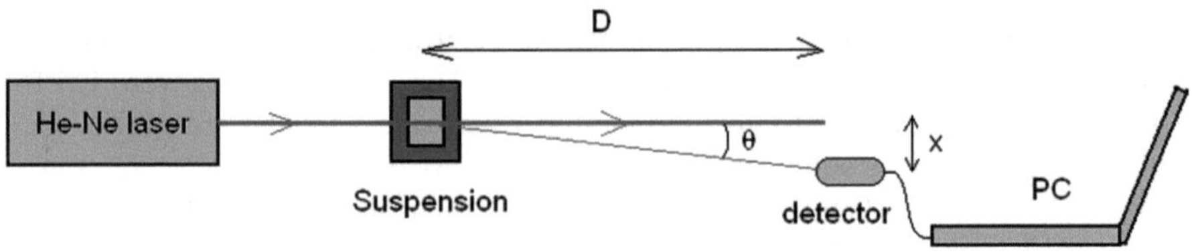

Fig. 1 A typical DLS experimental setup

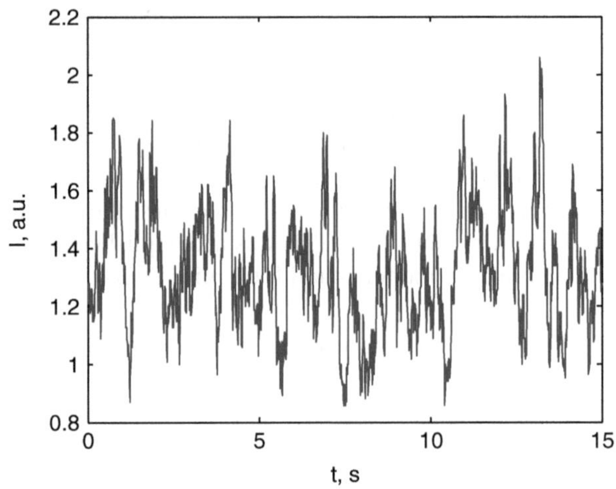

Fig. 2 A sequence of a time series recorded for sample lm9-3

A sequence of a time series recorded for sample lm9-3 is presented in Fig. 2. By subtracting the average intensity from the recorded time series and calculating the square of the intensity we obtain the power time series. The Fourier transform of the power time series is the power spectrum. We can compare the spectrum calculated from the experimental data with the theoretically expected spectrum, namely the functional form of the Lorentzian line $S(f)$ in Eq. (2).

$$S(f) = a_0 \cdot \frac{a_1}{(2\pi f) + a^2{}_1} \tag{2}$$

The Lorentzian line $S(f)$ has two free parameters a_0 and a_1 and is fit to the power spectrum using a non-linear minimization procedure to minimize the distance between the data-set and the line. We notice that a_0 enters linearly, thus only performs a scaling of the function in the range, which translates into a shift in the logarithmic representation. The a_1 parameter enters nonlinearly into the function. Its effect in the loglog scaled plot can approximately be described as a shift along the frequency axis. The possibility to fit the whole function is advantageous compared to the alternative method described in [19, 34, 35], where the $f_{1/2}$ (the frequency where half-maximal height is reached) was measured, since it takes more data points into account, thus increasing the quality of the fit.

Once the fit is completed and the parameters are found, the diameter of the SCs can be assessed as the double of the radius R. The radius can be derived as a function of the fitted parameter a_1 and other known quantities using Eq. (3):

$$R = \frac{2k_B T K^2}{6\pi \eta a_1} \quad \text{where} \quad K = \frac{4\pi n}{\lambda} \cdot \sin\left(\frac{\theta}{2}\right) \tag{3}$$

In Eq. (3) k_B is Boltzman's constant, T is the absolute temperature of the sample, η is the dynamic viscosity of the solvent, θ is the scattering angle, n is the refractive index of the scattering particles and λ is the wavelength of the laser radiation in vacuum.

As already mentioned the work described in this chapter was carried on using an experimental setup as described in Fig. 1. The wavelength was 633 nm, the light source was a He-Ne laser operating continuously and the power was 2 mW. The DLS experiment was carried on at 20°C. The cuvette-detector distance D was 0.615 m and x was 0.03 m making the scattering angle θ equal to $2°\ 47'\ 34''$. This is not typical for DLS where a bigger angle is chosen, usually 90°. The reason for choosing such a small angle is to shift the rollover point in the Lorentzian line towards smaller a_1 values, hence smaller frequencies, where the noise is considerably smaller.

A 900 s time series was recorded using the values of the parameters describes above. As the purpose of the experiment was to monitor the nanoparticle aggregation process, a program was used to slice the time series into chunks of data, using the desired time interval for each data chunk or slice, which is given as an input parameter. We should note however that using a small time span for each data slice will apparently increase the accuracy of the monitoring process but will decrease the precision of the Lorentzian function Eq. (2) fit as the amount of data to be fitted is smaller. Increasing the time span for each data slice will increase the precision of the fit, but will provide a poor information of the variation of the aggregates dimension in time. With this in mind, the time span of each time series slice was chosen to be 15 s.

Special care must be taken to start the recording simultaneously with the beginning of the dilution process and to produce a strong agitation and stirring of the aqueous suspension, otherwise the diffusion of the nanoparticles is slow compared with the time span of the experiment and the results might not be reproducible as different concentrations might occur in the beam area of the cuvette.

For the experiments described in this article a 10 × 10 × 40 mm quartz cuvete was used. First 0.01 mL concentrated nanofluid was introduced into a 5 mL seringe and after it 4.99 mL of deionised water was aspired in. At the time of deionised water aspiration, the time series recording was started. The mixture was injected fast into the cuvette producing turbulences. Right after that a certain fluid quantity was aspired and injected back. The procedure of aspiring and injecting fast the suspension was repeated three times in order to obtain a homogeneous suspension. The amount of aspired fluid was small enough that the laser beam remained under the upper surface of the fluid at all times. Anyway, due to the turbulences produced by the aspiration procedure the velocity of the SCs was different from the velocity of the natural Brownian motion and that can lead to false DLS particle size results. For this reason the first 7 s from the first slice of the time series was not processed as previously described.

Prior to the experiment the concentrated nanoparticle suspension was diluted in 25% citric acid, in order to prevent aggregation and a DLS time series was recorded. The time series was analyzed using the procedure described above. The PSD (*black line*) and the fitted Lorentzian line for the time series recorded on sample lm9-3 are presented in Fig. 3.

The parameters of the Lorentzian line found from the fit are: $a_0 = 2.27$ and $a_1 = 30.00$. Using Eq. (2) we found that the SCs have an average diameter of 18 nm. The PSD (*black line*) and the fitted Lorentzian line for the first time series, allotted to time $7+4 = 11$ s are presented in Fig. 4. The parameters of the Lorentzian line that produce the best fit are: $a_0 = 5.0034$ and $a_1 = 1.2615$. Using Eq. (3) we found that the SCs have an average diameter of 418 nm, considerably bigger than the average diameter of the SCs measured in the concentrated suspension.

The fitting procedure was repeated for all the consecutive time slices recorded during the experiment and the average diameter was calculated. The variation of the average diameter in time calculated using the modified version of the DLS particle sizing, as described above, is presented in Fig. 5.

We should notice, however, that the diameter we measured using the procedure described above, which is basically a DLS using nontypical values for the experimental parameters and for the detection system, is the hydrodynamic diameter, which is slightly different from the physical diameter. Moreover, the presence of particles with a wide diameter distribution changes the shape of the PSD departing it from the theoretical shape, as presented in Fig. 4.

Examining Fig. 5 we notice a very fast increase of the average aggregate diameter in time. Actually after 22.5 s since the dilution was initiated we can no longer

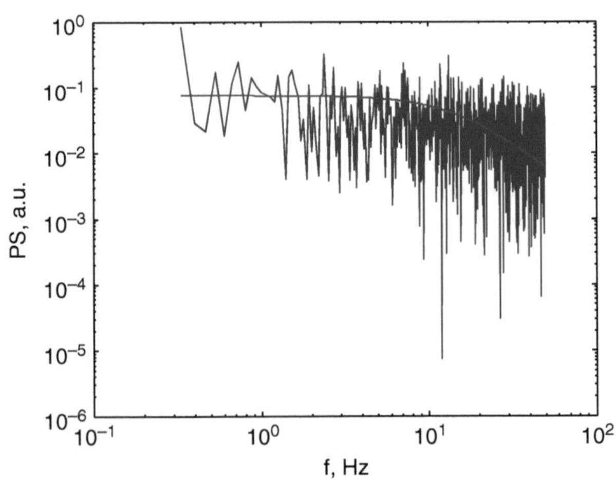

Fig. 3 The PSD (*black line*) and the fitted Lorentzian line (*gray*) for the time series recorded on sample lm9-3 diluted in citric acid

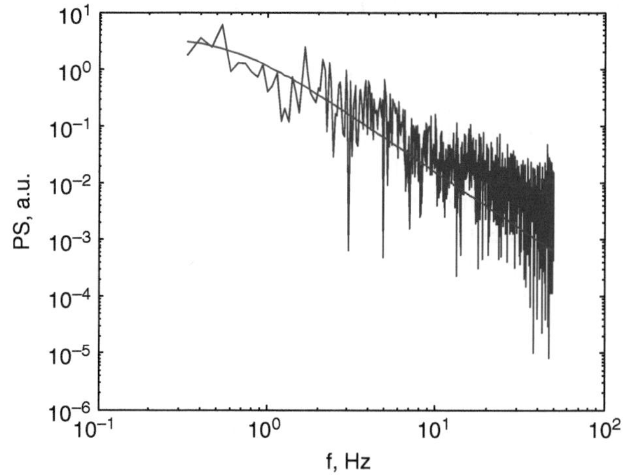

Fig. 4 The PSD (*black line*) and the fitted Lorentzian line (*gray*) for the time series slice allotted to time 11 s since the dilution was started on sample lm9-3

Fig. 5 The variation of the nanoparticle aggregates average diameter in time

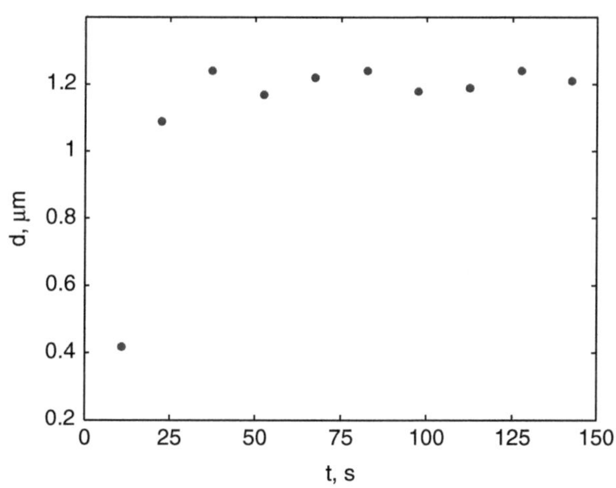

consider nanoparticles in suspension as the average measured diameter was bigger than 1 μm. As time passed the average diameter increasing rate decreased and a plateau can be noticed after 40 s. The average aggregate diameter remained around 1 μm; the variation around the average plateau value might be an artifact of the fitting procedure.

The next section presents the modified version of the static light scattering setup and data processing procedure.

4 Static Light Scattering Particle Sizing

Static light scattering (SLS hereafter) is a technique currently used in physical chemistry that measures the intensity of the scattered light to obtain the average molecular weight of a macromolecule like a polymer or a protein. Measurement of the scattering intensity at many angles allows calculation of the root mean square radius, also called the radius of gyration. By measuring the scattering intensity for many samples of various concentrations, the second virial coefficient can be calculated as well, as described in reference papers like [39–41]. For static light scattering experiments, a high intensity monochromatic light, usually a laser, is launched in a solution containing the macromolecules. One or many detectors are used to measure the scattering intensity at one or several angles.

A simplified experimental version of this technique, assisted by a computation procedure that can be used to monitor the size of the particles in suspension, in a range from 10 nm to 1.5 μm is presented further on.

Several empirical phase functions are frequently used to describe light scattering on a single scattering center; among them the Henyey–Greenstein phase function Eq. (3) and the two-parameters Gegenbauer kernel phase function are the most commonly used [23–25].

$$f(\mu) = \frac{1}{2}\frac{1-g^2}{\left(1-2\mu g+g^2\right)^{\frac{3}{2}}} \tag{4}$$

In Eq. (3) $\mu = \cos\theta, \theta$ is the polar scattering angle and $g = <\mu>$ is the scattering anisotropy parameter.

The Henyey–Greenstein phase function can not be measured directly. A detector has a finite transversal dimension *d* and can be seen from the active part of the cuvette as covering a certain solid angle, hence a polar angle interval $[\theta_1, \theta_2]$. The light intensity that can be measured using a detector is proportional to the integral of the phase function over the polar angle interval $[\theta_1, \theta_2]$, which is:

$$F(\mu_1,\mu_2) = \int_{\mu_1}^{\mu_2} f(\mu)d\mu = \frac{\left(1-g^2\right)}{g}\left[\frac{1}{\sqrt{1+g^2-2g\mu_1}} - \frac{1}{\sqrt{1+g^2-2g\mu_2}}\right] \tag{5}$$

where:

$$\mu_1 = \cos\theta_1, \quad \mu_2 = \cos\theta_2 \tag{6}$$

Moreover, the intensity that can be recorded using a detector is the result of the interference of the wavelets scattered by the scattering centers (SC hereafter), each wavelet having a different phase and amplitude in each location of the interference field. Thus an non-uniformly illuminated image is obtained, currently named speckled image, having a statistical distribution of the intensity over the interference field. The image changes in time as a consequence of the scattering centers complex movement of sedimentation and Brownian motion giving the aspect of "boiling speckles" [42, 43], therefore the signal recorded over a time interval must be averaged in order to compensate the "boiling speckles" fluctuations.

Coming back to the Henyey–Greenstein phase function Eq. (4), we can model light scattering on nanoparticle and nanoparticle clusters as light scattering on spherical particles having diameters in the range 0.050–15 μm and the refractive index equal to 1.5, a typical value for glassy substances. The spheres are considered to be in suspension in water, having the refractive index equal to 1.333.

Figure 6 presents the results of Mie calculations [44, 45] for *g* in the frame of the simple model described above, for different diameters in the range 0.15–2.5 μm, which is of interest for the experiments that were carried on. Examining Fig. 7 we notice a very fast increase of the g value in the very small diameter range and a plateau with small variations for spheres having the diameter bigger than 2.5 μm.

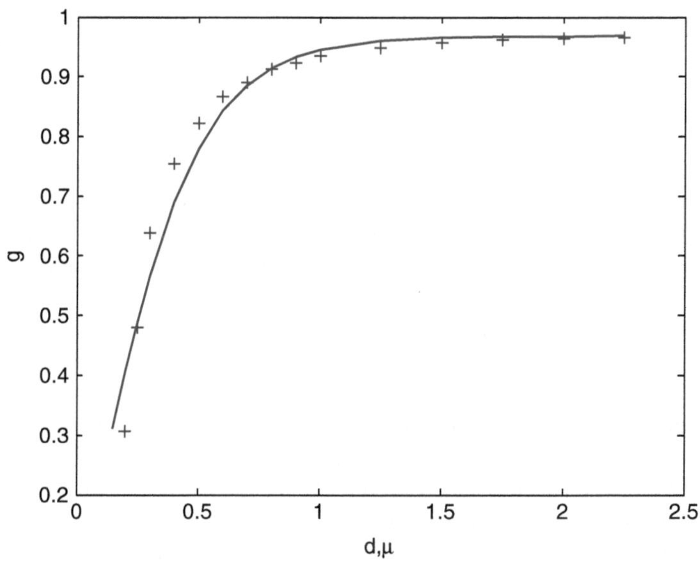

Fig. 6 The variation of the scattering anisotropy parameter g for magnetite spherical particles in aqueous suspension with the nanoparticle diameter. The data points are the calculated data and the *solid line* is the *fitted curve*, described by Eq. (7) using the parameters in Eq. (8)

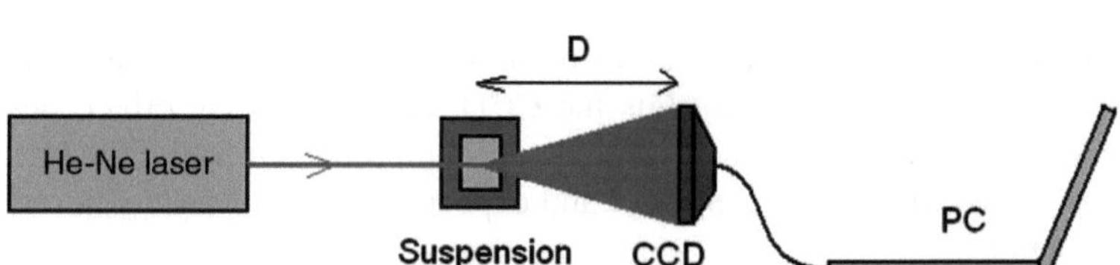

Fig. 7 The experimental setup for the modified SLS experiments, view from above

Several analytical function were fit on the calculated data presented in Fig. 6 and the preferred function is presented in Eq. (7), where d is the scatterer diameter, in μm, g is the dimensionless scattering parameter, a and b are the parameters to be determined by a least squares fit.

$$g(d) = a \cdot \tanh(b \cdot d) \tag{7}$$

The parameters that produced the best fit were:

$$a = 0.9722; \quad b = 2.1427 \tag{8}$$

Reverting Eqs. (7) and (8) we find Eq. (9) and this variation of the scattering parameter g with the scattering particle diameter was used in this work to assess the time variation of the nanoparticle aggregates in aqueous suspension:

$$d = \frac{1}{b} \cdot a \tanh\left(\frac{g}{a}\right) \tag{9}$$

The experimental setup consists of a He-Ne laser having the wavelength of 632 nm and a constant power of 2 mW, a cuvette, a sensitive detector, a CCD and a computer. First the detection system was calibrated. The calibration procedure provides the functional dependence of the grey level recorded by a certain cell on the CCD matrix with the light intensity. A schematic of the experimental setup is presented in Fig. 7.

The main goal of the experimental setup, as presented in [46] is to record the far interference field directly on the CCD and to capture a succession of bitmaps. Thus each individual cell becomes a detector and by extracting the light intensity recorded by a consecutive set of cells in a cell array, a light intensity versus scattering angle profile can be extracted. In order to accomplish this goal, D and the size of a cell on the CCD must be precisely measured. D was adjusted during the experiment in such a manner that the CCD cells covered the desired angular interval, 0.40–2.00°, therefore it was 8 cm. The cell on the CCD is square and the side of the cell is 5.83 μm.

A bitmap of the far-field scattered light on diluted nanofluid suspension is presented in Fig. 8. Examining it, we notice the speckle aspect. The resolution of each bitmap was 480×640 and the averages were performed on consecutive recording sequences of 5 s. A horizontal profile extracted from the bitmap area where the direct beam hits the CCD in Fig. 8 is presented in Fig. 9.

Examining Fig. 8 we notice a decrease of the intensity outside the saturated area where the direct beam hits the CCD, but the speckle effect, described in the previous section makes this type of profile improper for fitting the integral of the Henyey–Greenstein in Eq. (8) on it and expecting accurate results in respect of the g parameter.

In order to compensate the fluctuations that give the speckle aspect a movie was recorded for each sample subject to particle aggregation study. The framerate must

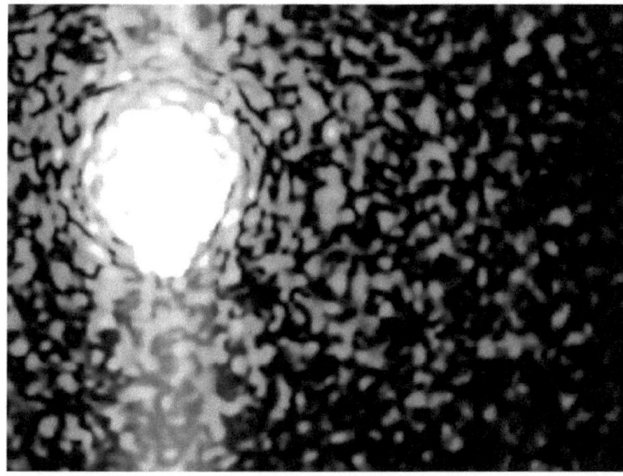

Fig. 8 A bitmap of the far field scattered light on diluted nanofluid

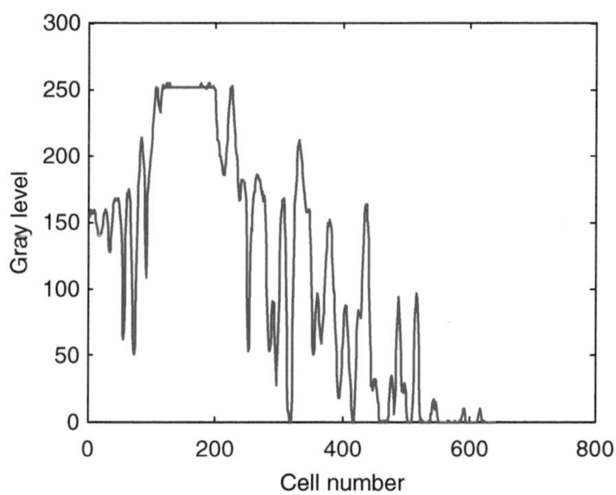

Fig. 9 A horizontal intensity profile extracted from the bitmap in Fig. 8, corresponding to the 240-th row

be adjusted accordingly to the size of the SCs. The bigger is the diameter, the slower is the fluctuation rate.

The movie was recorded using 30 frames per second. The movie is processed later on using a computer program written for this purpose [46]. The program reads sequences of successive frames. The number of frames in a sequence was chosen to be 150, therefore averages on time steps of 5 s were processed and for each sequence an average value of the *g* parameter was finally calculated.

From each frame of the movie the intensity levels recorded by an array of cells being less wide (high) as the spot, 80 pixels for the experiments reported here, and as long (wide) as the width of the CCD minus the saturated area, 420 for the experiments reported here, is extracted. Averages are performed on each individual sequence of frames and on each vertical set of 80 pixels. The light intensity is finally calculated using the calibration curve and an averaged intensity profile versus angle is produces for each sequence, hence time interval from the beginning of the

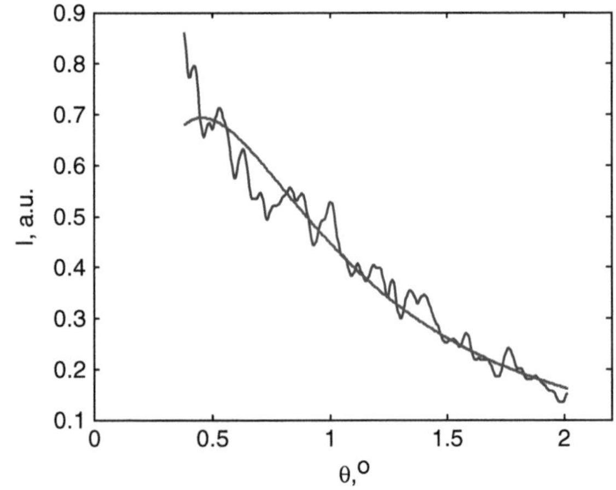

Fig. 10 The intensity profile for the time interval 65–70 s and the result of the Henyey Greenstein phase function fit, using Eq. (5), the smooth line

recording. The abscissa is expressed in angles. An averaged profile is presented in Fig. 10 revealing that the speckle fluctuations were considerably reduced.

As the program reads the whole movie and processes it, an averaged profile is produced for each sequence and allotted to the middle of the time interval it was extracted from.

Another program written for this purpose reads each averaged scattering profile in batch mode and fits the integral of the Henyey–Greenstein in Eq. (8) multiplied by a constant C proportional to the laser beam intensity on it. The program outputs the values of C and g that produces the best fit for each profile, thus the variation of the scattering parameter g with the time elapsed from the beginning of the experiment can be plotted and analyzed. The sample that was analyzed was a suspension of magnetite nanoparticles having citric acid as surfactant, manufactured as described in Sect. 1. The concentration was 0.17% volume ratio nanoparticles in deionized water at 20°C. The results of the preliminary experiments aiming to monitor the nanoparticle aggregation in aqueous suspension reveal that the aggregation process is very fast [46], therefore the experiment was conducted differently than in [46]. Instead of dissolving the concentrated nanofluid in a beaker and homogenizing it, the dilution was performed in situ, as described in Sect. 2. During all the operation the cuvette remained attached to the holder, the laser, the computer, the CCD and the data acquisition systems were in stand by. As soon as the 5 s operation was completed, recording was started, therefore a 5 s correction on the time axis was made on the plots in Figs. 11 and 12.

Figure 10 presents the average intensity profile for the 5 s sequence allotted to $t = 67.5$ s, and the result of the fit using Eq. (5), as described in this subchapter. We notice that the scattering parameter is 0.9497, corresponding to spherical particles with a diameter around 1.04 microns or bigger, according to Fig. 6.

The variation of the scattering parameter with the time elapsed from the beginning of the experiment, calculated using the procedure described in this section, is

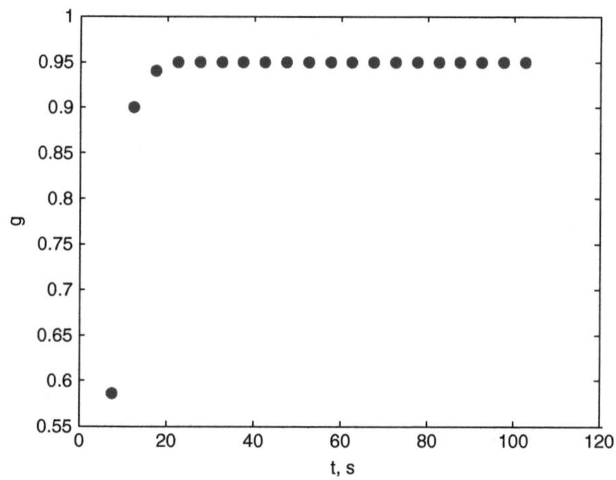

Fig. 11 The variation of the scattering parameter with time

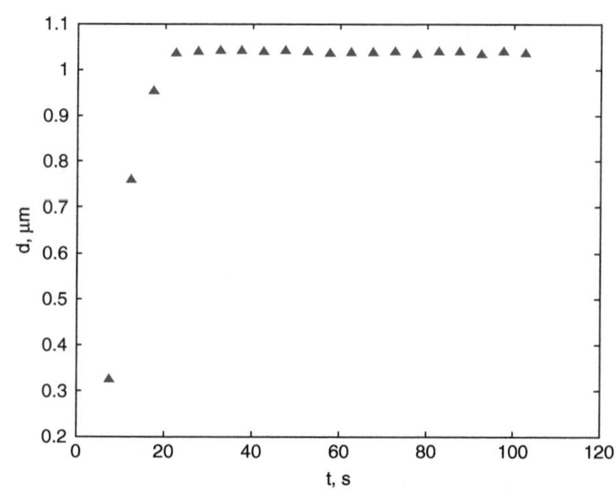

Fig. 12 The variation of the average diameter with the elapsed time

presented in Fig. 11. Examining Fig. 11 we notice a fast increase of the light scattering parameter g in time after nanoparticle dillution started followed by a plateau.

The fast increase can be explained by aggregates formation in the diluted nanofluid. The overall increase actually lasted for 15 s since dilution started.

The variation of the scattering parameter with the elapsed time appears different than the variation recorded during the preliminary experiments reported in [46]. In [46] recording started after 1 min and the scattering parameter apparently presents an increase followed by a decrease. The explanation lays in the scale of the g axis on the plot in [46], which ranges from 0.9814 to 0.9834, while in Fig. 11 the scale ranges from 0.55 to 1. During the preliminary experiments in [46] simply the fast increase was not recorded, as recording started at second 60 since the beginning of the dilution process. The small variations of the plateau in Fig. 11 are consistent with the variations in [46] and can be noticed when selecting a narrower range for the ordinate (g axis).

Once the variation of the anisotropy parameter with time is calculated, the average diameter can be assessed using Eq. (7) in Sect. 2. The variation of the average diameter with the elapsed time is presented in Fig. 12.

Examining Fig. 12 we notice a fast increase of the average diameter of the scattering centers, which for this sample are the aggregates. The average diameter of the light scatterers increased to 0.94 μm in less than 20 s from the beginning of the dilution and presents small fluctuations after the first 100 s, fluctuations that can be an artifact caused by the averaging process performed on a relative small number of frames, 150.

Increasing the number of frames that are averaged will reduce the speckle aspect, hence the apparent fluctuations in the calculated average diameter, but will cease to reveal the fast increase during the first seconds.

5 Discussion

In order to understand the significance of the average diameter previously mentioned and the variation of the average diameter in time, we should keep in mind that for small particles, comparable with the wavelength, Rayleigh approximation can be used to describe light scattering [47]. As the particle diameter d increases, at constant volume ratio, the nanoparticle number N should vary with the diameter d as:

$$N = \frac{V_{\text{nano}}}{\frac{4\pi}{3}\left(\frac{d}{2}\right)^3} \tag{10}$$

where V_{nano} is the total nanoparticle volume in suspension, a constant value. As light scattering on nanoparticles is a Rayleigh type scattering, the light intensity scattered by one individual particle is proportional to d^6 [30]. The average intensity scattered by all the nanoparticles in the sample and recorded at a constant angle is therefore proportional to d^3, thus increasing with the nanoparticle cluster diameter, as revealed by Eq. (8).

$$< I > \approx N \cdot < I(\theta) > \approx \left(\frac{d}{2}\right)^3 \tag{11}$$

We notice from Eq. (11) that the light intensity scattered by one cluster having a diameter in the range of microns is roughly 10^6 bigger that the intensity scattered by one nanoparticle having a diameter around 10 nm, therefore the far field landscape becomes dominated by light scattered by clusters, as soon as they appear.

The light scattering parameter g calculated using the procedure described in the previous section is actually the scattering parameter of light scattered by clusters, therefore the average diameter is different of the arithmetic average that can be calculated using TEM on a thin emulsion prepared from the nanofluid. Consequently both procedures described above are not sensitive in respect of measuring the amount of nanoparticles that turned into aggregates, but to reveal the presence of aggregates and to estimate the aggregation rate and the aggregates size.

Nevertheless, the experiment reveals that the procedure can be used for a semi-quantitative assessment of the time elapsed from the beginning of the dilution to the moment when cluster formation is completed in an aqueous suspension. Examining the results presented in Figs. 5 and 12 we notice that during the first 15 s a significant amount of nanoparticles turned into aggregates and the aggregate dimension was growing. After 20 s the aggregates dimension reached a plateau, which was found to be 1.2 μm by the modified DLS and 1.05 μm by the modified SLS. The further size increase can not be monitored using SLS, because the scattering parameter presents very small variations as the diameter increases further on, according to Fig. 6.

Another detail regarding the unconventional SLS particle sizing procedure described in this paper is that it is completely different of the already classical Dynamic Light Scattering (DLS), also known as Photon Correlation Spectroscopy (PCS) [25, 26], implemented on commercially available particle sizing devices. In the DLS technique, the dynamic information of the particles is derived from an autocorrelation of the intensity trace recorded during the experiment. The second order autocorrelation curve is generated from the intensity trace; it decays exponentially and the average decay rate is proportional, at constant angle and wavelength, with the diffusion coefficient of the particles in suspension, which, on its turn depends of the rheological properties of the base fluid and of the diameter of the particles in suspension. Therefore DLS provides information on the hydrodynamic diameter of the particles, not on the physical diameter. The modified SLS technique described in this paper does not use the time variation of the scattered light intensity but time averages, which are related to the intrinsic light scattering parameters of each individual particle, regardless whether it is in motion or not. Therefore, the SLS procedure described in this paper provides information on the physical diameter of the nanoparticles or aggregates, not on the hydrodynamic diameter, as DLS.

Another difference must be noted regarding the procedure described here. In Fig. 6, the plot of g versus diameter (theoretical curve) presents a plateau for diameters bigger that 2.5 μm, therefore this procedure is not suited for monitoring aggregates bigger than this value, while DLS is sensitive over a considerably wider range.

6 Conclusions

In this work two simple experimental procedures, assisted by a set of computer programs required to process data, are presented. The procedure consists of recording either a time series or an AVI type of movie, using a CCD, containing light scattered at small angles by suspensions over a time interval since the dilution process is initiated. The PSD is computed or the average light scattering intensity versus scattering angle is extracted and the light scattering parameter g is calculated using a least squares fit. The time variation of the average diameter of the aggregates in suspension is the output.

Using this procedure, we found that magnetite nanoparticles having citric acid as surfactant, in aqueous diluted suspension, 0.17% volume ratio, aggregate very fast. During the first 20 s, the aggregates dimension increases and the size reached 0.9 μm, which is extremely important when considering the nanoparticles for biomedical applications.

References

1. P. Vadasz, Heat conduction in nano-fluid suspensions. J. Heat Transfer **128**, 465–477 (2006)
2. U.S. Choi, Enhancing thermal conductivity of fluids with nanoparticles. ASME Fed. **231**, 99–103 (1995)
3. S.P. Jang, U.S. Choi, Role of Brownian motion in the enhanced thermal conductivity of nanofluids. Appl. Phys. Lett. **84**, 4316–4318 (2004)
4. W. Evans, J. Fish, P. Keblinski, Role of Brownian motion hydrodynamics on nanofluid thermal conductivity. Appl. Phys. Lett. **88**, 093116 (2006)
5. Y.M. Xuan, W. Roetzel, Conceptions for heat transfer correlation of nanofluids. Int. J. Heat Mass Transfer **43**, 3701–3707 (2000)
6. R. Prasher, P. Brattacharya, P.E. Phelan, Thermal conductivity of nanoscale colloidal solutions (Nanofluids). Phys. Rev. Lett. **94**, 025901 (2005)
7. K. Butter, P.H. Bomans, P.M. Frederik, G.J. Vroege, A.P. Philipse, Direct observation of dipolar chains in ferrofluids in zero field using cryogenic electron microscopy. J. Phys. Condens. Matter **15**, S1451–S1470 (2003). doi:10.1088/0953-8984/15/15/310
8. S.L. Tripp, S.V. Pusztay , A.E. Ribbe , A. Wei, Off-axis electron holography of self-assembled co nanoparticle rings, J. Am. Chem. Soc. **124**, 7914–7915 (2002)
9. Z. Xiao, C. Cai, X. Deng, Formation of robust mesoscopic ring structures by self-assembly of SiC13-terminated dendrimers. Chem. Commun. **16**, 1442–1443 (2001). doi:10.1039/b104306b
10. S.B. Clendenning, S.F. Bidoz, A. Pietrangelo, G. Yang, S. Han, P.M. Brodersen, C.M. Yip, Z. Lu, G.A. Ozin, I. Manners, Ordered 2D arrays of ferromagnetic Fe/Co nanoparticle rings from a highly metallized metallopolymer precursor. J. Mater. Chem. **14**, 1686–1690 (2004)
11. O.V. Salata, Applications of nanoparticles in biology and medicine. J. Nanobiotechnol. **2**(3), (2004). doi:10.1186/1477-3155-2-3
12. Q.A. Pankhurst, J. Connolly, S.K. Jones, J. Dobson, Applications of magnetic nanoparticles in biomedicine. J. Phys. D **36**, R167–R181 (2003)
13. F. Zhang, S.W. Chan, J.E. Spanier, E. Apak, Q. Jin, R.D. Robinson, I.P. Herman, Cerium oxide nanoparticles: size-selective formation and structure analysis. Appl. Phys. Lett. **80**(27), (2002). doi:10.1063/1.1430502
14. L.M. Lacava, B.M. Lacava, R.B. Azevedo, Z.G.M. Lacava, N. Buske, A.L. Tronconi, P. C. Morais, Nanoparticle sizing: a comparative study using atomic force microscopy, transmission electron microscopy, and ferromagnetic resonance. J. Magn. Magn. Mater. **225**(1–2), 79–83 (2001)
15. A.W. Hull, A new method of chemical analysis. J. Am. Chem. Soc. **41**(8), 1168–1175 (1919). doi:10.1021/ja02229a003
16. A.L. Patterson, The Scherrer formula for X-ray particle size determination. Phys. Rev. **56**(10), 978–982 (1939). doi:10.1103/PhysRev.56.978
17. D. Chicea, Coherent light scattering on nanofluids – computer simulation results. Appl. Opt. **47**(10), 1434–1442 (2008)

18. J.W. Goodman, Laser speckle and related phenomena, vol. 9, in *Series Topics in Applied Physics*, ed. by J.C. Dainty (Springer,New York,NY, (1984), pp. 13–19
19. J.D. Briers, Laser Doppler, speckle and related techniques for blood perfusion mapping and imaging. Physiol. Meas. **22**, R35–R66 (2001)
20. M. Giglio, M. Carpineti, A. Vailati, D. Brogioli, Near-field intensity correlation of scattered light. Appl. Opt. **40**, 4036–4040 (2001)
21. Y. Piederrière, J. Cariou, Y. Guern, B. Le Jeune, G. Le Brun, J. Lotrian, Opt. Expr. **12**, 176 (2004)
22. Y. Piederriere, J. Le Meur, J. Cariou, J.F. Abgrall, M.T. Blouch, Particle aggregation monitoring by speckle size measurement; application to blood platelets aggregation. Opt. Expr. **12**, 4596–4601 (2004)
23. D. Chicea, Speckle size and contrast measurement application in micron-size particle concentration assessment. Eur. Phys. J. Appl. Phys. **40**, 305–310 (2007). doi:10.1051/epjap:2007163
24. G.V.R. Born, Observations on the change in shape of blood platelets brought about by adenosine diphosphate. J. Physiol. **209**, 487–511 (1970)
25. Y. Ozaki, K. Satoh, Y. Yatomi, T. Yamamoto, Y. Shirasawa, S. Kume, Characterization of liposomes carrying von Willebrand factor-binding domain of platelet glycoprotein Ibα: a potential substitute for platelet transfusion. Anal. Biochem. **218**, 284–294 (1994)
26. K. Yabusaki, E. Kokufuta, Aggregation mechanism of blood platelets by time-resolved light scattering. Method Langmuir **18**, 39–45 (2002)
27. I. Turcu, S. Neamtu, C.V.L. Pop, Angular spreading of scattered light used to investigate platelets aggregation International Conference on Isotopic Processes PIM5 (2007)
28. I.V. Mindukshev, I.E. Jahatspanian, N.V. Goncharov, R.O. Jenkins, A.I. Krivchenk, A new method for studying platelets, based upon the low-angle light scattering technique. 1. Theoretical and experimental foundations of the method. Spectrosc. Int. J. **19**, 235–246 (2005)
29. I.V. Mindukshev, E.E. Ermolaeva, E.V. Vivulanets, E.Yu. Shabanova, N.N. Petrishchev, N.V. Goncharov, R.O. Jenkins, A.I. Krivchenko, A new method for studying platelets, based upon the low-angle light scattering technique. 2. Spectrosc. Int. J. **19**, 247–257 (2005)
30. B.H. Zimm, Molecular theory of the scattering of light in fluids. J. Chem. Phys. **13**, 141 (1945). doi:10.1063/1.1724013
31. D. Chicea, C.M. Goncea, On magnetic nanofluid synthesis and physical properties. Optoelectron. Adv. Mater. – Rapid Commun. **3**(3), 185–189 (2009)
32. W. Tscharnuter, in *Encyclopedia of Analytical Chemistry*, ed. by R.A. Meyers (Wiley, New York, NY, 2000), pp. 5469–5485
33. B.R. Ware, W.H. Flygare, The simultaneous measurement of the electrophoretic mobility and diffusion coefficient in bovine serum albumin solutions by light scattering. Chem. Phys. Lett. **12**, 81–85 (1971)
34. N.A. Clark, J.H. Lunacek, G.B. Benedek, A study of Brownian motion using light scattering. Am. J. Phys. **38**(5), 575–585 (1970)
35. S.B. Dubin, J.H. Lunacek, G.B. Benedek, Observation of the spectrum of light scattered by solutions of biological macromolecules. PNAS. **57**(5), 1164–1171 (1967)
36. B.J. Berne, R. Pecora, *Dynamic Light Scattering* (Wiley, New York, NY, 1976)
37. J.W. Goodman, *Statistical Optics* (Wiley Classics Library Edition, New York, NY, 2000)
38. E. Hecht, *Optics* (Addison-Wesley, New York, NY, 2001)
39. A. Einstein, Ann. Phys. (Leipzig) **33**, 1275–1298 (1910)
40. P. Debye, Light scattering in solutions. J. Appl. Phys. **15**, 338 (1944). doi:10.1063/1.1707436
41. B.H. Zimm, Molecular theory of the scattering of light in fluids. J. Chem. Phys. **13**, 141 (1945). doi:10.1063/1.1724013
42. M. Hammer, D. Schweitzer, B. Michel, E. Thamm, A. Kolb, Single scattering by red blood cells. Appl. Opt. **37**, 7410–7418 (1998)
43. M. Hammer, A.N. Yaroslavsky, D. Schweitzer, A scattering phase function for blood with physiological haematocrit. Phys. Med. Biol. **46**, N65–69 (2001)

44. S. Prahl's, Mie online calculator at, http://omlc.ogi.edu/calc/mie_calc.html, as of December 10, 2009
45. S.A. Prahl, M. Keijzer, S.L. Jacques, A.J. Welch, *SPIE proceedings of dosimetry of laser radiation in medicine and biology*. **5**, 102–111 (1989)
46. D. Chicea, Probing particle aggregation in aqueous suspensions by light scattering anisotropy measurements, presented at ROMOPTO 2009. J. Optoelectron. Adv. Mater. **12**, 152–158, (2010)
47. C.F. Bohren, D. Huffman, Absorption and scattering of light by small particles (Wiley, New York, NY, 1983)

Features of Classical and Quantum Fluid Flows Extending to Micro- and Nano-Scales

J.J. Niemela

Abstract This work presents some of the nano-*physics* that presents itself in the turbulent flows of classical and quantum fluids. It also points to the problems faced by experiments when attempting to make measurements at the extremely small scales which exist in these flows. The experimental focus is naturally on fluids at low temperature–in particular gaseous and liquid phases of the isotope ^{4}He– where both intense classical turbulence and quantum fluid flow can be studied.

1 Introduction

Turbulence flow in fluids is a model system for nonlinear systems with many degrees of freedom. In this regard, the advantage is that the equations of motion are presumably known and can under certain conditions be simulated precisely. Turbulent flows pose many difficulties with respect to numerical work precisely because of the large number of scales that need to be resolved, while analytic work is hampered by the intractability of the governing partial differential equations of motion, known collectively as the Navier-Stokes equations. This leaves much of the job to experimental work, and here there have been many advancements in recent years; largely this is due to the application of novel working fluids which allow much higher levels of turbulent activity to be obtained under controlled conditions in the laboratory. The most promising of these non-traditional fluids is low temperature helium, which will be described below. A recent review, from which this work is partially drawn, is that of Niemela and Sreenivasan [1].

As alluded to above, the range of length and time scales grows rapidly with increasing turbulent intensity and experiments must not only produce the turbulence, but also be able to make meaningful measurements of it at all relevant scales. Since the large scales are fixed, the increasing range of scales occurs at the low end, and is

J.J. Niemela (✉)
The Abdus Salam ICTP, Strada Costiera 11, 34014 Trieste, Italy
e-mail: niemela@ictp.it

A. Aldea, V. Bârsan (eds.), *Trends in Nanophysics,* Engineering Materials, 351
DOI 10.1007/978-3-642-12070-1_16, © Springer-Verlag Berlin Heidelberg 2010

limited only by an approach to the hydrodynamic limit (interestingly, an argument can be produced to show that this limit is a moving target that is difficult to reach, due to the fact that the effective mean free path decreases faster with increasing levels of turbulence than does the smallest scale in the flow [2]).

To motivate the discussion, we note that turbulence is widespread, existing in many different contexts, both natural and engineering. From the latter perspective it is important in pollutant dispersal within the atmosphere and in the design and motion of submarines, ships and aircraft and automobiles. In nature it is the norm rather than the exception for fluid flows, especially for large scale flows such as those occurring in the earth's atmosphere and oceans, the convection which takes place in the outer core and so forth. Further beyond our planet, turbulent flow plays a major role in the transport of energy in stars [3] and thus is a ubiquitous phenomena that needs little further justification for our attempts to understand it. As alluded to above, it can be considered a type of paradigm for other complex systems where there is a strong coupling between many degrees of freedom, such as fracture, earthquakes, and possibly even market fluctuations [4]. To better understand the turbulent flows we need first to consider a bit of background on equations used to describe the motion of fluids.

1.1 The Dynamics of Fluids

It is customary to express the governing equations of motion in an Eulerian coordinate system that is fixed in space. We will consider fluid "elements" which are in between molecular and macroscopic dimensions; i.e. they are considered like point particles except that they have a well-defined density ρ. Considering that these fluid particles move at a velocity $\mathbf{u}$, mass conservation can be written as follows:

$$\frac{\partial \rho}{\partial t} + \nabla \cdot (\rho \mathbf{u}) = \mathbf{0}, \tag{1}$$

or more simply as

$$\nabla \cdot \mathbf{u} = \mathbf{0} \tag{2}$$

in the case where the density can be considered constant within the flow. This simplification can be used even in the case where (small) density gradients actually *drive* the flow, as with buoyancy driven convection.

For a Newtonian fluid–i.e. one in which the stress tensor is linearly related to that of the strain– conservation of momentum is given by

$$\frac{D\mathbf{u}}{Dt} = -\frac{1}{\rho}\nabla p + \nu \nabla^2 \mathbf{u} + \mathbf{F_{ext}}, \tag{3}$$

where p is the pressure and $\nu = \mu/\rho$ the kinematic viscosity (here μ is the dynamic or "shear" viscosity). The derivative on the left-hand side of Eq. (3) is the so-called convective derivative $\frac{D}{Dt} = \frac{\partial}{\partial t} + \nabla \cdot \mathbf{u}$ and $\mathbf{F}_{\text{ext}}$ is a body force term which is could be related to forces such as gravity, rotation, or magnetic field, all of which are important in the majority of natural turbulent flows. The principal body force is that of gravity which gives rise to buoyant motions and convection. For this force there is merely one component, in the direction of gravity, and it can be written as $F_{\text{ext}} = g\alpha\Delta T$, where g is the acceleration of gravity, α is the isobaric coefficient of thermal expansion and ΔT is the temperature difference across a layer of fluid in the direction of gravity. Note that $\alpha\Delta T$ represents the relative variation in the fluid density across the layer.

For convective flows (the most common form of turbulence) it is generally assumed that the thermal driving does not affect the pressure (the fluid remains incompressible). This is the so-called Boussinesq approximation [5]. Energy conservation in the Boussinesq approximation is written as

$$\frac{\partial T}{\partial t} = \kappa \nabla^2 T - \mathbf{u} \cdot \nabla T \tag{4}$$

where κ is the thermal *diffusivity*, or thermometric conductivity, of the fluid at temperature T.

The three Boussinesq equations above fully describe fluid motion, determining the velocity, pressure and temperature, and presumably also describe turbulent flows. To be able to generalize the equations they must be non-dimensionalized and we do this by scaling all velocities in the flow by some characteristic velocity U, lengths by some characteristic length L, time by L/U, and pressure by ρU^2. In this case, Eq. (3) can be re-written in non-dimensional form as

$$\frac{\partial \mathbf{u}}{\partial t} + \mathbf{u} \cdot \nabla \mathbf{u} = -\nabla p + \frac{1}{\text{Re}} \nabla^2 \mathbf{u}, \tag{5}$$

where the so-called Reynolds number $\text{Re} = UL/\nu$ appears naturally as a consequence of the scaling of the variables. Here we have considered the simplest case of pressure driven flows without additional body force terms. Of course, initial and boundary conditions have to be specified and this leads to the infinite variety of fluid flows governed by such an equation with one free parameter. Specifically, if we have two flows that are geometrically similar, then in steady state they are also *dynamically* similar if the corresponding Re is the same for both. This is true regardless of the fluids used, the length scales (which could vary from astrophysical dimensions to nanoscales), or the velocities, viscosities, etc. Considering that the visible universe is composed mainly of fluids, Eq. (5) then describes most of what we can see. The existence of similarity parameters also means that scale model testing facilities such as wind tunnels or towing tanks can give results that are relevant to a full scale object of interest such as a commercial aircraft or a submarine. Physically, the Reynolds number is related to the ratio of the inertial forces (i.e., the term $\mathbf{u} \cdot \nabla \mathbf{u}$ in the convective derivative), and viscous forces (the term $\nu \nabla^2 \mathbf{u}$).

The Reynolds number is not the only possible similarity parameter: other parameters must be considered for example when external body forces are present. In particular, the inclusion of the buoyancy force into Eq. (3) yields a new similarity parameter namely the Rayleigh number

$$\mathrm{Ra} = g\alpha\Delta T L^3/\nu\kappa, \tag{6}$$

where g is the acceleration due to gravity, α, ν and κ, respectively, the isobaric thermal expansion coefficient, the kinematic viscosity and thermal diffusivity of the fluid, and ΔT and L are the characteristic temperature difference and the length scale of the flow (in this case the relevant scale is the vertical height of the fluid layer).

In physical terms, the Rayleigh number measures the ratio of the rate of potential energy release due to buoyancy to the rate of its dissipation due to thermal and viscous diffusion [6]. Another similarity parameter which arises as well in this case is the Prandtl number

$$Pr = \nu/\kappa \tag{7}$$

which can be expressed as the ratio of molecular time scales due to thermal diffusion ($\tau_\theta = L^2/\kappa$) and momentum diffusion ($\tau_v = L^2/\nu$).

1.2 Fluid Turbulence

While thermal turbulence is the more ubiquitous form, considering also astrophysical fluid flows, it is much simpler to consider isothermal, pressure-driven flows for which the Reynolds number is the only major similarity parameter. In this case, the Reynolds number defines the turbulence level; as we can appreciate from the discussion following Eq. (5), turbulence occurs when when inertia forces are large relative to the ability of the fluid to dissipate fluctuations. Turbulence is characterized by the existence of fluctuations on many length and time scales, sometimes referred to as "eddies", starting from the largest scale which cannot be larger than the size of the flow region, meaning the sample container in an experiment, and the smallest scale which is determined, as we shall see, by the Reynolds number, and which is the scale at which viscous dissipation becomes dominant and flow is converted into heat energy. A Reynolds number is not unique and can be defined for all size eddies, which have their own characteristic lengths and velocities. The energy of the large eddies is not lost to viscosity; it is instead transferred to other scales through nonlinear interactions in a cascade process (meaning interactions are more likely to occur between not-so-different scales). The energy in fact cascades from large scales to small scales since there it is dissipated. The smallest scale is then referred to usually as the *dissipation* or Kolmogorov scale η. The cascade process in a statistically steady state, i.e., where there is a steady input of energy at some large scale ℓ, results in a range of scales in k-space where the energy transfer occurs without viscous dissipation and this is commonly known as the *inertial* range, meaning precisely that–the range over which inertial forces outweigh dissipation or where the local Reynolds number is "large". The rate of energy dissipation per unit mass, ε, clearly must be the same as rate of energy input in steady state and is therefore independent of the viscosity. in other words, by reducing the fluid viscosity, we allow the cascade to continue to smaller scales where it can be dissipated (where the local Reynolds number becomes of order unity).

Dimensional analysis shows that

$$\eta = (\nu^3/\varepsilon)^{1/4} = \ell \mathrm{Re}_\ell^{-3/4}, \tag{8}$$

where $\mathrm{Re}_\ell = u\ell/\nu$ is the characteristic Reynolds number of the large scale. Thus for any given ℓ, the dissipation or Kolmogorov scale becomes increasingly smaller as the Reynolds number increases and so does the characteristic time scale. Thus strong turbulence can generate flows at arbitrarily small scales and in the micro- and nano-limit this poses severe requirements on measurement technology if all scales are to be resolved. As we shall see this has already become an issue in some laboratory measurements. Eq. (8) also tells us that the range of scales, namely, ℓ/η grows as $\mathrm{Re}^{3/4}$ or as $\mathrm{Re}^{9/4}$ in three dimensions. This leads to difficulties in numerical analysis: the state of the art [2] for fully resolved direct numerical simulation of turbulent flows is about $\mathrm{Re} \sim 10^4$ whereas oceanic and atmospheric flows are typically characterized by Reynolds number 3 orders of magnitude higher, as is also the case for flow about commercial jetliners in routine flight.

1.2.1 The Energy Cascade

It is customary to consider the energy cascade process in terms of the spectral wavenumber k. Here the energy spectrum is denoted by $E(k)$, where k is the wavenumber and $E(k)dk$ is the kinetic energy contained between k and $k + dk$. The Kolmogorov cascade process [7] is described by

$$E(k) = C_k \epsilon^{2/3} k^{-5/3}, \tag{9}$$

where C_k is a proportionality constant empirically determined [8, 9] to be of order unity (about 1.5).

In the inertial range, i.e. for scales much smaller than ℓ, but much larger than η, the turbulent flow can be considered to be roughly homogeneous and isotropic. Self-similarity in this range means that the turbulence at each scale is similar to that at any other. The local interaction picture of the cascade process implies that the local properties of the turbulence are independent of the conditions at the scales of energy injection and energy dissipation. It is for this reason that any universal statistical properties of turbulence, if they exist, are more likely to be found as $\mathrm{Re} \to \infty$, since the separation between the energy-injection scales and the dissipative scales is large.

2 Pushing Re in the Laboratory

The above discussion pointed to a need to attain the largest Reynolds numbers possible in laboratory experiments, namely that any general understanding of this complex flow derives from its statistic properties, since these are the only reproducible observables, and these properties should be be independent of the means in which the turbulence was generated (as well as the manner in which it is dissipated). Using the local interaction picture above, i.e., that at each scale the motion is the one *relative* to the immediate larger scales, which can be regarded as a steady background flow, it can then be possible to "lose" memory of the boundary conditions supplementing Eq. (5) and find universal properties common to all turbulent flows. Experimentally, we need to find a way to maximize Re under controlled conditions. This is considered below.

2.1 Helium as a Working Fluid

Maximizing the characteristic Reynolds number Re $= UL/\nu$ appearing in equation 5 can be accomplished experimentally by various means. Let us first take the characteristic velocity U: if we increase this too much we begin to introduce a new similarity parameter, namely the Mach number $M = U/c_s$ where c_s is the speed of sound in whatever fluid medium we have in mind. Thus U is not a good "knob" to turn. Let us take now the characteristic length L: here there is no fundamental limit as in the previous case. However, since Re is directly proportional to L we would need to increase it orders of magnitude to make a difference and now we have rather severe financial and technological limits. For this reason, the challenge of obtaining very large Re in academic laboratories has turned to searching for new working fluids in which the only remaining variable, the kinematic viscosity ν, is as small as possible. As a side remark, it should be noted that NASA's large transonic wind tunnel boosts Re by spraying liquid nitrogen into the airflow in order to increase the density (thus decreasing ν).

It is known that low temperature helium has the lowest kinematic viscosity of any known fluid and markedly lower that that of the commonly used test fluids air and water. Table 1 illustrates the difference between them.

In fact, we see that using helium an provide 2–3 orders of magnitude improvement in Re over the more common test fluids. Helium is not the only such fluid: it

Table 1 A comparison of kinematic viscosities for possible test fluids. The liquid phase of helium is evaluated at the saturated vapor pressure (SVP)

Fluid	$T(K)$	P(Bar)	$\nu(cm^2/s)$
Air	293	1	0.15
Water	293	1	0.01
Helium I (liquid)	2.2	SVP	1.8×10^{-4}
Helium gas	5.5	2.8	3.2×10^{-4}

should be noted that compressed air (approximately 200 bar) and compressed SF_6 (50 bar) can approach the values indicated for helium. In all cases, which fluid to use is based upon the advantages and limitations involved for a particular experiment. Nonetheless helium remains the fluid having the smallest kinematic viscosity and is in turn more versatile than the other two, and allows the possibility to examine quantum fluid behavior. For the purpose of this work, then, we will focus on helium.

Liquid helium exists all the way to absolute zero temperature, due principally to the zero point motion of this light atomic species and the lack of strong forces between the closed shell atoms. This is shown in the phase diagram of Fig. 1 where it can be seen that nearly 25 bar pressures are needed in order to solidify the liquid phase at absolute zero. There are three available fluids: helium gas, helium I, and helium II. Helium I is a classical viscous fluid, while helium II is the so-called superfluid phase and exists below the lambda point at around 2.2 K at the saturated vapor pressure [10] where the de Broglie wavelength of the atoms becomes comparable to the average inter-atom spacing. Indeed, helium II is related to a Bose-Einstein condensation and indeed exhibits macroscopic quantum effects; being a liquid, however, the non-negligible interactions between atoms gives it a somewhat different nature than a pure Bose-Einstein condensate.

If we consider temperatures that are low but finite then the motion of helium II can be described by a two-fluid hydrodynamic model; i.e., helium II flows as if it were composed of two interpenetrating and independent fluids, a superfluid associated with the quantum ground state of the system, with density ρ_s and velocity field v_s, and a normal component composed of the thermal excitations with density ρ_n and velocity field v_n. The ratio of superfluid to normal fluid density ρ_s/ρ_n varies smoothly from 0 at the lambda transition to 1 at the absolute zero of temperature. The superfluid component has zero entropy and zero viscosity, so all of the thermal energy is transported by the normal component leading to a number of interesting non-classical flows.

The two fluid model assumes that the curl of the superfluid velocity is zero and therefore that any flow is irrotational [12], a condition which would seem to rule out

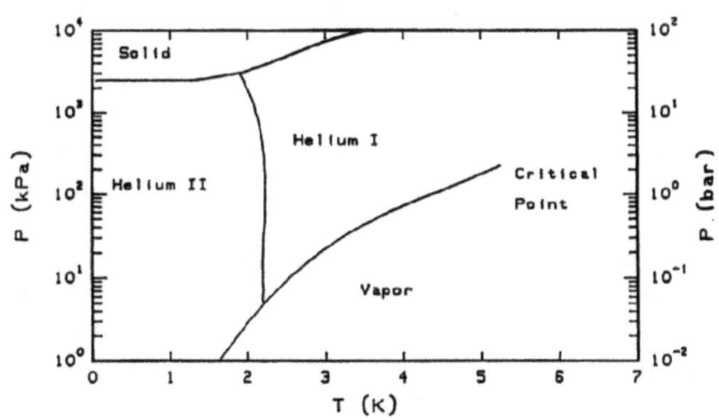

Fig. 1 Pressure-temperature phase diagram of helium. After Donnelly [11], 1991

turbulence completely. However, the circulation of the superfluid $\oint \mathbf{v}_s \cdot d\mathbf{r}$, can be non-zero in a so-called multiply-connected geometry. Such a geometry is provided by line vortices having an Angstrom-size core in which the superfluid density goes to zero (the core being defined as the length over which the superfluid density goes from zero to its nominal value). This circulation is quantized in units of h/m_4, where h is Planck's constant and m_4 is the mass of a helium atom. These *quantized* vortices have the same circulation (there is no evidence for multiply quantized vortices) and hence any rotational flow is made up of appropriate arrays of such single valued vortices. Turbulence, in particular, must be composed of a *tangle* of such vortices as has been discussed early on by Feynman [13]. To get quasi-*classical* turbulence in a superfluid (meaning to have a spectrum of scales in the flow) one then needs in addition to have the quantized vortices partially polarized so that *bundles* of single-quantized vortices can mimic flow on all scales, since a purely random tangle will have just one length scale, namely the average inter-vortex line spacing. A recent review of turbulent flows in quantum fluids [10] discusses the progress made in this area.

2.2 Superfluid Turbulence

In fact, empirically it was known that turbulence created in helium II appears identical to that produced in classical fluid including helium I. This was well illustrated in the work of Maurer and Tabeling [14] using a system of counter-rotating discs to create a highly turbulent flow in the layer of liquid helium contained between them. A local pressure probe was inserted into the fluid close to one of the rotating plates and the velocity fluctuations were deduced from the fluctuating pressure in such a way as to give the spectrum of turbulent energy [14]. What was striking about that experiment is that it was not possible to distinguish the fact that some measurements were made in the nearly pure superfluid phase; i.e. the measured energy spectra all showed identical Kolmogov $-5/3$ rolloff behavior (Eq. (9)) even though two of them were obtained for experiments conducted below the lambda point. In fact, one of the indistinguishable spectra was taken at a point on the saturated vapor pressure curve at low enough temperature where the residual normal fluid fraction was only a few percent!

Other experiments using towed grids of crossed tines to generate nearly homogeneous and isotropic turbulence in helium II showed similarly *classical* turbulence signatures [15–17].

So how does it happen? One thing to note is that the lowest temperature spectrum described was obtained for a fluid fraction which was small but not zero. If a quantized vortex line moves relative to the normal fluid it will produce a mutual friction force [18]. This mutual friction can provide coupling between the normal and superfluid in addition to providing a dissipation mechanism for the turbulent energy in the superfluid. Therefore it is not implausible that the measured energy spectra in the counter-rotating plate experiment could be similar. As alluded to above, this kind of coupling could take place as a result of local alignment of the axes of individual vortices producing bundles of vortices that can be oriented to mimic the eddies present in the normal fluid. Indeed, such polarization can easily be seen in the spinning of a container of helium II, in which all vortices will be aligned with their axes vertical to mimic solid body rotation [19].

All of this presumes that the flow takes place at scales larger than the average inter-vortex line spacing, below which it is clearly impossible to mimic classical eddies of any sort. It turns out that this is of the order one micron in most turbulent superfluid experiments, which is roughly the typical classical dissipation scale for the same flow. A cutoff scale, of course, does not exist in the flow of classical fluids, unless we consider the scale at which a hydrodynamic approximation loses validity (although we have mentioned above that such a cutoff might not be reached in turbulence). A related problem is the limit of zero normal fluid, where the viscosity of the entire fluid is zero and hence there is no dissipation mechanism. In either case, we need to understand how where the turbulent kinetic energy goes in the cascade process. A key to this comes from the early work of Scwarz [20] and Koplik and Levine [21] describing the reconnection that occurs as quantized vortex filaments approach one another. Such reconnections are evidently common in superfluid flows [22] and have been recently observed experimentally [23]. These reconnections lead to propagating helical or Kelvin waves along the cores [24] permitting the energy cascade to proceed beyond the level of the inter-vortex line spacing. Physically, the sharp kinks in any given reconnection process relax into Kelvin waves and continual reconnections lead to nonlinear interactions between the waves and to a cascade where at high frequency the waves can generate sound (phonons) and by this means dissipate the turbulent kinetic energy. This is illustrated in Fig. 2. A brief review of this has been produced [26] in the context of recent experiments [23] in which small hydrogen particles are attracted to the vortex cores which can then be visualized by their scattered light. While such visualizations are very important we still need to make quantitative measurements of the energy spectrum which can guide further

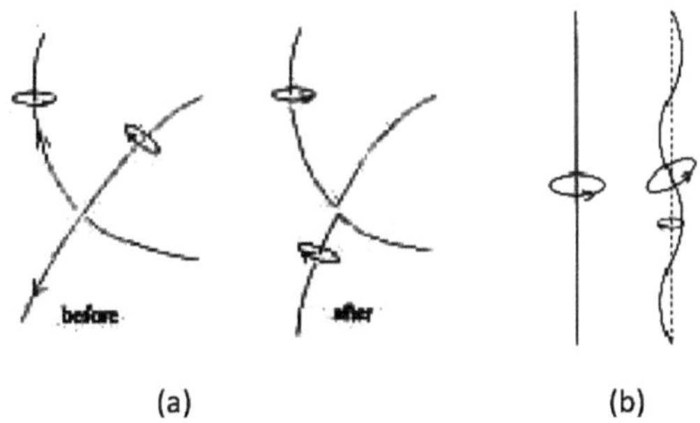

Fig. 2 (**a**) A reconnection event in which two vortex lines approach, break, and then reform, exchanging the broken parts of the lines. (**b**) Kelvin waves created by the reconnection which can produce phonons in the superfluid dissipating turbulent energy (*right*most vortex compared to the undisturbed vortex at its *left*) After Barenghi [25], 2007

theoretical work. Suitably small probes might be the helium atoms themselves, in the form of meta-stable molecules excited with laser pumping and which then decay emitting light through fluorescence [27].

2.3 Classical Fluid Turbulence

Many "classical" high-Reynolds number experiments have been reported using helium I or cryogenic helium gas. The latter phase has been more useful generally for turbulence experiments due to the ease of filling, lack of cavitation, and most importantly the great flexibility that exists in varying the Re and other parameters by operating in different parts of the phase diagram (Fig. 1).

One of the note-worthy experiments has already been discussed, namely the study of flow between two counter-rotating plates at low temperatures, but in this case using the gaseous phase of helium [28]. Here, unexpected transitions in the fourth-order moment of the velocity derivatives were found to occur at very high Re that were not observed in atmospheric measurements in the same range of Re (see the review by Niemela and Sreenivasan [1]). Without going into the details of the experiment, what is not clear is whether the velocity was resolved well enough at the smallest scales in the laboratory experiment; indeed for this experiment the Kolmogov scale was of the order of 1 μm at the highest Re which was somewhat smaller than the sensor used to probe the flow. This sensor was produced from a carbon fiber 7 μm in diameter (see Fig. 3). To make an effective 7 micron sensor, it was necessary to mask one such fiber with another stretched perpendicularly across, and then to evaporate gold over the entire exposed area of the fiber to provide a good electrical connection. The small masked area of the carbon fiber had high sensitivity at low temperatures and is shown schematically in Fig. 3. The possible new physics uncovered through carrying out a controlled laboratory experiment also points to the problem of resolving flow variables at small scales which is needed in this case. In the atmosphere, on the other hand, the Kolmogorov scales are large corresponding to the size of the largest scales in that flow (see Eq. (8)) so the same problem does

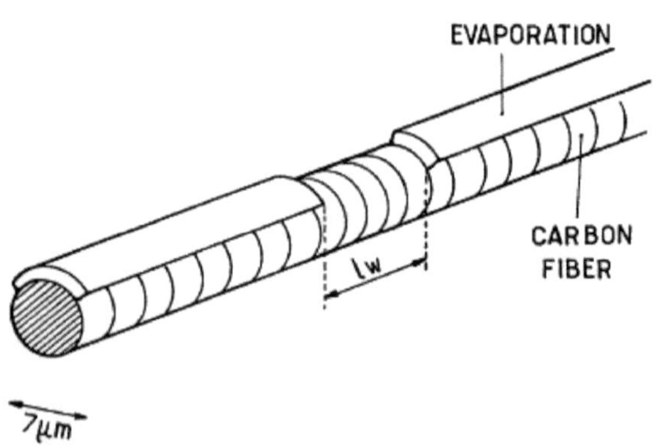

Fig. 3 Cryogenic hot wire probe, showing the active region for velocity measurements of size, 7 μm. After Belin et al. [28] 1998

not exist. However, this must be tempered with the lack of controlled experimental conditions, which limits the type of fundamental questions that can be posed and answered.

A considerable amount of work has been devoted to turbulent thermal convection, where a layer of fluid is heated from below, using cold helium gas [29–31]. The highest *Rayleigh* numbers (Eq. (6)) in any controlled laboratory experiment to date have been achieved in this way, up to $\mathrm{Ra} = 10^{17}$ using a large cryogenic apparatus with a sample space one meter in height [29]. These high Rayleigh numbers resulted from approaching the critical point of the gas where the isobaric thermal expansion coefficient diverges, and the thermal diffusivity consequently vanishes, leading to extremely large buoyant forcing (see Eq. (6)). The high Ra achieved are at the level of atmospheric flows and nearly at the estimated turbulent intensity of the Sun ($\mathrm{Ra} \sim 10^{21}$). The flexibility of the gas phase meant that 11 orders of magnitude of Ra could be achieved in a single apparatus, allowing scaling laws to be deduced accurately. Indeed, the resulting scaling law for the dimensionless heat transport has potential bearing on models used for stellar convection and provides a benchmark for high Ra (indeed, extrapolation of empirical observations in conventional fluid experiments over many orders of magnitude is difficult to justify and often can lead to erroneous conclusions).

Corresponding to these high Ra, the smallest scales again are in the micrometer range. The diffusive thermal boundary layers which form on the heated walls of the experiment, for example, are roughly $100\,\mu\mathrm{m}$ thick and it is over this thickness in which nearly all of the temperature drop occurs. Various plumes and thermal rise from these layers and have this characteristic dimension and their self-organization leads to a coherent mean circulation [33–35] which exhibits many interesting features. One of these is the random and spontaneous reversals in direction, where the PDF of the time spent in any one direction has a power-law roll-off. Interestingly, this turns out to be statistically identical to the PDF of medium energy solar flare lifetimes [36] (and perhaps not too surprisingly since the magnetic lines can be moved about due to the turbulent flow in the convective zone of the Sun). Again, as in the previous example, a limitation of these experiments is the ability to resolve all scales down to micron sizes with sensors that are robust. In particular, light scattering is usually not an option in the gas phase at low temperature but the techniques cited above for quantum flows [27] can work as well here.

Finally we mention the possibility of an engineering testing facility based on the flow of helium II. The same problems of resolution at the micron scale or below apply in this case too and robust sensors similar but smaller than those in fig. 3 would be needed. Figure 4 shows the schematic of an academic-sized helium tunnel that is capable of matching the Re of the largest wind tunnels in the world. Besides transducer development, scale model roughness requirements in such a small tunnel would be at the sub-micron level. One of the advantages of such a tunnel, besides its convenient size for academic use, is that powerful superconducting magnetic balance and suspension systems could be used to eliminate the need to attach fixed "stingers" to support scale models within the tunnel test section. They could also be used to measure the forces on the models directly. Of course, since the density of

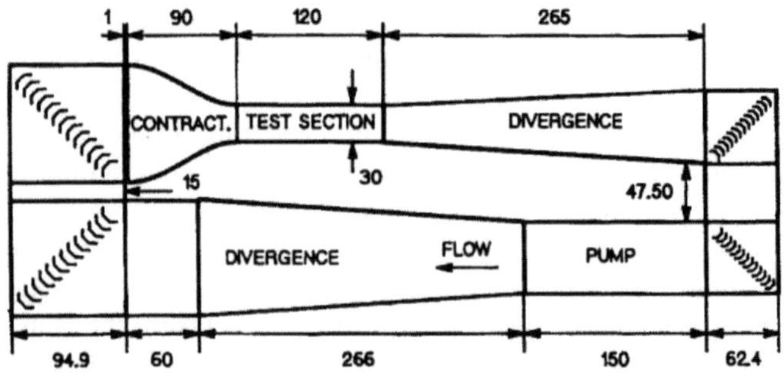

Fig. 4 Schematic of a helium tunnel with test section of only 30 cm capable of Re = $O(10^9)$. After Donnelly [32], 1998

helium is quite low compared to other fluids, the dynamic pressure on a model would reduced making such systems operate better at high Re. Such tunnels of course could also be used for basic research into the physics of ultra-high Re fluid flow, an area that was of much interest to Feynman and others at Caltech in which a small superfluid tunnel was conceived and abandoned nearly 50 years ago [11]. One of the reasons for not pursuing the tunnel was indeed the problem of measurement at low temperatures and at micron scales.

3 Discussion

We have discussed both classical and quantum turbulent fluid flows, the latter in the superfluid phase of ^{4}He. There are a number of reasons to consider turbulence in superfluids. The flow of helium II in the cooling circuits for the superconducting magnets used in large accelerator facilities such as LHC is turbulent and it is of great importance to understand the effectiveness of the heat transfer under those conditions. There is also an academic interest solely in complex quantum systems, where again, superfluid turbulence provides a model system where the equations of motion are known. A third motivation is that we still do not have a deep understanding of turbulence in classical fluids; indeed, this is often considered as one of the outstanding remaining problems in classical physics. Flow in a superfluid is in some sense simpler and somewhat different, so that a comparison of turbulence in classical and quantum fluids has the possibility of providing fresh insights into both.

As we have seen the range of scales in turbulent flows can be very large, growing as $Re^{9/4}$. Since the large scales are fixed, this large range is composed of ever smaller scales. Current turbulence experiments using low temperature helium generate length scales ranging from meters to micrometers at the highest Re and Ra. In quantum turbulence there is a cutoff scale associated with the inter-vortex line spacing and scales smaller than this exist only as propagating helical waves along the vortex cores, which themselves are the order of 1 Å in thickness. The small length scales pose technological challenges in terms of building robust transducers of the same effective size, if one wants to fully resolve the flow. Correspondingly, frequency response must be quite large – in the MHz range and beyond – and this presents an additional challenge. The use of low temperature helium means that all of the measurement technology has to survive and to be effective at low temperatures.

Measurement technology development, then, represents a parallel activity which is just as important as producing the large Re in the first place. Indeed, it is probably the biggest challenge at present to continued progress in understanding turbulence, either classical or quantum. Nature usually has surprises in store for those who look, and the flow at ultra-high Re, and in quantum flows, must be a rich area for exploration if the needed tools are developed. Certainly it is an interesting question to pose–and one which has been purely academic until recently–as to what new physics might be observed as the flow scales (from turbulence) become "small enough" to push the continuum approximation.

Acknowledgements Many thanks are due to my colleague K.R. Sreenivasan, who has contributed to much of the work reported here.

References

1. J.J. Niemela, K.R. Sreenivasan, J. Low Temp. Phys. **143**, 163 (2006)
2. U. Frisch, *Turbulence: The Legacy of A.N. Kolmogorov* (Cambridge University Press, Cambridge 1995)
3. E.M. Spiegel, Ann. Rev. Astron. Astrophys. **9**, 323 (1971)
4. B.B. Mandelbrot, Sci. Am. **280**, 50 (1999)
5. A. Oberbeck, *Annalen der Physik und Chemie* **7**, 271 (1879)
6. F.H. Busse, Rep. Prog. Phys. **41**, 1929 (1978)
7. A.N. Kolmogorov, Dokl. Akad. Nauk SSSR **30**, 9 (1941)
8. A.S. Monin, A.M. Yaglom, *Statistical Fluid Mechanics*, vol. 2. (MIT, Cambridge, MA 1975)
9. K.R. Sreenivasan, Phys. Fluids **7**, 2778 (1995)
10. W.F. Vinen, J.J. Niemela, J. Low Temp. Phys. **128**, 167 (2002)
11. R.J. Donnelly, in *High Reynolds Number Flows using Liquid and Gaseous Helium*, ed. by R.J. Donnelly (Springer 3 (1991)
12. D.J. Tritton, *Physical Fluid Dynamics* (Clarendon Press, Oxford 1988)
13. R.P. Feynman, in *Progress in Low Temperature Physics, Vol. 1*, edited by C.J. Gorter (Nother-Holland, Amsterdam, 1955)
14. J. Maurer, P. Tabeling, Europhys. Lett. **43**, 29 (1998)
15. J.J. Niemela, R.J. Donnelly, K.R. Sreenivasan, J. Low Temp. Phys. **138**, 537 (2005)
16. S.R. Stalp, J.J. Niemela, W.F. Vinen, R.J. Donnelly, Phys. Fluids **14**, 1377 (2002)
17. S.R. Stalp, L. Skrbek, R.J. Donnelly, Phys. Rev. Lett. **82**, 4831 (1999)
18. H.E. Hall and W.F. Vinen, Proc. Roy. Soc. A **238**, 204, 205 (1956)
19. R.J. Donnelly, *Quantized Vortices in Helium II* (Cambridge University Press, Cambridge 1991)
20. K.W. Schwarz, Phys. Rev. B **38**, 2398 (1988)
21. J. Koplik, H. Levine, Phys. rev. Lett. **71**, 1375 (1993)
22. M. Tsubota, T. Araki, adn S.K. Nemirovskii, Phys. Rev. B **62**, 11751 (2000)
23. M.S. Paoletti, M.E. Fisher, K.R. Sreenivasan, D.P. Lathrop, Phys. Rev. Lett. **101**, 154501 (2008)

24. B.V. Svistunov, Phys. Rev. B **52**, 3647 (1995)
25. C.F. Barenghi, in *Vortices and Turbulence at Very Low Temperatures*, ed. by C.F. Barenghi, Y.A. Sergeev (Springer, New York, NY 2008), p. 1 [Amazon] [WorldCat]
26. J.J. Niemela, Physics **1**, 26 (2008) http://physics.aps.org/articles/v1/26
27. W. Guo, J.D. Wright, S.B. Cahn, J.A. Nikkel, D.N. McKinsey, Phys. Rev. Lett. **102**, 235301 (2009)
28. F. Belin, J. Maurer, P. Tabeling, H. Williame, in *Flow at Ultra-high Reynolds and Rayleigh Numbers, A Status Report*, Vol. 345 ed. by R.J. Donnelly, K.R. Sreenivasan (Springer, New York, NY 1998)
29. J.J. Niemela, L. Skrbek, K.R. Sreenivasan, R.J. Donnelly, Nature **404**, 837 (2000)
30. J.J. Niemela, K.R. Sreenivasan, J. Fluid Mech. **481**, 355 (2003)
31. J.J. Niemela, K.R. Sreenivasan, Phys. Rev. Lett. **100**, 184502 (2008)
32. R.J. Donnelly, in *Flow at Ultra-high Reynolds and Rayleigh Numbers, A Status Report*, vol. 1, ed. by R.J. Donnelly, K.R. Sreenivasan (Springer, New York, NY 1998)
33. J.J. Niemela, L. Skrbek, K.R. Sreenivasan, R.J. Donnelly, J. Fluid Mech. **449**, 169 (2001)
34. K.R. Sreenivasan, A. Bershadskii, J.J. Niemela, Phys. Rev. E **65**, 056306 (2002)
35. R.C. Hwa, C.B. Yang, S.Bershadskii, J.J. Niemela, K. R. Sreenivasan, Phys. Rev. E **72** 066308 (2005)
36. A. Bershadskii, J.J. Niemela, K.R. Sreenivasan, Phys. Lett. A **331**, 15 (2004)

Index

A. Aldea, V. Bârsan (eds.), *Trends in Nanophysics*, Engineering Materials,
DOI 10.1007/978-3-642-12070-1, © Springer-Verlag Berlin Heidelberg 2010

N

Q

MIX
Papier aus verantwortungsvollen Quellen
Paper from responsible sources
FSC® C105338

If you have any concerns about our products,
you can contact us on
ProductSafety@springernature.com

In case Publisher is established outside the EU,
the EU authorized representative is:
Springer Nature Customer Service Center GmbH
Europaplatz 3, 69115 Heidelberg, Germany

Printed by Libri Plureos GmbH
in Hamburg, Germany